DISPERSAL

CENTRE NATIONAL
DE LA RECHERCHE
SCIENTIFIQUE

National
Science
Foundation
Celebrating 50 Years

IUBS

DISPERSAL

Edited by

JEAN CLOBERT
ETIENNE DANCHIN
ANDRÉ A. DHONDT
JAMES D. NICHOLS

OXFORD
UNIVERSITY PRESS

*This book has been printed digitally and produced in a standard specification
in order to ensure its continuing availability*

OXFORD
UNIVERSITY PRESS

Great Clarendon Street, Oxford OX2 6DP

Oxford University Press is a department of the University of Oxford.
It furthers the University's objective of excellence in research, scholarship,
and education by publishing worldwide in

Oxford New York

Auckland Bangkok Buenos Aires Cape Town Chennai
Dar es Salaam Delhi Hong Kong Istanbul Karachi Kolkata
Kuala Lumpur Madrid Melbourne Mexico City Mumbai Nairobi
São Paulo Shanghai Taipei Tokyo Toronto

Oxford is a registered trade mark of Oxford University Press
in the UK and in certain other countries

Published in the United States
by Oxford University Press Inc., New York

© Oxford University Press 2001

The moral rights of the author have been asserted

Database right Oxford University Press (maker)

Reprinted 2003

ISBN 0-19-850659-7

Contents

Foreword   ix
Peter Waser
List of contributors   xiii
Introduction   xvii
Jean Clobert, Jerry O. Wolff, James D. Nichols, Etienne Danchin,
and André A. Dhondt

Part I. Measures of dispersal: genetic and demographic approaches

  1. Methods for estimating dispersal probabilities and related
      parameters using marked animals   3
      Robert E. Bennetts, James D. Nichols, Jean-Dominique Lebreton,
      Roger Pradel, James E. Hines, and Wiley M. Kitchens
  2. Genetic approaches to the estimation of dispersal rates   18
      François Rousset
  3. How to measure dispersal: the genetic approach.
      The example of fire ants   29
      Kenneth G. Ross
  4. Dispersal in pikas (*Ochotona princeps*): combining genetic and
      demographic approaches to reveal spatial and temporal patterns   43
      Mary M. Peacock and Chris Ray
  5. Invasion fitness and adaptive dynamics in spatial
      population models   57
      Régis Ferrière and Jean-François Le Galliard

Part II. Why disperse? Habitat variability, intraspecific interactions, multi-determinism, and interspecific interactions

  6. On the relationship between the ideal free distribution and
      the evolution of dispersal   83
      Robert D. Holt and Michael Barfield
  7. The landscape context of dispersal   96
      John A. Wiens
  8. Dispersal, intraspecific competition, kin competition and
      kin facilitation: a review of the empirical evidence   110
      Xavier Lambin, Jon Aars, and Stuart B. Piertney
  9. Inbreeding, kinship, and the evolution of natal dispersal   123
      Nicolas Perrin and Jérôme Goudet
 10. Inbreeding versus outbreeding in captive and wild populations
      of naked mole-rats   143
      M. Justin O'Riain and Stanton Braude

11. Multiple causes of the evolution of dispersal   155
     Sylvain Gandon and Yannis Michalakis
12. Parasitism and predation as causes of dispersal   168
     Wolfgang W. Weisser, Karen D. McCoy, and Thierry Boulinier
12a. Dispersal and parasitism   169
     Thierry Boulinier, Karen D. McCoy, and Gabriele Sorci
12b. The effects of predation on dispersal   180
     Wolfgang W. Weisser

Part III. Mechanisms of dispersal: genetically based dispersal, condition-dependent dispersal, and dispersal cues

13. The genetic basis of dispersal and migration, and
     its consequences for the evolution of correlated traits   191
     Derek A. Roff and Daphne J. Fairbairn
14. Condition-dependent dispersal   203
     Rolf A. Ims and Dag Ø. Hjermann
15. Proximate mechanisms of natal dispersal: the role of
     body condition and hormones   217
     Alfred M. Dufty, Jr. and James R. Belthoff
16. Habitat selection by dispersers: integrating proximate and
     ultimate approaches   230
     Judy A. Stamps
17. Public information and breeding habitat selection   243
     Etienne Danchin, Dik Heg, and Blandine Doligez

Part IV. Dispersal from the individual to the ecosystem level: individuals, populations, species, and communities

18. Dispersal, individual phenotype, and phenotypic plasticity   261
     *Courtney J. Murren, Romain Julliard, Carl D. Schlichting,
     and Jean Clobert*
19. Dispersal and the genetic properties of metapopulations   273
     Michael C. Whitlock
20. Population dynamic consequences of dispersal in
     local populations and in metapopulations   283
     Ilkka Hanski
21. Dispersal in antagonistic interactions   299
     Minus van Baalen and Michael E. Hochberg
22. The properties of competitive communities with
     coupled local and regional dynamics   311
     *Nicolas Mouquet, Gerard S.E.E. Mulder, Vincent A.A. Jansen,
     and Michel Loreau*

Part V. Perspectives

23. The evolutionary consequences of gene flow and
    local adaptation: future approaches   329
    Nick H. Barton
24. Perspectives on the study of dispersal evolution   341
    Ophélie Ronce, Isabelle Olivieri, Jean Clobert,
    and Etienne Danchin
25. Dispersal in theory and practice: consequences for
    conservation biology   358
    David W. Macdonald and Dominic D.P. Johnson

References   373
Index   443

Foreword

In 1922, the American ornithologist Joseph Grinnell noted that 'accidentals', defined as individuals sighted outside the normal range of a species, are not accidental at all. Instead, they are a 'regular thing, to be expected ... the process is part of the ordinary evolutionary program' (Grinnell 1922, p. 374). Grinnell then went on to foreshadow many of the themes that drive the current resurgence of interest in dispersal: its role in the colonization (or recolonization) of unoccupied areas of suitable habitat, the stochastic nature of rare, long-distance movements and the difficulty in detecting them, the interplay between innate and environmental factors in triggering dispersal, the apparent contradiction between the low survival of individual dispersers and the great importance that such movements may have in species persistence. In his list of population consequences of dispersal, he even presaged current interest in dispersal's role in biotic responses to global warming: 'barriers of climate are continually moving about over the earth's surface ... populations are by them being herded about ... it is the rule for the population, by means of [dispersing] individuals, to keep up with the receding barrier and not only that but to press the advance' (p. 379). He also ventured a description of the *causes* of dispersal: 'individuals ... occupying a restricted habitat may be likened to the molecules of a gas in a container which are continually beating against one another and against the confining walls, with resulting pressure outwards ... [and furthermore] the number of individual units is being augmented ... at each annual period of reproduction, with correspondingly reinforced outward pressure' (p. 377).

So what have we learned about dispersal in the 80 years since Grinnell's essay? Or, for that matter, in the 70 years since Sewall Wright (1931) outlined the importance of dispersal (and of restricted dispersal) to the very process of evolution?

This book provides a variety of answers. One conspicuous change has been in the effort to quantify. This is apparent for both patterns and ideas. Trying to work out what would now be referred to as incidence functions for California bird species, Grinnell mused: 'I attempted to carry out the figures, which seem to behave according to some mathematical formula; but when I came to deal with 3/5 of an occurrence I decided it was profitless to go on' (p. 375). Population ecologists these days are not so shy! Although we may not yet agree how best to extract it, this book makes it clear that crucial information about the dispersal process can be extracted from the distribution of individuals among habitat patches, from the distribution of dispersal distances, and from the spatiotemporal patterns of sightings and recaptures.

The mathematization is even clearer with regard to ideas about the evolution of dispersal. Grinnell ventured: 'it is obvious that the interests of the individual are sacrificed in the interests of the species'; perhaps wisely, he left it at that – what if Grinnell had realized how complex an issue this actually is! In this book, a cornucopia of models brings the latest in abstract formulations of fitness and spatial dynamics to bear on the ultimate causes of dispersal. In so doing, the models expose how complex a phenomenon dispersal actually is. Should dispersal be modelled as a fixed or a conditional trait?

Should it be viewed from the parent's or the offspring's point of view? In our models, must we incorporate the fact there is feedback between the evolution of dispersal tendencies and local rates of inbreeding or kin competition? Should we include the fact that the fitness of dispersers is likely to be density and frequency dependent, or the fact that the world is stochastic rather than deterministic?

Perusing this book, some may mirror the cautious view expressed by François Rousset with regard to genetic estimators of migration rate: 'How much this field has progressed is not clear' (chapter 2). Indeed, we are almost as limited as was Grinnell with regard to data. We know that rates of dispersal between populations must be influenced by distance and by the nature of the habitat matrix, but have few measurements of those effects. We know that the mortality rate of dispersers is a crucial parameter for many models, both demographic and evolutionary, but mortality during dispersal remains extremely difficult to measure. We know that dispersal is often a conditional decision, but rarely know what aspects of an individual's internal state or of its ecological or demographic surroundings trigger that decision.

But here is where the menu of ideas is fresh and exciting. Estimation of emigration rates from mark–recapture statistics dates back at least to C.H.N. Jackson, who, however, left some aspects of the argument to higher authority: 'Prof R.A. Fisher informs me that dispersal rate varies, approximately, inversely as the periphery [of the study area]' (Jackson 1938, p. 244). Sixty years later, the ubiquity of microchips and the approaches described in this book allow us (among other things) to test Professor Fisher's assumptions with the sophisticated new generation of mark–recapture estimators. The huge storehouse of information now accessible with hypervariable genetic markers is launching more realistic and powerful 'indirect' methods to investigate interpopulation movement. In this book are chapters that focus on under-investigated components of the dispersal process, showing what can be gained if we focus not on emigration but on where individuals choose to settle, or on their behaviour as they pass through unsuitable habitats. There are synthetic models that predict dispersers' sex ratios as a function of the combined effects of kin competition and inbreeding avoidance, superseding the 'endless discussions concerning the role of inbreeding and sex-specific competition as alternative explanations for the evolution of dispersal' (Ronce *et al.*, chapter 24). There are chapters suggesting fascinating links between dispersal and predation, dispersal and life history evolution, dispersal and species coexistence.

Because the range of ideas and systems discussed in this book is so broad, readers might well question whether dispersal is a unitary phenomenon. Academics, like other species, disperse (Fig. 1), and the dispersal of academics seems parallel in many regards to other cases of dispersal discussed in this book. Suppose we could determine what mix of academic kin competition and inbreeding avoidance leads to the observed dispersal rates, what patterns of patch choice, interpatch movement and demographic stochasticity lead to the observed dispersal distributions, or what patterns of academic dispersal maximize population persistence. Would we expect the answers to be generalizable to other species? Each reader will have to decide whether a grand unified theory of dispersal is on the horizon.

And what if dispersal is not a unitary phenomenon? Then the value of this book lies elsewhere. Like Grinnell's accidentals, dispersing academics (and the ideas that they carry with them) are surely beneficial to biologists as a species. This book provides an

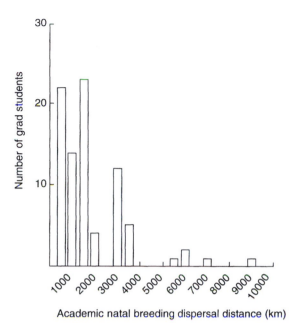

Figure 1. The distance between the natal site (where the degree was received) and the first breeding site (the location of the first job) for Biological Sciences PhD students at Purdue University, USA, 1980–1990. The distribution parallels those exhibited by many other species, and illustrates, for example, the significant opportunity for kin interaction among short-distance dispersers and the presence of rare long-distance dispersers with the potential to influence the large-scale persistence and evolution of ideas. Variation in habitat quality is evident (the peak at 3000–4000 km represents the North American West Coast, the gap between 4000 and 5000 km the Atlantic Ocean). Population biologists are over-represented in the tail of the distribution, but it is unclear whether this is a genotypic effect or an effect of condition.

intellectual dispersal opportunity for an unusually broad range of ideas about the biological dispersal process. Not all of these ideas may survive. However, we can expect that those that do will (in Grinnell's terms) 'press the advance', and colonize important new patches of the conceptual landscape.

Peter Waser

Contributors

Aars, Jon
Department of Zoology, University of
Aberdeen, Tillydrone Avenue, Aberdeen,
AB24 2TZ, UK.
e-mail: Jon.Aars@abdn.ac.uk

Barfield, Michael
Department of Ecology and Evolutionary
Biology, University of Kansas, Lawrence,
Kansas 66045, USA.
e-mail: barfield@ukans.edu

Barton, Nick H.
Institute of Cell, Animal and Population
Biology, University of Edinburgh,
King's Buildings, Edinburgh EH9 3JT, UK.
e-mail: n.barton@uk.ac.ed

Belthoff, James R.
Department of Biology, Boise State University,
1910 University Drive, Boise,
Idaho 83725, USA.
e-mail: jbeltho@boisestate.edu

Bennetts, Robert E.
Station Biologique de la Tour du Valat,
Le Sambuc, 13200 Arles, France.
e-mail: bennetts@tour-du-valat.com

Boulinier, Thierry
Laboratoire d'Ecologie, UMR 7625,
Université Pierre et Marie Curie, Case 237,
7 Quai St. Bernard,
75252 Paris Cedex 05, France.
e-mail: tboulini@snv.jussieu.fr

Braude, Stanton
Biology Department, Washington University,
One Brookings Drive, Box 1137, St. Louis,
MO 63130, USA.
e-mail: braude@biology.wustl.edu

Clobert, Jean
Laboratoire d'Ecologie, UMR 7625
Université Pierre et Marie Curie, Case 237,
7 Quai St. Bernard,
75252 Paris Cedex 05, France.
e-mail : jclobert@snv.jussieu.fr

Danchin, Etienne
Laboratoire d'Ecologie, UMR 7625
Université Pierre et Marie Curie, Case 237,
7 Quai St. Bernard,
75252 Paris Cedex 05, France.
e-mail: edanchin@snv.jussieu.fr

Dhondt, André A.
Laboratory of Ornithology, Cornell
University, Ithaca, NY 14850, USA.
e-mail: aad4@cornell.edu

Doligez, Blandine
UPMC, Laboratoire d'Ecologie, Bât. A7,
Case 237, 7 Quai St. Bernard,
75252 Paris Cedex 05, France.
e-mail: bdoligez@snv.jussieu.fr

Dufty, Alfred M., Jr.
Department of Biology,
Boise State University,
1910 University Drive, Boise,
Idaho 83725, USA.
e-mail: adufty@email.boisestate.edu

Fairbairn, Daphne J.
Department of Biology, Concordia University,
1455 de Maisonneuve Blvd. West, Montreal,
Quebec H3G 1M8, Canada.
e-mail: fairbrn@vax2.concordia.ca

Ferrière, Régis
Laboratoire d'Ecologie, UMR 7625,
Ecole Normale Supérieure, 46 Rue d'Ulm,
75230 Paris Cedex 05, France.
e-mail: ferriere@horus.biologie.ens.fr

Gandon, Sylvain
Centre d'Etudes sur le Polymorphisme des
Microorganismes, UMR CNRS-IRD 9926,
Equipe: Evolution des Systèmes Symbiotiques,
IRD, 911 Avenue Agropolis, BP 5045,
34032 Montpellier Cedex 1, France.
e-mail: sgandon@snv.jussieu.fr

Goudet, Jérôme
Institut de Zoologie et d'Ecologie Animale,
Bât. Biologie, Université de Lausanne,

CH 1015 Lausanne, Switzerland.
e-mail: jerome.goudet@ie-zea.unil.ch

Hanski, Ilkka
Metapopulation Research Group,
Department of Ecology and Systematics,
University of Helsinki, PO Box 17,
Arkadiankatu 7,
FIN-00014 Helsinki, Finland.
e-mail: ilkka.hanski@helsinki.fi

Heg, Dik
Zoological Laboratory,
Kerklaan 30,
PO Box 14,
9750 AA Haren,
The Netherlands.
e-mail: d.h.heg@biol.rug.nl

Hines, James E.
U.S. Geological Survey,
Biological Resources Division,
Patuxent Wildlife Research Center,
Laurel, Maryland 20708, USA.
e-mail: Jim_Hines@usgs.gov

Hjermann, Dag Ø.
Department of Biology, University of Oslo,
PO Box 1050 Blindern, N-0316 Oslo, Norway.
e-mail: hjermann@math.iuo.no

Hochberg, Michael E.
Institut des Sciences de l'Evolution,
UMR 5554, Université de Montpellier
2 (CC 065), 34095 Montpellier, France.
e-mail: hochberg@isem.univ-montp2.fr

Holt, Robert D.
Natural History Museum and
Center for Biodiversity Research,
University of Kansas,
Lawrence, Kansas 66045, USA.
e-mail: predator@falcon.cc.ukans.edu

Ims, Rolf A.
Division of Zoology, Department of Biology,
University of Oslo, PO Box 1050 Blindern,
N-0316 Oslo, Norway.
e-mail: r.a.ims@bio.uio.no

Jansen, Vincent A.A.
School of Biological Sciences, Royal Holloway,
University of London, Egham TW20 0EX, UK.
e-mail: vincent@lyapunov.zoo.ox.ac.uk

Johnson, Dominic D.P.
Wildlife Conservation Research Unit,
Department of Zoology,
University of Oxford, South Parks Road,
Oxford OX1 3PS, UK.
e-mail: dominic.johnson@zoo.ox.ac.uk

Julliard, Romain
C.R.B.P.O., Muséum National d'Histoire
Naturelle, 55 Rue Buffon,
F–75005 Paris, France.
e-mail: julliard@mnhn.fr

Kitchens, Wiley M.
U.S. Geological Survey,
Biological Resources Division,
Florida Cooperative Fish and
Wildlife Research Unit,
University of Florida,
Gainesville, Florida 32611, USA.
e-mail: KitchensW@wec.ufl.edu

Lambin, Xavier
Department of Zoology,
University of Aberdeen,
Tillydrone Avenue,
Aberdeen AB24 2TZ, UK.
e-mail: X.Lambin@abdn.ac.uk

Lebreton, Jean-Dominique
CEFE CNRS, BP 5051,
34033 Montpellier Cedex 1, France.
e-mail: Lebreton@srvlinux.cefe.cnrs-mop.fr

Le Galliard, Jean-François
Laboratoire d'Ecologie, UMR 7625,
Université Pierre et Marie Curie, Bât. A7,
Case 237, 7 Quai St. Bernard,
75252 Paris Cedex 05, France.
e-mail: galliard@snv.jussieu.fr

Loreau, Michel
Laboratoire d'Ecologie, UMR 7625,
Ecole Normale Supérieure, 46 Rue d'Ulm,
F-75230 Paris Cedex 05, France.
e-mail: loreau@ens.fr

Macdonald, David W.
Wildlife Conservation Research Unit,
Department of Zoology,
University of Oxford,
South Parks Road, Oxford OX1 3PS, UK.
e-mail: david.macdonald@zoo.ox.ac.uk

McCoy, Karen D.
Laboratoire d'Ecologie, UMR 7625,
Université Pierre et Marie Curie, Case 237
7 Quai St. Bernard,
75252 Paris Cedex 05, France.
e-mail: kmaccoy@snv.jussieu.fr

Michalakis, Yannis
Centre d'Etudes sur le Polymorphisme des
Microorganismes, UMR CNRS-IRD 9926,
Equipe: Evolution des Systèmes Symbiotiques,
IRD, 911 Avenue Agropolis, BP 5045,
34032 Montpellier Cedex 1, France.
e-mail: michalakis@cepm.mpl.ird.fr

Mouquet, Nicolas
Laboratoire d'Ecologie, UMR 7625,
Ecole Normale Supérieure, 46 Rue d'Ulm,
75230 Paris Cedex 05, France.
e-mail: mouquet@ens.fr

Mulder, Gerard S.E.E.
Institute of Evolutionary and Ecological
Sciences, Leiden University, PO Box 9516,
NL-2300 RA Leiden, The Netherlands
e-mail: mulder@rulsfb.leidenuniv.nl

Murren, Courtney J.
Department of Ecology and Evolutionary
Biology, U-43, University of Connecticut,
Storrs, CT 06269, USA.
e-mail: cmurren@utk.edu

Nichols, James D.
U.S. Geological Survey, Biological Resources
Division, Patuxent Wildlife Research Center,
Laurel, Maryland 20708, USA.
e-mail: Jim_Nichols@usgs.gov

Olivieri, Isabelle
Institut des Sciences de l'Evolution,
Université Montpellier II, cc65, Place Eugène
Bataillon, 34095 Montpellier Cedex 5, France.
e-mail: olivieri@isem.isem.univ-montp2.fr

O'Riain, M. Justin
Laboratoire d'Ecologie, UMR 7625,
Université Pierre et Marie Curie, Case 237,
7 Quai St. Bernard,
75252 Paris Cedex 05, France.
Present address:
Zoology Department,
University of Cape Town,
Private Bag,

Rondebosch, 7700,
Afrique du Sud.
e-mail: joriain@botzoo.uct.ac.za

Peacock, Mary M.
Biological Resources Research Center,
Department of Biology, University of Nevada,
Reno, NV 89557–0015, USA.
e-mail: mpeacock@scs.unr.edu

Perrin, Nicolas
Institute of Ecology, University of Lausanne,
CH-1015 Lausanne, Switzerland.
e-mail: nicolas.perrin@ie-zea.unil.ch

Piertney, Stuart B.
Department of Zoology, University of
Aberdeen, Tillydrone Avenue,
Aberdeen, AB24 2TZ, UK.
e-mail: S.Piertney@abdn.ac.uk

Pradel, Roger
CEFE CNRS, BP 5051,
34033 Montpellier Cedex 1, France.
e-mail: pradel@srvlinux.cefe.cnrs-mop.fr

Ray, Chris
Biological Resources Research Center,
Department of Biology,
University of Nevada,
Reno, NV 89557–0015, USA.
e-mail: cray@unr.edu

Roff, Derek A.
Department of Biology, McGill University,
1205 Dr Penfield Ave., Montreal,
Quebec H3A 1B1, Canada.
e-mail: droff@bio1.lan.mcgill.ca

Ronce, Ophélie
Institut des Sciences de l'Evolution,
Université Montpellier II, cc65,
Place Eugène Bataillon,
34095 Montpellier Cedex 5, France.
e-mail: ronce@isem.isem.univ-montp2.fr

Ross, Kenneth G.
Department of Entomology, University of
Georgia, Athens, GA 30602–2603, USA.
e-mail: kenross@arches.uga.edu

Rousset, François
Laboratoire Génétique et Environnement,
Institut des Sciences de l'Evolution,
Université Montpellier II,

34095 Montpellier Cedex, France.
e-mail rousset@isem.univ-montp2.fr

Schlichting, Carl D.
Department of Ecology and Evolutionary
Biology, U-43, University of Connecticut,
Storrs, CT 06269, USA.
e-mail: schlicht@uconnvm.uconn.edu

Sorci, Gabriele
Laboratoire d'Ecologie, UMR 7625,
Université Pierre et Marie Curie,
Case 237, 7 Quai St. Bernard,
75252 Paris Cedex 05, France.
e-mail: gsorci@snv.jussieu.fr

Stamps, Judy A.
Section of Evolution and Ecology,
University of California,
Davis, CA 95616, USA.
e-mail: jastamps@uc.davis.edu

van Baalen, Minus
Institute for Biodiversity and Ecosystem
Dynamics, Section of Population Biology,
University of Amsterdam, PO Box 94084,
1090 GB Amsterdam, The Netherlands.
Present address:
Laboratoire d'Ecologie, UMR 7625
Université Pierre et Marie Curie, Case 237,
7 Quai St. Bernard,

75252 Paris Cedex 05, France.
e-mail: mvbaalen@snv.jussieu.fr

Waser, Peter
Department of Biological Sciences,
Purdue University, W. Lafayette,
IN 47907, USA.
e-mail: pwaser@bilbo.bio.purdue.edu

Weisser Wolfgang W.
Zoology Institute, University of Basel,
Rheinsprung 9, 4051 Basel, Switzerland.
Present address: Institute of Ecology,
Friedrich Schiller University,
Dornburger Strasse 159,
07743 Jena, Germany.
e-mail: wolfgang.weisser@uni-jena.de

Whitlock, Michael C.
Department of Zoology, University of
British Columbia, Vancouver,
BC V6T 1Z4, Canada.
e-mail: whitlock@zoology.ubc.ca

Wiens, John A.
Department of Biology and
Graduate Degree Program in Ecology,
Colorado State University,
Fort Collins, CO 80523, USA.
e-mail: jaws@lamar.colostate.edu

Introduction

Jean Clobert, Jerry O. Wolff, James D. Nichols,
Etienne Danchin, and André A. Dhondt

We will begin with a thought experiment. What if:

- Australia was not isolated from the other continents and the Bering Strait had never been closed
- Christopher Columbus had not reached America and Moses had not crossed the Red Sea
- Flemish and Italian painters of the Middle Ages had not interacted
- No wars had occurred during the twentieth century
- The French disliked hunting and all countries were equally economically healthy
- Royal families promoted outbreeding

The question is: would the world still look the same?

All the above events were affected by a common process, which is best described as 'movement'. One of the most studied yet least understood concepts in ecology and evolutionary biology is the movement of individuals, propagules, and genes. The motivations, causes, and consequences of movement are major evolutionary forces that are affected by social systems, growth forms, seasonality, habitat, and a host of other ecological and behavioural parameters. Movement has consequences for individuals as well as for populations and communities, and its effects on inclusive fitness are ultimately the selecting forces for dispersal, migration, exploration, and other types of movement that affect the distribution, abundance, and dispersion of individuals.

Movement takes different forms (Dingle 1996). During the conference that led to this book, we realized that 'movement' means different things to different investigators and in different disciplines, as well as for different taxa. Instead of trying to cover all the possible meanings of this term (which would be difficult if not impossible) we attempt to develop an overall operational definition and then leave the authors of each chapter with the responsibility of modifying this definition for their own purposes. For the general purposes of this book, we focus on two types of movement: **natal dispersal**, i.e. *the movement between the natal area or social group and the area or social group where breeding first takes place*; and **breeding dispersal**, i.e. *the movement between two successive breeding areas or social groups*. We include the social group in our definition because part of the theoretical framework that explains the motivation for dispersal involves social interactions. We focus on these types of movement because they are the most likely to result in gene flow. However, the distinction between these and other types of movement is not clear-cut, and other forms of movement may also interact with dispersal; this will be reflected in some parts of the book (Roff and Fairbairn, chapter 13; Wiens, chapter 7).

Why another book on dispersal? Apart from the fact that dispersal is probably the most important life history trait involved in both species persistence and evolution, we believe that the field of dispersal studies has matured to the point where different approaches are converging towards a common paradigm. In these proceedings, we avoid theoretical and methodological debate and concentrate our efforts on generality and on developing this unifying paradigm.

Over the last 10 years, the methods used to address spatial problems, either directly or indirectly, have matured and have been refined such that the accuracy, precision, and types of questions that can be addressed have all improved. There has been a parallel evolution in the fields of demography (Bennetts *et al.*, chapter 1) and genetics (Rousset, chapter 2; Ross, chapter 3). Furthermore, important progress has been made in measuring population growth rate and defining fitness in fragmented populations (Ferrière and Le Galliard, chapter 5). In particular, the recent surge of adaptive dynamics models opens the possibility of merging three related factors: individual interactions localized in space, population dynamics, and evolution. Although we still need methods that link genetic and demographic data, progress is being made in integrating information from demography and genetics with our perception of the role of dispersal in metapopulation functioning (Peacock and Ray, chapter 4). However, to understand the consequences of movement, we need to understand first the cause of dispersal, and then the mechanisms underlying movement, dispersal, and habitat selection.

Kin interaction, inbreeding avoidance, local mate and resource competition, and temporal and spatial heterogeneity are all forces that guide the evolution of dispersal. The influence of spatial heterogeneity on the evolution of dispersal, which has been emphasized recently, is contingent upon the existence of condition-dependent dispersal, and shares many of the ideas developed in the context of Fretwells' (1972) Ideal Free Distribution (Holt and Barfield, chapter 6; Wiens, chapter 7; Stamps, chapter 16). Habitat heterogeneity, limited resources, and condition-dependent dispersal all indicate that competition (between conspecifics, members of the same sex, or kin) plays a key role in the evolution of dispersal (Lambin *et al.*, chapter 8). However, the roles of these different forms of competition need to be measured, and we still lack sufficient evidence of the frequency of certain types of competition (including their co-occurrence within a single species) and their effectiveness. For example, philopatry carries a potential cost in terms of kin competition, as well as the potential for mating with relatives. Because they are so inextricably related, it follows that inbreeding, kin competition, and sex-specific habitat requirements cannot be separated from one another (O'Riain and Braude, chapter 10) when assessing the costs and benefits of dispersal (Perrin and Goudet, chapter 9).

This shows that we need a multi-causal approach, including biotic and abiotic factors, in order to understand the interplay between the numerous factors that have been shown individually to promote the evolution of dispersal. Such an approach is shown in chapter 11 (Gandon and Michalakis). Spatial heterogeneity can be induced by variation in habitat quality as well as in social structure. Quality is thus reflected not only by resource level, but also by densities of competitors, parasites, and predators. Many species of hosts and parasites (Boulinier *et al.*, chapter 12a) as well as prey and predators (Weisser, chapter 12b) respond to one another's spatial distributions, and, in some species, dispersal has been shown to respond to the level of parasitism or predation.

In many cases, dispersal is condition dependent and influenced by numerous factors. However, identifying mechanisms that directly affect dispersal of individuals has been difficult. One way to test and possibly verify discrete mechanisms is by studying the genetic, physiological, and behavioural components by which a dispersal phenotype is produced. In numerous insects, the production of a winged dispersal morph has a genetic basis with no apparent environmental influence (Roff and Fairbairn, chapter 13). In other insects, however, environmental determination of dispersal phenotypes seems to be the rule (Weisser, chapter 12b). The selective factors that promote condition-dependent *versus* condition-independent dispersal are not always clear, but the existence of non-zero spatial and environmental autocorrelation is likely to be the key to this process. Whereas migration appears to be strongly genetically determined (Roff and Fairbairn, chapter 13), dispersal is frequently condition dependent (Ims and Hjermann, chapter 14). In addition, the dispersal phenotype may be open to environmental influences quite early in development (including grand-maternal effects). Steroid hormones appear to play a prominent role in dispersal decisions (Dufty and Belthoff, chapter 15). The extent to which these hormones, acting alone or in interaction with other categories of hormones such as leptin (a hormone linked to conditions influencing dispersal), actually shape the development of the dispersal phenotype remains to be seen.

What cue triggers the decision to disperse? The existence of condition-dependent dispersal implies that some cues are available in the environment and that they provide reliable information on its current or future quality, for example with respect to breeding. Dispersal is a three-step process involving: a decision to leave the natal (breeding) site, a transience phase, and the selection of a new breeding habitat. Although we have some evidence of the cues used in the first and last steps, we still know little about the transience phase (Wiens, chapter 7), or about how information gathered during the entire process interacts to produce optimal habitat selection (Stamps, chapter 16). Relying on passive information delivered by conspecifics (especially their reproductive success) is a cheap way of assessing environmental quality, although the feasibility of this strategy will depend on environmental autocorrelation and, to a lesser extent, on the species' life cycle (Danchin *et al.*, chapter 17). Further research on these topics is needed to link physiology, development, and behaviour better to understand dispersal and settlement decisions (Ronce *et al.*, chapter 24).

A study of the evolution of dispersal would not be complete without an understanding of the consequences of this movement on gene flow and adaptation, as well as on the attributes and fitness of individual life history traits and on population and community dynamics. Accounting for more realism in description and models of dispersal and population structure has profound effects on effective population size, the probability of fixation of beneficial alleles, genetic variance, and the potential for local adaptation (Whitlock, chapter 19). Incorporating precise details of dispersing and resident individuals or landscape structure will have an important impact on predictions about population persistence and evolution, to the extent that we cannot yet make any general statements (Barton, chapter 23). Indeed, individuals that disperse frequently are not a random population subset. Dispersers often show particular genotypes or phenotypic traits which may render them more efficient at colonizing or at integrating into a new population (Murren *et al.*, chapter 18). In most theoretical models, however, immigrant fitness is predicted to be lower than that of residents (especially because of the costs of dispersal), even though there is no empirical support for this assumption. Although comparisons between resident and immigrant fitness are complicated by technical and theoretical problems, it is still possible to study the interplay between dispersal motivation, individual characters, and fitness (Ronce *et al.*, chapter 24).

It is now widely accepted that dispersal plays a major role in population dynamics, especially through density-dependent emigration and colonization (Hanski, chapter 20). As with information acquisition, spatio-temporal variability and autocorrelation may be needed to understand the demographic consequences of dispersal in terms of which strategy should be favoured (Danchin *et al.*, chapter 17) and of stabilizing and synchronizing local population dynamics. In addition, metapopulation persistence and contour appear to be dependent on both demographic and genetic effects, but the interaction between these factors as well as their feedback on the evolution of dispersal is still poorly understood (Hanski, Barton, and Ronce *et al.*, chapters 20, 23, and 24 respectively).

If we have difficulty in understanding the consequences of dispersal for population dynamics and the evolution of other traits and syndromes, we should be prepared to face even more difficulties in understanding its role in antagonistic interactions (van Baalen and Hochberg, chapter 21) and in community organization (Mouquet *et al.*, chapter 22). Until recently, dispersal in host–parasite and prey–predator interactions has been considered to be random. Incorporating condition dependence into the dispersal of the different components of these systems has revealed surprisingly complex patterns and problems, to the extent that they put into question previous results on the stability of such systems and on their feedback on evolutionary outcomes, especially the co-evolution of dispersal rates of interacting species (van Baalen and Hochberg, chapter 21). The co-evolution of dispersal rates among interacting species may be the major force shaping dispersal patterns of some species. Given these results, it is hardly surprising that dispersal is crucial in explaining species coexistence in some communities (Mouquet *et al.*, chapter 22). The spatial dimension is a further component of the species niche, where some species can trade local competition for spatial exploitation, giving rise to the concept of meta-community. The interplay between local processes, regional biodiversity, and ecosystem functioning is an exciting new area where the ecosystems response to habitat fragmentation and global change needs to be investigated (Macdonald and Johnson, chapter 25).

In this volume, we are looking for unifying principles to develop a paradigm for applying dispersal theory to all taxa and levels of biological organization so that it may be applicable to evolutionary biology as well as conservation biology. Our hope is that these contributions will stimulate further research into the causes and consequences of dispersal evolution. We are far from having a global understanding of the role of space in individual, species, and community dynamics and/or evolution (Barton, chapter 23; Ronce *et al.*, chapter 24), or a concise understanding of the role of space in conservation biology (Macdonald and Johnson, chapter 25). The workshop that gave rise to this book was designed to encourage interaction between individuals and disciplines, to provide a general overview of other scientists' work, ideas, and methods. We originally planned to have more than 40 chapters, in order to cover as many taxa and aspects of dispersal as possible. This proved to be too ambitious. We have chosen to concentrate on reviews and more theoretical approaches, with a limited number of empirical examples.

The contributors were asked to use examples from as many taxonomic groups as possible, but some taxa are obviously under-represented (e.g. plants and marine systems) owing to the editors' bias. However, it should be relatively simple to extend and transpose the material discussed in most chapters on to other systems.

The conference that led to this book was held in Roscoff (France) from 23 April to 1 May 1999. It was organized as a CNRS–NSF joint seminar. We thank the National Science Foundation (NSF, USA), the Département des Sciences de la Vie and the Direction des Relations Internationales of the Centre National de la Recherche Scientifique (CNRS, France) for their generous support for the Roscoff conference. This book was published with the financial support of the International Union of Biological Sciences (IUBS) and related to a major element of Diversitas, an International Scientific Program on Biodiversity co-sponsored by International Union of Biological Sciences (IUBS), United Nations Educational, Scientific and Cultural Organization (UNESCO), International Council of Scientific Unions (ICSU), Scientific Committee on Problems of the Environment (SCOPE), International Geosphere-Biosphere Programme (IGBP) and International Union of Microbial Societies (IUMS).

This book would not have appeared without the help of Robert Barbault, CNRS Directeur Adjoint des Sciences de la Vie, who supported the project from its very origin. We particularly thank Françoise Saunier, without whom the Roscoff conference would not have taken place and the book would not have been produced. Molly Brewton accepted the difficult task of editing every chapter. Monique Avnaim helped in various steps of the organization of the conference and the editing process. Marie-Bernadette Tesson was our financial manager. André Toulmond, Director of the Roscoff Biological Station, kindly welcomed the conference to the station. Our contact in Roscoff, Nicole Sanséau, helped with the organization of the conference. Several of our students, Clotilde Biard, Blandine Doligez, Jean-François Legalliard, and Sandrine Meylan, kindly helped in the day-to-day organization of the conference. N.H. Barton, P. Beerli, N.C. Bennett, S. Bondrup-Nielsen, T. Boulinier, S.P. Carroll, C. Combes, D. Couvet, J.L. Dickinson, F.S. Dobson, J.A. Endler, B.J. Ens, G.W. Esch, L. Fahrig, R. Ferrière, S. Gandon, S. Geritz, C. Godfray, J. Goudet, I. Hanski, D.H. Heg, M.H. Hochberg, R.D. Holt, R.A. Ims, V.A.A. Jansen, H.B. John-Alder, R. Julliard, W.D. Koenig, X. Lambin, J.D. Lebreton, P.H. Lundberg, M. Massot, E. Matthysen, D.E. McCauley, I. Michalakis, A.P. Møller, P.J. Morin, T.A. Mousseau, I. Olivieri, N. Perrin, K.H. Pollock, O. Ronce, K.G. Ross, F. Rousset, F. Sarrazin, C.D. Schlichting, C.J. Schwarz, J. Shykoff, G. Sorci, J.A. Stamps, P.D. Taylor, T.J. Valone, M. van Baalen, M.J. Wade, P.M. Waser, M.C. Whitlock, J.A. Wiens, J.C. Wingfield, and J.O. Wolff were solicited as referees of the various chapters. Their work greatly contributed to the scientific level of the book. We would like to thank Cathy Kennedy, Senior Editor in Biology at Oxford University Press, for her interest in the project. Lastly, we thank the authors for attending the conference, for their diligence in preparing their manuscripts, and for their patience with the editorial staff. We hope that you will enjoy reading this book as much as we have enjoyed producing it.

Part I

Measures of dispersal:
genetic and demographic approaches

1

Methods for estimating dispersal probabilities and related parameters using marked animals

Robert E. Bennetts, James D. Nichols, Jean-Dominique Lebreton,
Roger Pradel, James E. Hines, and Wiley M. Kitchens

Abstract

Deriving valid inferences about the causes and consequences of dispersal from empirical studies depends largely on our ability reliably to estimate parameters associated with dispersal. Here, we present a review of the methods available for estimating dispersal and related parameters using marked individuals. We emphasize methods that place dispersal in a probabilistic framework. In this context, we define a dispersal event as a movement of a specified distance or from one predefined patch to another, the magnitude of the distance or the definition of a 'patch' depending on the ecological or evolutionary question(s) being addressed. We have organized the chapter based on four general classes of data for animals that are captured, marked, and released alive: (1) *recovery data*, in which animals are recovered dead at a subsequent time, (2) *recapture/ resighting data*, in which animals are either recaptured or resighted alive on subsequent sampling occasions, (3) *known-status data*, in which marked animals are reobserved alive or dead at specified times with probability 1.0, and (4) *combined data*, in which data are of more than one type (e.g., live recapture and ring recovery). For each data type, we discuss the data required, the estimation techniques, and the types of questions that might be addressed from studies conducted at single and multiple sites.

Keywords: capture–recapture, capture–resighting, dispersal, Heisey–Fuller estimator, Kaplan–Meier estimator, logistic regression, movement, parameter estimation, proportional hazards model, radio-telemetry, recovery data

Introduction

The movement of organisms from one location to another is integral to the study of both evolution and population dynamics. The study of evolution has focused primarily on the two fundamental determinants of fitness: reproductive rates and survival probabilities. Moving from one location, or, conversely, staying behind, can strongly affect both of these parameters (e.g., Parker and Stuart 1976). In the study of population ecology, all population change results from births, deaths, immigration, and emigration. There has been a relatively long and rich history of scientific investigation of the birth and death processes. In contrast, the process of dispersal (i.e., immigration and emigration) and its resulting spatial patterns constitute one of the biggest gaps in our knowledge of ecological dynamics (Wiens, chapter 7).

Until recently, available methodology often imposed limitations on our ability to derive inferences regarding many questions related to dispersal (North 1988; Nichols 1996). However, given the methodological advances over the past several years, it is our belief that the availability of suitable methods should now rarely pose a significant limitation on our ability to derive inferences about dispersal. Rather, the burden of deriving inference has shifted to the researcher's ability to implement well designed studies that are focused on relevant questions about the evolution and the population dynamic consequences of dispersal. Here, we review the methodological tools currently available for deriving inference about dispersal based on data from observations of marked animals over time and space.

This volume focuses on natal and breeding dispersal, and many chapters define dispersal in this context; however, the methods for estimating dispersal-related parameters may be equally applicable to other kinds of movement and migration. In the context of a probabilistic framework, we define a dispersal event as a movement of a specified distance or from one predefined patch to another. The magnitude of the distance or the definition of a 'patch' depends on the ecological or evolutionary questions being addressed. Dispersal is also a behavioural process which has a beginning (departure from a site), an intermediate stage (searching for or moving to a new site), and an end (settlement in a new site). Important ecological or evolutionary questions can arise from each of these stages, individually or in combination. For example, one may wish to examine factors influencing 'stay or leave' decisions (van Baalen and Hochberg, chapter 21), which would require a focus on the probabilities of departing (or staying at) a given site. Or interest may be on the process of selecting a new site, which may warrant a focus on how many dispersing animals 'stop' and 'sample' a given site, or how long they remain at an intermediate site (Kaiser 1995). One also may be interested in combinations of stages, such as the conditions of the 'departure' site relative to those of the 'settled' site, thus warranting a focus on the probabilities of moving from one specific site to another, perhaps in relation to time- and/or space-dependent covariates.

The methods available for estimating parameters associated with dispersal have been expanding rapidly, particularly over the past decade. Recent reviews of these advances have been made by Nichols (1996), in the context of migration, and by Nichols and Kaiser (1999) in a more general movement context. In order to complement, rather than duplicate, these reviews, the extent to which we discuss different methods is quite uneven. In some cases, where the methods have been well covered in previous reviews, we present only a superficial treatment. In contrast, we provide a somewhat more thorough coverage of methods (e.g., known status models) that have been covered less extensively in earlier reviews (Nichols 1996). Finally, it is not our intention that the chapter provide the reader with a full understanding of the methods we present. Such an understanding will undoubtedly require substantial further reading of the literature. Rather, we hope to introduce the reader to a basic 'toolbox' which will enable a wide variety of ecological and evolutionary questions about dispersal to be addressed.

Data types and associated estimation methods

A common feature of all the methods we describe is that they use data on observations of animals that can be individually identified. In most cases, animals initially will be physically captured and a mark applied. Subsequent encounters can involve recaptures or resightings of live animals, or recovery of dead animals. It may also be possible to avoid capture altogether if individuals can be photographed and identified from natural markings (e.g., dolphins, Slooten *et al.* 1992; tigers, Karanth and Nichols 1998), or from other markings not applied by investigators (manatees, Langtimm *et al.* 1998). Similarly, studies are currently underway that exploit the ability to identify individual animals using genetic markers from hair, feather, or faecal samples.

We have organized this chapter based on four general classes of reobservation data for animals that are captured, marked, and released alive: (1) *recovery data*, in which animals are recovered dead at a subsequent time, (2) *recapture/resighting data*, in which animals are either recaptured or resighted alive on subsequent sampling occasions, (3) *known-status data*, in which marked animals are reobserved alive or dead at specified times with probability 1.0, and (4) *combined data*, in which data are of more than one type (e.g., live recapture and ring recovery).

Within each major data type, we have also distinguished whether data were collected at single versus multiple study sites, as this is a primary determinant of what movement parameters can be estimated (Nichols 1996). Regardless of the types of data that are collected, when sampling is done at a single site, it limits the parameters that can be estimated to those related to that site. For example, it is sometimes possible to estimate parameters such as departure probabilities or residence time using data collected at a single site. However, it is not possible to estimate rates of dispersal between specific study sites when the sites of interest are not all sampled. In contrast, multi-site studies enable a much broader spectrum of parameters to be estimated. However, this broader potential comes with the cost that studies may be more difficult to orchestrate and require a substantial amount of data, such that the number of estimable parameters may be limited or estimates may have low precision.

Recovery data

Data description

Recovery data are obtained from animals that have been marked (usually with rings) on k occasions, released alive, and recovered dead at a subsequent time (Brownie *et al.* 1985). Such data are typically collected for exploited species in conjunction with annual harvests (e.g., waterfowl and fish). A key distinction between recovery data and the others we discuss is that recovery of the tag occurs as a result of the animal's death; thus, a given animal is encountered only once after initial marking.

Estimation methods

We know of no methods for estimating dispersal parameters using recovery data from a single site, although it is possible to estimate some parameters relevant to dispersal using recovery data from multiple sites (Fig. 1.1; see also Nichols 1996). The distances between the locations where animals are marked and where they are recovered have been considered for use in addressing questions about natal dispersal (van Noordwijk 1995), but spatially variable detection probabilities make this a difficult problem.

Schwarz and Arnason (1990) developed a set of models for migratory animals that are marked in one location (e.g., migratory birds on the breeding grounds) and then

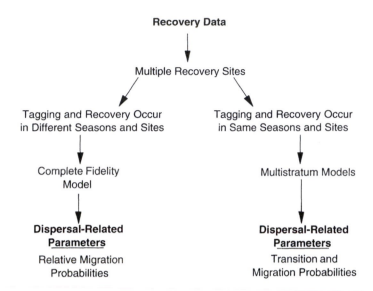

Figure 1.1. Schematic representation of methods for estimating dispersal-related parameters, using recovery data.

recovered in different locations (e.g., the wintering grounds). These models differ in their assumptions about fidelity to areas of ringing or recovery. In the case of their 'complete fidelity model', where birds return to the same breeding and wintering areas throughout their lifetime, survival probabilities corresponding to different recovery areas can be estimated. Such estimates could be useful in addressing questions about fitness consequences of movement to different locations.

If marking and recovery occur during the same season of the year, and thus at the same set of locations (e.g., waterfowl on wintering areas), multistratum models can be used to estimate area-specific probabilities of survival and movement (Hilborn 1990; Schwarz 1993; Schwarz *et al.* 1993). These estimated probabilities should be useful in testing hypotheses about relative rates of dispersal, based on estimates of habitat 'quality' or on predictions of expected fitness (Fretwell and Lucas 1970; Fretwell 1972) in the different locations.

Recapture/resighting data

Data description

Recapture/resighting data are obtained from animals that have been marked on k occasions, released alive, and either recaptured or resighted alive at a subsequent sampling period (often the same period during which marking occurs). Thus, in contrast to recovery data, animals may be observed on more than one occasion after initial marking.

Estimation methods

Inferences about movement based on single-location studies are necessarily limited (Fig. 1.2). However, it is possible to estimate the number of immigrant individuals

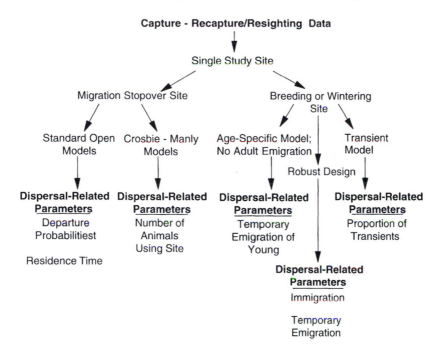

Figure 1.2. Schematic representation of methods for estimating dispersal-related parameters, using capture–recapture/resighting data from a single site.

entering a population between two sampling periods (Nichols and Pollock 1990; Pollock *et al.* 1990) using an age-specific modification of Pollock's (1982) robust design. Estimation of the relative contributions of immigration and *in situ* reproduction to population growth (Nichols and Pollock 1990) could be useful in addressing questions about the status of areas as sources or sinks (Pulliam 1988).

An open-population modelling approach developed by Pradel *et al.* (1997) permits estimation of the proportion of transients among newly captured individuals under an assumption of equal capture probabilities. Parameters associated with different forms of temporary emigration can be estimated under specific sampling designs and associated models (e.g., Clobert *et al.* 1994a; Kendall *et al.* 1997; Schwarz and Stobo 1997). It is also possible to address questions about departure probabilities and residence times at locations used temporarily by animals during movements (Lavee *et al.* 1991; Kaiser 1995). This approach might be adapted to address questions about how dispersing animals sample potential sites for future settlement.

Studies conducted at multiple sites can provide estimates of several parameters that should be directly relevant to addressing questions about the evolution and population-dynamic consequences of dispersal. The original multisite model developed by Arnason (1973) has been extended recently by Hestbeck *et al.* (1991), Brownie *et al.* (1993), Nichols *et al.* (1993), and Schwarz *et al.* (1993) (Fig. 1.3). Multisite models permit location-specific differences in capture probability and yield estimates of location-specific survival probabilities and probabilities of movement between all pairs of locations.

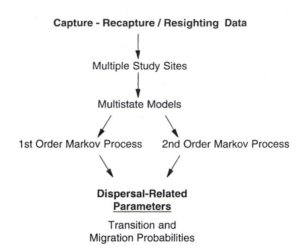

Figure 1.3.   Schematic representation of methods for estimating dispersal-related parameters, using capture–recapture/resighting data from multiple sites.

Such location-specific estimates are needed in order to investigate population-dynamic consequences of dispersal (Hanski, chapter 20).

Movement probabilities can be modelled directly as functions of external covariates, providing many opportunities for addressing questions about the ecology and evolution of dispersal (Nichols and Kendall 1995; Spendelow *et al.* 1995). For example, predictions of habitat selection models (Fretwell and Lucas 1970) and of models on the evolution of dispersal (McPeek and Holt 1992) and assumptions underlying stepping-stone models of population genetics (e.g., Crow and Kimura 1970) can be addressed directly within the structures of multistate models (Nichols and Kendall 1995; Spendelow *et al.* 1995). Movement probabilities can also be modelled as functions of covariates associated with individual animals, permitting additional interesting questions to be addressed. Sex-specific movement probability estimates can be used to test predictions about sex-biased dispersal emerging from inbreeding avoidance (e.g., Perrin and Goudet, chapter 9) and other hypotheses. Expansion of the state space to include not only location, but also temporally variable, individual-level state variables (e.g., body size, reproductive activity) permits tests of hypotheses about condition-dependent dispersal (Ims and Hjermann, chapter 14).

Recent work (J.-D. Lebreton *et al.*, unpublished) on multisite recruitment models (where recruitment refers to an attempt to breed for the first time) has focused on the estimation of parameters such as prebreeding survival probabilities and age-specific breeding probabilities for animals produced in, and recruited to, different subpopulations. This work includes three different types of multistate models corresponding to three field situations differing in biological features and corresponding sampling approaches. All three sampling situations involve capture–recapture/resighting on the breeding grounds during the breeding season each year. Animals are marked as young (just hatched, born, or fledged), and older animals can be resighted and/or recaptured.

The three specific sampling situations differ with respect to (1) presence of nonbreeders on the breeding grounds and (2) variability in the age of first reproduction.

In the first sampling situation, both breeders and nonbreeders are present on the breeding grounds and can be detected and their breeding state determined. Age at first reproduction can be variable such that we envision age-specific breeding probabilities. We use a full age-specific multistate approach in which the state space is defined by both location (the different sampled subpopulations) and reproductive activity (breeding or nonbreeding).

Parameters estimated under this approach will include local survival probabilities for each age–state combination. Imposition of constraints on these parameters can be used to test hypotheses about age specificity of survival, and about variation in survival associated with different locations and reproductive states. The other model parameters of primary biological interest are the transition probabilities reflecting both movement (from one subpopulation to another) and transitions between breeding states (breeding to nonbreeding and vice versa). Examples of relevant questions include: is the probability of moving related to breeding state (e.g., do nonbreeders have higher movement probabilities than breeders)? Are breeding state transitions associated with location (e.g., is the probability of making the transition from nonbreeder to breeder similar for the different locations or higher for locations of better habitat)? Nonbreeder-to-breeder transitions for the younger age classes can be used to address questions about recruitment. Age-specific probabilities of making this transition on natal locations versus non-natal locations will be especially relevant to questions about the roles of different subpopulations as sources or sinks.

In the second sampling situation, nonbreeders are not present on the breeding grounds, so young animals cannot be recaptured or resighted until they initiate breeding activity. Animals are assumed to remain breeders (or at least to exhibit 'adult' breeding probabilities) for all years following the year of initial reproduction, and age at first reproduction is assumed to be constant, such that all animals begin breeding and become exposed to recapture/resighting efforts at the same age (e.g., breeding probability is 0.0 for all ages less than some threshold, m, and either 1.0 or some unknown adult breeding probability for animals of age m and older). Estimated parameters include location-specific prebreeding survival probability (the probability that a young animal will survive until age m) and adult survival probability. Movement probabilities for prebreeders will reflect the probability that a young animal produced at some location r became a breeder at age m at location s, given that the animal survived the prebreeding period and reached age m. These movement parameters for prebreeders may be useful in distinguishing source and sink subpopulations.

In the third sampling situation, nonbreeders are again not present on the breeding grounds, so young animals cannot be recaptured or resighted until they initiate breeding activity. Animals are also assumed to remain breeders (or at least to exhibit 'adult' breeding probabilities) for all years following the year of initial reproduction. However, in this situation, the age at first reproduction is not assumed to be constant, such that animals may begin breeding and become exposed to recapture/resighting efforts at different ages. We thus consider different age-specific probabilities of entering the adult breeding population. After the initial breeding effort, animals are assumed to breed with probability 1.0, or with some unknown adult breeding probability. The modelling approach requires specification of the first possible age of reproduction, k, and of a maximum age, m, after which all animals are assumed to breed with probabilities characteristic of adults. In addition, all animals older than the first possible age of reproduction, k, are assumed to exhibit adult survival probabilities, regardless of whether or not they have become breeders. The modelling approach used here can be viewed as a multisite extension of the approach used by Clobert *et al.* (1994a).

Age-specific transition probabilities between the unobservable prebreeder component of the population and the observable breeder component are estimated directly. These transition probabilities reflect the probability that a prebreeder begins breeding, conditional on survival until the age in question. We might expect these probabilities generally to increase with age until age m, at which they become 1.0. These age-specific probabilities to the breeder state are also referenced to locations, reflecting probabilities associated with young produced on a specific area and recruited to the breeding population of another (or the same) breeding location. Comparisons of age-specific breeding probabilities associated with recruitment into (or from) different locations should permit inferences about the relative contributions of different areas to metapopulation dynamics.

In addition to the above modelling, recent research on methods for studying recruitment (natal dispersal) in multisite (metapopulation) systems has also extended to reverse-time approaches (Pollock *et al.* 1974; Nichols *et al.* 1986, 2000; Pradel 1996). These approaches involve reversing the time order of capture–recapture/resighting data and applying usual multistate models such as those of Brownie *et al.* (1993). We estimate unconditional transition probabilities reflecting the probability that an animal of age class j at location r in time $i + 1$ was a member of age class k at location s in time i. If the models involve multiple age classes, then the robust design (Pollock 1982) is required in order to estimate capture probabilities for the initial age class.

This approach permits decomposition of the animals present at location r, time $i + 1$, into components corresponding to surviving adults and young from the previous time period, i, from the same location (r) and from the other sampled location, s. In addition, the proportion of animals that were immigrants to the study system can also be estimated. Thus, hypotheses about the relative contributions to population growth of movement, both within the metapopulation and from outside sources, can be addressed directly. Questions about relative source and sink subpopulations, and about possible temporal variation in source–sink status, can be directly addressed as well.

Known-status data

Data description

The distinguishing feature of known-status data is that a marked animal can presumably be located at will. Known-status sampling is primarily obtained using radio-telemetry, although we could envisage other methods (e.g., radio-isotope tracers) that might enable the status of an animal to be determined at any given time. From a statistical standpoint, having a known status implies that a detection parameter (i.e., probability of observing the animal given that it is present in the sampled area) does not need to be estimated; it is assumed to be 1.0. From an information standpoint, it is possible to know whether an animal moved, when it moved, and, often, where it moved. The resulting data are thus viewed as complete histories (with respect to the event of interest, dispersal in our case) of individual animals over the duration of the study.

In some studies using radio-telemetry, status is not always known. In the case of temporary or intermittent failure to detect individuals (i.e., the detection probability is < 1.0), it may be advisable to treat the data in a capture–recapture context, rather than as known-status. By doing so, the detection probability is explicitly incorporated into the likelihood (Pollock *et al.* 1995). When an animal cannot be detected for a long period, after a given point in time the data are 'censored' (i.e., the animal is removed from the sample) (Lee 1980). Censoring does not bias the resulting estimates provided that it is independent of the status of the animal (Pollock *et al.* 1989a).

Estimation methods

The estimation methods that we consider initially are all based on an analogy between dispersal and mortality, as events that may (or may not) occur over a given time interval. Dispersal and mortality models using known-fate data can be viewed using any of three common functions, which are related through simple transformations (Lee 1980; Pollock *et al.* 1989a; White and Garrott 1990). The *fidelity function*, $F(t)$, (analogous to the *survival function*) (Fig. 1.4) is the (unconditional) probability that dispersal has not occurred between an initial time t_0 and a subsequent time t. Thus, it defines the cumulative probability that an animal has not dispersed since the start of the study. The

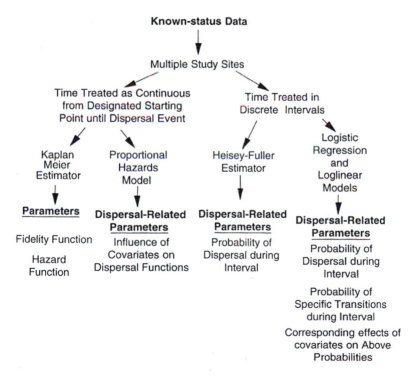

Figure 1.4. Schematic representation of methods for estimating dispersal-related parameters, using known-status data from multiple sites.

probability density function, f(t), for dispersal is defined as the limit of the probability that the event of interest (dispersal) occurred over a very short interval of t to $t + \Delta t$ (Lee 1980). The *hazard function*, h(t), is defined as the conditional probability of the event occurring during a short interval of t to $t + \Delta t$, conditional on the animal having been available (i.e., alive at the initial location) at the start of the interval (Lee 1980; Pollock *et al.* 1989a; White and Garrott 1990). The hazard function corresponds directly to the risk that dispersal will occur over a given time interval (Allison 1995). Models often differ in what is assumed about the underlying hazard function (see below).

Heisey–Fuller estimator. The Heisey–Fuller model (Heisey and Fuller 1985) was developed as an extension of an estimator of nest success described by Mayfield (1961, 1975) and applied to telemetry data by Trent and Rongstad (1974). The basic model estimates daily dispersal rate (d) as the proportion of the number of dispersal events observed (deaths in the original context of mortality) to the total number of animal observation days, where each animal observed for one day equals one observation day. The overall dispersal rate for the period of interest (say, K days) is then estimated as the complement of the estimated probability of not dispersing: $1 - (1 - d)^K$. Thus, it is assumed that the dispersal probability is the same for each day of the period of interest; in other words, that the hazard function is constant over the period. Common environmental phenomena (e.g., weather events) likely to influence dispersal could easily result in violations of this assumption. When such an assumption is not reasonable, the period of interest can be subdivided and a different daily rate estimated for each period, or, in some cases, alternative methods that do not assume a constant hazard function (e.g., Kaplan–Meier) may be more appropriate.

Heisey and Fuller (1985) originally proposed their approach primarily for comparison of cause-specific mortality rates. In this context, dispersal events to different locations could be treated as competing risks, for which the Heisey–Fuller model may be applicable (Nichols 1996).

The Kaplan–Meier estimator. The Kaplan–Meier (KM) estimator, or product limit estimator, is nonparametric in that no assumptions are made about the underlying hazard function (Kaplan and Meier 1958; Miller 1983). It is also robust and easy to compute. The primary disadvantage of the Kaplan–Meier approach is that it is asymptotically less efficient than parametric methods, such that, when censoring is high, estimates may be less precise than those obtained using a parametric fidelity function (Miller 1983; Efron 1988).

The KM estimator is an extension of a binomial estimator, and thus is closely related to other methods we describe (e.g., logistic regression) (Efron 1988). Data required for the estimator are the times elapsed between the beginning of a study and the observed times of dispersal. The probability of not dispersing over a given time interval is estimated as the proportion of the sample population that did not disperse out of the total sample available (i.e., alive and status known) to disperse during that interval. The fidelity function for any arbitrary time t is estimated as the cumulative product of not dispersing from the beginning of the study (time $t = 0$) through time t. Pollock *et al.* (1989b) developed a simple extension of the KM estimator to allow for new individuals to be added to an ongoing study (i.e., staggered entry), a situation common with radio-telemetry data. Detailed descriptions of this estimator, including variance estimators, can be found in Cox (1972, 1975), Cox and Oakes (1984), Pollock *et al.* (1989a,b), and White and Garrott (1990).

Log rank tests (Savage 1956; Cox and Oakes 1984) have typically been used to compare survival (fidelity) functions. However, Tsai (1996) recently compared the performance of the log rank test to several alternatives and found that the relative performance of the different tests depended greatly on the properties of the hazard function and sample sizes.

Example analysis. Because we believe that the KM estimator has considerable potential for estimating dispersal and we have seen no examples of its use for dispersal in the literature, we present a brief example here in the context of natal dispersal of snail kites (*Rostrhamus sociabilis*). Radio transmitters were attached at the time of fledging (approximately 35 days post-hatching). A juvenile was considered to have dispersed when it was located alive in a wetland other than its natal one. We considered a wetland to be distinct if it was separated by a physical barrier (e.g., ridge or primary levee), such that it was under a different hydrological regime than adjacent wetlands, either through natural or managed control (Bennetts and Kitchens 1997). The time of dispersal was estimated as the midpoint between the time of a bird's previous location, its natal wetland, and the time of its first location outside of its natal wetland. We considered dispersal in relation to the time of fledging (thus as a function of age), although we could just as easily have considered it relative to date. Birds were censored either if we were unable to locate their radio signal or if they were known to have died prior to dispersal from their natal area. We were also interested in whether dispersal from lake habitats differed from that from marsh habitats. Lakes are hydrologically more stable, which may influence the reliability of food resources. Thus, we predicted that juvenile birds might be less likely to disperse from lakes during their first year. We used log rank tests to test differences among dispersal functions.

We attached 117 radio transmitters on juvenile snail kites over a 3-year period from 1992 to 1994. As indicated by the fidelity function (Fig. 1.5), 75% of the juvenile kites dispersed from their natal wetland during their first year post-fledging. The hazard function illustrates that all dispersers moved within the first 220 days, with a pulse of dispersal at about 30 days. The results were consistent with our prediction that the probability of dispersal during the first year was lower ($\chi^2 = 6.22$, df $= 1$, $P = 0.01$) from lakes (0.60, ± 0.09 SE) than from marshes (0.84, ± 0.06 SE).

Methods for including covariates

Proportional hazards model. The proportional hazards model, sometimes called *Cox regression*, was introduced by Cox (1972, 1975) as a means of evaluating time-dependent covariates on survival functions. This model has since been applied to situations analogous to dispersal, for example, whether or not military cadets stay or leave service (e.g., Morita *et al.* 1989), and, more recently, to the dispersal of animals (Conroy *et al.* 1996). The proportional hazards model is semiparametric in that it assumes an underlying hazard function, but does not specify its form (Conroy *et al.* 1996). Coefficients (β_i) of explanatory variables are tested in a regression context for their effect on dispersal. The model incorporates covariates in such a way that it is assumed that the covariates act upon the baseline hazard function in a multiplicative manner. Thus, individual

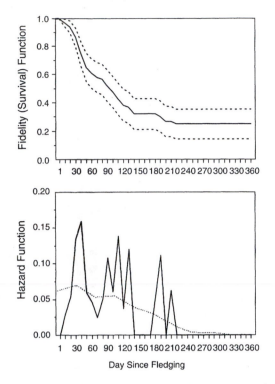

Figure 1.5. Kaplan–Meier estimates of the fidelity function (±SE) (top) for natal dispersal of juvenile snail kites during their first year post-fledging. Also shown is the corresponding hazard function for these estimates (bottom), including both the raw function (solid line) and the same function smoothed (dashed line).

hazard functions are assumed to be proportional to one another (Lee 1980; Pollock *et al.* 1989a; White and Garrott 1990), an assumption that may commonly be violated in natural populations. Conroy *et al.* (1996) found the proportional hazards model to have low bias and good confidence interval coverage when applied to estimation of movement rates, but its properties certainly warrant further consideration, particularly with respect to the proportional hazards assumption.

Logistic regression. When known-status data are collected over time intervals defined by discrete sampling periods (i.e., the status of each animal is ascertained at each sampling period, but not necessarily between samples), it is reasonable to consider the use of logistic regression (Cox 1970; Lee 1980; Hosmer and Lemeshow 1989). For this approach, the probability that an animal disperses during interval t to $t + 1$ is considered a binomial parameter, π, and the odds ratio for dispersal is the ratio between π and $1 - \pi$ (the probability of not dispersing). Then, evaluation of the risk factors influencing dispersal (or covariates associated with it) can be considered within a generalized linear modelling framework using a logit link function (e.g., Bennetts *et al.* 2000).

In this context, it becomes quite reasonable to consider factors that influence whether animals 'stay' or 'leave' a given site. Conroy *et al.* (1996) recently applied logistic regression to estimate specific transition (movement) probabilities among habitats.

Based on a Monte Carlo simulation of this application, Conroy *et al.* (1996) found that, for estimates of movement rates, bias was low but confidence interval coverage was poor. Both bias and confidence interval coverage were poor for assessment of a habitat effect on specific transitions. Thus, additional work on the properties of logistic regression in this context is probably warranted.

Combined data types

The combination of different data sources permits estimation of additional parameters relevant to dispersal (Fig. 1.6; also see Nichols 1996; Nichols and Kaiser 1999). For example, Burnham (1993) developed a model for single-site studies that uses both site-specific capture–recapture/resighting data and range-wide ring recovery data. These data permit estimation of true survival and the probability of permanent emigration (permanent movement away from the site of the capture–recapture/resighting sampling efforts) (see Szymczak and Rexstad 1991). M.S. Lindberg *et al.* (unpublished work) have extended Burnham's (1993) approach to include the estimation of temporary emigration (following Kendall *et al.* 1997) using the robust design. Barker (1995, 1997) considered data from single-site recapture/resighting sampling in combination with range-wide tag recoveries and ancillary resighting observations. These models also permit direct estimation of survival probability and the probability of permanent emigration (Barker 1995, 1997). Barker (1995) outlined extensions of his approach to multi-site studies, and the general approach of Lindberg *et al.* (unpublished) can be extended to multi-site studies as well. Such extended models would permit the estimation of the usual between-site movement probabilities and location-specific survival probabilities, as well as

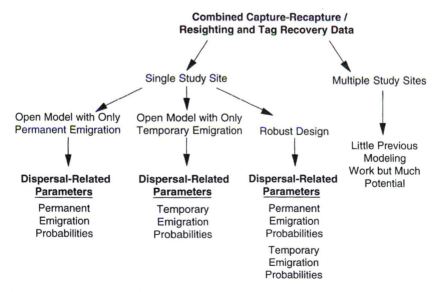

Figure 1.6. Schematic representation of methods for estimating dispersal-related parameters, using a combination of capture–recapture/resighting and tag–recovery data.

estimates of probabilities associated with permanent and temporary movement out of the study system.

Discussion

This volume contains many interesting predictions, questions, and hypotheses about the evolution and consequences of dispersal behaviour. Although there has clearly been much interesting and useful empirical work done on dispersal-related topics (e.g., reviews of Roff and Fairbairn, chapter 13; Stamps, chapter 16; Lambin *et al.*, chapter 8; Ims and Hjermann, chapter 14), we are struck by the relatively small number of instances of usage of state-of-the-art approaches to the estimation and modelling of data on animal movement. The large number of interesting theoretical questions and model-based predictions presented in this volume would seem to present enormous opportunities for important empirical testing. We believe that the strongest inferences will be obtained from tests based on estimation models that formally and explicitly deal with the employed sampling situations (e.g., with detection probabilities smaller than 1.0). These estimation methods should prove useful, not only for more descriptive studies of dispersal, but also for experimental approaches yielding much stronger inferences.

It is very clear that gene frequency data available through the use of molecular genetic techniques carry information about dispersal (e.g., Peacock and Ray, chapter 4; Rousset, chapter 2; Ross, chapter 3). However, it is not nearly so clear exactly how such data can be used to obtain estimates of dispersal probabilities (Rousset, chapter 2). In fact, this estimation problem falls into the general class of notoriously difficult problems involving attempts to infer process from a description of pattern (e.g., efforts to draw inferences about community dynamics from a description of community pattern; Strong *et al.* 1984).

It seems to us that there should be some potential for combining direct estimation methods based on marked animals (described above) with genetic migration models based on molecular data. We can envisage the development of joint likelihoods containing both sorts of data. The data on marked individuals would be modelled using the methods described above (e.g., multisite capture–recapture models), with location-specific dispersal probabilities incorporated as parameters in the estimation models. Genetic data from the multiple study locations could then be modelled under alternative sets of assumptions about migration (e.g., island model versus isolation by distance). These models for the genetic data would also contain parameters for migration that would be constrained to be either equal to, or proportional to (e.g., if the movement probability parameters based on direct estimation of marked animals include only movement and do not necessarily correspond to reproducing animals responsible for gene flow), the movement parameters for the models based on marked animals. We would thus have alternative joint likelihoods, with the same likelihood component for marked animals appearing in all alternative models, but different models for the genetic data corresponding to the competing hypotheses about gene flow and population structure.

Such joint likelihoods could serve two potentially useful purposes. First, the data for marked animals could be used to help distinguish between the competing genetic models, using likelihood ratio tests or model selection metrics such as Akaike's Information Criterion (see Akaike 1973; Burnham and Anderson 1998) to assess formally which genetic model best matched the data on marked animals. Second, given the most appropriate genetic model, estimation of migration probabilities would be accomplished by bringing both sources of data (observed movements of marked animals and location-specific gene frequency data) to bear upon the single estimation problem. Such an approach seems potentially useful, although we recognize that it is conditional on the development of reasonable models that relate migration parameters to data on genetic pattern, and that such development may be problematic (Rousset, chapter 2).

Acknowledgements

This chapter was greatly improved by the helpful comments of Romain Julliard, Kenneth Pollock, and Carl Schwarz.

2

Genetic approaches to the estimation of dispersal rates

François Rousset

Abstract

This chapter reviews current methods for the estimation of dispersal rates from genetic data, based on models of neutral and selected markers. The causes of their inferred failures are discussed, with a more detailed investigation of difficulties raised by long-distance dispersal. It is emphasized that studies at a local geographical scale are more likely to yield valuable estimates. It should be possible to define more efficient and reliable methods, particularly when it is possible to sample from most subpopulations connected by dispersal at a local scale.

Keywords: backward dispersal, dispersal distance, F statistics, maximum likelihood, island model, isolation by distance, comparison of genetic and demographic estimates, long-distance dispersal, robustness, neighbourhood, wave of advance, autocorrelation statistics, assignment procedures

Introduction

A number of theoretical analyses have concluded that various evolutionary phenomena – say, whether local adaptation will occur or not – depend on specific parameters of dispersal such as the fraction of individuals entering a subpopulation each generation (m) or the second moment of dispersal distance (σ^2), generally in combination with selection parameters or with subpopulation size or density parameters. These dispersal parameters refer to natal and breeding dispersal rates: they describe dispersal between reproductive events. These are also backward dispersal rates: for a given breeding individual, one considers the position where its parents reproduced. Thus, passive dispersal of seeds to unsuitable habitats, for example, is not considered.

Many studies attempt to estimate such parameters using genetic markers. Perhaps the first study of this kind was that of Dobzhansky and Wright (1941), using lethals in populations of *Drosophila pseudoobscura*. They concluded (p. 49) that 'we entertain no illusions regarding the lack of precision in the results obtained. At best it is hoped that the order of magnitude of certain variables effective in population dynamics has been arrived at. But at a certain stage of the development of a scientific subject even such rough estimates may be useful to guide further work.' How much this field has progressed is unclear. Studies of population structure guide future enquiries even if they do not provide an *estimate* of dispersal. However, the current routine use of genetic methods to 'estimate dispersal' still has weak theoretical foundations and few demonstrated successes. This short review will discuss current methods, results, prospects, and concerns about such an approach. Because it is of increasing interest for ecological studies to identify immigrants individually and to know their origin, I will also describe some recent attempts in this direction. A more extended description of the statistical techniques used in these different methods may be found in Rousset (2000b).

Some methods

I will first review some of the methods proposed to estimate the dispersal parameters σ or m. I will emphasize those methods that are based on a well-defined theoretical and statistical framework, and that have been tested by comparison with demographic estimates. The classical F_{ST} techniques are based on one-time samples of supposedly neutral alleles from several subpopulations; other methods using the same information will be presented. Then, clines of selected markers will be considered. Inference from such clines requires their existence and information on the selection process affecting the markers. Some other methods will be discussed briefly.

Neutral differentiation

Allelic identity and F statistics

F_{ST} is the quantity that Wright initially devised for his theory of adaptation in subdivided populations (see Provine 1986, for a historical account). Since this early work F_{ST} has generated a considerable literature. It is best defined as $F_{ST} \equiv (Q_w - Q_b)/(1 - Q_b)$ where Q is the probability of identity in state of pairs of genes either within (Q_w) or between (Q_b) subpopulations. Here, identity in state is simply whether the alleles are similar according to the nature of the data. For example, microsatellite alleles are identical in state when allele size is identical if data consist of allele sizes, and they are identical in state when their sequence is identical if data consist of allele sequences.

Given a relationship between F_{ST} values and demographic parameter values, such as subpopulation size and dispersal rates, one may deduce an estimate of demographic parameter values from an estimate of F_{ST}. The usual example is the infinite island model with subpopulations of N adults and backward dispersal rate m, for which $F_{ST} \approx 1/(1 + 4Nm)$, so that an estimate of Nm may be deduced from an estimate of F_{ST}. A straightforward way of estimating F_{ST} is to estimate each of the probabilities of identity and to compute the ratio of estimates $\hat{F}_{ST} \equiv (\hat{Q}_w - \hat{Q}_b)/(1 - \hat{Q}_b)$. Note that this is essentially what the most used analysis-of-variance estimator of F_{ST} (Cockerham 1973; Weir and Cockerham 1984) does.

While the *nhi502@abdn.ac.uk* approach is conceptually straightforward, it does not claim efficiency. I will present attempts to go beyond this method, both in terms of the range of models considered and the efficiency of estimation.

F_{ST} analogues

Analogues of F_{ST} have also been defined taking into account microsatellite allele size (Slatkin 1995; Michalakis and Excoffier 1996). Several methods have also been devised specifically for the analysis of sequence data. They have in common the assumption of a linear relationship between a measure of sequence divergence (or size divergence for microsatellites) and the coalescence time of alleles. This may be used to construct measures analogous to F statistics in their definition (pairwise comparisons of genes) and in their quantitative interpretation (e.g. Excoffier *et al.* 1992). Different alternatives have been proposed for the analysis of sequence data. One may construct an allele phylogeny and count the minimal number of dispersal events compatible with this phylogeny (Slatkin and Maddison 1989; Hudson *et al.* 1992). An estimate of Nm is then deduced from numerically established relationships between this minimal number of events and dispersal rates. Wakeley (1998) derived an estimator of Nm based on the total length of the genealogy of n sampled sequences from d sampled demes and on the length of the genealogy of d sequences, one from each of the d demes. He presented some simulations suggesting that this estimator performs better than the moment method.

F_{ST} analogues that 'use additional information from the data', such as measures devised to take into account allele size for microsatellite data, may have poor statistical properties in comparison to more conventional F statistics, when the assumptions involved in 'making use of additional information' do not exactly hold and even in some cases when they do hold (as shown in Slatkin 1995, Table 1, for $Nm = 10$). In the latter case, it thus appears that what constitutes 'information' does not always conform to *a priori* intuition. Other examples are known in genetic data analysis (e.g. Ewens 1972). A strength of likelihood theory (see, for example, Cox and Hinkley 1974, for a sound exposition) is that it defines what is information and makes optimal use of it under some conditions. Some likelihood approaches are discussed below.

Likelihood approaches

Legitimately, there is increasing interest in likelihood methods for the estimation of demographic parameters from genetic data. Such attempts have been hampered by both theoretical and practical problems. These problems include the difficulty of computing the likelihood of a sample under theoretical models and the difficulty of finding the maximum likelihood estimate given the likelihood. Two slightly different algorithms have been proposed to compute likelihoods under coalescent approximations. One was originally described by Griffiths and Tavaré (1994). Using a variant of this algorithm Nath and Griffiths (1996) have implemented a maximum likelihood estimator of Nm using large subpopulation, low dispersal approximations to island models. Extensions of this approach are described by Bahlo and Griffiths (2000) and incorporated in the GENETREE software (http://www.maths.monash.edu.au/~mbahlo/mpg/gtree.html). The other is described by Beerli and Felsenstein (1999) and incorporated in the LAMARC software (http://evolution.genetics.washington.edu/lamarc.html). They used it to investigate estimation of several parameters in a two-subpopulations model with different subpopulation sizes, N_1 and N_2, and asymmetrical dispersal rates, m_1 and m_2. For sequence data it was found that the method performed generally better than an *ad hoc* F_{ST} analogue for sequence data, for the estimation of $N_i m_i$, $N_i \mu$, and m_i/μ, where μ is a measure of mutation.

An alternative approach has been taken by Rannala and Hartigan (1996). They used the diffusion approximation to the infinite island model, which gives the distribution of gene frequencies in populations, from which the likelihood of samples can be deduced. They considered a 'pseudo' maximum likelihood estimate (MLE) of Nm, an approximation for the MLE which is expected to be asymptotically as efficient as the MLE but is much easier to compute than the MLE. They investigated the efficiency of this estimator in an island model with low dispersal rates and found it better than the efficiency of an Nm estimator based on Cockerham's analysis of variance (ANOVA) estimator of F_{ST}. A program is available at http://allele.bio.sunysb.edu/software.html. Like all the models reviewed here, the likelihood model rests upon the assumption of demographic equilibrium; Rannala and Hartigan's (1995) alternative derivation of the same distribution for a model of subpopulations of variable size rests upon the assumption that the backward migration rate is independent of subpopulation size.

Tufto *et al.* (1996) have proposed a multinormal approximation to the distribution of gene frequencies in a set of subpopulations connected by an essentially arbitrary dispersal pattern. Actually, the parameters of this multinormal distribution can be expressed in terms of probabilities of identity, in the same ways as F_{ST} parameters. The status of these approximations is unclear. For example, they are not compatible with the diffusion approximation to the infinite island model. Nevertheless, Tufto *et al.* have provided some simulations in which the estimates show little bias.

Isolation by distance

A specific class of alternatives to the island model is the isolation by distance models. These models embody the hypothesis that dispersal occurs preferentially between nearby subpopulations. Such models were first discussed by Wright (1943, 1946), but a rigorous formulation is available only for the lattice models and is from Malécot (1950). Some analyses of the nearest neighbour ('stepping-stone') case were given by Malécot (for a one-dimensional model) and by Weiss and Kimura (1965). Nagylaki (e.g. 1976) and Sawyer (1977) gave results for more general dispersal distributions.

The literature has had difficulties making sense of these analyses. One trap is that F_{ST} values have too often been identified with probabilities of identity by descent within subpopulations. A more coherent interpretation of the models was developed by Slatkin (1991, 1993), who studied approximations for F_{ST} as a function of geographical distance. Another difficulty is that most authors have followed some of Wright's intuitions in assuming that F_{ST} values were determined by a 'neighbourhood' size. It is hard to make sense of such ideas (see Rousset 1997, for discussion). Rather, it is instructive to look at the pattern of F_{ST} values through a series of dispersal models showing the transition from a stepping-stone to an island model (Fig. 2.1). In particular, this comparison shows that σ^2 is not the only feature of the dispersal distribution that determines pairwise F_{ST} values. What σ^2, or more precisely the product $D\sigma^2$ where D is the population density, approximately determines is the increase of differentiation of pairwise F_{ST} values with distance.

Thus, in principle, it is possible to estimate $D\sigma^2$ from the increase of pairwise F_{ST} values with distance. A regression method along this line is described in Rousset (1997) and incorporated in the GENEPOP software (Raymond and Rousset 1995a; ftp://ftp.cefe.cnrs-mop.fr/genepop/). It may be easy to improve on this estimation method for

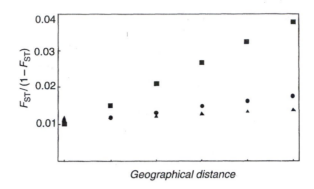

Figure 2.1. Comparison of expected differentiations for a fixed backward dispersal rate m and increasing values of σ. In a two-dimensional habitat, an approximately linear relationship is expected between $F_{ST}/(1 - F_{ST})$ and the logarithm of distance. The figure is drawn for the geometric dispersal model (see Rousset, 1997, p. 1228, for details), with an axial migration rate $m \equiv d/2 = 1/3$, and $\sigma^2 = 1$ (■), 5 (●) or 22 (▲). For increasing σ, the island model with total migration rate $1 - (1 - m)^2$ is approached.

$D\sigma^2$ by considering only a family of dispersal distributions where m is completely determined by σ^2; in this case F_{ST} values, not simply their increase with distance, will contain information about σ^2. But such improvements will not be robust to incorrect specification of the dispersal distribution.

Auto-correlation methods are basically descriptive methods which fall out of the task of estimating dispersal, but there have been some attempts to turn them into an estimation method of $D\sigma^2$. Epperson and Li (1997) established numerical relationships between some autocorrelation statistics and $D\sigma^2$ values for a two-allele model with dispersal between adjacent subpopulations only. The generality of such relationships for the models described above remains to be investigated. Since these statistics are constructed from pairwise comparisons of genes or genotypes, they are somewhat analogous to F statistics and there should be essentially the same information in such statistics as in estimators of pairwise F_{ST} values (see also Hardy and Vekemans 1999).

Assignment methods

Given samples from most or all subpopulations of a set of subpopulations exchanging individuals, it is possible to use an assignment procedure to find the origin of an immigrant (e.g. Paetkau *et al.* 1995; Rannala and Mountain 1997). It would then be possible to estimate dispersal rates between all pairs of subpopulations. The performance of such an approach has been little investigated up to now (Cornuet *et al.* 1999) but, given the appropriate data, it could be more efficient than other methods and would allow estimation of parameters not estimable by other methods, particularly in cases where dispersal has been rare or nonexistent up to the last few generations.

Clines of selected markers

This method considers gene frequency clines that are maintained, either by different selection pressures in geographically adjacent areas or by selection against hybrid genotypes between two taxa ('tension zones'). Theoretical models predict the shape of clines, notably the steepness in the cline centre, as a function of the σ^2 parameter and of one or several selection coefficients (for tension zones: Bazykin 1969; Barton 1979; for selection variable in space: Nagylaki 1975). Additional information on dispersal and selection is given by linkage disequilibria between loci. Such methods may be therefore be applied only when multilocus clines are observed, and depend on assumptions specific to each case study (e.g. external information on recombination rates and epitasis between loci). Given this substantial drawback, the method is still applicable to many situations of intrinsic interest for evolutionary studies.

This approach differs in several ways from analyses based on neutral models. First, it estimates a dispersal parameter (specifically σ), rather than its product with some population density parameter. Second, drift is neglected relative to selection: gene frequencies in the different subpopulations are fixed values, functions of the parameters defining the demography and the selection regime, not random variables as in the neutral models. The likelihood function is then given by multinomial sampling in each subpopulation. Thus, estimating the parameters is conceptually straightforward and relatively easily tailored to the specific demography and selection regime of different organisms, at least when only a few loci subject to selection need be considered and when the expected frequencies of each genotype are computed directly using recursion equations for specific values of the parameters to be estimated (e.g. Lenormand *et al.* 1998). Finally, the approach seems promising in terms of precision. The main practical difficulty is to find the maximum likelihood estimates: it is necessary to estimate many parameters simultaneously, and there is no perfect algorithm to estimate them. The currently used algorithms are variants of the Metropolis algorithm (including simulated annealing: Kirkpatrick *et al.* 1983). No software seems to be available. Various alternatives to such numerical procedures have been considered previously in the analysis of few-loci systems (Mallet *et al.* 1990), and approximations may still be required for cases involving a large number of loci (e.g. Szymura and Barton 1991; Sites *et al.* 1995; Porter *et al.* 1997). Some of these approximations make the maximum likelihood status of resulting estimates sometimes dubious. For example, models of linkage disequilibrium for panmictic populations have been used as statistical models for linkage disequilibria generated by selection.

Successes and failures

In the formulation of any of the above methods, assumptions are made relative to, for example, the temporal stability or, alternatively, the specific demographic scenarios involving fluctuations in population size. Numerical studies of the robustness of estimators when such assumptions are not valid may miss some important factor. Thus, it is important to test the methods against independent estimates of dispersal parameters obtained by demographic methods such as mark–recapture. Several surveys have concluded the existence of some rough agreement, but many have indicated discrepancies in specific cases (e.g. Hastings and Harrison 1994; Slatkin 1994; Ward *et al.* 1994; Koenig *et al.* 1996, for reviews). While many studies have compared their estimates of genetic differentiation to *a priori* expectations based on various arguments, few have compared them to independent estimates obtained by statistical analysis of demographic observations from the same populations. Below, I exclude studies that have observed that an '*Nm* estimate' was larger than the number of individuals dispersing as far as the subpopulations compared, and concluded the existence of a discrepancy between the different estimates. Such a result is expected under the isolation by distance models, and does not allow further quantitative assessment. The few remaining studies include those of Waser and Elliott (1991), though no estimate is deduced from their genetic analysis, Johnson and Black (1995), Sites *et al.* (1995), and Rousset (1997, 2000a), the latter reanalysing three data sets, one from Long *et al.* (1986), Wood (1987), and Wood *et al.* (1985), the others from Johnson and Black (1995) and Waser and Elliott (1991). The dispersal parameter considered in all cases was σ or $D\sigma^2$. Of these studies, Waser and Elliott (1991) and Johnson and Black (1995) concluded discrepancies between the two approaches, while the others concluded relative agreement, with estimates differing by a factor of 2. The conclusion therefore depends on the methods of analysis. It must be noted that these comparisons are based on local patterns of genetic differentiation, the maximum distance between samples being 20σ.

3. Interpretation of failures

Discrepancies between genetic and demographic estimates have been attributed to several different causes. The dispersal parameters in the genetic models are backward dispersal rates (defined from considering where the parent of an adult was), while demographic studies may estimate rates with different interpretations, such as a forward dispersal rate (considering where juveniles go). However, backward dispersal rates may be estimated in some demographic studies. The genetic models have some other obvious deficiencies, the list of which is well known.

The neutrality of genetic markers is often assumed. This assumption is less likely to hold as one considers larger areas because of habitat-related selection. The act of faith of assuming neutrality of allozyme markers is likely to lose its charm as microsatellites are increasingly used, but has already been replaced by the act of faith of assuming neutrality of microsatellites. It should be noted that locus-specific patterns of differentiation can themselves give some evidence for non-neutrality of the markers used, as discussed previously for F_{ST} values (e.g. Beaumont and Nichols 1996; Ross *et al.* 1999).

A second assumption is demographic stability. Some types of demographic instability, such as range expansions, have a more lasting impact on estimates of dispersal as one considers larger areas (e.g. Slatkin 1993; Hardy and Vekemans 1999), but the impact of other types of demographic fluctuations is more complex and cannot be summed up so neatly (e.g. Pannell and Charlesworth 1999). In fact, demographic instability raises questions about the very aims of the study of dispersal, not simply about particular estimation methods: if dispersal permanently varies in time, what would be the interesting dispersal parameters to estimate?

A third category of assumptions relates to mutation rates and mutational processes. Some recent literature has discussed the effect of mutational processes on the value of F_{ST} and some other measures (e.g. Slatkin 1995) and attempted to construct statistics taking into account supposed mutational processes at microsatellite loci. Later work has shown that for F statistics a high mutation rate is a more important feature than the supposed mutational processes, because the quantitative impact of a 10^3 variation in mutation rate is much larger than the variation resulting from different mutational processes. In particular, homoplasy itself is not a problem for analyses based on F statistics (Rousset 1996), while mutation rate may affect the spatial scale within which low mutation approximations can be used. High mutation rates will not be so much a concern as long as one focuses on local processes (involving subpopulations exchanging individuals at rates m higher than the mutation rate, or at distances $r \ll \sigma/\sqrt{2\mu}$; see Rousset 1997, for details), which should often be the case when one is studying processes of local adaptation.

What may have been most critical to many inferred discrepancies between direct and indirect estimates is an inadequate use of the models. At a local scale, the classical formulas of the island model are surprisingly accurate approximations for much more general dispersal models (Fig. 2.1; see also Hudson 1998, and Rousset 1999, for a discussion of models with spatial heterogeneity of local density and of local dispersal rates), yet these approximations are derived under the assumption of low dispersal rates between well-identified subpopulations and may be of dubious value for many studies. The problem in comparing 'Nm estimates' with the number of individuals exchanged between the subpopulations compared has been noted above. In the isolation by distance models, it has often been assumed that σ^2 plays a role analogous to m in the island model, in contradiction with the theory reviewed above. It is clear that many of the reported discrepancies between direct and indirect estimates of 'dispersal' may actually reflect misinterpretations of the theory of F statistics.

The question of the robustness of the analyses to deviations from assumptions remains, and becomes more pressing, when additional assumptions are necessary to 'take into account more information from the data'. The robustness of likelihood estimators, when assumptions about past demographic fluctuations are not valid, has not been addressed, since the application of likelihood methods is still in its infancy, but some insight can be gained by considering a slightly different problem. Nordborg and Donnelly (1997) considered the estimation of the selfing rate, using sequence data, and compared moments methods based on the comparison of pairs of genes (analogous to the F_{ST} methods described above) with more full-blown likelihood methods making use of Griffiths and Tavaré's (1994) coalescent algorithms. They show that the likelihood function is made of two parts, one corresponding to a 'fast process' and depending on recent selfing rates, and one corresponding to a 'slow process' which depends on all other assumptions of the model. The estimation of selfing rates was not much improved by an approach based on the total likelihood for realistic sample sizes, and the improvement is questionable, since it depends on assumptions relative to the slow process. An important conclusion from their analysis is that one does not necessarily gain much in seeking more 'information' in the data, such as making use of supposed relationships between sequence divergence and coalescence times. In the aim of robustness, a safe approach would use likelihood models for the fast process only. For the infinite island model, such a model might be given by Wright's distribution of gene frequencies (as used by Rannala and Hartigan 1996) since it is a low mutation approximation common to different mutation models.

Long-distance dispersal

In some cases, discrepancies between genetic and demographic estimates have been attributed to the failure of 'direct' (demographic) estimates of dispersal rates to detect long-distance dispersal (e.g. Mallet *et al.* 1990; Barton and Gale 1993; Koenig *et al.* 1996). Long-distance dispersal would be missed by demographic studies but would have an impact on genetic estimates. However, it is not clear why long-distance dispersal would not be 'missed' by the genetic methods as well. Indeed, one may question the logic of indirect methods for the estimation of σ when the value of σ is determined in large part by a few individuals dispersing long distance. This point has been little discussed in the literature, and I will illustrate it by numerical examples of the effects of long-distance dispersal on clines of selected alleles, and on neutral differentiation in a linear habitat as measured by F_{ST}. I focus on 'rare' long-distance dispersal, as specified below, because abundant long-distance dispersal would be apparent in the demographic studies. In these examples I consider that the probability m_k ($k \neq 0$) of dispersing by k spatial units is proportional to $1/k^n$ for some n. n measures how dispersal probability decreases with distance, and distributions with larger n have comparatively lower long-distance dispersal. When n goes down to 3, σ becomes infinite (although the dispersal distribution looks 'Gaussian'; Fig. 2.2a), and the kurtosis becomes infinite as soon as $n \leq 5$. Thus, when n goes down to 3, such distributions represent cases where most dispersal is local but where the small fraction of long-distance dispersal contributes in large part to σ values. The consequence of rare long-distance dispersal on the accuracy of theoretical approximations in terms of σ may be investigated by considering low values of n. For both neutral differentiation and one-locus clines maintained by selection against heterozygotes, the effect of long-distance dispersal is seen only for $n < 3.5$ (Fig. 2.2b).

For $n = 3.01$ the maximum slope in the centre of the cline is 5.68 times larger than as given by this theoretical approximation. Likewise, local neutral differentiation is greater than expected from the theoretical results summarized above (Fig. 2.2c).

The main conclusion is the not-unexpected fact that local genetic differentiation is affected by local dispersal patterns: a low level of long-distance dispersal has little effect on spatial patterns of genetic differentiation. In fact, by ignoring those individuals that go beyond a more or less arbitrarily chosen distance, one can compute a σ value for the truncated dispersal distribution and compute theoretical approximations according to the parameters of the truncated distribution. These approximations will better match the exact pattern. In using such a truncation we may approximate what is effectively done in comparisons of genetic and demographic estimates of dispersal: individuals dispersing at long distances are ignored, and matches between both estimates may reflect a match between local processes only. Conversely, it is unclear whether we can explain discrepancies between the two approaches by long-distance dispersal. Whether long-distance dispersal is important will depend on whether one is interested in testing models by comparisons of different methods of estimating dispersal as it affects local processes, or whether one is really interested in estimating σ without distinguishing local and long-distance processes. In the latter case, the genetic methods described above are surely not appropriate.

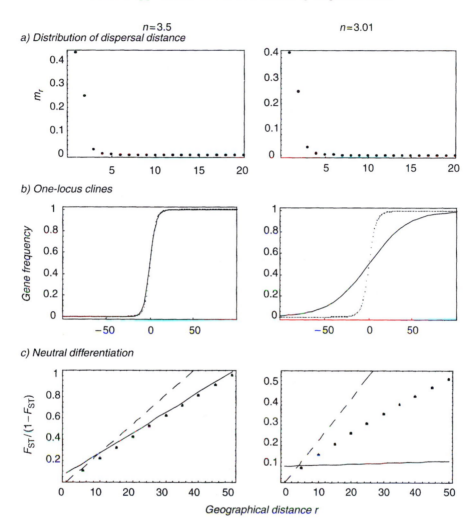

Figure 2.2. Effects of increasing levels of long-distance dispersal. There is more long-distance dispersal when $n = 3.01$ than when $n = 3.5$ (see text for details). From top to bottom, the dispersal distribution, gene frequency clines at a locus subject to underdominance, and neutral differentiation in a one-dimensional (elongated) habitat. In (b) the exact cline (dots) is compared with expectations according to Bazykin's (1969) formula, $p = 1 + \exp(-r\sqrt{2}s/\sigma)^{-1}$, for selection $s = 0.05$. In (c) the exact expectation (dots) is compared with the expected increase of differentiation according to σ for the exact distribution of dispersal (solid line) and to σ for the distribution truncated to its $k \le 7$ terms (dashed line).

Demographic 'waves of advance' have occasionally been used to estimate σ (e.g. van den Bosch *et al.* 1992; Metz and van den Bosch 1995), and waves of advance of genes may in principle be used in the same way. The theoretical approximations in terms of σ are less robust to long-distance dispersal than in the above processes (see Mollison 1991), raising more concern about the value of this method for many organisms.

Practical advice?

What practical advice could be given is limited by the scanty and piecemeal nature of the theoretical background, and it remains that the most widely used approaches give estimates of dispersal parameters only in combination with population density/size parameters. Moreover, one is often interested in evolutionary processes that cannot be understood through the simplest models, such as processes that would depend on asymmetries in gene flow between different habitats or on demographic fluctuations. There has been little discussion of the appropriate parameters to estimate in such cases, and of how to estimate them.

The above discussion of robustness issues suggests that one should focus on local processes: to estimate σ from populations spread over hundreds of kilometres or more when most dispersal occurs within 1 km is indeed dubious, yet this is a common practice. The choice of markers deserves little comment; for many reasons microsatellite markers have been adopted for genetic studies. It is likely that the efficiency of statistical analyses is optimal for some intermediate level of variability and that such optima may be best attained with microsatellite loci. As noted above, their high mutation rates may not be a major problem when one is interested in local processes such as studies of local adaptation.

One may expect that, for estimating dispersal parameters, the F_{ST} methods will be superseded by some likelihood methods. As pointed out in our discussion of the selfing rate example, it is not obvious which is the best likelihood model to consider. One may obtain additional information from, say, the mutation model, but this raises questions about the validity of the mutation models used. Alternatively, it may turn out that likelihood inferences will be roughly independent of the assumed mutation model, but this may also imply that not much information is to be deduced from mutational processes. Finally, data allowing comparisons of estimates obtained by demographic methods will be most needed. This is the only way to test the validity of genetic methods which, however elaborate and robust they may be, remain based on specific classes of demographic and genetic models.

In contrast to the techniques reviewed above, there has been almost no attempt to develop techniques specifically applicable to situations in which one can sample most subpopulations connected by migration at a local scale, perhaps because such sampling is not possible in most population genetic studies. With such sampling, however, it should be possible to estimate migration rates between the different subpopulations, using likelihood models more similar to the models of mark–recapture studies rather than likelihood models assuming specific past demographic scenarios.

Acknowledgements

I thank T. Lenormand and the participants of the Roscoff meeting for discussion, C. Chevillon for reading the manuscript, and P. Beerli for sharing unpublished information. My work on population structure has been supported by Service Commun de Biosystématique de Montpellier, by the PNETOX, and by the Entente Interdépartementale pour la Démoustication du Languedoc Roussillon (no. 95.162). This is publication ISEM 00-051.

3

How to measure dispersal: the genetic approach. The example of fire ants

Kenneth G. Ross

Abstract

Dispersal and gene flow were studied in the fire ant *Solenopsis invicta* by combining population genetic analyses and natural history information. In introduced and native populations of this ant, local genetic structure is evident in the polygyne social form (multiple queens per nest) but not the monogyne form (single queen per nest), a pattern particularly evident for the mitochondrial DNA (mtDNA). This result is predicted from differences in the reproductive behaviour of queens of the two forms. Moreover, the social forms are genetically differentiated where they occur in sympatry, a pattern that, again, is most pronounced for the mtDNA. Such between-form differentiation is expected if queens are unable to cross social boundaries. Detailed analyses of between-form gene flow using the mtDNA and non-neutral nuclear markers suggest that such gene flow occurs via only one of four possible routes (one not involving queens). The results reveal that social behaviour can have important consequences for dispersal and gene flow patterns, and that the genetic structure of wild populations can be accurately predicted when genetic and natural history information are appropriately combined.

Keywords: fire ants, gene flow, genetic structure, mtDNA, nuclear markers, reproductive strategies, sex ratios, social behaviour, social organisms, *Solenopsis invicta*

Introduction

Dispersal is a central concept in population and evolutionary biology. Dispersal is a key component of life history strategies, and as such it is linked to such population features as density regulation and colonizing ability. Dispersal is of special interest in population genetics because it usually results in gene flow, the movement and integration of genes from one population into another. Gene flow not only modulates the action of selection and genetic drift, for instance by altering effective population sizes and the availability of selectable genetic variation (Slatkin 1985a), it determines in large measure the genetic structure of populations, defined as patterns in the spatial distribution of genetic variation (Barton and Clark 1990). The causal relationship between dispersal, gene flow, and genetic structure means that detailed analyses of genetic structure can be used to decipher the nature of dispersal and gene flow regimes in wild populations (Bohonak 1999). This task is aided increasingly by the use of a wide assortment of genetic markers and sophisticated analytical procedures for describing and interpreting patterns of genetic structure.

Dispersal takes on special significance in social organisms because the existence of social behaviour, as well as variation in the form that it takes, is expected to influence dispersal. In a most basic sense, persistent social behaviour is not possible unless dispersal is delayed or abandoned. For instance, cooperative breeding in vertebrates, and maternal care in plant-feeding bugs require that the young remain with their nuclear family for an extended period (Emlen 1991; Tallamy and Schaefer 1997). Social behaviour can entail the complete loss of dispersal, perhaps in one sex only, if group members are tolerant of additional helpers or reproductives and social mechanisms exist for their permanent integration (Koenig *et al.* 1992; Bourke and Franks 1995). Importantly, variation in patterns of dispersal is often linked with variation in social organization within species. In killer whales (*Orcinus orca*), for example, two distinctive forms of social organization based on different foraging strategies exist, and gene flow regimes within and between the forms are influenced by this social polymorphism (Hoelzel 1998). In ants, colony queen number commonly varies within species, and dispersal strategies of both sexes are correlated with this central feature of social organization (Bourke and Franks 1995; Ross and Keller 1995a). Thus, the origin and elaboration of social behaviour typically involves coincident changes in dispersal behaviours and local gene flow regimes. Such changes in gene flow regimes, in turn, influence subsequent rates and patterns of social evolution by altering the distribution of genetic variance within and between social groups (Wade and Breden 1987).

The fire ant *Solenopsis invicta* is a useful model for studying dispersal and gene flow patterns and relating this information to particular social attributes. Several features of *S. invicta* make it appropriate for such studies. This ant is common in its introduced range in the USA, and it is easily collected and manipulated in the laboratory. Furthermore, a large number of genetic markers of diverse types have been developed for the species (Ross *et al.* 1999). Most important, naturally occurring variation in the social behaviour exists in both the native and introduced ranges (Ross and Keller 1995a; Ross *et al.* 1996). Finally, an extensive literature covering all areas of fire ant biology constitutes a rich backdrop of information to support wide-ranging population studies (reviewed in Lofgren *et al.* 1975; Vinson and Greenberg 1986; Ross and Keller 1995a; Tschinkel 1998). Much of this literature deals with the social and reproductive biology, making it possible in some cases to generate clear predictions regarding gene flow patterns and population genetic structure. Thus, a central theme of the research summarized in this chapter is the use of genetic data to test predictions about dispersal and gene flow derived from basic knowledge of fire ant natural history. This point is emphasized because studies are often undertaken on organisms for which there is insufficient information to explain observed genetic structure reasonably in terms of known features of the dispersal and reproductive biology.

Social biology of *Solenopsis invicta*

S. invicta, a native of South America, was introduced to the USA in the 1930s. Since that time, this pest insect has become widespread throughout the American South (Lofgren 1986). Fire ants are highly eusocial insects that form populous colonies housed in nests in the soil. Three 'castes' exist: the workers, which are wingless, and the winged queens and males. Workers, although obligately sterile, play major roles in virtually all facets of fire ant reproduction. The workers determine which diploid (female) larvae are sexualized (induced to develop as queens rather than workers), and they decide which of the sexualized brood are reared to adulthood and fed to maturity as young adults (Vargo and Fletcher 1986). Also, the workers decide how many and which types of adult queens will be allowed to function as active egg-layers in the colony (Fletcher 1986; Keller and Ross 1993a; Ross and Keller 1998). Finally, workers presumably initiate and coordinate the processes of colony movement and budding (a process that gives rise to daughter colonies). Because workers are intimately involved in these reproductive activities, this caste potentially has a major impact on dispersal and gene flow patterns even though its members are completely sterile.

Queens possess wings early in life (an obvious adaptation for dispersal), but the wings are shed at the onset of egg laying (Vargo and Laurel 1994); subsequent dispersal by queens is limited to movement along foraging galleries that connect nests under the soil surface. Fire ant queens mate early in adult life with a single male (Ross *et al.* 1993). Mating apparently takes place predominantly in the air during mass-mating flights including sexuals from numerous colonies; passive dispersal of both sexes by wind can be extensive during such flights (Markin *et al.* 1971). Mating may also occur within or on the queens' parent nests in some cases, but this has not been demonstrated directly.

Males, which are permanently winged, live relatively briefly and have a limited behavioural repertoire involving mostly activities directly related to mating. Males apparently can mate only once (Ball and Vinson 1983). As is true throughout the order Hymenoptera, fire ant males are normally haploid, with each individual male producing genetically identical haploid sperm.

A characteristic of fire ant social biology central to understanding their dispersal strategies and genetic structure is the existence of two types of social organization within the species. In the **monogyne** social form, colonies consist of a single egg-laying queen and her offspring, these simple families occupying a single nest at a given time. In the **polygyne** social form, colonies consist of multiple queens (two to several hundred in the introduced range) and their offspring, and these diffuse families typically inhabit multiple nests connected by underground galleries (Vargo and Porter 1989; Zakharov and Tompson 1998). Polygyne nestmate queens are closely related in native populations but not in introduced populations, indicating that the dispersal and reproductive biology of this form have evolved during colonization (Ross 1993; Ross *et al.* 1996).

As is true for many other ants exhibiting similar social polymorphisms, the monogyne (M) and polygyne (P) forms of *S. invicta* differ in several important reproductive and social characteristics other than colony queen number (Table 3.1). For instance, mating takes place only in high-altitude (> 100 m) mating swarms in the M form, whereas the site of mating is apparently more variable in the P form – in high-altitude swarms, and probably also in low-altitude (approximately 2–5 m) swarms and within or near the parent nest (Porter 1991; Ross *et al.* 1996; DeHeer *et al.* 1999; Goodisman *et al.* 2000). Modes of colony multiplication also differ between the forms. In the M form, new colonies originate exclusively by means of newly mated queens founding colonies in the absence of workers (independent founding). This mode of colony multiplication may also occur in the P form, but more commonly a parent colony gives rise directly to a daughter colony by means of budding, a process of coordinated movement of workers

Table 3.1.   Differences in reproductive and social biology between the monogyne (M) and polygyne (P) social forms of *S. invicta* in its introduced (USA) range

	Monogyne (M) form	Polygyne (P) form
Site of mating	High-altitude aerial swarms	High- or low-altitude aerial swarms; within or near parent nest[a]
Mode of colony multiplication	Independent founding by newly mated queens	Budding
Site of initiation of queen egg laying	Incipient colony	Established polygyne colony or bud
Energy reserves of young winged queens	Extensive	Usually meagre
Operational sex ratio	Close to one	Strongly female biased

[a]Good evidence exists for mating within or near the parent nest in the polygyne form in the native range in South America (Ross *et al.* 1996), but evidence for this phenomenon in the introduced range is circumstantial (Porter 1991).

and queens from the parent nest to a nearby site where a new nest is constructed (Vargo and Porter 1989). Because of the different modes of colony multiplication, the context in which queens begin their reproductive lives differs between the two forms, with M queens becoming functional egg-layers in incipient colonies they found (workers absent) and P queens typically seeking adoption as new egg-layers in existing polygyne colonies or buds (workers present). Linked to these reproductive differences are differences in the amounts of energy reserves accumulated by young queens (Keller and Ross 1993b). Monogyne queens accumulate extensive fat and glycogen stores to support flight and oogenesis, and to provide nutriment for their first brood in the absence of workers. Polygyne queens typically accumulate far smaller amounts of these reserves, consistent with reduction or elimination of the mating flight and with the presence of workers to feed the queens as they initiate oogenesis in established societies. A final reproductive difference between the forms in the introduced range is that the operational sex ratio is close to unity in the M form but strongly female biased (approximately $6:1$) in the P form (Vargo 1996).

Many of the reproductive and social differences between the two forms of *S. invicta* appear to have relatively simple genetic explanations. For instance, the characteristic differences in queen energy reserves are associated with genotype at the locus *Gp-9* (Ross and Keller 1998; DeHeer *et al.* 1999). Because the amount of energy reserves influences queen dispersal tendency in this and other ants (Sundström 1995; DeHeer *et al.* 1999), variation at *Gp-9* is expected to exert a profound influence on dispersal and gene flow regimes *within* each form. Variation at *Gp-9* also affects worker tolerance of nestmate queens. Colonies containing workers with the typical polygyne genotype ($Gp-9^{Bb}$) will tolerate multiple queens, but only if these queens possess this same genotype; conversely, colonies consisting solely of workers with the typical monogyne genotype ($Gp-9^{BB}$) will tolerate only a single queen, and only if she also possesses the monogyne genotype. Thus, there is genetic incompatibility between the forms based on the *Gp-9* genotype, and variation at this gene is expected to influence the extent and patterns of gene flow *between* as well as within the two forms.

The differences in operational sex ratio between the forms in the introduced range also appear to have simple genetic underpinnings. The strongly female-biased ratios in the P form stem from a loss of allelic variation at the sex-determining locus during the introduction of *S. invicta* to the USA (Ross *et al.* 1993). Reduced sex allele diversity means high levels of sex locus homozygosity and, consequently, frequent production of sterile diploid males (fire ants have complementary sex determination, which requires high levels of sex allele diversity to ensure that most diploids become females; Cook and Crozier 1995). Sterile diploid males, which are found in abundance only in the P form,[1] are apparently raised in lieu of fertile haploid males, thus biasing the operational sex ratio toward females in this social form (Ross and Keller 1995a).

Knowledge of these features of the reproductive and social biology of the two social forms of *S. invicta* can be used to generate predictions about dispersal and gene flow within and between the forms. These predictions then can be tested empirically using formal population genetic analyses.

Predictions of dispersal, gene flow, and genetic structure in introduced *S. invicta*

Given the differences between the two forms in their sites of mating and modes of colony reproduction (Table 3.1), queen vagility and dispersal potential are predicted to be lower in the P form than the M form. This should lead to relatively strong local mitochondrial DNA (mtDNA) structure in the P form but not in the M form (Table 3.2).[2] Depending on the extent to which queen-mediated local gene flow is restricted in the P form (as well as on patterns of dispersal of the male mates of these queens), significant local structure may also be detected at nuclear markers in the P form, but is not expected in any case in the M form (Table 3.2).

Dispersal and gene flow between the social forms of *S. invicta* can occur via four routes (Ross and Shoemaker 1993; Shoemaker and Ross 1996), two involving queens

Table 3.2. Summary of expected patterns of hierarchical genetic structure in introduced (USA) populations of *S. invicta* based on features of the reproductive and social biology

mtDNA
- *Local (among-site) differentiation:* strong in P form, absent in M form
- *Between-form differentiation:* strong
- *Between-region differentiation:* modest to large

Nuclear DNA
- *Local (among-site) differentiation:* detectable in P form, absent in M form
- *Between-form differentiation:* weak but detectable
- *Between-region differentiation:* modest

[1]Diploid males are not found in the M form because the independently founded colonies of this form fail to survive the founding stage if they produce diploid males (Ross and Fletcher 1986).
[2]In fire ants, as in most animals, the mtDNA is maternally inherited (Shoemaker and Ross 1996), so the distribution of variation at this genome is affected by female but not male dispersal.

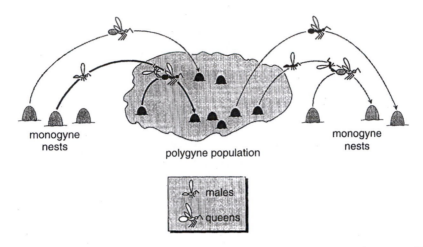

monogyne
nests

polygyne population

monogyne
nests

males

queens

Figure 3.1. Potential routes of dispersal and gene flow between the two social forms of *S. invicta*. The upper two routes involve queens becoming egg-layers in nests of the alternative form, and the lower two involve males mating with queens of the alternative form. The only route predicted to be common in the introduced (USA) range is that involving monogyne males mating with polygyne queens (heavy arrow).

becoming egg-layers in nests of the alternative form, and two involving males mating with queens of the alternative form (Fig. 3.1). Differences between the forms in characteristic *Gp-9* genotypes, coupled with the genotype-specific effects of *Gp-9* on worker behaviour and queen phenotype, lead to two predictions regarding queen-mediated gene flow between the forms. First, it appears that queens are not tolerated by workers of the alternative form because of *Gp-9* genotypic incompatibilities (M queens are not accepted into P colonies, and P queens are rarely are accepted into M colonies; Ross and Keller 1998). Thus, existing colonies are unlikely to obtain new or replacement queens produced by the alternative social form. Second, P queens generally have insufficient reserves to found single-queen colonies independently (associated with their typical *Gp-9* genotype). These facts suggest that queens do not often become reproductives in nests of the alternative form. That is, queens are unable to cross the social boundaries and serve as effective agents of between-form gene flow. This is expected to lead to strong differentiation between sympatric social forms at the mtDNA (Table 3.2).

The strongly female-biased operational sex ratio in the P form, coupled with the even sex ratio in the M form and strong dispersal ability of M males, also lead to two predictions regarding male-mediated routes of gene flow between the forms. M males are expected to mate commonly with P queens, but the converse pattern, P males mating with M queens, is expected to be rare (Ross and Shoemaker 1993; Ross and Keller 1995b). The occurrence of frequent inter-form nuclear gene flow via the former route is expected to dampen differentiation between the forms promoted by a lack of queen exchange across social boundaries, leading to the prediction of weak (but perhaps still significant) nuclear differentiation between sympatric social forms (Table 3.2).

At larger (regional) spatial scales that far surpass typical dispersal distances of these insects, some effect of isolation by distance is expected to be revealed in the form of significant differentiation at the mtDNA and nuclear markers (Table 3.2). The mtDNA may show relatively greater differentiation, because of the smaller effective population size for this genome than for the nuclear genome and its correspondingly greater susceptibility to random differentiation through the effects of drift. On the other hand, because the introduced populations of *S. invicta* are not at equilibrium, any single set of markers may not yield evidence for isolation by distance, especially at large geographical scales where structure equilibrates most slowly (Porter and Geiger 1995).

Footprints of dispersal and gene flow in *S. invicta*

Patterns of genetic structure predicted from knowledge of the biology of *S. invicta* (Table 3.2) may be compared to observed patterns of structure derived using data from numerous genetic markers. Genetic data have been obtained from ants collected from large numbers of nests, sampled at scales relevant for testing the predictions in both the native and introduced ranges (Shoemaker and Ross 1996; Ross and Shoemaker 1997; Ross *et al.* 1997, 1999). The most comprehensive data set comes from samples collected from 535 nests of the two social forms in two states in the introduced range, Georgia and Louisiana, with ants of each form sampled from multiple sites in Georgia (see Ross *et al.* 1999). Data were obtained from 28 presumably neutral nuclear markers as well as the mtDNA.[3] The present discussion will focus on analyses of these samples from the introduced range but, where appropriate, results from studies of native (Argentina) populations also will be presented (Ross *et al.* 1997). The extent of structure (differentiation) at various levels is described by means of hierarchical and single-level F_{ST} analyses (Weir and Cockerham 1984), with the significance of differentiation at specific levels or locations evaluated by means of exact tests, permutation tests, and/or bootstrapping of F_{ST} values over nuclear loci (Excoffier *et al.* 1992; Raymond and Rousset 1995b).

Full hierarchical F_{ST} analyses

The hierarchical partitioning of total mtDNA variance for the samples from the introduced range is illustrated in Fig. 3.2. More than a third of the total mtDNA variance, a significant proportion, occurs among sites within the forms. This result is consistent with the strong local differentiation expected within the P form, based on the dispersal and reproductive biology (Table 3.2). Between-form mtDNA differentiation also is substantial and statistically significant, a result predicted as well. No between-state (between-region) differentiation is evident from the hierarchical F_{ST} analysis, although exact tests reveal strong differentiation between the ants in Georgia and Louisiana when the between-form and among-site levels are disregarded. This example reveals how hierarchical F_{ST} analysis, by partitioning the total metapopulation variance among multiple levels, can mask the absolute amount of differentiation that exists at higher levels (see Ross *et al.* 1999, for discussion).

---

[3]The markers include eight allozyme loci, five codominant Randomly Amplified Polymorphic DNA (RAPD) loci, seven microsatellite loci, eight dominant RAPD loci, and a 4-kilobase segment of the mtDNA surveyed using restriction enzyme fragment analysis (Ross *et al.* 1999).

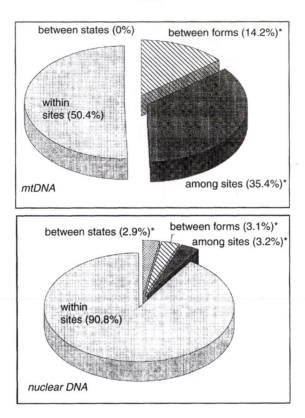

Figure 3.2. Partitioning of genetic variance in introduced (USA) populations of *S. invicta* at the mtDNA (top panel) and nuclear DNA (bottom panel) based on hierarchical F_{ST} analyses. The three levels of structure examined were: between states (Georgia and Louisiana), between social forms within states (monogyne and polygyne), and among sampling sites within forms. Asterisks indicate a statistically significant structure at a given level of structure. (Modified after Ross *et al.* 1999.)

The full hierarchical analysis of nuclear variation in the introduced range yields a very different pattern from that found for the mtDNA (Fig. 3.2), as expected given the different predictions for the two genomes at each level of structure (Table 3.2). Nuclear differentiation is relatively modest but significant at each of the three hierarchical levels investigated. As was true for the mtDNA, this result is consistent with the predictions in Table 3.2. However, a more critical evaluation of the match between predicted and observed structure for both genomes requires closer scrutiny of the genetic structure within each state and form, as well as in other populations from the native range, as follows.

Local genetic structure

Considering local structure within each of the two forms in Georgia, the predicted patterns of strong and highly significant local mtDNA differentiation in the P form and the absence of such mtDNA differentiation in the M form are observed (Fig. 3.3). Moreover, a very similar set of patterns is seen in a native population from Corrientes

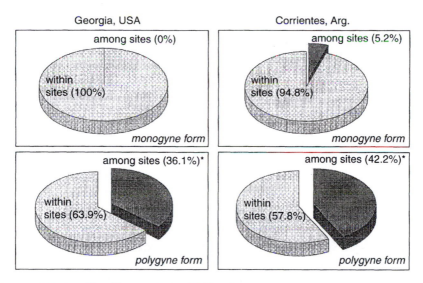

Figure 3.3. Partitioning of local (among-site) mtDNA variance in the two social forms of *S. invicta* based on single-level F_{ST} analyses. Asterisks indicate statistically significant differentiation among sites. Results are shown for an introduced population (Georgia, left panels) and a native population (Corrientes, right panels).

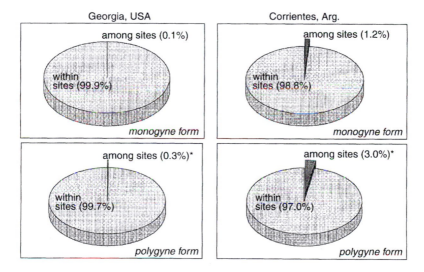

Figure 3.4. Partitioning of local (among-site) nuclear DNA variance in the two social forms of *S. invicta* based on single-level F_{ST} analyses. Asterisks indicate statistically significant differentiation among sites. Results are shown for an introduced population (Georgia, left panels) and a native population (Corrientes, right panels). Data were obtained for 28 loci in the introduced population and 15 loci in the native population.

Province in Argentina (Fig. 3.3). At the nuclear markers, among-site differentiation is low in both social forms in Georgia but, importantly, structure at this level is statistically significant in the P form but not in the M form (Fig. 3.4). Again, similar patterns are seen in the native Corrientes population (Fig. 3.4). Thus, restricted local queen dispersal in the P form relative to the M form leads to consistent and predictable differences between the forms in local genetic structure, using markers of both the mitochondrial and nuclear genomes. That it is differences in queen rather than male dispersal that are most important is suggested by the much stronger differences in structure between the forms at the mtDNA marker than at the nuclear markers.

Between-form differentiation

Pronounced and highly significant mtDNA differentiation exists between the sympatric social forms in Georgia, and similar patterns are observed in the introduced Louisiana

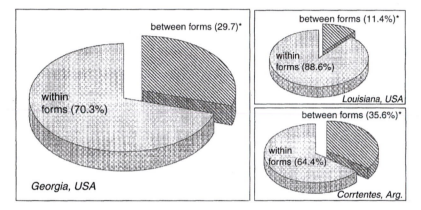

Figure 3.5. Partitioning of between-form mtDNA variance in *S. invicta* based on single-level F_{ST} analyses. Asterisks indicate statistically significant differentiation between the social forms. Results are shown for introduced (Georgia and Louisiana) and native (Corrientes) populations.

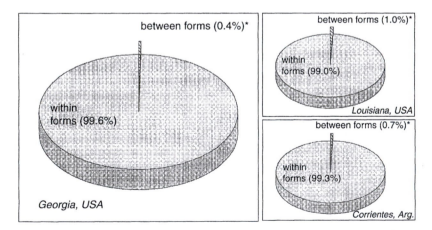

Figure 3.6. Partitioning of between-form nuclear DNA variance in *S. invicta* based on single-level F_{ST} analyses. Asterisks indicate statistically significant differentiation between the social forms. Results are shown for introduced (Georgia and Louisiana) and native (Corrientes) populations. Data were obtained for 27 loci in the introduced populations and 15 loci in the native population.

and native Corrientes ants as well (Fig. 3.5). In Georgia, significant between-form differentiation persists even after among-site differentiation within the P form is taken into account (Ross *et al.* 1999). At the nuclear markers, differentiation between the forms in Georgia is weak but significant, with virtually identical patterns seen in the introduced Louisiana and native Corrientes populations (Fig. 3.6). Thus, the hypothesized inability of queens to become egg-layers in nests of the alternative social form, coupled with the hypothesized male-mediated gene flow via at least one route, led to distinctive and predictable patterns of between-form structure at the two genomes (see Avise 1994, for other examples of sex-specific dispersal patterns registered as differential structure at the mtDNA and nuclear genomes).

Measurements of between-form gene flow in *S. invicta*

Measurements of gene flow in wild populations, expressed in terms of the genetically effective migration rate ($N_e m$), are often obtained from values of F_{ST} using the island model of Wright (Slatkin 1985a). Another approach to measuring $N_e m$ uses allele trees reconstructed for non-recombining genes in combination with coalescence theory (Slatkin and Maddison 1989). Both approaches have been used to quantify gene flow between sympatric social forms in native populations of *S. invicta* (Ross *et al.* 1997). The results suggest that nuclear gene flow between the forms is sufficiently high to preclude their evolutionary divergence in the absence of strong form-specific selection ($9 < N_e m < 25$), but that queens may not play an important role as agents of this between-form gene flow (for the mtDNA, $0.1 < N_e m < 7$).

Such gene flow measurements cannot be made in introduced *S. invicta* populations because the equilibrium assumptions underlying these approaches are violated. Nonetheless, qualitative and quantitative statements regarding the amounts of between-form gene flow occurring via the four possible routes can be obtained using gene frequency data from the mtDNA and from non-neutral nuclear markers. Inspection of the mtDNA haplotype frequencies in the two social forms in Georgia (Table 3.3) gives an indication of the importance of each of the two queen-mediated routes of interform gene flow. One haplotype (C) is common in the P form but absent in the M form, consistent with a complete lack of between-form gene flow by means of P queens becoming egg-layers in M nests. Two other haplotypes (B and D) have a moderate combined frequency in the M form but a very low frequency in the P form, consistent with rare or no between-form gene flow by means of M queens becoming egg-layers in P nests.

Use of non-neutral nuclear markers to measure male-mediated gene flow between the forms relies on the fact that strong form-specific selection generates very different allele frequencies in the forms at such markers (Ross 1997). Two such non-neutral loci exhibiting divergent allele frequencies are *Pgm-3* and *Gp-9* (Table 3.3). The frequency at which queens mate with males of the alternative form can be estimated by simply comparing the allele frequencies of the mates of each type of queen with the allele frequencies of wild-caught males of each form (Ross and Shoemaker 1993; Shoemaker and Ross 1996; Ross 1997). When this was done for M queens in Georgia, it was found that males with which these queens mated possessed the *Pgm-3*[a] allele at a frequency almost identical to that estimated for wild-caught males sampled from M nests (Shoemaker and Ross 1996).

Table 3.3. Frequencies of mtDNA haplotypes and of alleles at two non-neutral nuclear loci in the two social forms of *S. invicta* in its introduced range (Georgia, USA)

	Monogyne (M) form	Polygyne (P) form
mtDNA haplotypes		
A	0.87	0.48
B + D	0.13	0.01
C	0.00	0.51
Pgm-3 alleles		
A	0.76	0.39
a	0.24	0.61
Gp-9 alleles		
B	1.0	0.50
b	0.0	0.50

Data for the mtDNA are from winged queens from 149 M nests and egg-laying queens from 356 P nests (Ross and Shoemaker 1997). Data for *Pgm-3* and *Gp-9* are from winged queens from 466 M nests and egg-laying queens from 427 P nests (Shoemaker and Ross 1996; Ross 1997).

The same result was found for the gene *Gp-9*. This is consistent with the conclusion that M queens mate exclusively with males produced by M colonies. When this approach was employed for P queens in Georgia, it was found that the frequency of the *Pgm-3ª* allele among the mates of P queens was intermediate to that estimated for wild-caught males of the two social forms, although it was more similar to the frequency characteristic of M males. Simple extrapolation suggests that around 85% of the P queens studied mated with M males, an estimate very similar to that obtained using data from *Gp-9* (Ross 1997). These assessments of between-form gene flow corroborate the conclusions that are based on the natural history and on patterns of genetic structure, in suggesting that only one of four possible routes of such gene flow, M males mating with P queens, is likely to be important in the introduced range (Fig. 3.1). In native populations, which are more likely to approach equilibrium conditions, formal analyses of cytonuclear associations may eventually prove useful in revealing patterns and magnitudes of between-form gene flow (e.g., Goodisman and Asmussen 1997).

Discussion

The results reviewed in this chapter bear on several important points concerning the assessment of dispersal, gene flow, and genetic structure. A first point is that a genetic approach employing a predictive framework based on the natural history of the study organism can be a powerful tool for inferring patterns of dispersal and gene flow in the wild. However, this approach requires not only access to detailed basic biological information, but also the availability of numerous well-characterized and complementary genetic markers, the implementation of carefully conceived and biologically relevant sampling schemes, and the ability to obtain large numbers of samples. Clearly, the approach can be taken only in organisms that have been the subject of considerable previous study, which argues for the importance of intensive long-term studies on model organisms with appropriate attributes.

A second point is that social behaviour can have important, and sometimes unanticipated, consequences for dispersal and gene flow, owing to the interrelationships between social and reproductive behaviours in social animals. In fire ants, for example, specific social tolerance behaviours displayed by workers determine which queens will be accepted as new egg-layers, and workers are intolerant of queens bearing *Gp-9* genotypes characteristic of the alternative form. Thus, the social behaviours of the workers, obligately sterile individuals, profoundly influence patterns of local gene flow between sympatric social types. This example also illustrates clearly the distinction between dispersal and gene flow, which often is not fully appreciated by population biologists. Queens are likely to disperse commonly to nests of the alternative form, yet this dispersal does not result in their incorporation into the local breeding unit. Organisms in which group living is mandatory but social barriers impede the incorporation of immigrating individuals into social groups may provide many dramatic examples of the distinctiveness of dispersal and gene flow (e.g., O'Riain and Braude, chapter 10).

A final point is that substantive issues exist regarding the best statistical approaches to adopt when analysing and interpreting genetic structure. F_{ST}, long the workhorse statistic for describing genetic structure (see Rousset, chapter 2), summarizes the average extent of differentiation between all units sampled at a given level and is subject to some arbitrariness stemming from the selection of sampling scales; thus, it can mislead with respect to both the generality and location of observed differentiation (e.g., Lewontin 1972; Urbanek *et al.* 1996). These problems have been addressed in the fire ant work by analysing patterns of structure in several different areas in the native and introduced ranges, and by designing meaningful sampling schemes based on knowledge of fire ant biology. It is now widely appreciated that F_{ST} can yield biased views of structure when used with highly polymorphic markers such as microsatellites (Charlesworth 1998; Pannell and Charlesworth 1999; Balloux *et al.* 2000), especially when considering situations involving large spatial scales, long periods of divergence, and low gene flow. Analogues of F_{ST} that incorporate information on presumed evolutionary relationships of alleles (such as R_{ST} of Slatkin 1985a) may not provide unbiased descriptions of structure if the assumed mutational models do not correspond well with the actual mutational processes generating the polymorphisms (Balloux *et al.* 2000). These latter problems are avoided to some extent in introduced fire ants because the use of allele relationships is unwarranted (because of non-equilibrium conditions) and the times for divergence are small, but in other systems, including native fire ants, the choice of what statistic to use to partition genetic variance can influence the results substantially (Ross *et al.* 1997; Goodman 1998).

All *relative* measures of structure that partition variance among levels suffer from the drawback that their values are sensitive to total metapopulation diversity; thus, for instance, changes in gene flow patterns that reduce total diversity can inflate F_{ST}, even if there is no increase in absolute levels of between-population divergence (Pannell and Charlesworth 1999). Hierarchical analyses using relative measures can pose additional problems: while having the advantage of pointing to particular biological causes of variation at each level, hierarchical analyses have the possible disadvantage of masking substantial absolute differentiation at higher sampling levels (see example above for mtDNA differentiation between the states; also Rousset 1999). For these reasons, methods that partition total genetic diversity always should be used in conjunction with absolute measures of differentiation, such as exact or G tests (Raymond and Rousset 1995b; Lugon-Moulin *et al.* 1999). Finally, alternative methods of describing genetic structure, such as those based on spatial autocorrelation (Smouse and Peakall 1999), may be appropriate supplements in some systems to the more widely employed methods described in this chapter.

Acknowledgements

I thank François Rousset, Jim Nichols, and two anonymous reviewers for comments on an earlier draft of this chapter. The work presented was supported by grants from the National Geographic Society and the U.S. National Science Foundation.

4

Dispersal in pikas (Ochotona princeps): combining genetic and demographic approaches to reveal spatial and temporal patterns

Mary M. Peacock and Chris Ray

Abstract

Demographic studies of dispersal lack the spatial and temporal breadth necessary to characterize the full patterns of movement, settlement and mating within and between populations. Genetic studies can provide some of these missing data on mating patterns and gene flow, allowing us to address certain evolutionary questions. However, genetic inference of dispersal is not yet an exact science, and there is no one-to-one correspondence between patterns of gene flow and dispersal. In order to determine the details of dispersal, it is useful to use demographic data to guide genetic inference. This paper uses dispersal data from a small, territorial lagomorph to explore several examples of the potential interplay between genetic and demographic data, from pattern detection through hypothesis testing.

Keywords: demographic stochasticity, environmental stochasticity, competition for resources, DNA fingerprinting, genetic effective population size, genetic inference, meta-population dynamics, inbreeding avoidance, inbreeding

Introduction

Our ability to test dispersal hypotheses depends directly on our ability to characterize dispersal patterns, yet many difficulties can frustrate our attempts to characterize dispersal. Besides the difficulty of measuring individual movements, there are problems with estimating dispersal from raw data. Discrepancy between true and estimated dispersal can undermine hypothesis testing. This paper describes our attempts to improve interpretation of dispersal patterns by using a combination of demographic and genetic inference.

A gap between true and estimated dispersal patterns can develop during data collection. Many studies rely on direct (demographic or 'observational') methods for measuring dispersal, such as mark–recapture. The raw data gathered using direct methods generally do not represent the full distribution of individual movements. Mark–recapture studies are spatially and temporally restricted, with limited ability to reveal long-distance movements such as dispersal among populations (Smith and Ivins 1983; Dobson 1994; van Vuren and Armitage 1994; Koenig *et al.* 1996). Consequently, direct methods provide data mainly on the within-population component of dispersal. Various models can be used to extrapolate full dispersal patterns from mark–recapture data, but these models rely upon assumptions that often cannot be tested (see Bennetts *et al.*, chapter 1).

The limitations of direct methods may prevent definitive tests of hypotheses concerning the evolutionary mechanisms driving dispersal, such as competition for mates, competition for resources, and inbreeding avoidance. To progress beyond mere debate about these mechanisms, we must have full information on dispersal, mating patterns, and parentage. Accurate identification of parentage is crucial not only for identifying mating patterns, but also to determine the actual dispersal distances of juvenile animals. Questions concerning inbreeding and outbreeding should also be evaluated against the background of genetic variation found within the population. Patterns of genetic variation within and among populations can be used to infer long-term patterns of gene flow. These indirect (genetic or 'molecular') methods offer a temporal perspective on dispersal that is unavailable from direct methods.

Unfortunately, genetic inference can fail. Because genetic patterns reflect effective gene flow over the long term, they may not reflect current dispersal. On the other hand, stochasticity can play a role in the genetic patterns observed at any one place and time. 'Historical accidents' such as founder events (Wright 1940) can temporarily alter the genetic landscape, changing expectations about long-term gene flow and effective dispersal. Genetic inference is further complicated by the fact that different population structures, with different patterns of local growth and dispersal, can result in similar genetic patterns. For example, both a large, panmictic population and a collection of small, isolated populations can maintain high overall genetic diversity (Wright 1931). The caveat in this example is that a subdivided population may not maintain high genetic diversity if there is frequent local extinction (Wright 1940). In order to infer dispersal patterns from genetic data, we must know enough about dynamics and dispersal to choose an appropriate model for the system under study (Felsenstein 1982).

The benefits of combining direct and indirect methods are best illustrated by examples in which both methods have been employed sequentially. In the bird literature, molecular approaches prompted a major revision of mating and dispersal models that were developed through observation (Burke *et al.* 1989; Westneat 1990; van Vuren and Armitage 1994; Hughes 1998; Piertney *et al.* 1998). A similar revision is underway in mammalian models (Webb *et al.* 1994; Favre *et al.* 1997; Girman *et al.* 1997; Peacock 1997). Here, we use the case of one small mammal, the pika (*Ochotona princeps*), to compare the results of studies based on purely demographic methods with those that also embrace genetic inference. This paper explores the evolution of our understanding of pika dispersal to date, from purely demographic studies, through molecular research, to the development of population genetic models that are informed by demographic data. This example clarifies the necessity of integrating demographic and genetic approaches throughout the process of dispersal research, from model development through testing.

Pika natural history

Pikas are small lagomorphs (approximately 150 g) found throughout Asia and North America. North American populations are found in western alpine habitats (Smith and Weston 1990) on patches of talus (broken rock). Pikas use the subterranean portion of the talus as a refuge from both predators and thermal extremes. Lagomorphs do not hibernate, and overwinter activity requires that pikas maintain a high metabolic rate to counteract heat loss. This high metabolic rate leads to summer heat stress, which may curtail dispersal (Smith 1974b).

Adult pikas are individually territorial and rarely disperse from established territories (Tapper 1973; Smith and Ivins 1983; Peacock 1997). Successful juveniles obtain a territory during the summer of their birth, and reproduce as yearlings the following spring (Smith and Weston 1990). Territory ownership is important for overwinter survival, and only territorial individuals are known to reproduce. Each territory centres on one or more 'haypiles' or caches of vegetation stored for overwinter consumption. The carrying capacity of a talus patch is a simple function of patch size, easily determined by maximum haypile densities, which vary little across the range of North American pikas (Tapper 1973; Smith 1974a, 1978; Southwick *et al.* 1986). Haypile sites are relatively permanent, so that the number of unused sites indicates the number of available territories on a patch (Peacock 1997). Average territory size is equivalent in males and females (Smith and Ivins 1984; Peacock 1997).

Talus is patchily distributed and patches support relatively small populations (less than 100 individuals; Tapper 1973; Smith 1978). Smaller pika populations experience frequent local extinction (Smith and Gilpin 1997; Moilanen *et al.* 1998), and even large populations may succumb to extreme weather conditions (Smith 1978). Low allozyme diversity in *O. princeps* has been attributed to repeated genetic bottlenecks and a concomitant reduction in N_e, which suggests widespread extinction–recolonization dynamics (Hafner and Sullivan 1995).

Behavioural observations suggest that pikas are facultatively monogamous (Smith and Ivins 1983), meaning that polygyny is prevented only by ecological constraints (Kleiman 1981). Males do not appear capable of either monopolizing essential resources to attract multiple females or defending groups of widely dispersed females (Smith 1987). Females often initiate two litters per season, but on average only two young from the first litter are weaned successfully (Smith 1978; Smith and Ivins 1983). The first litter is conceived prior to snowmelt, and these juveniles appear above ground in early summer. Most successful immigration into populations occurs in the fall or spring months (Tapper 1973; Smith 1987).

Dispersal and mating patterns inferred from demographic methods

Most pika studies have used mark–recapture and mark–resight methods to quantify movement, and observational data on social interaction among adults to infer mating patterns and parentage of juvenile animals (Millar and Zwickel 1972; Tapper 1973;

Smith and Ivins 1983, 1984). These observational data suggest several provocative conclusions. First, juvenile settlement patterns are primarily philopatric (i.e., settlement is on the natal or closest available territory; Tapper 1973; Smith and Ivins 1983; Brandt 1985). Second, adults on neighbouring territories form monogamous mated pairs (Smith and Ivins 1983). Third, pikas that disappear from a population generally die rather than disperse successfully (Smith 1987), a conclusion based on relatively low observed immigration into fully marked populations (Smith and Ivins 1983, 1984). In light of the high adult survivorship observed in this species (47–53% per year; Millar and Zwickel 1972; Smith 1978; Peacock 1997), these conclusions suggest that pikas inbreed regularly with close relatives. In fact, it has been predicted that pika populations are composed of clusters of highly related individuals, and that local populations are genetically isolated and genetically depauperate (Smith 1993).

A re-evaluation of dispersal patterns: results from genetic and demographic methods

In order to test hypotheses of limited dispersal and close inbreeding in pikas, recent studies have combined genetic and demographic data to infer dispersal and mating patterns (Peacock 1995, 1997; Peacock and Smith 1997a,b). These studies were designed to address two primary questions. First, what is the pattern of juvenile dispersal, and which of several general dispersal models does this pattern support? General dispersal models include competition for resources (Dobson 1982), competition for mates (Greenwood 1980; Dobson 1982; Koenig *et al.* 1996), and inbreeding avoidance (Moore and Ali 1984). See Johnson and Gaines (1990) for an excellent review of these models. Second, does spatial subdivision of the population influence dispersal and mating patterns? Two similar studies were undertaken, in populations with contrasting spatial structure (one more patchy than the other). Demographic data from these studies were also used to build genetic simulation models designed to test the effects of population dynamics and spatial structure on genetic parameters (Ray 1997). These simulation models complete the triumvirate of genetic inference: after observing genetic patterns and inferring demographic processes, these models demonstrate whether the observed genetic patterns could result from alternative demographic scenarios.

Study populations and methods

The first study focused on two pika populations ('Pipet Tarn' and 'Cabin slope') that occupy large and relatively continuous talus patches at high elevation (3170 m) in the Sierra Nevada mountain range (Mono County, California, USA). These populations experience environments typical for pikas throughout most of North America. The second study focused on an atypical population occupying a highly fragmented collection of habitat patches at low elevation (2550 m). This fragmented habitat was formed from the tailing piles of abandoned gold mines in the town of Bodie (also in Mono County), which lies 35 km from other large pika populations in the Sierra Nevada. The mine tailings were colonized circa 1900 (Severaid 1955; Peacock and Smith 1997a; Smith and Gilpin 1997). The Bodie population marks the distributional boundary for pikas along the eastern Sierra Nevada, due to its high summer temperatures (Smith 1974b).

From each population, a large sample of adult and juvenile pikas was live-trapped, weighed, sexed, and tagged for individual identification (virtually all individuals within a certain segment of each population were trapped each year). Ear tissue was collected for genetic analysis. Multilocus minisatellite DNA fingerprints were used to examine relatedness between mated pairs, genetic similarity among pikas within and between populations, gene differentiation between disjunct populations (F_{ST}), population and metapopulation heterozygosity (H) and effective size (N_e) (Lynch 1991). Fingerprinting analyses were also used to determine parentage, so that the true dispersal patterns of juvenile animals could be compared with territory availability. An internal size standard (lambda) was run in each lane to facilitate gel-wide comparisons. Comparisons were made only within gels. On gels without internal size standards, comparisons were made only if DNA fingerprints were within three lanes of one another. Territory occupants were identified by physical association with established haypiles; usually, the occupant was observed in the act of caching forage at the haypile site. The total number of territories within a patch was determined from the total number of established haypile sites.

Dispersal in continuous habitat

Juvenile settlement patterns were studied over a 4-year period within a 40-territory segment of the Pipet Tarn population (Peacock 1997). All territory openings (vacant haypile sites) were mapped at the beginning of each field season. The settlement pattern of marked juveniles whose natal territories had been identified by genetic exclusion analysis was overlaid on to territory availability for each year. Individuals were sampled less systematically from the Cabin slope population, to provide information on the rate of genetically effective dispersal between populations separated by 2 km of typical alpine pika habitat.

Within-population patterns

A philopatric settlement pattern was observed similar to that reported in previous studies on pika dispersal; by the end of each summer, most vacant territories were filled by juveniles born on the study site. Twenty-eight of the 30 juveniles born at Pipet Tarn dispersed from their natal territory to the closest or one of the closest vacant territories. This pattern was significantly different from a random settlement pattern ($P < 0.05$, two-tailed Monte Carlo randomization test).

One potential effect of philopatry is fine-scale genetic structuring within a population (Shields 1983). Despite a philopatric settlement pattern, a genetic isolation-by-distance analysis showed no evidence of genetic subdivision within the Pipet Tarn population. There was no relationship between the band-sharing score calculated for each pair of territorial adults and distance between territories ($R^2 = 0.031$). Although mated pairs tended to be neighbours (modal number of territories separating mated pairs was zero), average band-sharing between mates (0.410, $N = 11$; SD $= 0.062$; range 0.25–0.477) was well below that for first-order relatives ($N = 44$ band-sharing comparisons; 95% confidence interval (CI), 0.547–0.784). Despite a philopatric settlement pattern, no clusters of highly related individuals were apparent, and the pikas did not appear to be mating, on average, with highly related individuals.

The lack of genetic structure within the Pipet Tarn population appeared to be due to a high rate of turnover. On average, 34% of territory occupants were replaced each winter by unmarked individuals. The majority (63%) of juveniles born at Pipet Tarn that settled territories during their natal summer died overwinter and were replaced by immigrants of unknown origin. Of the 40 individuals recruited into the adult population over a 4-year period, only 11 were identified as natal residents of Pipet Tarn. Natal recruits had either zero, one, or two unattributable bands when compared with putative parents ($N = 6$) or band-sharing with at least one adult that was a resident in the year they were born ($N = 5$) that fell well within the distribution for first-order relatives (mean band-sharing $= 0.609$, SD $= 0.078$; range 0.483–0.785). Putative parents were identified using data on spatial overlap and were considered the biological parents if the juvenile had zero unattributable bands or, in the presence of one or two unattributable bands, the juvenile had a sibling with no unattributable bands (Peacock 1997). All juveniles with either one or two unattributable bands with respect to their putative parents had band-sharing values with both putative parents that fell well within the distribution for first-order relatives. Genetic similarities below 0.45 occurred only between juveniles and a putative parent when more than two bands could not be attributed to either social parent.

Among-patch movement

Four juvenile immigrants settled in the Pipet Tarn population during the summer months, and 25 unmarked individuals settled overwinter. Forty-eight per cent of the overwinter immigrants had band-sharing scores with at least one resident adult in the Pipet Tarn population that fell well within the distribution for first-order relatives. Either (a) these immigrants and their purported first-order relatives originated from the same external population, or (b) these immigrants originated from within or near the Pipet Tarn population. The latter explanation would be consistent with the philopatric settlement pattern seen during the summer months.

The band-sharing scores between the remaining immigrants and Pipet Tarn residents suggest that these individuals immigrated from within 2 km, the distance between the Pipet Tarn and Cabin slope populations. The average band-sharing score between immigrants and Pipet Tarn residents lay within the 95% CI of the distribution of band-sharing scores between residents of Pipet Tarn and Cabin slope populations (Fig. 4.1; 95% CI, 0.100–0.525). If immigrants were dispersing from distances greater than 2 km, the band-sharing score between immigrants and Pipet Tarn residents should be lower than that between Pipet Tarn and Cabin slope residents. As expected, the mean band-sharing score between immigrants and Pipet Tarn residents was well above the highest band-sharing score between Pipet Tarn and Bodie residents (Fig. 4.1; $N = 13$ immigrants; mean $= 0.444$; SD $= 0.048$; range 0.339–0.517).

Band-sharing scores between adults from Pipet Tarn ($N = 17$) and Cabin slope ($N = 10$) were used to estimate gene flow between disjunct talus patches within the Sierra

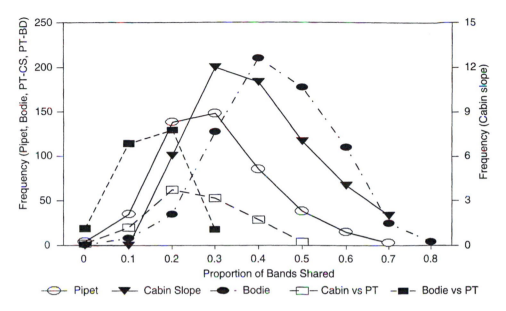

Figure 4.1. The distribution of multilocus minisatellite band-sharing scores for pairs of adults drawn from Pipet Tarn (open circles; $N = 659$ comparisons), Cabin slope (filled triangles; $N = 42$), and Bodie (filled circles; $N = 697$), and the distribution of band-sharing scores from Pipet Tarn–Cabin slope pairs (open squares; $N = 169$), and Pipet Tarn–Bodie pairs (filled squares; $N = 280$). Dotted lines are used for visual emphasis of related data, and do not represent interpolation between points.

Nevada (Lynch 1991). The result ($F_{ST} = 0.105$) suggests that there is gene flow between these populations. In fact, during the 4-year study (approximately one generation), three individuals in the Pipet Tarn population had band-sharing scores with Cabin slope residents that fell within the distribution for first-order relatives.

Although the mean band-sharing score among residents within Pipet Tarn (0.359, SD $= 0.133$) was higher than that reported for unrelated individuals in many species considered outbred, it was much lower than that reported for *highly* inbred species such as Naked Mole Rats, Channel Island Foxes, and inbred mice (≥ 0.80) (Gilbert *et al.* 1990; Reeve *et al.* 1990). Because juvenile pikas were recruited into their natal population, the background band-sharing includes unrelated individuals and individuals of various degrees of relatedness. The mean band-sharing among adult residents also varied among years (0.315–0.437), which suggests movement into and out of the population. Average heterozygosity estimate ($H = 0.709$) was high and consistent with estimates for species considered outbred (Westneat 1990; Hoelzel and Dover 1991; Stephens *et al.* 1992; Scribner *et al.* 1994).

Conclusion

Despite philopatric settlement patterns, the study populations were neither genetically isolated nor highly inbred. These data suggest that juveniles sometimes move long distances. This movement pattern could be created by competition for territories, where juveniles maturing early in the season settle philopatrically, and those maturing later must travel further to find a vacant territory. Thus, these data support the resource competition model of dispersal (Dobson 1982).

Dispersal in highly fragmented habitat

The Bodie population has been described as a metapopulation (Smith and Gilpin 1997; Hanski 1998; Moilanen *et al.* 1998), because populations on individual patches experience frequent extinction and recolonization. Most patches are small, but one relative 'mainland' in the northern portion of the site supports approximately 50 territories. Recent demographic declines at Bodie are restricted to the southern patches (Smith and Gilpin 1997), suggesting that the subpopulation on the 'mainland' may produce excess migrants that support the northern portion of the population (but see Moilanen *et al.* 1998). Despite the demographic evidence for recolonization and dispersal, it has been suggested that subpopulations at Bodie should be genetically isolated if the philopatric settlement pattern observed in continuous habitat is exacerbated by habitat fragmentation (Smith 1987). The Bodie metapopulation is also assumed to be isolated from other pika sources, being 35 km from other known populations (Smith and Gilpin 1997).

Dispersal patterns were studied within 17 habitat patches in the northern portion of the study area, including the 'mainland' (Peacock and Smith 1997a). Dispersal within and among patches was assessed using independent mark–resight and molecular studies, similar to those described above. In addition, multilocus DNA fingerprinting was used to identify parent–offspring pairs located on separate patches. The resident patch of the mother was considered the natal patch of the juvenile.

Dispersal

Territory turnover rate was higher at Bodie (approximately 62%) than that reported for pika populations found in continuous habitat (37–53%) (Millar and Zwickel 1972; Smith 1978; Peacock 1997). Higher turnover represents more opportunities for juveniles to settle philopatrically. Indeed, only two of 47 juveniles with known parentage dispersed further than the closest available territory. Juveniles born on the 'mainland' settled on territories available within that patch ($N = 4$). Of 11 juveniles that dispersed between patches, nine emigrated from fully occupied patches (with no vacant territories). Eight of these juveniles settled on the closest patch with vacancies. The ninth juvenile dispersed 396 m, passing at least four intervening patches with available territories. The remaining two juveniles dispersed from natal patches with vacancies. One settled on the closest available patch, while the other dispersed 210 m, passing at least two intervening patches with available territories. The majority of unmarked adults (68%; $N = 13$) found on study patches at the beginning of each field season had band-sharing scores with at least one adult on the patch that fell within the distribution for first-order relatives. These data suggest that these adults were recruited on to their natal patches.

Genetic structure

The philopatric settlement pattern observed in this population implies a genetically structured population with high levels of relatedness within patches and within clusters of neighbouring patches. To test for genetic structure at several spatial scales, mean band-sharing scores were compared for pairs of resident adults, where the pairs were 'spatially referenced', or separated by certain physical distances. Pairs were drawn from (a) within a patch, (b) nearest neighbouring patches, (c) patches separated by intermediate distances (400–500 m), and (d) patches separated by the width of the study area (700–800 m). In each instance, the mean band-sharing score of all spatially referenced pairs was compared to a distribution of mean band-sharing scores determined by sampling at random from the population. The sample size for each randomized mean was equal to the sample size for each spatially referenced mean. Spatial genetic structure would be evident if the mean band-sharing score between spatially referenced pairs lay outside the 95% CI of a randomly generated score distribution. Despite a predominantly philopatric settlement pattern, there was little evidence for genetic structure within years, and no persistent genetic structure across years.

Genetic diversity

Mean band-sharing among first-order relatives (0.638) and among adult residents (0.483) was higher at Bodie than Pipet Tarn (first-order relatives 0.603 and adult residents 0.336). However, heterozygosity (Bodie $H = 0.736$; Pipet Tarn $H = 0.709$) was virtually identical. Although individuals tended to mate with neighbours (females 87%, males 78%), band-sharing scores between mated pairs did not differ from a pattern of random mating ($P = 0.025$).

Conclusion

Despite the supposed isolation and fragmentation of the Bodie population, it has maintained genetic variation comparable to that of non-isolated populations in less fragmented habitat. These results call into question either the isolation of the population or the effect of fragmentation and local extinction on genetic diversity, or both. Models designed to quantify the potential for genetic diversity in this metapopulation are summarized below.

A lesson from failed genetic inference

The Bodie population has been touted as the best mammalian example of a 'classical' metapopulation (Smith and Gilpin 1997; Hanski 1998; Moilanen *et al.* 1998). Although the largest patch may harbour 50 individuals, this population is small enough to experience strong genetic drift, and cannot be a source of lasting genetic diversity. Several classical population genetic models predict genetic depauperacy in metapopulations with frequent local extinctions, due to the homogenizing effect of repeated colonizations or founder events (Wright 1940; Slatkin 1977; Maruyama and Kimura 1980; Ewens 1989; Gilpin 1991; Barton and Whitlock 1997; Hedrick and Gilpin 1997;

Whitlock and Barton 1997). Yet this pika metapopulation exhibits wide genetic diversity, with heterozygosity as great as that found in another large and well-mixed population. Based on the genetic data and the predictions of the above models, there appears to be no reason to suspect frequent local extinctions and a metapopulation structure. However, the demographic history of this system includes the founding of the metapopulation from a small number of individuals less than 40 generations ago, and repeated founder events in almost every local population since that time (Smith 1974a; Smith and Gilpin 1997).

In order to explore the potential for maintenance of genetic diversity in this metapopulation, a more detailed demographic model was constructed (Ray 1997). The model incorporated the founding and subsequent dynamics of the metapopulation and tracked the fate of a set of neutral alleles assumed present at the founding event. The model included no mutation, but allowed local genetic drift, founder drift during local colonizations, and the potential for immigration from a large allele source external to the metapopulation. Details of local population growth, extinction, and colonization rates were based on rough estimates gathered from demographic studies of this metapopulation (Smith 1974a; Peacock and Smith 1997a; Smith and Gilpin 1997). These estimates were varied in a sensitivity analysis, to determine their impact on model predictions. In addition to allowing for demographic stochasticity, local extinctions were assumed to be a logarithmic function of local population size, according to expectations for extinctions due to environmental stochasticity in growing populations (Goodman 1987; Lande 1993; Mangel and Tier 1993). Local growth was geometric, truncated at a ceiling determined by patch size (Smith 1974a). Because there was little evidence for genetic structure in this system, dispersal was not spatially structured.

The model predicts that the effective metapopulation size (including the initial founder event and no subsequent immigration) is near 40. It is unlikely that such a small population could retain the observed heterozygosity ($H = 0.736$), suggesting that this is not an isolated metapopulation.

On the other hand, the model predicts the current effective size of this metapopulation is approximately two-thirds the census size (Table 4.1). This result has an intuitive explanation: a highly subdivided metapopulation experiencing uncorrelated local extinctions is analogous to a single large population of individuals experiencing uncorrelated deaths, and loses diversity at a similarly slow rate (Ray 2000). Also, fluctuations in metapopulation size decrease with subdivision, much as fluctuations due to a birth–death process decrease with the number of individuals in the population. This pika metapopulation is highly subdivided, with fewer than 400 individuals distributed among over 60 local populations (Smith and Gilpin 1997). Therefore, we should expect it to maintain relatively high levels of genetic diversity, despite frequent local extinctions.

Sewall Wright pointed out long ago that population subdivision can raise effective population size, due to independent genetic drift in local populations. He also pointed out that local extinction can lower effective population size. Extension of his models to incorporate demographic details from the pika system, or for a more general case (Fig. 4.2), reveals that the effects of population subdivision may compensate for the effects of local extinction. As subdivision increases, the heterozygosity retained by a metapopulation may approach that retained by a single large and constant population of the same total size (Fig. 4.2). This result was presaged by Whitlock and Barton (1997),

Table 4.1. Diversity at generation 50 in metapopulations patterned after the Bodie pika metapopulation

n_{init}	p_{init}	Occ_{ave}	N_{50}^{meta}	R_{gen}	Nm	P_e	H_0	H_{meta}	N_e^{meta}	$H_{'single\ large'}$
Closed metapopulations										
2	1	0.456	203	3	0.67	0.493−0.016	0.5	0.235 (0−0.496)	33	0.442 (0.268−0.500)
		0.593	264		0.58	0.367−0.011		0.262 (0−0.499)	39	0.454 (0.340−0.500)
		0.569	253		0.22	0.176−0.006		0.270 (0−0.499)	41	0.452 (0.325−0.500)
6	2	0.442	197		0.65	0.493−0.016		0.257 (0−0.499)	38	0.439 (0.265−0.500)
		0.578	258		0.57	0.367−0.011		0.280 (0−0.499)	43	0.453 (0.329−0.500)
		0.591	263		0.23	0.176−0.006		0.291 (0−0.499)	46	0.454 (0.340−0.500)
257	34	0.562	250	2	0.83	0.493−0.016	0.97	0.702 (0.525−0.828)	78	0.876 (0.826−0.910)
		0.612	273		0.60	0.367−0.011		0.834 (0.726−0.897)	166	0.885 (0.840−0.918)
		0.570	254		0.11	0.176−0.006		0.868 (0.782−0.914)	225	0.879 (0.828−0.913)
Open metapopulation, receiving approximately two unique alleles per 10 generations										
2	1	0.408	182	3	0.41	0.335−0.010	0.5	0.328 (0−0.911)	60	

n_{init} is the initial population size on each patch. p_{init} is the initial number of patches occupied. Occ_{ave} is the average metapopulation occupancy. N_{50}^{meta} is average metapopulation size at generation 50. R_{gen} is the per-generation geometric growth rate. P_e is given for the smallest ($N=2$) and largest ($N=50$) population sizes. Nm is the average number of immigrants per patch. N_e^{meta} is the effective metapopulation size, estimated from H_0 and H_{meta}. In the last column, diversity at generation 50 is given for unsubdivided populations with the given H_0 and constant size equal to N_{50}^{meta}. Numbers in parentheses are 95% confidence intervals. Note that N_e^{meta} estimated from minisatellite data ranges from 19 to 43 (Peacock and Smith 1997a), in good agreement with models parameterized with reasonable demographic estimates, shown in italic type. These models predict lower absolute heterozygosity than detected with minisatellite data, because they do not incorporate mutation. This discrepancy is overcome by comparing observed and predicted N_e^{meta}. N_e^{meta} is solely a function of the rate of decline in H. Note also that N_e^{meta} is high for a more contemporary parameterization of the metapopulation (shown in normal type), which now occupies a large percentage of the metapopulation. Despite frequent local extinctions, the contemporary metapopulation loses diversity nearly as slowly as does the equivalent ideal population.

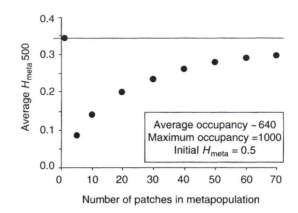

Figure 4.2. The heterozygosity retained after 500 generations in metapopulations modelled with different degrees of subdivision. Each metapopulation holds a maximum of 1000 individuals. The patterns of subdivision range from five populations (of 200 individuals each) to 70 populations (of 14–15 individuals each). This generalized model is similar to that described for the Bodie pika metapopulation, except that all patches are of similar size. Initial heterozygosity was 0.5, with two alleles at equal frequency, both within and across populations. Dots represent the heterozygosity measured across populations (H_{meta}) after 500 generations, averaged over 1000 simulations. Average metapopulation occupancy was 640 individuals throughout simulations, due to frequent local extinctions. Local extinction probabilities per time step ranged from 0.1 to 0.03, depending on the size of local populations, which were subject to both demographic and environmental stochasticity (see text). Colonizing propagules were drawn at random from occupied patches, and contained two individuals each. Recolonized populations doubled to carrying capacity. Local populations experienced founder drift and drift at reproduction. There was no mutation or selection in these models. The solid line represents the heterozygosity maintained after 500 generations in an ideal (unsubdivided, unfluctuating) population of 640 individuals, with $H_0 = 0.5$. Similar results are obtained for models of highly polymorphic loci (Ray 2000).

who pointed out that metapopulation size is ultimately a function of variance in individual reproductive success. Variance in reproductive success due to local extinction–recolonization dynamics is minimal when local populations hold few individuals. If local populations held just one individual, variance in reproductive success due to local extinction could be identical to that in an ideal population experiencing demographic stochasticity.

The conclusion that a metapopulation may maintain high levels of diversity despite frequent local extinction has important implications for the application of genetic inference in dispersal studies. We cannot assume that high diversity precludes extinction–recolonization dynamics (cf. Gilpin 1991). Before embracing any such generalities, we must analyse all the potential relationships between demography and genetic structure. The analysis given here is but one example of the need to incorporate demographic realism into metapopulation genetic models.

Discussion

From model development through data collection and interpretation, dispersal research should integrate demographic and genetic approaches. Models that predict the effects of dispersal on genetic structure must incorporate sufficient demographic detail to make accurate predictions. Ironically, we need sufficient information about population demography in order to choose the appropriate genetic model with which to infer demography.

Luckily, we can squeeze a lot of genetic inference out of a little demographic data. For example, by observing the pattern of territory vacancy and juvenile settlement within a population and the pattern of genetic relatedness within and between populations, we may discover the pattern of dispersal at a range of spatial scales. In the case of the pika, juveniles appear to search for the nearest available habitat, dispersing when necessary. However, securing a philopatric territory during the natal summer does not guarantee success, as these settlers are commonly replaced overwinter. Genetic data reveal that these replacements often involve long-distance dispersers. Dispersal success is also evident in the lack of genetic population structure and inbreeding. The data from both Pipet Tarn and Bodie populations support a resource competition model (Dobson 1982), where the resource is territory. The number and spacing of vacancies is variable in time, so dispersing individuals will move various distances to find openings. Purely demographic studies may not reveal the dispersal success of individuals forced to move out of their natal populations. Purely genetic studies may not reveal the pattern of territory vacancy. Using a combined approach reveals the mechanics of the dispersal process.

After revealing the mechanics of dispersal, we can advance to hypotheses concerning its evolution. At this stage, incorporating data from highly variable genetic markers allows definitive tests of inbreeding hypotheses. Genetic similarity among known first-order relatives can be used to assess levels of relatedness between mated pairs (Packer and Pusey 1993; Girman *et al.* 1997; Peacock and Smith 1997b). The degree of genetic similarity between mated pairs is a result of mating pattern, but also the result of long-term dispersal and gene flow. Populations with little or no immigration can lose genetic variation over time through drift, so that random mating and 'inbreeding' have similar fitness consequences. Inbreeding 'avoidance' can be assessed only in the context of population genetic structure. Genetic data are similarly important for revealing mating patterns predicted by other dispersal hypotheses. For example, male-biased dispersal has often been interpreted as competition for mates (Greenwood 1980; Dobson 1982; Koenig *et al.* 1996). However, identification of extra-pair copulations using molecular markers has shown that subordinate males can adopt mating strategies that overcome the social mating system (see Hughes 1998). Data on parentage, mating patterns, and genetic population structure, coupled with observed dispersal and immigration, reveal contemporary and historical patterns that form the necessary background for testing dispersal hypotheses.

Particularly revealing are studies that combine genetic and demographic analyses to follow up on predictions that were developed using direct methods alone. Analyses of genetic population structure in wild dog packs demonstrated that, although most dispersal occurs between neighbouring packs, occasional long-distance movement introduces local genetic variation and reduces inbreeding (Girman *et al.* 1997). Demographic data on dispersal in lions indicated little opportunity for close inbreeding, and genetic data revealed that even moderate levels of inbreeding were uncommon (Packer and Pusey 1993). As expected from demographic data, natal philopatry in male red grouse and Greater White-toothed shrews has resulted in significant population divergence;

however, genetic studies show that female dispersal mediates gene flow and reduces inbreeding and loss of genetic variation within populations (Favre *et al.* 1997; Piertney *et al.* 1998). In each case, genetic methods reveal patterns that would be nearly impossible to uncover using direct methods alone.

Despite the large number of studies on dispersal in birds and mammals, there is little consensus on what explains the evolution of dispersal in these taxa (Moore and Ali 1984; Dobson 1985; Johnson and Gaines 1990; Stenseth and Lidicker 1992). The lack of a discernible pattern may be due in part to the limitations of direct methods. As genetic inference is applied to dispersal questions, more general evolutionary patterns may emerge. But progress through genetic inference may be hindered by the application of models with simplistic assumptions. There is a real need for further exploration of the interplay between demographic and genetic structure, especially in spatially complex systems (Hastings and Harrison 1994; Wang and Caballero 1999; Ray 2000).

Acknowledgements

Part of this work was supported by NSF Dissertation Improvement Grant DEB-9520820 and separate Awards for Research in College Sciences (ARCS) scholarships to C. Ray and M.M. Peacock. Support is not endorsement.

5

Invasion fitness and adaptive dynamics in spatial population models

Régis Ferrière and Jean-François Le Galliard

Abstract

Disentangling proximate and ultimate factors of dispersal and assessing their relative effects requires an appropriate measure of fitness. Yet there have been few theoretical attempts to define fitness coherently from demographic 'first principles', when space-related traits such as dispersal are adaptive. In this chapter, we present the framework of adaptive dynamics and argue that *invasion fitness* is a robust concept accounting for ecological processes that operate at the individual level. The derivation of invasion fitness for spatial ecological scenarios is presented. Spatial invasion fitness involves the effect of neighbours on a focal individual, mediated by coefficients analogous to relatedness coefficients of population genetics. Spatial invasion fitness can be used to investigate the joint evolution of dispersal and altruism, two traits that both have a direct influence on, and are strongly responsive to, the spatial distribution of individuals. Our deterministic predictions of dispersal and altruism evolution based on spatial invasion fitness are in good agreement with stochastic individual-based simulations of the mutation–selection process acting on these traits.

Keywords: adaptive dynamics, altruism, dispersal, invasion fitness, lattice models

Introduction

Even in homogeneous habitats, *spatial fluctuations* of population size arise inevitably as a result of demographic stochasticity, and *spatial correlations* build up from the imperfect mixing of individuals, induced by their limited range of dispersal (Tilman and Kareiva 1997; Dieckmann *et al.* 2000). As a consequence, selective forces acting on the life history traits of individuals are neither uniform nor independent across space. Dispersal propensity (in the broad sense of natal dispersal and breeding dispersal) is therefore a pivotal component of the individuals' phenotype, for it is both a target of selection and a primary factor in spatial fluctuations and correlations in the selective regime (Ferrière *et al.* 2000).

Since the seminal work of Hamilton and May (1977), we know that the avoidance of competition with related individuals is an important factor in explaining the evolution of dispersal. It has recently been argued that dispersal probabilities evolving under the sole effect of kin competition provide a null model against which to assess the relative importance of alternative selective forces, as predicted by more elaborate kin selection models (Ronce 1999). In kin selection theory based on diallelic, haploid genetics, the commonly used measure of fitness is *invasion fitness*, that is, the *per capita* growth rate of a mutant when rare. For pairwise interactions involving an 'actor' and a 'recipient', the definition of invasion fitness involves the *relatedness* of the recipient to the actor (Grafen 1979), for which the correct definition is the probability that the recipient is a mutant (Day and Taylor 1998). However, this assumes that the altered phenotype of a mutant has no effect on that probability and therefore does not change relatedness. Obviously, this does not hold true when the phenotypic traits under consideration, like dispersal, modify the distribution of individuals across space. Furthermore, modelling a mutant's initial rarity requires some care in spatial models (Rousset and Billiard, 2000), for the population size is locally finite everywhere, and the initial number of mutants may not be regarded locally as infinitesimal.

The purpose of this chapter is to provide a modelling framework that allows us to investigate the evolutionary dynamics of adaptive, continuous traits, while accounting explicitly for both the reciprocal effects of these traits on the spatial distribution of individuals, and for the effects of the spatial heterogeneity of selective pressures on the traits' evolutionary dynamics. In section 2, we provide a general argument that the notion of invasion fitness is appropriate to capture 'first' demographic principles operating at the level of individuals, and to describe the long-term evolutionary dynamics of adaptive life history traits (Metz *et al.* 1992, 1996; Dieckmann and Law 1996; Geritz *et al.* 1997, 1998). We then present, in section 3, van Baalen and Rand's (1998) extension of the notion of invasion fitness to spatially heterogeneous populations. *Spatial invasion fitness* is derived from first demographic and behavioural principles operating at the level of individuals and their nearby neighbours. In non-spatial populations, where individuals are assumed to be constantly well mixed and interactions occur at random between them, invasion fitness can be obtained as the Malthusian growth rate of a simple birth-and-death process (Ferrière and Clobert 1992; Metz *et al.* 1992; Ferrière and Gatto 1995). In contrast, when interactions develop locally and dispersal is limited to neighbourhoods, the process of mutant growth should be modelled by keeping track of spatial statistics that describe local population structures beyond global densities. The theory of *correlation equations* (Matsuda *et al.* 1992; Morris 1997; Rand 1998) provides the appropriate mathematical tools. Under certain assumptions about habitat structure and the model's mathematical properties, invasion fitness can then be obtained as the dominant eigenvalue of a matrix (van Baalen and Rand 1998), just as one would recover the population growth rate of a simple Leslie model (Caswell 1989). In the spatial setting, the matrix involved contains demographic parameters that depend upon the local spatial structure of the population.

In section 4 we use this framework to investigate the joint evolution of dispersal and altruistic behaviour. The evolution of dispersal and the evolution of altruism have been the focus of two rather independent lines of research that trace back to the seminal work of Hamilton (1964). Yet there are serious reasons for trying to merge these lines. With limited dispersal, individuals are likely to interact with relatives, and kin selection models would then predict the evolution of altruism. Yet neighbours do not only interact socially: they compete with one another as well. Thus, clustering of relatives may not be sufficient for sociality to evolve. A dose of dispersal might be needed, so that a locally successful strategy can be exported throughout the resident population. Co-adaptive changes in dispersal and social behaviour may thus be expected (J.F. Le Galliard, R. Feuieré and U. Dieckmann, unpublished manuscript).

Adaptive dynamics and the concept of invasion fitness

We will first introduce the basics of a general and coherent mathematical theory of Darwinian evolution which aims at describing the evolutionary dynamics of adaptive, continuous traits. This *adaptive dynamics theory* (founding papers are Metz *et al.* 1992, 1996; Dieckmann and Law 1996; Geritz *et al.* 1997) satisfies three important requirements:

- Adaptive dynamics are modelled as a macroscopic description derived from microscopic mechanisms. Selective pressures are set by ecological mechanisms operating at the 'microscopic' level of individuals.
- Adaptive dynamics incorporate the stochastic elements of evolutionary processes, arising from the random process of mutation and from the extinction risk of initially small mutant populations.
- Adaptive dynamics describe evolution as a dynamical process, identifying potential evolutionary endpoints, and among them those that, indeed, are attractors for the traits' dynamics.

In this section, we present a brief overview of the principles of adaptive dynamics modelling, to show that a consistent measure of fitness arises naturally from the description of microscopic processes underlying ecological interactions (the reader should refer to Marrow *et al.* (1992), Metz *et al.* (1992), Dieckmann and Law (1996), for a more thorough treatment). In the following sections, we shall see how to derive this fitness measure for a class of spatial population models where the individual probability of dispersal is one of the adaptive traits under consideration.

The canonical equation of adaptive dynamics

We consider a closed population of a single species. Individuals are characterized by a suite of adaptive, quantitative traits that define their *phenotype*. They reproduce and die at rates that depend upon their phenotype and their environment, including external factors as well as their own congeneric population. Haploid inheritance is assumed, and there is a non-zero probability for a birth event to produce a mutant offspring, that is, an individual that differs from its parent in one of the traits. Individuals interact with one another, and the process of selection determines changes in the abundance of each phenotype through time.

Direct individual-based models accounting for the stochasticity of birth, death, and mutational events could be run to study how the distribution of phenotypes present in the population evolves through time. The theory of adaptive dynamics was developed as an alternative to intensive computer calculations, to provide a handy, deterministic description of the stochastic processes of mutation and selection.

Adaptive dynamics models rest on two basic principles (Metz *et al.* 1996): mutual exclusion, 'in general two phenotypes x and x' differing only slightly cannot coexist indefinitely in the population'; and time-scale separation, 'the time scale of selection is much faster than that of mutation'. Thus, one may regard the adaptive dynamics as a trait substitution sequence. Each step occurs at a rate equal to the probability $w(x'/x)$

per unit time for a specific phenotype substitution, say x' substituted to x. The so-called *canonical equation* of adaptive dynamics then describes how the mean of the probability distribution of trait values in the evolving population changes through time. If we keep using x to denote this mean, the canonical equation reads (Dieckmann and Law 1996):

$$\frac{d}{dt} x = \int (x' - x) \cdot w(x'|x) \, dx' \tag{2.1}$$

where the integral sum is taken over the whole range of feasible phenotypes.

Following on the traditional view of the evolutionary process as a hill-climbing walk on an adaptive landscape (Wright 1931), we seek to recast the canonical equation into the form

$$\frac{d}{dt} x = \eta(x) \cdot \frac{\partial}{\partial x'} W(x',x) \Big|_{x'=x} \tag{2.2}$$

where the coefficient $\eta(x)$ would scale the rate of evolutionary change, and $W(x',x)$ would rigorously define the measure of fitness of individuals with trait value x' in the environment set by the bearers of trait value x. This mathematical derivation ought to be underpinned by a biologically consistent description of the mutation–selection process.

Mutant invasion rate as a measure of fitness

To recast the canonical equation (2.1) into the form of equation (2.2), we first expand $w(x'|x)$ as the product of a mutation term and a selection term. To keep notations simple, we shall restrict ourselves to the case where phenotypes are characterized by a single trait. The mutation term is the probability per unit time that the mutant enters the population. It involves four multiplicative components: the *per capita* birth rate $b(x)$ of phenotype x, the fraction $\mu(x)$ of births affected by mutations, the equilibrium population size n_x of phenotype x, and the probability of a mutation step size $x' - x$ from phenotype x. The selection term is the probability that the initially rare mutant gets to fixation. Under the assumption that the population is *well mixed*, we can neglect the effects of the mutant density on the demographic rates of the mutant and resident populations. Let us denote the *per capita* birth and death rates of the rare mutant in a resident population of phenotype x by $b(x',x)$ and $d(x',x)$. Then, the difference $b(x',x) - d(x',x)$ measures the mutant invasion rate, that is, the *per capita* growth rate of initially rare mutants, hereafter denoted by $s(x',x)$. The theory of stochastic birth-and-death processes (e.g. Renshaw 1991, pp. 39–41) shows that the probability that the mutant population escapes initial extinction starting from size 1 is zero if $s(x',x) < 0$, and is approximately equal to $s(x',x)/b(x',x)$ otherwise.

Altogether we obtain:

$$w(x'|x) = \mu(x) \cdot b(x) \cdot n_x \cdot M(x,x' - x) \cdot \frac{[s(x',x)]_+}{b(x',x)} \tag{2.3}$$

where n_x denotes the equilibrium population size of phenotype x. The quantity $[s(x',x)]_+$ is equal to $s(x',x)$ if $s(x',x) > 0$, and to zero otherwise; this means that only advantageous mutants, with a positive invasion rate, have a non-zero chance of becoming established. Up to first order in the mutation step size $x' - x$ we have also:

$$\frac{s(x',x)}{b(x',x)} \approx \frac{1}{b(x)} \cdot (x' - x) \cdot \left. \frac{\partial s}{\partial x'} \right|_{x'=x} \tag{2.4}$$

where we have used $s(x,x) = 0$ since the population of phenotype x is at demographic equilibrium. If we assume the mutational process to be symmetrical, and denote the variance of the mutation distribution M by $\sigma^2(x)$, we can insert equation (2.3), together with equation (2.4), into equation (2.1) and compute the integral to obtain (Dieckmann and Law 1996):

$$\frac{d}{dt} x = \left[\mu(x) \cdot \frac{\sigma^2(x)}{2} \cdot \hat{n}_x \right] \cdot \left. \frac{\partial s}{\partial x'} \right|_{x'=x} \tag{2.5}$$

which conforms precisely to equation (2.2).

According to this deterministic approximation of adaptive dynamics, the evolutionary rate $\eta(x)$ of equation (2.2) is given by the bracketed product which encapsulates the influence of mutation. Most importantly, this derivation identifies the mutant invasion rate $s(x',x)$ as the appropriate measure of fitness denoted by $W(x',x)$ in equation (2.2). Therefore, we call $s(x',x)$ the mutant *invasion fitness*.

Invasion fitness, stability and evolutionary branching

The *selection derivative* (Marrow *et al.* 1992), $\partial s/\partial x|_{x'=x}$, determines the direction of adaptive change. When the selection derivative is positive (or negative), an increase (or a decrease) of the trait value x will be advantageous in the vicinity of the resident trait value. Phenotypes that nullify the selection derivative are called *evolutionary singularities* and represent potential endpoints for the evolutionary process. Yet careful inspection of stability properties of evolutionary singularities is required before conclusions can be drawn about the adaptive dynamics in their vicinity (Geritz *et al.* 1998):

- If invasion fitness reaches a local maximum at an evolutionary singularity, then this singularity is an *evolutionarily stable strategy* (ESS), in the classical terminology of evolutionary biology.
- An ESS need not be attainable: if the selection derivative increases near the ESS, any evolutionary trajectory starting nearby will actually be repelled away from the ESS. In this case, the ESS also is an *evolutionary repellor*.
- Conversely, a singularity may attract evolutionary trajectories and yet correspond to a fitness minimum. In this perhaps most remarkable case, selection is initially stabilizing and drives the population to a point where ecological interactions turn the selective regime into a disruptive one, and dimorphism evolves. This phenomenon is known as *evolutionary branching*. The canonical equation for adaptive dynamics provides an approximate model for evolutionary trajectories heading to a branching phenotype, but obviously fails to capture the population's further evolutionary dynamics.

Spatial invasion fitness in homogeneous habitats

One conclusion to be drawn from the previous section is that the derivation of invasion fitness must be underpinned by an ecological model for the population dynamics. The definition of a fitness measure as a function of space-related traits therefore requires that spatial structure and local interactions are both incorporated into the underlying ecological model.

Spatial population models

Spatial models fall into two main categories, depending on the continuous versus discrete structure of the habitat. Traditional models for continuous space (*reaction–diffusion models*; see Okubo 1980) run into serious biological inconsistencies, like the assumption that infinitely many 'nano-individuals' may live in arbitrarily small areas. It is only recently that two new types of mathematically sound and biologically consistent models have been derived. *Hydrodynamics limit models* are spatially explicit; akin to reaction–diffusion equations, they involve correction terms that account for local interactions and dispersal (Durrett and Levin 1994). *Moment equations* are spatially implicit; they describe the dynamics of the statistical moments of the distribution of individuals in space (Bolker and Pacala 1999; Dieckmann and Law 2000). For modelling spatial population processes over discrete space, there is a long tradition of *metapopulation models* (Levins 1969; Hanski and Gilpin 1997; Hanski 1999a, and references therein). Classical models of metapopulations are not truly spatial in the sense that they do not involve the notion of neighbourhood; dispersal is global, and all dispersing individuals, irrespective of their location, are mixed into a common pool before being redistributed to patches.[1] *Stepping-stone models* (Kendall 1948; Kingman 1969; Renshaw 1986) assume that a set of finite populations is distributed on a regular lattice of patches. Dispersal takes place between neighbouring patches. In the field of population genetics, stepping-stone models usually assume that all patches are saturated to their carrying capacity (Malécot 1948, 1975; Kimura 1953). *Lattice models* (Matsuda *et al.* 1992; Morris 1997; Rand 1998) have been developed recently as another tool for modelling population dynamics in discrete space. Lattice models prescribe the possible locations of individuals on a network of sites, each site hosting at most one individual. There is no saturation assumption: all sites need not be occupied. Local interactions and local dispersal occur between any site and its neighbourhood of connected sites. Like moment equations, lattice models are spatially implicit, and they aim at describing neighbour–range spatial correlations.

When it comes to deriving a measure of invasion fitness from these ecological models, operational results are scant. So far, no rigorous invasion criterion has been established for models of hydrodynamics limits or moments. Invasion fitness in metapopulations has been worked out by Olivieri *et al.* (1995) and, in greater generality, by Metz and

---

[1]For the sake of completeness, we should mention the so-called *two-patch* or *n-patch models* frequently used (possibly overused) to describe local population regulation by means of simple nonlinear density dependence (like the Ricker map). For examples and corresponding references, see chapter 3 in Hanski (1999a). Unfortunately, as they treat the densities of local populations as continuous variables, they have to rely on the rather unsatisfactory premise that local population size is infinite.

Gyllenberg (unpublished manuscript). However, as we have already pointed out, such models do not account for limited dispersal, and therefore address spatial processes in a rather special way. The study of interacting populations, using stepping-stone models, remains very limited. Only lattice models have led to a rigorous mathematical definition of invasion fitness in space (van Baalen and Rand 1998), and it is models of this type that we shall consider further in the rest of this chapter.

Modelling the spatial dynamics of population lattices

The population is distributed over an infinite network, or lattice, of connected sites (Fig. 5.1). A site contains at most one individual. Interactions (social, competitive, parasitic, etc.) may occur only between individuals that inhabit connected sites, and movement may occur only from a given site to a connected site. This has the important consequence that the spatial scale is the same for dispersal and interactions. For simplicity, we shall assume that each site is connected to the same number (n) of neighbouring sites (e.g. a regular lattice). Each site is in one of a limited number of possible states: empty, or occupied by an individual of one out of N possible types. The configuration of the whole lattice is given by the states of all sites. The lattice configuration changes as a result of two types of events potentially affecting any site during any short time interval: birth or immigration of an individual from a neighbouring site, and death or emigration of the individual occupying a site. In general, dispersal (emigration–immigration) is not restricted to the newborn class.

We aim to describe the temporal dynamics of the frequencies of sites that are empty and sites that are occupied by any given phenotype (Matsuda *et al.* 1992; Rand 1998). The probability that the state of a site changes depends not only on its current state but also on the state of neighbouring sites, for two different reasons. On the one hand, dispersal and birth are local events whose realization is conditional on the availability of empty sites in the neighbourhood. The likelihood that an individual in a given site moves or exports its offspring is proportional to the frequency of empty sites in its neighbourhood. On the other hand, local interactions with neighbours will affect the birth rate and death rate of any focal individual. For example, individuals might negatively affect one another's birth rate through local competition for food. In this case, the birth rate could be seen as a decreasing function of the number of neighbours.

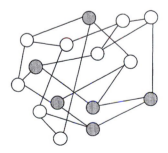

Figure 5.1. Example of a random lattice. Each site is randomly linked to a fixed number n of other sites. Here $n = 3$. Dark circles are occupied sites; open circles are empty sites.

Therefore the *frequency of sites in state i* among all sites in the lattice, p_i, must depend on the neighbourhood structure, as described by a second-order statistic for the distribution of the configurations of all pairs of nearest-neighbour sites. The dynamics of pair configurations depend in turn on the state of triplets including the pairs' neighbours, and so on. A full description of the lattice dynamics eventually requires an infinite hierarchy of statistics, each one describing the spatial structure on a particular scale (sites, pairs, triplets, and so on) in relation to the next one (Morris 1997). To make a model tractable, one has to choose a particular scale of description, and make appropriate approximations to close the exact, infinite system at that scale. This means that the frequencies of configurations beyond the chosen spatial scale are estimated from the frequencies of configurations up to that scale. No mathematical procedure is currently available to systematically identify the scale at which the system should be closed and the closure procedure that should be applied in order to obtain the best approximation of the dynamics of the infinite-dimensional model. This will depend on the particular model under consideration and on the biological motivation guiding the analysis (Morris 1997; Dieckmann and Law 2000).

Our aim is to describe the dynamics of lattices at the most local scale, that of pairs of nearest neighbours. Pair-dynamics models can account for the effect of spatial correlations that arise at a local scale and vanish quickly, although they are not concerned with the development of large-scale spatial structures. It should be noticed that, at least for regular lattices, one may straightforwardly recover the frequencies of sites in the various states (i.e., the p_i values) simply by adding the appropriate pair frequencies. Pair-dynamics models offer a handy compromise between the need to incorporate and describe some of the spatial complexity of the population dynamics, and the aim of deriving useful analytical results on population equilibrium and invasion conditions. The pair-dynamics approach has been used to construct appropriate correlation equations for plant dynamics models (Harada and Iwasa 1994; Satô and Konno 1995), spatial games (Morris 1997; Nakamaru *et al.* 1997), social interactions (Matsuda *et al.* 1992; Harada *et al.* 1995; van Baalen and Rand 1998), and epidemic models (Keeling 1996; Morris 1997). In the case of a spatial game on a regular lattice, however, Morris (1997) showed that the pair-dynamics description could fail dramatically. Then, moving up to the triplet dynamics is often sufficient to obtain a substantial improvement in the closure accuracy.

From individuals to pair-dynamics and correlation equations

We define p_{ij} as the frequency of pairs of nearest-neighbour sites, one being in state i and the other in state j. Such a pair is denoted by (i, j), and the frequency p_{ij} is calculated over all pairs[2] in the lattice. We shall take four heuristic steps in order to derive the so-called *correlation equations*; that is, a set of non-linear differential equations that describe the lattice dynamics at the spatial scale of pairs. The four steps are:

(1) Write the rates of local events for anchored pairs. An *anchored pair* is one that contains a given site z occupied by an individual in a specified state i. By definition,

---

[2]Note that the pairs are symmetrical, which implies $(i,j) = (j,i)$.

local pair events affect anchored pairs, and are triggered by a site event at the anchored site z (see Fig. 5.2). Four local events have to be considered (see below for details).

(2) Average the rates of local events for anchored pairs calculated at step 1 over all sites z in state i.

(3) Calculate the rate of change of the frequency of all (i,j) pairs by bookkeeping all possible transitions of anchored pairs that may create or destroy an (i,j) pair.

(4) Apply an appropriate closure procedure designed to approximate all statistics involving triplets in terms of statistics for pairs.

Notations are introduced in Table 5.1. (See Morris (1997) and Rand (1998), for a rigorous account of all mathematical details involved.)

Step 1: Transition rates for anchored pairs

We define the anchored pair $(i \in z, j \in z')$ to be the pair spanning the sites located at z and z', and hosting a type i individual in site z while site z' is in state j. We consider the four local events that can affect such a pair as a result of an individual event occurring at z (Fig. 5.2): a birth event at z when j is the empty state; two mortality events affecting the i individual at z, differing in the presence or absence of an individual at z'; a dispersal event from z to z', assuming z' to be empty. The individual birth rate, death rate, and dispersal rate involve three additive components: (1) an intrinsic, baseline rate that may depend on the individual's phenotype, (2) an interaction term that measures the effect of neighbours, and (3) a cost term that depends on the individual's phenotype. To calculate the rate of local events, we must introduce the number $n_{k:ij}(z)$ of neighbouring sites in state k next to the z site of an anchored pair $(i \in z, j \in z')$. We simply add the contributions to the event rate affecting the i individual at z resulting from all possible configurations of the neighbourhood of site z. The *per capita* rate of the birth and dispersal local events should be scaled by ϕ, the inverse neighbourhood size. This reflects the fact that a birth or dispersal event affecting, at a given rate, a focal individual that belongs to n pairs, will affect any of these pairs at a rate n times slower; in contrast, a death event at

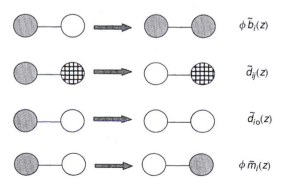

Figure 5.2. The four local pair events and their rates. Open circles are empty sites. Each dark circle is occupied by a type i individual. Hatched circles are in state j. See text for notations and explanations.

Table 5.1.   Variables and parameters of the lattice model

N	Set of all phenotypes present in the population
z	Generic notation for the location of a site in the graph
i, j, k	Generic notations for different site states
p_i	Frequency of sites in state i among all sites (*site frequency*)
p_{ij}	Frequency of (i,j) pairs among all pairs of sites (*pair frequency*)
$q_{i:j}$	Probability that, next to a site in state j, there is a site in state i (*aggregation coefficient*)
$q_{i:jk}$	Probability that, next to a site in state j in a (j,k) pair, there is a site in state i
n	Number of neighbouring sites to any given site (constant)
$n_{k:ij}(z)$	Number of sites in state k in the neighbourhood of a type i at site z, in a (i,j) pair
ϕ	Probability of randomly generating a connection with (or 'probability of making a connection at random with') all sites that could potentially be connected to that site ($\phi = 1/n$).
$b_i(z)$	Intrinsic *per capita* birth rate at location z
$d_i(z)$	Intrinsic *per capita* death rate at location z
$m_i(z)$	Intrinsic *per capita* dispersal rate of type i at location z
$E_{ij}^b(z)$	Additive effect (competition, cooperation) on the *per capita* birth rate of a type i individual located at z, induced by interaction with a type j individual located in the neighbourhood
$E_{ij}^d(z)$	Additive effect (competition, cooperation) on the *per capita* death rate of a type i individual located at z, induced by interaction with a type j individual located in the neighbourhood
$C_i^b(z)$	Cost of type i strategy, decreasing the birth rate of a type i individual located at z
$C_i^d(z)$	Cost of type i strategy, decreasing the death rate of a type i individual located at z

z will concomitantly affect all n pairs containing z. Altogether, this yields the following rates for each of the transitions depicted in Fig. 5.2:

$$\phi \tilde{b}_i(z) = \phi \left(b_i(z) + \sum_{k \in N} E_{ik}^b(z) n_{k:io}(z) - C_i^b(z) \right) \tag{3.1a}$$

$$\tilde{d}_{ij}(z) = d_i(z) + E_{ij}^d(z) + \sum_{k \in N} E_{ik}^d(z) n_{k:ij}(z) + C_i^d(z) \tag{3.1b}$$

$$\tilde{d}_{io}(z) = d_i(z) + \sum_{k \in N} E_{ik}^d(z) n_{k:io}(z) + C_i^d(z) \tag{3.1c}$$

$$\phi \tilde{m}_i(z) = \phi m_i \tag{3.1d}$$

Notice that, for the sake of simplicity, we have assumed that the intrinsic dispersal rate $m_i(z)$ of any focal individual was merely equal to the intrinsic dispersal rate. There is, however, no conceptual predicament entailed by extending the model and making dispersal conditional on the neighbourhood composition (Rand 1998).

Step 2: Averaging transition rates for anchored pairs over the lattice

Assuming that the lattice is homogeneous, we can take the intrinsic rates, the interaction effects, and the costs of interaction to be independent of the location z of any focal individual, and set $b(z) \equiv b$, $d(z) \equiv d$, $m(z) \equiv m$, $E_{ij}^b(z) \equiv E_{ij}^b$, $E_{ij}^d(z) \equiv E_{ij}^d$, $C_i^b(z) \equiv C_i^b$, and $C_i^d(z) \equiv C_i^d$. Transition rates for anchored pairs given by equations (3.1) are still influenced by the local configurations of the lattice, through the neighbourhood structure terms $n_{k:ij}(z)$, which depend on the location z. *Local fluctuations* caused by demographic stochasticity induce spatial variations in the neighbourhood structure. If we knew at any time the state of every site z, then we could calculate each $n_{k:ij}(z)$ and obtain all transition probabilities for each anchored pair. However, the large number of sites makes this endeavour hopeless. Instead, we aim to derive average transition rates for anchored pairs across the lattice. We first compute an average measure of the neighbourhood structure, $\bar{n}_{k:ij} = \sum n_{k:ij}(z)/|i|$, calculated as the total number $|i|$ of sites in state i becomes very large; the sum is taken over all sites z that host a type i individual belonging to an (i,j) pair. Likewise, we define $q_{k:ij}$ as the average proportion of sites in state k in the neighbourhood of a site in state i within a (i,j) pair; in other words, $q_{k:ij}$ is the conditional probability of having a site in state k in the vicinity of a site in state i, given that one of the latter's neighbouring sites is in state j. Since a focal site in an anchored pair is connected to $(n-1)$ sites outside that pair, we have $\bar{n}_{k:ij} = (n-1)q_{k:ij}$. This averaging procedure applied to all local pair-events rates, equations (3.1), eventually yields the following average rates:

$$\phi \bar{b}_i = \phi \left(b_i + \sum_{k \in N} E_{ik}^b (n-1)q_{k:io} - C_i^b \right) \tag{3.2a}$$

$$\bar{d}_{ij} = d_i + E_{ij}^d + \sum_{k \in N} E_{ik}^d (n-1)q_{k:ij} + C_i^d \tag{3.2b}$$

$$\bar{d}_{io} = d_i + \sum_{k \in N} E_{ik}^d (n-1)q_{k:io} + C_i^d \tag{3.2c}$$

$$\phi \bar{m}_i = \phi m_i \tag{3.2d}$$

Step 3: Pair transition rates and equations for pair-dynamics

To compute the transition rates for all possible pairs, we have to complete the bookkeeping of all local pair events that may create or destroy any given pair, and use the average rates given by equations (3.2). This is done in Box 1 for one particular type of pair, in the case of a lattice where there are three possible states for a site: empty, or occupied by one of two types. Once all pair transition rates are available, it is straightforward to assemble a system of differential equations that govern the temporal dynamics of pair frequencies. It turns out that the combinations of rates that enter these equations can be simplified by making use of the following composite rates (van Baalen and Rand 1998):

$\alpha_{ij} = (1 - \phi)(\bar{b}_i + \bar{m}_i)q_{i:oj}$ is the rate at which type i enters a pair (o,j) with $j \neq i$,

$\beta_i = \phi \bar{b}_i + (1 - \phi)(\bar{b}_i + \bar{m}_i)q_{i:oi}$ is the rate at which type i enters a pair (o,i),

$\delta_{ij} = \bar{d}_{ij} + (1 - \phi)\bar{m}_i q_{o:ij}$ is the rate of loss of type i from (i,j) pairs.

We shall refer to these equalities as equations (3.3a), (3.3b), and (3.3c), respectively. It is also convenient to introduce the auxiliary parameter $\alpha_{ij}' = (1 - \phi)(\bar{b}_i + \bar{m}_i)$.

Box 1. Derivation of pair dynamics

We consider a dimorphic population with two types of individuals, x and y. We perform the bookkeeping of all possible transitions and their rates that may create or destroy (x,o) pairs. The frequency of this pair is affected by six potential events, which can be grasped easily by mere graphical depiction (Fig. 5.B1; also see van Baalen and Rand 1998). The rate of each transition is computed by summing the appropriate average rates of local pair events.

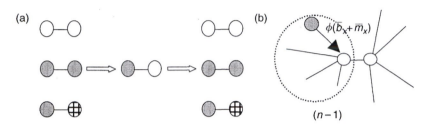

Figure 5.B1. How local pair events affect the pair (x,o). (a) All possible transitions that may create and destroy the focal pair (in the middle). (b) An example of a local pair event showing how a pair (x,o) can be created from a pair (o,o): reproduction or dispersal occurs in an anchored pair that belongs to the neighbourhood of one of the empty sites of the focal pair. This happens at rate $\phi(\bar{b}_x + \bar{m}_x)$ for each of the $(n-1)q_{x:oo}$ possible anchored pairs under consideration.

Pairs (x,o) are created by:

- the transition from (o,o), as illustrated in Fig. 5.B1. There are, on average, $(n-1)q_{x:oo}$ anchored pairs (x,o) whose empty site belongs also to a pair (o,o); the empty pair (o,o) will be turned into an (x,o) pair by reproduction at the local pair event rate ϕb_x, and by dispersal at the rate $\phi \bar{m}_x$.
- the transition from (x,x), either due to death at rate $\bar{d}_{xx}$ to movement towards a neighbouring site. In the latter case, there are $(n-1)q_{o:xx}$ anchored pairs that may undergo the corresponding transition, each at an average local pair event rate $\phi \bar{m}_x$.
- the transition from (x,y), which is calculated in a similar way.

Pairs (x,o) are destroyed by:

- the transition to (o,o), due to death at rate $\bar{d}_{xo}$ or to dispersal. Again, we calculate the number of anchored pairs where this transition may take place to be $(n-1)q_{o:ox}$, and for each of them the transition occurs at the rate $\phi \bar{m}_x$.
- the transition to (x,x), due to reproduction within this pair at rate $\phi \bar{b}_x$ or to a reproduction or dispersal event involving an x neighbour. The latter transition involves $(n-1)q_{x:ox}$ (x,o) anchored pairs, which are affected by a local birth event at rate and $\phi \bar{b}_x$ by a local dispersal event at rate $\phi \bar{m}_x$.
- Likewise, the transition to (x,y) involves $(n-1)q_{y:ox}$ anchored pairs (y,o), undergoing local birth at rate $\phi \bar{b}_y$ and local dispersal at rate $\phi \bar{m}_y$.

Collecting all these transition rates together, and using the notation $\bar{\phi} = (n-1)\phi$, we finally obtain the following rate of change for the pair frequency p_{xo}:

$$\frac{dp_{xo}}{dt} = (\bar{b}_x + \bar{m}_x)\bar{\phi}q_{x:oo}p_{oo} + (\bar{d}_{xx} + \bar{\phi}\bar{m}_x q_{o:xx})p_{xx} + (\bar{d}_{yx} + \bar{\phi}\bar{m}_y q_{o:yx})p_{xy}$$
$$- (\phi\bar{b}_x + \bar{d}_{xo} + \bar{\phi}\bar{m}_x q_{o:xo} + \bar{\phi}(\bar{b}_x + \bar{m}_x)q_{x:ox} + \bar{\phi}(\bar{b}_y + \bar{m}_y)q_{y:ox}) \qquad \text{(B1.1)}$$

Step 4: Closing the system

The equations for pair frequencies obtained at step 3 involve the conditional probabilities q_{kij}. This implies that the system is not closed: the frequencies of pairs depend on the frequencies of triplets, and, to avoid a cascade of dependency on even more complex configurations, the frequencies of configurations involved beyond pairs have to be approximated from the pairs. Finding an accurate approximation amounts to solving the 'closure problem' posed by the dynamical system under concern.

The general form of such a *pair approximation* can be written as $q_{k:i}$, the probability that there is a site in state k next to a site in state i, plus an error term capturing an estimation bias due to local fluctuations (Morris 1997). Different pair approximations have been developed, reflecting different ways of correcting for the neighbourhood structure (Matsuda *et al.* 1992; van Baalen 2000), the lattice regularity (Morris 1997), and the distribution of local fluctuations (Morris 1997). *Ad hoc* corrections accounting for the population clustering pattern have also been proposed (Satō *et al.* 1994). In general, we can safely assume that an infinite random lattice, or a more regular lattice with weak aggregation, will produce a small bias. The *standard pair approximation* (Matsuda *et al.* 1992) precisely equals the bias to zero and therefore reads $q_{k:ij} \cong q_{k:i}$. It has been challenged against individual-based simulations in a number of models corresponding to various biological situations (Matsuda *et al.* 1992; Harada and Iwasa 1994; Satō and Konno 1995; Kubo *et al.* 1996; Nakamaru *et al.* 1997). The match is often very good, but sometimes devastatingly bad. In such cases, moving up the description level to the spatial scale of triplets can suffice to improve matters substantially (Morris 1997). Satō *et al.* (1994), Harada *et al.* (1995), Morris (1997), Ellner *et al.* (1998), and van Baalen (2000) have investigated the alternative path of deriving better pair approximations.

Here, we shall content ourselves with the standard pair approximation and apply it to equations (3.2) and (3.3). This yields:

$$\phi \bar{b}_i = \phi \left(b_i + \sum_{k \in N} E_{ik}^b (n-1) q_{k:i} - C_i^b \right) \tag{3.4a}$$

$$\bar{d}_{ij} = d_i + E_{ij}^d + \sum_{k \in N} E_{ik}^d (n-1) q_{k:i} + C_i^d \tag{3.4b}$$

$$\bar{d}_{io} = d_i + \sum_{k \in N} E_{ik}^d (n-1) q_{k:i} + C_i^d \tag{3.4c}$$

and

$$\alpha_{ij} = (1 - \phi)(\bar{b}_i + \bar{m}_i) q_{i:o} = \alpha_i \tag{3.5a}$$

$$\beta_i = \phi \bar{b}_i + (1 - \phi)(\bar{b}_i + \bar{m}_i) q_{i:o} \tag{3.5b}$$

$$\delta_{ij} = \bar{d}_{ij} + (1 - \phi)\bar{m}_i q_{o:i} \tag{3.5c}$$

These approximate expressions can be inserted into the system of differential equations for pair frequencies written with exact pair transition rates; see equation (B1.1). If there is a single phenotype x in the population (resident phenotype), the dynamics of pairs obey the following system of so-called *correlation equations* (Rand 1998):

$$\begin{pmatrix} dp_{ox}/dt \\ dp_{xx}/dt \end{pmatrix} = \begin{pmatrix} \alpha'_x q_{o:o} - (\beta_x + \delta_{xo}) & \delta_{xx} \\ 2\beta_x & -2\delta_{xx} \end{pmatrix} \begin{pmatrix} p_{ox} \\ p_{xx} \end{pmatrix} \tag{3.6}$$

The equilibrium state of the population, fully characterized by $q_{o:x}$ and $q_{o:o}$, may then be found by solving the system $dp_{ox}/dt = 0$ and $dp_{xx}/dt = 0$.

Spatial invasion fitness

We now have the modelling machinery in place to tackle the calculation of invasion fitness, that is, a measure of the population growth rate of a mutant (phenotype y) introduced at low frequency in the resident population where only phenotype x is present. When two strategies x and y are represented in the population, there are six possible types of pairs. A simple bookkeeping procedure is applied to all possible transitions of these pairs, and the rates defined by equations (3.3) are used to construct a system of correlation equations (3.7):

$$\begin{pmatrix} dp_{ox}/dt \\ dp_{xx}/dt \\ dp_{oy}/dt \\ dp_{yx}/dt \\ dp_{yy}/dt \end{pmatrix} = \begin{pmatrix} \alpha'_x q_{o:o} - (\beta_x + \alpha_y + \delta_{xo}) & \delta_{xx} & 0 & \delta_{yx} & 0 \\ 2\beta_x & -2\delta_{xx} & 0 & 0 & 0 \\ 0 & 0 & \alpha'_y q_{o:o} - (\beta_y + \alpha'_y q_{y:o} + \delta_{xo}) & \delta_{xy} & \delta_{yy} \\ 0 & 0 & \alpha_x + \alpha'_y q_{x:o} & -(\delta_{xy} + \delta_{yx}) & 0 \\ 0 & 0 & 2\beta_y & 0 & -2\delta_{yy} \end{pmatrix} \begin{pmatrix} p_{ox} \\ p_{xx} \\ p_{oy} \\ p_{yx} \\ p_{yy} \end{pmatrix} \tag{3.7}$$

The mutant rate of growth, denoted by $s(y,x)$, can be obtained by summing up the last three equations of system (3.5):

$$\frac{dp_y}{dt} = \frac{dp_{oy}}{dt} + \frac{dp_{xy}}{dt} + \frac{dp_{yy}}{dt} = s(y,x)p_y \tag{3.8}$$

which, after some algebra, simplifies into:

$$s(y,x) = \bar{b}_y q_{o:y} - \bar{d}_y \tag{3.9}$$

where

$$\bar{b}_y = b_y + (n-1)E^b_{yx}q_{x:y} + (n-1)E^b_{yy}q_{y:y} - C^b_y \tag{3.10}$$

$$\bar{d}_y = d_y + nE^d_{yx}q_{x:y} + nE^d_{yy}q_{y:y} - C^d_y \tag{3.11}$$

Rearranging terms, we obtain the final expression:

$$s(y,x) = \left[(b_y - C^b_y)q_{o:y} - d_y - C^d_y\right] + \left[(n-1)E^b_{yx}q_{o:y} - nE^d_{yx}\right]q_{x:y}$$
$$+ \left[(n-1)E^b_{yy}q_{o:y} - nE^d_{yy}\right]q_{y:y} \tag{3.12}$$

This expression bears an interesting relationship to the notion of *direct* or *neighbour-modulated fitness* (Hamilton 1970; Frank 1998). Direct fitness is defined by summing the fitness effects on an individual caused by all the phenotypes of neighbours (including the individual itself). Likewise, spatial invasion fitness is obtained by adding the effects on a focal mutant of a resident or mutant neighbour, weighted by the probability that the focal individual is neighboured by a resident or a mutant individual.

Further analysis based on spatial invasion fitness as defined by equation (3.12) requires that we solve equation (3.7) for $q_{y:y}$, $q_{x:y}$, and $q_{o:y}$. This can be done numerically by following the algorithmic recipe outlined in Box 2, or even analytically in the simplest cases (Matsuda *et al.* 1992).

Box 2. A numerical recipe to compute spatial invasion fitness

The expression of mutant population growth rate depends on the spatial statistics $q_{o:y}$, $q_{x:y}$, and $q_{y:y}$, which *a priori* vary over time. Yet the so-called *relaxation property* of the system entails that the statistics $q_{o:y}$, $q_{x:y}$, and $q_{y:y}$, converge very fast to equilibrium values, compared with the slow growth or decline of the system variables $p_{o:y}$, $p_{x:y}$, and $p_{y:y}$ (Matsuda *et al.* 1992, our simulations). Therefore, to obtain a measure of spatial invasion fitness, we may write an auxiliary system of differential equations for the variables $q_{o:y}$, $q_{x:y}$, and $q_{y:y}$ only, solve it for equilibrium, and insert the result into equation (4.1).

The numerical derivation of this auxiliary system relies on the initial rarity of the mutant in the resident population. This, by definition, means that $q_{y:o} = 0$. This property allows us to write a closed model for the mutant pair dynamics, using the 3×3 lower-right block M of the transition matrix which appears in equation (3.5):

$$\frac{\mathrm{d}p_y}{\mathrm{d}t} = M(\bar{q}_y)\bar{p}_y \quad \text{with } \bar{p}_y = (p_{oy}, p_{xy}, p_{yy}) \tag{B2.1}$$

Using the relations $\mathrm{d}p_y/\mathrm{d}t = s(y,x)p_y$ and $\bar{p}_y = p_y\bar{q}_y$, we can further transform this system into

$$\frac{\mathrm{d}\bar{q}_y}{\mathrm{d}t} = \left[M(\bar{q}_y) - s(y,x)I \right]\bar{q}_y \tag{B2.2}$$

(I is the 3×3 identity matrix.) At equilibrium, $\mathrm{d}\bar{q}_y/\mathrm{d}t = 0$, and the spatial statistics $\bar{q}_y$ are obtained by solving (numerically, or analytically in the simplest cases) the non-linear system $M(\bar{q}_y)\bar{q}_y = \lambda\bar{q}_y$, which involves four unknowns ($q_{o:y}$, $q_{x:y}$, $q_{y:y}$, and the corresponding eigenvalue λ) and three equations, along with the constraint $q_{o:y} = 1 - q_{x:y} - q_{y:y}$. Solving for λ at the same time yields the numerical value of the spatial invasion fitness $s(y,x)$ (4.1).

Application: coadaptation of dispersal and altruism

Empirical work has stressed the importance of spatial structure and spatial processes for the evolution of dispersal (Hanski, chapter 20; Ims and Hjermann, chapter 14;

Ronce *et al.*, chapter 24). Coadaptation of other life history components is also expected to have a decisive influence on the evolution of dispersal, because of physiological and/or genetic correlations (Ronce *et al.*, chapter 24; Roff and Fairbairn, chapter 13) or behavioural alternatives (see Lambin *et al.*, chapter 8, for a discussion of the joint adaptation of dispersal, competition, and cooperation). There is an urgent need for theory to incorporate these empirical facts.

The purpose of this section is to take a step forward in that direction. We make use of the framework of lattice population models to investigate the joint evolution of dispersal and social behaviour, while accounting explicitly for local interaction and dispersal processes. More specifically, our main objectives are (i) to identify selective pressures acting on these traits, (ii) to make predictions about their relative effects on the direction of evolution, and (iii) to relate them quantitatively to basic individual and interaction traits. The material presented here provides a short review of analyses expounded in Le Galliard (1999), J.-F. Le Galliard, R. Ferrière, and U. Dieckmann (unpublished manuscript), and R. Ferrière and J.-F. Le Galliard (unpublished manuscript).

Model assumptions

We focus on two adaptive components of the individual's phenotype: dispersal and altruism (Table 5.2). The former trait is measured by the dispersal rate m. The altruistic trait is measured by the total investment in altruism u and the amount u/n of help an actor individual may distribute over its neighbourhood. This amount additively affects any recipient's intrinsic birth rate. Note that this is a simplified description of altruism, because individuals will have a total potential amount u to give and will always give the same amount of help per neighbour, whatever the number of receivers. In the biological realm, this would mean the absence of any kind of strategical distribution of altruism.

Both traits are costly to the bearer. A linear model for the cost of dispersal m is assumed, whereas the cost of altruism scales algebraically with the amount of total investment u (Table 5.2). The total cost is subtracted from the intrinsic birth rate. The costs of dispersal and altruism are paid unconditionally, irrespective of the movement actually performed by the individual and the average amount of help actually given to the neighbourhood. A representative biological instance would be an organism where both dispersal and altruism imply an initial ontogenetic shift towards a fixed physiological or morphological state that would determine the individual lifetime investment of dispersal and altruism. This state would permanently impact the birth rate. An example of this might be the case of a dispersal structure (O'Riain *et al.* 1996).

Table 5.2.   Specific variables and parameters of the model

b	Intrinsic *per capita* birth rate ($b = 2.0$)
d	Intrinsic *per capita* death rate ($d = 1.0$)
m	Intrinsic *per capita* dispersal rate (*adaptive trait*)
u	Intrinsic *per capita* investment in altruism rate (*adaptive trait*)
κu^{γ}	Cost of altruism, decreasing the birth rate
νm	Cost of dispersal, decreasing the birth rate

Starting from the general model presented in section 3, we make two simplifying assumptions on our way to derive the measure of spatial invasion fitness: the intrinsic birth and death rates are independent of the phenotype, and costs and benefits impact the birth rate only.

Referring to notations introduced in Table 5.1, this means that $b_i \equiv b$, $d_i \equiv d$, $E_{ij}^d \equiv 0$, $C_i^d \equiv 0$. We use the notation $C(u,m)$ to designate the total cost associated with investment in altruism u and dispersal rate m, $C(u,m) = \kappa u^\gamma + \nu m$ (Table 5.2). Parameters κ and ν measure the sensitivity of the costs of altruism and dispersal. The parameter γ further indicates how the sensitivity of the cost of altruism varies with the degree of altruism. A high value of γ means that the cost of altruism increases slowly with the degree of altruism when the degree of altruism is low, and becomes more sensitive to altruism as the degree of altruism increases.

Adaptive dynamics of dispersal and altruism

Spatial invasion fitness s follows from the general model of equation (3.10) and is given here by:

$$s \equiv \left(b + u(1 - \phi)q_{x:y} + u'(1 - \phi)q_{y:y} - C(u',m')\right)q_{o:y} - d \tag{4.1}$$

where $x = (u,m)$ denotes the resident phenotype; $y = (u',m')$, the mutant phenotype. The canonical equation (2.5) reads:

$$\frac{d}{dt}\begin{pmatrix} u \\ m \end{pmatrix} = \begin{pmatrix} \eta \cdot \dfrac{\sigma^2}{2} \cdot p_x \cdot \dfrac{\partial s}{\partial u'}\Big|_{u'=u} \\ \eta \cdot \dfrac{\sigma^2}{2} \cdot p_x \cdot \dfrac{\partial s}{\partial m'}\Big|_{m'=m} \end{pmatrix} \tag{4.2}$$

where η and σ^2, respectively, denote the mutation rate and the mutation step variance, which we assume to be the same for both traits and independent of the current phenotypic mean. Making use of the facts that the resident population is at equilibrium and that the mutant is little different from the resident, a first-order approximation of spatial invasion fitness reads:

$$s/q_{o:y} \cong [d/q_{o:x}^2 - (1 - \phi)u](q_{o:y} - q_{o:x}) + (1 - \phi)q_{y:y}(u' - u) - [C(u',m') - C(u,m)] \tag{4.3}$$

(R. Ferrière and J.-F. Le Galliard, unpublished manuscript). This expression clearly identifies three components of selection operating on dispersal and altruism. The first term in the right-hand side of equation (4.3) quantifies the pressure for opening free space in an individual's neighbourhood. It is stronger under more crowded conditions (i.e., when $q_{o:x}$ is low), and increases with the intrinsic death rate d: when mortality is low, there is little selective advantage to be gained from opening space by reducing altruism or increasing dispersal. Also, this pressure is opposed if the resident degree of altruism, u, is sufficiently high, since then it pays off to interact with more neighbours, regardless of their altruism phenotype (resident or mutant). The second term in equation (4.3) expresses the pressure for increased altruism; it is stronger under more aggregated mutant conditions. The third term measures the pressure for reducing the direct costs of dispersal and altruism. By following the numerical recipe for the calculation of aggregation coefficients and spatial invasion fitness (Box 2), explicit analytical expressions for $q_{y:y}$ and $q_{o:y}$ can be obtained. It is thus possible to write each component of selection as a function of individual parameters.

In general, when the evolution of one trait alone is considered, the adaptive dynamics of the trait are monotonous and converge to a point attractor. This attractive point corresponds to a singularity of the adaptive dynamic, that is, a point where the selection derivative vanishes. A mutant appearing around this phenotype value is actually counterselected and cannot invade (Fig. 5.3). The pattern of stabilizing selection is well explained by the relative effects of conflicting pressures. Focusing on the case of dispersal, we can see that at low dispersal the predominant selective pressure is induced by local competition for space; reduced aggregation is favoured, and this selects for higher dispersal rates. As dispersal increases, the intensity of the opposed selective pressure induced by the cost of dispersal also increases. An intermediate equilibrium value is reached at which both pressures exactly compensate each other. Numerical analysis of the dispersal rate at this attractor suggests that its value is mainly sensitive to the parameter v, which scales the cost of dispersal.

We now consider the coadaptive dynamics of dispersal and altruism. The selective gradient respective to either trait vanishes along the corresponding isocline (Fig. 5.4), which is the set of evolutionary singularities obtained for this trait, for each possible value of the other trait. Both isoclines cross at the singularity of the coadaptive dynamics, denoted by (m^*, u^*).

When the cost of altruism is high and very sensitive to a change in the degree of altruism, the singularity is always a stable node (J.-F. Le Galliard, R. Ferrière, and

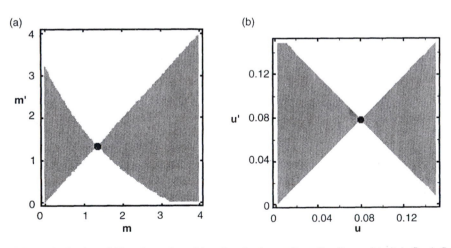

Figure 5.3. Pairwise invasibility plots when either the altruism trait or the dispersal trait is fixed. Spatial invasion fitness (see equation 4.1) is positive in the shaded region. (a) Evolution of dispersal for fixed altruism ($u = 0.1$). (b) Evolution of altruism for fixed dispersal ($m = 0.5$). In both cases, there is a unique evolutionary singularity that is attracting and evolutionarily stable. Parameter values: $n = 4$, $b = 2.0$, $d = 1.0$, $\gamma = 2.0$, $\kappa = 1.0$, $v = 0.1$.

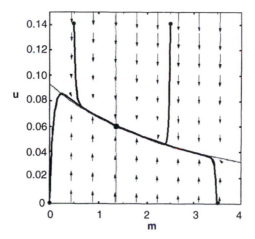

Figure 5.4. Coadaptive dynamics of dispersal and altruism. Predictions from the canonical equation (4.2). Thin lines are evolutionary isoclines. The four thick lines are examples of evolutionary trajectories, starting from four different ancestral states. The crossing point of the isoclines gives the singularity, which is attracting and evolutionarily stable. Parameter values: as in Fig. 5.3.

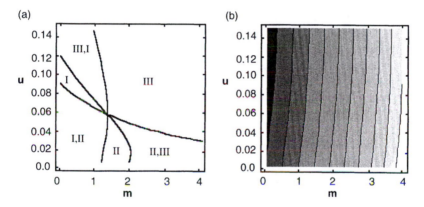

Figure 5.5. (a) Zero-contour lines of the components of selection along adaptive trajectories. In each of the six delineated regions, positive pressures are indicated. Component I, pressure for reducing local competition for space; component II, pressure for increased altruism under aggregated conditions; component III, pressure for reducing the direct costs of dispersal and altruism. (b) Spatial aggregation, shown as a contour plot of the aggregation coefficient q_{xx} for a pure population of phenotype x. Parameter values: as in Fig. 5.3.

U. Dieckmann, unpublished manuscript). The dispersal rate still converges monotonously to the singularity, but explaining the adaptive dynamics in the two-dimensional trait space now requires that we consider how the three selective pressures interplay. This can be done by identifying the sign of each selection component evaluated locally in the direction of adaptation (Fig. 5.5). For example, the four trajectories (1–4) depicted in Fig. 5.4 can be interpreted in this way. (1) Starting from low dispersal and low altruism, mutants that invest more in altruism and dispersal are initially favoured (selection components I and II are positive): being more altruistic is advantageous because the level of aggregation is high; being slightly more mobile is also beneficial for it yields more free space in the neighbourhood. In a second phase of the dynamics, mutants dispersing more are selected for (selective component I is positive); this reduces spatial aggregation and therefore promotes invasion by less altruistic phenotypes. (2) Initially, dispersal is low and altruism is high. Only the first phase of the adaptive dynamics differs. Here, the adaptive dynamics begin with the reduction of the cost of altruism and the reduction of local competition for space (components III and I are positive). (3) Starting with a high dispersal–low altruism phenotype, selection favours an increase in altruism and a decrease in dispersal: at low altruism, mutants with lower dispersal rates pay a significantly reduced cost (component III is positive), and the benefit of more altruism in a population that develops more aggregation dominates the cost of increased local competition (component II is positive). (4) Finally, when ancestral dispersal and altruism are high, the selective pressure for reduced costs dominates (component III is positive) and drives the system all the way down to the singularity where both traits stabilize.

Revisiting Hamilton's rule

Hamilton (1964) formulated his famous rule, according to which, if an actor expresses a behaviour that costs him C offspring and increases by B the number of individuals related to the actor, this behaviour is selected for if $Br > C$. There has been much debate over the interpretation of the fitness costs C, benefits B, and relatedness r that make Hamilton's rule work, and by which this rule can be generalized for more complex ecological scenarios.

Defining and measuring relatedness in spatially structured populations is a long-standing problem of population genetics (Malécot 1948; Rousset and Billiard 2000).

The spatial invasion condition provides a natural definition of relatedness as a measure of phenotypic correlation between neighbours (Frank 1998; van Baalen and Rand 1998). When altruistic and selfish individuals are identical in their basic demographic rates (b, d, and m), altruists with phenotype $y = (u',m)$ can invade non-altruists with phenotype $x = (0,m)$ if:

$$u' \cdot (1 - \phi) \cdot q_{y:y} > \kappa \cdot u'^{\gamma}, \qquad (4.4)$$

that is, we have recovered a variant of Hamilton's rule in which $B \equiv (1 - \phi)u'$, $C \equiv \kappa u'^{\gamma}$, and the coefficient of relatedness r is given by:

$$r = q_{y:y}$$

As already mentioned, $q_{y:y}$, and therefore r, can be computed from the invasion matrix (Box 2). This coefficient r estimates how much of an altruist's environment consists of other altruists, an interpretation that is consistent with the findings of Day and Taylor (1998). The precise interpretation of B, C, and r in Hamilton's rule, however, is dependent on the details of demographic processes operating in the population. For example, van Baalen and Rand (1998) noted that, if the cost of altruism is incurred as an increased mortality rate instead of a decreased birth rate, for zero dispersal the invasion condition of altruists in a selfish population becomes:

$$u' \cdot (1 - \phi) \cdot q_{y:y} > (b/d)\kappa \cdot u'^{\gamma}. \tag{4.6}$$

This provides another version of the Hamilton's rule where the cost C is recovered as the cost of altruism corrected for intrinsic birth and death rates. Other variants of the spatial Hamilton's rule, where relatedness similarly depends on local demographic processes, have been established by Ferrière and Michod (1995, 1996) for the invasion of cooperation in a spatial iterated Prisoner's Dilemma.

How kin selection models handle relatedness is usually problematic (Day and Taylor 1998; Rousset and Billiard 2000). This is not to suggest that kin selection is not the ultimate cause of the evolution of altruism in viscous populations, as Hamilton (1964) originally asserted it, but that measuring inclusive fitness as defined in kin selection models may not correctly predict the evolutionary dynamics of social traits when selection is density dependent. Using spatial invasion fitness, Hamilton's principle is recovered as an emergent property of the model. This backs up Nunney's (1985) statement that group selection acts when there is positive preferential association of common phenotypes, but that kin selection is the only form of group selection that is able to maintain altruism.

Does spatial invasion fitness rightly predict evolutionary dynamics?

Although our coevolutionary model of dispersal and altruism incorporates salient features of the ecological and evolutionary processes (including density dependence, demographic stochasticity, and evolutionary feedback), it remains underpinned by several critical simplifications. We assume an infinite lattice size, and describe the dynamics of local densities by making use of the standard pair approximation (Morris 1997). The derivation of the fitness measure relies on the small frequency of mutants when they appear and on the relaxation approximation that they instantaneously build up a characteristic invasion structure that may serve as a vehicle for their potential spread (Dieckmann and Law 2000). Furthermore, the deterministic description of the adaptive dynamics gives an approximation for the mean path of the stochastic mutation selection process (Dieckmann and Law 1996), which itself already entails averaging over an infinite number of realizations.

Notwithstanding all this, the properties of stochastic simulations are remarkably well captured by the deterministic predictions (Fig. 5.6; J.-F. Le Galliard, R. Ferrière, and U. Dieckmann, unpublished manuscript, for a more thorough comparison). The positions of the isoclines and the attracting singularity (m^*, u^*) remain nearly unchanged. Overall trends of stochastic trajectories are predicted correctly by the deterministic model. Wilder fluctuations in trait values, involving the repeated rise and fall of altruism, are observed nearer to the singularity, as the selection gradient tends to weaken there. In our case, these complex regimes in the degree of altruism, which have received some attention elsewhere (Doebeli and Knowlton 1998), are best explained by genetic drift in regions of low selection pressure across the phenotypic space.

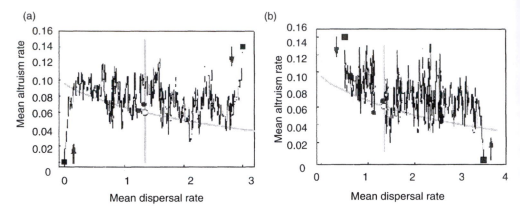

Figure 5.6. Mean trajectory of ten individual-based simulations of dispersal and altruism evolutionary dynamics. Thick grey lines are isoclines predicted by the canonical equation (4.2) (see Fig. 5.4). Black squares indicate initial states. (a) Simulations of two trajectories starting at $(m,u) = (0,0)$ and $(m,u) = (3,0.14)$, respectively. (b) Simulations of two trajectories starting at $(m,u) = (0.5,0.14)$ and $(m,u) = (3.5,0.0)$, respectively. The stochastic trajectories, although rather jerky near the convergence state at the isoclines intercept (open circles), hit rather close to it (black circles) after following closely the deterministic path predicted by spatial invasion fitness (see Fig. 5.4). For both traits, the mutation rate is 10^{-2} and the mutation variance is 10^{-2}. Time at the end of simulations: 500,000 time units. Lattice size: 900 sites.

Concluding remarks

Defining invasion fitness for spatial ecologies is no trivial matter. Starting from demographic and behavioural processes operating at the individual level and locally between close neighbours, the invasion exponent of a simple system of correlation equations for the dynamics of a mutant population provides a tractable solution to this problem. The notion of spatial invasion fitness allows one to *derive*, rather than postulate, an explicit relationship between distinct components of selection on the one hand, and the characteristics of the individuals and their interactions on the other. Numerical simulations of individual-based models confirm that the resulting spatial invasion fitness correctly predicts the dynamics of the stochastic mutation–selection process. On the empirical side, Rainey and Travisano's (1998) experiments on the evolution of polymorphism in bacteria have shown that invasion fitness measured in spatially heterogeneous populations successfully predicts the maintenance of morph diversity. In contrast, the destruction of local structures developed in the course of population growth alters the phenotypes' invasion fitnesses and modifies the eventual phenotypic composition of the population.

The mathematical derivation of spatial invasion fitness proceeds by averaging over space the transition rates of pairs. This amounts to looking at the local structure of the mutant population as homogeneously replicated across the whole (infinite) lattice. The non-homogeneous distribution of the pairs containing mutants, induced by the finite size of the mutant population and the non-typical clustering pattern that may develop at the earliest stage of invasion, may also require us to incorporate correction terms to our measure of spatial invasion fitness. There may be an interesting parallel to be drawn with the theory of evolutionary games in continuous space. In this context, the initial clustering of mutants entails that fitness should be defined not from space averages of individual traits, but as the speed at which the front of a mutant cluster moves forward and propagates mutants through space (Hutson and Vickers 1992; Ferrière and Michod 1995, 1996; Ellner *et al.* 1998).

We have used the notion of spatial invasion fitness to model the joint adaptive dynamics of dispersal and altruism. Even without further corrections for more subtle spatial effects, spatial invasion fitness appears to give very consistent predictions on how these two behavioural traits coevolve. The analysis of this particular model underlines three important and general achievements of adaptive dynamics based on the notion of spatial invasion fitness. First, it unravels the interplay of the ecological (spatial) dynamics of a population and the evolutionary dynamics of the individual traits. Spatial self-structuring shapes the selective pressures, which in return may alter the aggregation pattern. Here we have seen that a high degree of spatial aggregation is not a prerequisite for, but rather a consequence of, the joint evolution of altruism and dispersal. Second, this analysis underlines important transient effects. A state of high dispersal or high altruism may be maintained transitorily, up to the point where the direction of selection changes or even reverts. In general, this means that variations, under the same environmental conditions and for the same species, of adaptive traits may be explained by different ancestral states and by the observation of populations at different points in time in their evolutionary history. Finally, this approach allows us to separate out distinct components of the selective regime and to express these components in terms of individual traits and characteristics of the population aggregation structure. In practice, there is potential here to predict how the selective pressures should equilibrate to produce patterns observed empirically, and how dispersal-related traits may respond to the experimental manipulation of each component of the selective regime.

Acknowledgements

Initially, we planned to co-author this chapter with Denis Couvet, whose expertise in population genetics would have been critical to achieve a complete reinterpretation of traditional kin selection models in a spatially explicit context. Obviously the shot is longer than we thought, and we are still a long way away from the huge integration of spatial ecology and population genetics. Yet discussions with Denis were most useful to delineate the critical issues and pitfalls in this enterprise. He deserves very grateful acknowledgement. We also owe many thanks to Ulf Dieckmann, who kindly wrote the program that was used to produce the individual-based simulations presented in section 4 and is contributing substantially to our ongoing research project on the coadaptive dynamics of altruism and dispersal. Many thanks also to Mats Gyllenberg, Laurent Lehman, Nicolas Perrin, François Rousset, and Minus van Baalen for stimulating discussions and invaluable comments on a previous draft of this chapter; to Molly Brewton for her most careful editing of our manuscript; and to the Editors of this volume for their constant support and patience. Part of this work has been supported by the Adaptive Dynamic Network (IIASA, Laxenburg, Austria).

Part II

Why disperse?
Habitat variability, intraspecific interactions, multi-determinism, and interspecific interactions

6

On the relationship between the ideal free distribution and the evolution of dispersal

Robert D. Holt and Michael Barfield

Abstract

The ideal free distribution (IFD), first proposed by Fretwell and Lucas, describes how a pattern, the distribution of abundances and fitnesses across habitats, emerges from a process: evolution moulding the habitat selection strategies of individuals. Key ecological mechanisms assumed in the IFD include localized density dependence and the freedom of individuals to choose habitats so as to maximise fitness. The central prediction of the IFD is that fitnesses are equilibrated across space. Models for the evolution of dispersal in spatially heterogeneous but temporally constant environments also often predict fitness equilibration. Given combined temporal and spatial heterogeneity, there is no simple characterisation of fitness. Nonetheless, numerical studies suggest that the evolution of dispersal can produce a distribution characterised by approximate equilibration of local fitness measures (e.g., geometric means of local reproductive success, spatial reproductive values) among habitats, at least as a useful rule-of-thumb. However, there are clear counter-examples, particularly involving the use of sink habitats, or forces such as kin competition. The concept of fitness equilibration represented by the IFD describes a widespread but by no means universal outcome of the evolution of dispersal in heterogeneous environments.

Keywords: ideal free habitat selection, evolution in variable environments, spatial reproductive value

Introduction

Almost three decades ago, Steve Fretwell and Henry Lucas developed a set of models to predict animal distributions, based upon the premise that organisms select habitats to maximise fitness (Fretwell and Lucas 1970; Fretwell 1972). Their simplest model led to a pattern they dubbed the 'ideal free distribution' (IFD). Roughly speaking, at an IFD individuals in different habitats have equal average fitness. Ideal free theory occupies a central position in behavioural ecology (Rosenzweig 1985; Brown 1998; Fryxell and Lundberg 1998). But what does it have to do with dispersal? The topic of dispersal includes (though not exclusively) trans-generational spatial flows among habitats. In contrast, habitat selection theory focuses more closely on within-generation patterns of habitat utilisation. Our aim is to highlight conceptual linkages between the IFD and the evolutionary theory of dispersal. We use numerical studies to suggest that the core conclusion of ideal free theory, namely that organisms should be distributed such that fitnesses are equilibrated across space, at times pertains to the evolution of dispersal.

However, in many reasonable circumstances, the evolution of dispersal does not lead to an IFD.

A summary of ideal free habitat selection theory

It is useful to summarise the basic ideal free model (for reviews of habitat selection theory see Rosenzweig 1985, 1991; Kacelnik *et al.* 1992; Tregenza 1995; Morris 1994; Brown 1998). For simplicity, consider an environment with two habitats. Fretwell (1972) assumed that habitat 'suitability' for an individual would decrease with the abundance of conspecifics in each habitat. By 'suitability', Fretwell meant something closely related to individual fitness, as suggested by this quote: 'The suitability of a habitat is a reflection of the average genetic contribution of resident adults to the next generation and must be closely related to the average lifetime production of reproducing offspring in the habitat' (Fretwell 1972, p. 106).

Fitness is most easily characterised in species with the simple life history of discrete, synchronised generations. We assume such a life history and concentrate upon natal dispersal. As noted by Michod (1999, p. 50), it is crucial to use absolute fitnesses when concerned with evolution in spatially structured populations. We use 'local fitness' to denote absolute fitness or selective value (combining viability and fecundity into expected number of gametes produced by a newborn zygote; Roughgarden 1979) of an individual residing in a given habitat. Let $F_i(N_i)$ be local fitness in habitat i, given N_i conspecific individuals there, and assume F_i declines with N_i (see Stamps, chapter 16, for implications of positive density dependence). Carrying capacity of habitat i is the local population size $K_i > 0$ at which $F_i(K_i) = 1$; if no K_i exists, the habitat is an intrinsic sink (Holt 1985; Pulliam 1988), where the population cannot persist without immigration. Given the behavioural assumptions that individuals are identical, perfectly assess current habitat suitability, move without cost, and do not interfere with one another's movement, ideal free theory predicts that: (i) as long as $F_1 > F_2$, individuals in habitat 2 should leave and reside in habitat 1; (ii) if both habitats are occupied, local densities should thus be adjusted such that $F_1 = F_2$ (Fretwell 1972, pp. 86–87). If the system reaches demographic equilibrium, at which average fitness equals unity, density in each habitat must match the carrying capacity of that habitat (Holt 1985). At a demographic equilibrium, no sink habitats should be occupied in an IFD. The statement that average local fitnesses are equilibrated across space at an evolutionary equilibrium appears to be the most general statement of what is meant by an 'ideal free *distribution*'. This is to be distinguished from assuming that organisms show ideal free *behaviour*, those simplifying assumptions made by Fretwell and Lucas in their original model (for a list of these assumptions, see Tregenza 1995, p. 285). If the ideal free assumptions are violated, populations of course may exhibit distributions for which local fitnesses are not spatially equilibrated. Stamps (chapter 16) points out many ways in which organisms may be neither ideal in their habitat selection, nor free in their dispersal.

There is not a one-to-one correspondence between ideal free behaviour and an ideal free distribution. Individual organisms may not be ideal or free at the level of individual behavioural traits and decision rules, yet populations may nonetheless converge on an

IFD (Houston and McNamara 1988). Hugie and Grand (1998) note that, even with unequal competitors and non-ideal free movements, the original IFD model 'may be sufficient to approximate the distributions of animals'. Conversely, organisms may not have an IFD despite having appropriate behaviours, for instance due to unstable resource–consumer interactions (Abrams 1999). Another important complication is that the local fitness measure used by Fretwell and Lucas may inadequately predict evolution in heterogeneous environments. For instance, Brown (1998) shows that, if individuals allocate time between habitats differing in risk of predation, overall fitness is a non-linear function of local fitnesses. At the evolutionarily stable state (ESS), the prevailing habitat use strategy does not equalise total fitness returns from the two habitats, but rather marginal fitness returns from each habitat.

Comparable themes arise in the evolution of dispersal. As noted elsewhere in this volume (e.g., Ims and Hjermann, chapter 14), dispersal behaviour may occur uncon-ditionally, for example an individual's dispersal rate may be independent of the local environment, but vary by genotype, or instead may vary by habitat type, local density, or individual condition. An 'ideal free' disperser might be defined as an individual that moves between habitats only if, by so doing, it increases its realised fitness. We will show that, even if dispersers are not ideal free, a population may in some circumstances evolve toward a distribution that is approximately ideal free.

That the evolution of dispersal may have an 'ideal free' interpretation has been touched on in previous literature (Holt 1985). For instance, Levin *et al.* (1984) studied a model for evolution in a parameter D, the fraction of individuals that disperse from natal sites. In summarising their results they 'conjecture that the [evolutionarily stable value for D] is determined by the equilibration of long-term expectations from dispersing and non-dispersing individuals'. Lemel *et al.* (1997) suggest that a unifying rule for the evolution of dispersal is that 'the dispersal rates which permit the spatial homogeniza-tion of fitnesses are ESSs'. In the following pages, we focus on the suggestion that the evolution of dispersal tends towards an equilibration of fitnesses among habitats. Abstractly, dispersal is an 'assortative parameter', determining (in part) the numbers of individuals in different 'classes', which in the case of dispersal are local habitats. Slatkin (1978a) demonstrated that, given density- and frequency-dependent interactions in a population, natural selection acts on assortative parameters so 'as to equilibrate the fitnesses of those classes' and noted that this generalisation may fail in temporally varying environments.

The concept of 'fitness equilibration' should be viewed in the same spirit, we suggest, as 'fitness maximisation' in classical evolutionary genetics. In many circumstances, and in particular when individual relative fitnesses are density and frequency independent, the outcome of natural selection can be characterised as hill-climbing in an adaptive topography, or as the maximisation of a fitness function (Roughgarden 1979). This characterisation provides a compact description of the outcome of evolutionary dynamics in a wide variety of situations, and its failure in other situations (e.g., with frequency-dependent selection) highlights important features of evolution. The chapters in this volume span diverse evolutionary causes of dispersal (Ronce *et al.*, chapter 24). We will see that fitness equilibration does characterise some situations and some causal mechanisms for the evolution of dispersal, but in others fitness equilibration is not expected.

The ideal free distribution and dispersal evolution in temporally constant environments

We first return to Fretwell's original model and cast it in a form appropriate for studying dispersal. Genetic variation is clonal and affects only dispersal rules. Dispersal strategies are defined by fixed dispersal rules, independent of population size (a non-'ideal' behaviour); however, dispersal rates may be conditional on habitat type. We first assume that the population settles into a stable equilibrium, and then consider temporal variation. Different dispersal rules give different realised fitnesses for a genotype, for a given array of local fitnesses. We consider two distinct movement scenarios: 'scalar' movement and localised dispersal.

Scalar model

The simplest model is a 'scalar' model (terminology after Tuljapurkar and Istock 1993). Examples of the use of scalar models to study dispersal evolution include Metz *et al.* (1983), Levin *et al.* (1984), and Holt (1997b). Assume that after reproduction all newborns in a population of size N enter a general dispersal pool and enter each of two habitats with a fixed probability. Thus, individuals are non-'ideal'. Genetic variability in the propensity to utilise different habitats permits dispersal to evolve. Assume the initial population uses habitat 1 with probability p. Expected fitness of an individual is $W = pF_1(N_1) + (1-p)F_2(N_2)$, where $N_1 = pN$, and $N_2 = (1-p)N$. We assume the population is at demographic equilibrium, so $W = 1$; hence, either $F_1 = F_2 = 1$ (so the equilibrium population $N_i^* = K_i$), or if, say, $F_1 > 1$ then $F_2 < 1$ (thus, $N_1^* < K_1$ and $N_2^* > K_2$).

A rare invading clone with a different pattern of habitat choice has an overall fitness of $W' = p'F_1(N_1) + (1-p')F_2(N_2)$. (The densities in this case are summed over both clones in a habitat.) It immediately follows that, if $F_1 > 1 > F_2$, then, given $p' > p$, the invading clone increases in frequency, whereas, if $p' < p$, the invading clone declines towards extinction; selection obviously favours increased use of whichever habitat provides higher fitness. A necessary condition for an evolutionary equilibrium with no ongoing directional selection on habitat choice is $F_1 = F_2 = 1$. In other words, an evolutionarily stable state for dispersal requires the population to exhibit an IFD, even though individuals making up the population do not have ideal free behaviours. Such an equilibrium could be monomorphic or a mix of different dispersal types, with relative abundances adjusted such that the IFD holds (McPeek and Holt 1992).

At a joint demographic and evolutionary equilibrium, each habitat is at its respective carrying capacity. Natal dispersal involves organisms born in one habitat that settle in the other. For dispersal not to affect local population size, immigrants should equal emigrants in each habitat. If p is the average propensity of an individual to use habitat 1, then the number of immigrants into habitat 2 is $I = K_1(1-p)$, and the number of emigrants there is $E = K_2p$. Setting $I = E$ implies $p = K_1/(K_1 + K_2)$. A pattern of equalised immigration and emigration has been called 'balanced dispersal' (empirical examples include Doncaster *et al.* 1997; Diffendorfer 1998). Given balanced dispersal, if there is an inverse relation between local population size and local dispersal probability, local populations can be at carrying capacity, yet dispersal can still be ongoing (McPeek and Holt 1992; Lemel *et al.* 1997).

Partial mixing

The above 'scalar' model assumes the population mixes each generation, so that a single number characterises habitat use, and a single fitness measure describes the iteration of numbers through time. More generally, one expects populations in different habitats to be partially decoupled. For populations with discrete generations occupying multiple habitats, stage-structured matrix models (Tuljapurkar and Caswell 1997) provide a natural framework for analysing population and evolutionary dynamics (Holt 1996). Consider the following stage-structured model (where 'stage' = 'habitat') for a single species with discrete generations, local density dependence, and fixed habitat-specific dispersal rates, where dispersal (defined simply as movement between habitats) occurs a single time in an individual's lifetime:

$$\begin{bmatrix} N_1(t+1) \\ N_2(t+2) \end{bmatrix} = \begin{bmatrix} (1-e_{12})F_1(t) & e_{21}F_2(t) \\ e_{12}F_1(t) & (1-e_{21})F_2(t) \end{bmatrix} \begin{bmatrix} N_1(t) \\ N_2(t) \end{bmatrix} \tag{1}$$

Here $N_i(t)$ is abundance and $F_i(t)$ is realised local fitness (surviving reproductive offspring) in habitat i at generation t. Dispersal is defined by e_{ij}, the *per capita* rate of movement of newborn individuals from habitat i to habitat j. Local fitness may decrease with increasing local density (summed over clones, if there is more than one clone). If there is no density dependence in habitat j, we assume $F_j < 1$, so habitat j is intrinsically a sink. We assume no direct cost of dispersal. Dispersal may be conditional by habitat but is not directly responsive to density; individuals are not 'ideal', because they may move from high to low fitness habitats.

To examine the evolution of dispersal, we permit clones that are identical in within-patch performance but differ in their propensity to move among habitats to compete with one another. We assume local densities are sufficiently great that kin competition can be ignored (Taylor and Frank 1996; see Lambin *et al.*, chapter 8). Several authors have used models of this basic form to examine the evolution of dispersal (Hastings 1983; Holt 1985; McPeek and Holt 1992; Holt and McPeek 1996; Doebeli and Ruxton 1997; Lemel *et al.* 1997). A general (albeit abstract) measure of fitness in populations whose dynamics are defined by a stage-structured matrix process is the dominant Lyapunov exponent of that process (Metz *et al.* 1992). In model (1), if the local fitnesses are fixed quantities (i.e., we ignore density dependence and temporal variability for a moment), the transition matrix has fixed elements, and the dominant Lyapunov exponent is simply the dominant eigenvalue of the transition matrix (Metz *et al.* 1992). The dominant eigenvalue for model (1) that provides a fitness measure is:

$$W = \frac{1}{2}\left[(1-e_{12})F_1 + (1-e_{21})F_2 + \sqrt{Q}\right] \tag{2}$$

where

$$Q = [(1-e_{12})F_1 + (1-e_{21})F_2]^2 - 4(1-e_{12}-e_{21})F_1F_2$$

For a population to be stable, one or both local fitnesses must be density dependent. We assume this is the case, so the F_i values are adjusted such that $W = 1$; this requires that either $F_1 = F_2$, or if, say, $F_1 > 1$, then $F_2 < 1$. A novel clone, when sufficiently rare relative to the resident clone, experiences density dependence mainly from the resident clone; its asymptotic growth rate when rare is given by expression (2), with F_i being the same local fitnesses experienced by the resident clone, but e_{ij} being different between clones. With partial mixing, in contrast to the 'scalar' model, fitness in a spatially heterogeneous environment is a strongly non-linear combination of local fitnesses. Nonetheless, the evolution of dispersal tends in the same direction, towards an IFD. Rather than dwelling on algebraic details, the salient results are as follows.

(i) $F_1 > 1 > F_2$. If the dispersal parameters are free to vary, then $dW/de_{12} < 0$ and $dW/de_{21} > 0$; selection favours avoidance of the habitat with lower fitness and movement towards the habitat with higher fitness. If dispersal rates are constrained to be equal ($e_{12} = e_{21} = e$), then $dW/de < 0$. With unconditional dispersal, in a spatially heterogeneous habitat, one expects populations to evolve by selection towards very low dispersal rates (Hastings 1983; Holt 1985) at which each habitat will be at its respective carrying capacity, and fitness will converge on unity across space.

(ii) $F_1 = F_2 = 1$ (i.e., each habitat is at its respective carrying capacity). If habitats have unequal carrying capacities, an IFD is compatible with ongoing dispersal, given 'balanced dispersal' (Doncaster et al. 1997), i.e., dispersal that is asymmetrical and inversely related to carrying capacity: $K_1/K_2 = e_{21}/e_{12}$ (see also Lemel et al. 1997). McPeek and Holt (1992) note this for a special case; a simple demonstration of this necessary relationship for an equilibrium comes from adding and subtracting the two equations in (1), and evaluating them at equilibrium, with each patch at its respective K.

Sink habitats and the non-equilibration of local fitnesses

So, in constant environments, the evolution of dispersal rates, given non-ideal individuals with fixed dispersal propensities, tends toward an IFD for the population as a whole. But a rich body of theory shows that a principal selective factor favouring dispersal is temporal heterogeneity (e.g., Levin et al. 1984; McPeek and Holt 1992). How does temporal variation affect the ideal free property of the population distribution? Analysing adaptive evolution in variable environments is a challenging problem (Haccou and Iwasa 1995), particularly in structured populations, in large measure because there is no simple characterisation of fitness (Tuljapurkar 1990). To complicate matters further, given local density dependence, dispersal alters local population size (see Hanski, chapter 20), in turn modifying fitnesses and thus selection on dispersal; in models of spatially structured populations in fluctuating environments, one cannot solve for densities through time and thus cannot analytically characterise temporal variation in fitness. The feedback between demographic functioning of populations and the selective pressure on dispersal is at the core of ESS analyses of dispersal, yet analytically finding this ESS is usually impossible.

For non-dispersing organisms with discrete generations, the appropriate measure of fitness is the geometric mean of fitness through time (Cohen 1993; Haccou and Iwasa 1995; Yoshimura and Jansen 1996). In the following paragraphs, we report numerical studies of evolution of conditional dispersal strategies in the above scalar and matrix models. We assumed two basic spatial scenarios: a source–sink system, with density dependence in the source, or two habitats with unequal positive carrying capacities. Previous studies along these lines (e.g., McPeek and Holt 1992; Holt 1997b) have explored the issue of the evolution of dispersal *per se*. Here, we complement this earlier work by interpreting the evolutionary equilibrium in terms of spatial equilibration of two proxies for local fitness: geometric mean *local* fitness and spatial reproductive value (see Holt 1996).

Consider first the above 'scalar' model, but now assume that fitness fluctuates among generations in one or both habitats (due either to temporal variation in parameters or to unstable dynamics), so that the average fitness in generation t is $W = pF_1(N_1(t)) + (1 - p)F_2(N_2(t))$. K.A. Schmidt, J.A. Earnhardt, J.S. Brown and R.D. Holt (unpublished manuscript) show that if both habitats are used, the ESS value for p is the solution of:

$$\sum_k \frac{q_k(F_1 - F_2)}{[pF_1 + (1 - p)F_2]} = 0 \tag{3}$$

where q_k is the probability that the kth type of generation occurs. If local fitnesses are constant, we recover the earlier result that, at the ESS, local fitnesses are equilibrated across space. But if fitnesses vary temporally, this need not hold. Consider the case where habitat 2 is a sink habitat with constant fitness $F_2 < 1$, whereas habitat 1 is a source with fluctuating fitness. Holt (1997b) and Jansen and Yoshimura (1998) demonstrate that partial use of the sink is selectively advantageous if the source is sufficiently variable in fitness, so that in some generations the source has a lower expected fitness than does the sink; this adaptive use of a sink is a 'bet-hedging' strategy (Seger and Brockmann 1987). A limiting case is for the source to alternate cyclically between good and bad years, with fitness being zero in bad years and positive (but density dependent) in good years. Nonetheless, partial dispersal to the sink can permit persistence (Jansen and Yoshimura 1998); by assuming overall geometric fitness is unity (which can be achieved by adjusting population size, and thus density-dependent fitness, in the source in good years), after manipulating equation (3), one can show that the ESS for use of the source habitat is the solution of a quadratic, $F_2^2 p^2 + 2(1 - F_2^2)p + (F_2^2 - 1) = 0$. When F_2 is near 1, p is near 0 (most individuals should avoid the source habitat), whereas when F_2 is near zero, p converges on $1/2$. (If maximal source fitness during good years does not permit overall geometric fitness to be unity, then the population goes extinct.) Geometric mean local fitness in the source is zero; geometric mean fitness in the sink is a constant between zero and 1.

This example shows that it is certainly *not* always true that dispersal evolves so as to equalise local fitnesses (here interpreted as the geometric mean of local fitness) across space, when one habitat is an intrinsic sink (adding negative density dependence to the sink does not affect this qualitative conclusion; R.D. Holt, unpublished results). If utilisation of sink habitats were part of a bet-hedging strategy, one would not expect to see balanced dispersal, with immigrants equalling emigrants (Doncaster *et al.* 1997; Diffendorfer 1998), or an IFD.

However, if both habitats have positive carrying capacities (i.e., neither is an intrinsic sink) and do not suffer local extinctions (i.e., all $F_i(t) > 0$), numerical studies suggest that at the ESS for dispersal the realised geometric mean of local fitness is often approximately equalised between the two habitats. We were surprised at this outcome. In our simulations, we examined a range of assumptions about temporal variation (e.g., uniform versus normal distributions) and the functional form of local density dependence. Particular forms for density dependence examined include the exponential logistic $[F_i(t) = \exp(r_i(t)(1 - N_i(t)/K_i(t)))]$, the Ricker model $[F_i(t) = \exp(r_i(t) - d_i(t)N_i(t))]$ and a flexible phenomenological model of density dependence $[F_i(t) = R_i(t)/(1 + N_i(t)^{d_i(t)})]$, where $N_i(t)$ is the summed density over all dispersal types within a patch in generation t. (The quantities $R_i(t)$ and $\exp(r_i(t))$ are maximal rates of increase in generation t; $d_i(t)$ measures the strength of density dependence; and $K_i(t)$ is the carrying capacity of habitat i.)

Figure 6.1 shows a representative example. Geometric mean local fitnesses were calculated over 10 000 generations for populations monomorphic in p (for 101 values of p uniformly distributed in [0, 1]). The ESS was then determined by clonal competition. In all cases examined, as long as each habitat patch had a positive carrying capacity in each generation, we found that the population at the ESS exhibited a distribution that was close to ideal free, as measured by realised geometric mean local fitnesses. Comparable results arise in the more general model with localised dispersal (model [1]). Again, there appears to be a qualitative difference between a system with sink habitats, and one in which all habitats have positive carrying capacities. For the source–sink system, if the population persists and is initially restricted entirely to the source, then regardless of the amount of temporal variation in fitness, very small rates of localised dispersal coupling the sink and source are disfavoured. This can be readily shown analytically for a cyclical environment of period 2. However, larger rates of dispersal can be favoured, even if small rates of dispersal are not. To find the ESS pattern of dispersal, we allowed a large

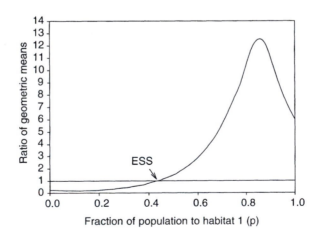

Figure 6.1. Relative geometric mean fitness, for two habitats coupled with scalar dispersal. In the example shown, local fitnesses were described by $F_i(t) = R_i(t)/(1 + N_i(t)^d)$. The density-independent growth parameter for each habitat was given by a sequence of independent draws from normal distributions, with a mean in habitat 1 of $R_1 = 4$, a mean in habitat 2 of $R_2 = 6$, and a standard deviation for both of 0.5. For both habitats, $d = 2$. (The simulations discarded zero and negative values of R_i; in our simulations with these parameters this never occurred.)

Table 6.1. Source–sink dynamics with unstable source and localised dispersal.
A. Winning dispersal rates. Local fitness in the source is given by the Ricker model.
B. Geometric mean local fitnesses for the evolutionarily stable dispersal rates

R_{sink}	0.3		0.4		0.5		0.6		0.7	
A. Dispersal rates of persisting clone Standard deviation of r										
	e_{12}	e_{21}	e_{12}	e_{21}	e_{12}	e_{21}	e_{12}	e_{21}	e_{12}	e_{21}
0.5	0.00	1.00	0.00	1.00	0.02	1.00	0.08	1.00	0.16	1.00
0.6	0.00	1.00	0.00	1.00	0.08	1.00	0.18	1.00	0.26	1.00
0.7	0.00	1.00	0.04	1.00	0.18	1.00	0.28	1.00	0.34	1.00
0.8	0.00	1.00	0.16	1.00	0.30	1.00	0.38	1.00	0.44	0.88
0.9	0.25	1.00	0.38	1.00	0.48	0.86	0.56	0.72	0.62	0.63
B. Geometric mean local fitness of persisting clone in source Standard deviation of r										
0.5	1.000		1.000		0.997		0.977		0.943	
0.6	1.000		1.000		0.993		0.972		0.941	
0.7	1.000		0.999		0.987		0.966		0.934	
0.8	1.000		0.993		0.979		0.956		0.920	
0.9	0.999		0.986		0.966		0.936		0.894	

number of clones to compete, spanning all feasible values of the two dispersal rates. Table 6.1 presents a typical example (the table assumes Ricker density dependence in the source, with mean $r_1 = 1.5$, $d = 1$, and random normal deviates in r_1). At low sink fitness, and low fitness variance in the source, the evolutionarily stable state of the population is no dispersal, leaving the sink unoccupied. Dispersal can be favoured, given higher sink fitness or larger temporal variance in source fitness. Winning clones always disperse at a higher *per capita* rate from the sink to the source than in the reverse direction. In the source, there is often incomplete dispersal for each generation, whereas typically all individuals born in the sink immediately disperse back into the source.

Table 6.1 also shows the long-term geometric mean local fitnesses of the winning clones. As in the scalar model, natural selection does *not* equilibrate geometric mean fitness among habitats; at the ESS for dispersal, the source typically has a higher realised geometric mean *local* fitness than does the sink. In short, evolution of dispersal does not necessarily lead to an IFD, as measured by *local* (geometric mean) fitness, when the use of intrinsic sink habitats is favoured by high temporal variance in fitness in source habitats.

By contrast, as in the scalar model, if both habitats have positive carrying capacities and dispersal is localised, numerical studies suggest that the geometric mean of local fitnesses in the two habitats is typically approximately equalised at the ESS. A characteristic example is shown in Fig. 6.2, which shows a surface describing the ratio of geometric mean fitnesses for two habitats as a function of dispersal rates. The straight line defining the intersection of this surface with the plane of equal geometric mean fitnesses gives the set of dispersal rates that equilibrate geometric mean fitness; the dot is the ESS arising from competitive trials. At the ESS, geometric mean local fitnesses are very nearly equal across space. It is also clear from Fig. 6.2 that fitness equilibration does not capture all the important features of dispersal evolution. The line of intersection

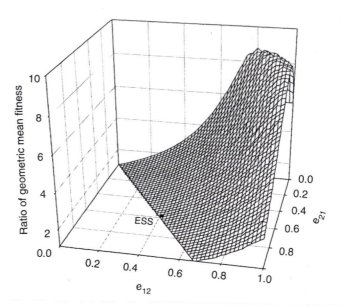

Figure 6.2. Relative geometric mean fitnesses in two coupled patches, as a function of dispersal rates for localised dispersal. These simulations used the expression $R_i(t)/(1 + N_i(t)^{d_i(t)})$ for local fitnesses. The values for $R_i(t)$ were drawn from a uniform distribution, with a mean of 4 for habitat 1 and a mean of 6 for habitat 2, with a range in both cases of width 6. Independent uniform deviates between 1 and 2 were used for both the $d_i(t)$ values. Thus, the populations experienced very high levels of variation in local growth rate. The ESS was found from clonal competition trials (including pairwise competition and trials with an initially uniform distribution of all feasible dispersal clones – increments in dispersal rates of 0.02 separated clonal dispersal values). Geometric mean fitness was evaluated over 100 000 generations, after an initialisation period of 10 000 generations, with initial densities at carrying capacity in the two patches. The dot indicates the ESS for dispersal rates. At the ESS, geometric mean fitness is approximately equalized in the two habitats.

describes an equivalency class of dispersal rates, where each combination leads to fitness equilibration. When one allows these clones to compete, a unique winner emerges. Thus, there are evolutionary forces operating that are not encapsulated by this fitness equilibration rule; otherwise, these clones would be selectively neutral relative to one another.

We were puzzled by finding that geometric mean local fitness was approximately equilibrated in these simulations; given dispersal, the fitness measure relevant to predicting the course of evolution is much more complex than just the realised geometric mean fitness in each habitat. Moreover, using this measure revealed to us small (but real) discrepancies from the ESS. Fitness equilibration by this measure provides a fairly accurate 'rule-of-thumb' describing the outcome of dispersal evolution, but it is reasonable to ask whether there are other, more accurate, measures that would reveal true equilibration (if such exists).

A broad consensus has emerged that the most general definition of fitness for a structured population with non-linear dynamics in a variable environment is the dominant Lyapunov exponent of a series of random matrices, where matrix elements describe transitions among different classes or subpopulations (Tuljapurkar 1990; Metz et al. 1992; Ferrière and Gatto 1995). With limited dispersal, localised density dependence, and fluctuating environments, however, analytical characterisation of this measure of fitness can be very difficult (Tuljapurkar 1997). In the absence of an analytical expression for this true fitness measure, we conjectured that evolution might tend to equalise the realised *spatial reproductive values* of each habitat. In stage-structured models with fixed coefficients (Caswell 1989), the reproductive value of a given stage is the relative contribution of an individual in that stage to eventual population size. In a spatially heterogeneous environment, 'habitat' corresponds to 'stage'. The spatial reproductive value of each habitat is the relative contribution of that habitat to the entire population (Holt 1996; Rousset 2000c). Given dispersal, an individual in one focal habitat leaves descendants across a number of habitats. The fitnesses of those descendants must enter into the calculation of the reproductive value ascribed to the focal habitat. One can abstractly define a stochastic reproductive value (Tuljapurkar 1990, 1997), but, in practice, simulation studies are required to assess reproductive value in fluctuating environments.

We assessed spatial reproductive value using neutral genetic markers. First, for any given parameter set, competitive trials determined ESS dispersal rates. Then, with just the dominant clone, the system was run for 100 generations, at which time an additional population was introduced at low density to each habitat, with dispersal parameters and

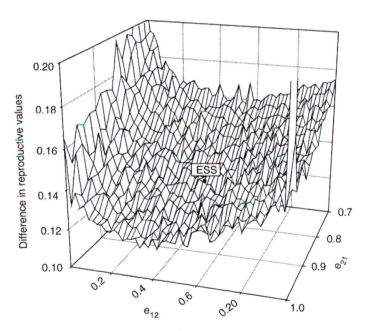

Figure 6.3. Differences in spatial reproductive value between source and sink, as a function of dispersal rates, given localised dispersal. The general protocol for evaluating reproductive value is described in the text. Simulations were carried out with local fitnesses described by the Ricker model in the source and a constant growth rate < 1 in the sink. $R_{sink} = 0.9$. The source $r(t)$ was drawn from a normal distribution, with mean $r = 1.5$, $\sigma = 0.5$, and $d = 1$. The winning dispersal strategy is $e_{12} = 0.48$, $e_{21} = 0.86$ (indicated by dot). At the ESS, spatial reproductive values are not equalised, and the winning strategy does not minimise the differences in spatial reproductive values between habitats.

local fitnesses identical to the resident clone. After 100 further generations, the total number of descendants from each new population was divided by the initial propagule size to give the reproductive value for the habitat in which that population was introduced. Because populations are fluctuating, this value can be sensitive to initial conditions and the exact pattern of environmental fluctuations; we carried out 100 000 runs for each parameter set and averaged the resulting reproductive values.

Again, numerical studies revealed a dichotomy between scenarios in which one patch was a sink and those in which both habitats had positive carrying capacities. In the source–sink system, at evolutionary equilibrium, realised reproductive value of the source did not usually equal that of the sink; Fig. 6.3 is a typical example. By contrast, given positive carrying capacities, reproductive values were approximately equalised across habitats at the evolutionary equilibrium (not shown). Our studies suggest that this equilibration among habitats is approximate, rather than exact. What impresses us is that, even in cases that involve very pronounced temporal variation in local fitnesses, equilibration of these fitness proxies provides a useful rule-of-thumb for the evolutionary outcome, deviating by less than 1–2% from the actual ESS. Thus, even though we assume organisms that are not ideal, the population as a whole converges on a near-IFD at the evolutionary equilibrium for dispersal.

Discussion and conclusions

Given that all habitats have a positive carrying capacity, our numerical studies suggest that the evolution of dispersal tends to produce an approximate IFD, at which the quantity equilibrated across space is one of two proxies for fitness: spatial reproductive value or geometric mean local fitness. It will be an important task in future work to assess the generality of these conclusions. We are intrigued by the observation that the geometric mean *local* fitness is, to a reasonable approximation, equilibrated at the ESS for dispersal. This has practical implications: it is more likely that one could devise a reasonable estimate for local geometric mean fitness in given habitats (for a concrete example, see Boyce and Perrins 1987) than that one could directly measure spatial reproductive values – not to mention dominant Lyapunov exponents.

Our results suggest that the equilibration of local fitness measures across space can often describe the outcome of the evolution of dispersal. They also show, however, that the evolution of dispersal need not equilibrate fitnesses, even approximately, if some habitats are sinks. Natural selection favours utilisation of sink habitats if source habitats have high temporal variance in fitness (Holt 1997b; Jansen and Yoshimura 1998). Organisms with high variance in fitness in high-quality habitats might be expected not to exhibit an IFD because they are more likely to include low-quality but stable sink habitats in their habitat repertoire.

The models we examined have many simplifying assumptions. For instance, we assumed that each individual has a fixed propensity to disperse, which could vary by habitat, but in no other way. Several authors in this volume (e.g., Ims and Hjermann, chapter 14; Stamps, chapter 16; Murren *et al.*, chapter 18) discuss the importance of factors such as individual plasticity, the dependence of dispersal upon internal conditions, and density-dependent dispersal. It would be useful to assess fitness equilibration across habitats, given more realistic assumptions about the proximal causes, conditional dependence, and costs of dispersal. Two assumptions in our models were that there was no direct cost of dispersal and that once a disperser moved it would have the same expected fitness as a non-dispersing resident. Stamps (chapter 16) suggests that evolution has favoured a wide range of mechanisms to reduce the costs of settling into new habitats. However, it is unlikely that such mechanisms will completely eliminate dispersal costs and/or differences between residents and immigrants. Incorporating these realistic features of dispersal could alter an expectation of fitness equilibration at evolutionary equilibrium.

In the models discussed above, dispersal evolved due to spatio-temporal variation in fitness, in populations large enough to consider abundance to be a continuous variable. Yet fitness variation is only one of several potential factors important in the evolution of dispersal (Johnson and Gaines 1990). Fitness equilibration is unlikely to be an evolutionary outcome for many alternative mechanisms. For instance, with highly localised density dependence and very few individuals interacting within patches, demographic stochasticity promotes the evolution of dispersal; clones with zero dispersal walk randomly to extinction in patches with low carrying capacities, leaving behind only dispersing clones (Travis and Dytham 1998). Moreover, kin competition becomes important; movement by a single individual alters fitnesses in both natal and recipient habitats, changing inclusive fitness (Taylor and Frank 1996). These two effects of individual discreteness favour intermediate dispersal rates, even in constant environments. Hamilton and May (1977) showed that this mechanism operates even if dispersal is costly. If some of this cost arises because dispersers land in low-quality sink habitats, one will observe sustained spatial variation in local fitness. Thus, kin competition (see Perrin and Goudet, chapter 9; Gandon and Michalakis, chapter 11) is an important driver of the evolution of dispersal that is not likely to lead to fitness equilibration among habitats. An absence of equilibration is also likely if parents rather than offspring govern dispersal (Murren *et al.*, chapter 18). Moreover, observing a non-ideal free distribution is compatible with a variety of evolutionary causes of dispersal; for example, such a pattern could arise from 'bet-hedging' use of sink habitats, or from kin selection with localised competition.

Despite these limitations in the scope of fitness equilibration, we believe it fair to conclude that Fretwell and Lucas's concept of the IFD provides an insight that proves surprisingly useful for characterising the population-level outcome of evolution of dispersal rates, even among organisms that are not particularly ideal in their individual habitat selection behaviour. Fitness equilibration appears to be most likely when considering movement among habitats, none of which are intrinsic sinks, and when the prime driver of dispersal evolution is spatio-temporal variation in fitness, rather than factors such as kin competition.

Acknowledgements

We acknowledge support by the National Science Foundation and gratefully appreciate the patience of the editors, the useful critiques of the reviewers, and insights from Steve Fretwell about the historical origins of ideal free theory.

7

The landscape context of dispersal

John A. Wiens

Abstract

Dispersal is usually considered in terms of movement from one location and the consequences of arriving at other locations. Movement between these locations is often viewed as a linear process, on which the characteristics of the intervening area have little effect. The area between the beginning and ending of dispersal, however, is not a featureless matrix of unsuitable habitat, but is instead a richly textured mosaic of patches of different shapes, sizes, arrangements, and qualities: a landscape. Movement through such a mosaic is not likely to be linear, and *how* an individual traverses the landscape then becomes important. Different locations in a landscape are associated with different costs and benefits, and the movement of an individual determines, and is determined by, the combination of costs and benefits it encounters. The propositions that movement during dispersal may depend on the structure of landscapes, and that these landscape effects can influence population distribution and the probabilities of movement among locations, are supported by several studies that have used experimental model systems (EMSs) to document and model individual movements in fine-scale mosaics. These studies also illustrate the importance of the scales on which organisms perceive landscape structure. Assessing how landscapes may influence dispersal probabilities therefore requires a consideration of the interaction between the scale of landscape pattern and the scale on which organisms respond to that pattern. Landscapes change over time as a consequence of natural processes, such as succession, disturbance, or human land use. Habitat fragmentation has received particular attention because of its potential effects on the probability that dispersal will link otherwise isolated components of a metapopulation. This probability is a consequence of the intersection of the dispersal–distance function of a population with the frequency distribution of distances among suitable habitat patches in a landscape. The effects of fragmentation may be altered either by changes in the dispersal behaviour of individuals or by changes in the landscape mosaic. Landscape structure clearly can influence the dispersal process, but its consideration introduces additional complexity into evaluations of dispersal and its consequences. Determining how much of this complexity must be considered to develop an adequate understanding of dispersal requires that we learn much more about how dispersal actually occurs in spatially complex landscape mosaics. Experimental model systems and spatially explicit, individual-based simulation models both have considerable potential, but limitations. Until we have a better understanding of the dispersal process, however, we should not assume that we can ignore the details of the movement–landscape interaction.

Keywords: landscape ecology, beetles, movement, dispersal mechanisms, cost–benefit functions, scale, patch dynamics, metapopulations, fragmentation

Introduction

Dispersal is one of the most important, yet least understood, features of ecology, population biology, and evolution. It determines the probability that an individual currently 'here' will later be 'there', and as a consequence be exposed to different opportunities and risks. It gives populations, communities, and ecosystems their characteristic texture in space and time.

Dispersal of individuals from a location generates a dispersal–distance function (Fig. 7.1) that is characteristic of a particular species, stage of the life cycle, sex, environment, and time. Much of what is interesting and important about dispersal follows from the form of such functions. What is 'interesting and important', however, depends upon one's perspective. To an empiricist, the immediate question may be whether the function is real. Is the decrease in numbers of dispersing individuals with distance from the source accurately portrayed, or is it biased by changes in detection or recapture probability with increasing distance? A modeller may seek the simplest mathematical equation that captures the essence of the dispersal function. An evolutionary biologist might be interested in why the function has the form it does, what is the adaptive value of dispersing or the effect on fitness of dispersing various distances? To a population ecologist, the importance of such functions lies in their consequences, such as source–sink or metapopulation dynamics, while a population geneticist would focus on how dispersal affects gene flow and the genetic structure of populations. A behavioural ecologist might ask what factors motivate individuals to initiate dispersal or how social factors affect dispersal. The diversity of these approaches leads to ongoing debates about what qualifies as 'dispersal' and how it should be defined (Shields 1987; Stenseth and Lidicker 1992). Nonetheless, all of these facets of dispersal are obviously 'interesting and important'.

Despite their differences, these approaches all focus either on what causes dispersal or on its consequences – the beginning and ending of the dispersal process. Implicitly or explicitly, they adopt the view of dispersal depicted in Fig. 7.2A: individuals leave a given

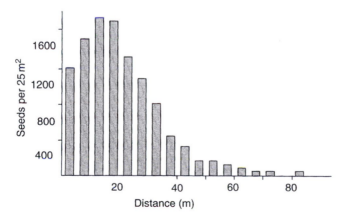

Figure 7.1. A typical dispersal–distance function, showing the density of seeds of *Tachigalia versicolor* in relation to the distance travelled from the parent plant (after Augspurger and Katijama 1992).

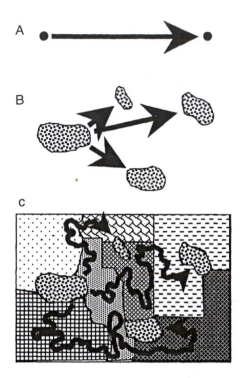

Figure 7.2. Dispersal pathways, as considered by (A) most ecological and evolutionary models of dispersal and by mark–recapture analyses; (B) patch–matrix models, such as island biogeography or metapopulation theories; and (C) an incorporation of landscape mosaic structure and patch-specific movement pathways.

patch or population, cross a gap, and (somehow) end up later in another patch or population. Both decisions, when to leave and when to stop, vary among species and individuals. Whether an individual leaves a patch, for example, may depend on the mode of dispersal, genetic predispositions to disperse, local population density, habitat change, age, or reproductive status, among other factors. The decision to stop may involve various elements of habitat selection or patch choice, such as conspecific attraction, habitat quality, or physiological factors (e.g. exhaustion of energy reserves).

Most of the contributions to this volume deal in one way or another with these leaving or arriving phases of the dispersal process – the causes and consequences of dispersal. Here I instead address what happens along the way and how a consideration of landscape structure and dynamics may both enrich and complicate our understanding of the dispersal process.

The landscape perspective

The simplest and arguably most efficient way to disperse is to follow a straight-line pathway from the origin to a stopping place (Fig. 7.2A). This view of dispersal is fostered by mark–recapture studies, in which the linear distance between the marking location and the recapture location provides an empirical measure of dispersal distance (Fig. 7.1). Such dispersal–distance functions can also be generated by diffusion or reaction–diffusion models (Okubo 1980; Andow *et al.* 1990; Turchin 1998), which build upon random-walk algorithms to produce estimates of the rate of spread of individuals from a point. Temporal and spatial autocorrelation can be incorporated by modelling movement as a correlated random walk, which produces directionality in dispersal and anisotropy in the diffusion pattern (Schippers *et al.* 1996; Zollner and Lima 1999). More complex patch–matrix situations are generally modelled as variations on this source–matrix–target theme (Fig. 7.2B). Thus, island biogeography theory (MacArthur and Wilson 1967) or metapopulation and source–sink models of population dynamics (Pulliam 1988; Hanski and Gilpin 1997) consider the linear distance of an island or subpopulation from a source to be a key determinant of colonization and extinction probabilities. In such models, the matrix is featureless and 'distance' is measured as a linear value, a 'gap' to be crossed. Indeed, both empirical studies (Grubb and Doherty 1999) and models (With 1999; With and King 1999) of gap-crossing focus on the size of the gaps but not their composition.

All of these approaches therefore assume that individuals either move from one place to another in straight lines or that any deviations from linear movement are unimportant. Functionally, the environment through which individuals disperse has no structure and, consequently, risks during dispersal (e.g. mortality) are uniformly distributed and environmental 'resistance' to dispersal (e.g. the diffusion coefficient) is spatially constant. The dispersal environment is homogeneous. Real environments, however, are heterogeneous, and this heterogeneity can affect both the likelihood that individuals will undertake dispersal and the ways in which they move. Models of the evolution of dispersal, for example, generally converge on the conclusion that selection should favour reduced dispersal in spatially variable environments (Gillespie 1975; Hastings 1983; Levin *et al.* 1984; Holt 1985; see Johnson and Gaines 1990). This is because passive diffusion will, on average, result in individuals moving to less favourable environments, as most individuals already occur in the most favourable habitats (note that this result is contingent on the evolution of optimal patch choice). Temporal variation, on the other hand, should promote the evolution of increased dispersal rates, as it enhances the likelihood of finding a suitable location as the quality of the present location deteriorates over time.

According to these models, then, spatial heterogeneity may not by itself promote the evolution of dispersal. Such models, of course, deal with the adaptiveness of leaving a location, usually determined by the fitness consequences of remaining in a patch versus the probability of finding a better patch (i.e. the 'leaving' and 'arriving' phases of dispersal). Once the evolutionary decision has been made that dispersal is adaptive, however, individuals must confront the spatial heterogeneity of their surroundings in order to disperse successfully. Part of this heterogeneity is simply the spatial variance or uncertainty in habitat conditions or survival probabilities, but there is more to 'heterogeneity' than variability alone (Wiens 2000). The structural configuration of patches, corridors, ecotones, patch boundaries, and the like gives ecological landscapes a spatial pattern and texture. Given this structure, *how* individuals move becomes important, and the realized route of movement of individuals through such a complex mosaic is likely to be anything but linear or random (Fig. 7.2C).

One way to view the landscape mosaic is cartographically, as the spatial arrangement of mappable properties such as patches, boundaries, corridors or matrix (Wiens *et al.* 1993; Forman 1995). Because they differ in internal structure, habitat patches may vary in their resistance to movement (their viscosity), and dispersing individuals may therefore travel at different rates and over different distances depending on which patch types they encounter. Boundaries between patches may exhibit different permeabilities, impeding movement to varying degrees (Haddad 1999). Corridors or landscape connectivity may facilitate movement, increasing dispersal distances and producing an overall directionality to movements (Bennett 1999; Haddad 1999). Moreover, because landscapes are open systems (Pickett *et al.* 1994; Wiens 1995a, 1999), what happens in one patch is influenced by its surroundings: patch context is important. As I shall develop below, the structure of a landscape depends on which kind of organism is viewing it, so the 'map' must be adjusted to the organism.

One may also view landscapes functionally, by considering variations among landscape elements in terms of risks and rewards, a cost–benefit perspective. There is a rich tradition of modelling the evolution of dispersal in terms of costs and benefits to individual dispersers. In the Fretwell–Lucas (1970) model, for example, dispersal and patch choice are dictated by the relative quality of the patches, which changes over time as the distribution of population density across the landscape changes (see also Herzig 1995; Ruxton and Rohani 1998).

This model, however, assumes that all patches are equally accessible to individuals and that the patches exist in an ecologically neutral matrix. But both the patches and the 'matrix' are anything but neutral. An individual residing in particular patches or passing through the landscape mosaic will encounter different mating opportunities, resource levels, predation risks, physiological stresses, and the like. The balance of these costs and benefits changes as an individual moves from place to place. Consequently, a landscape can be viewed not as patches of differing quality contained in a matrix, but as a cost–benefit surface in which there are 'peaks' where benefits outweigh costs and 'valleys' where costs exceed benefits (Wiens 1997a; cf. Wright 1932). Because 'costs' and 'benefits' also differ among organisms, the form of this cost–benefit surface also varies among species, sexes, or age classes.

Optimization provides one way of thinking about how individuals of a given species, sex, or age might move through such a cost–benefit landscape. Confronted with contours of varying costs and benefits, an optimally dispersing individual might be expected to move among the high-benefit (or low-cost) locations, 'keeping to the high ground' of the cost–benefit surface. Because the distribution of peaks, ridges, and valleys on such a surface is likely to be nonrandom (mirroring the spatial patterning of the real landscape), the dispersal pathway traced by such an optimal disperser would be nonlinear and tortuous (Fig. 7.2C). In this setting, a straight-line movement pathway would normally be suboptimal. On the other hand, there are situations in which the details of the cost–benefit landscape topography may be relatively unimportant. For an extreme habitat specialist, for example, patches may be either suitable or unsuitable. If such an organism undertakes dispersal, variations in the costs of the different components of the matrix may not matter much; all that matters is reaching a suitable patch. Depending on the size, density, and dispersion of suitable patches in the landscape, a straight-line pathway may be the best way of doing this. Linear movement may also be favoured if dispersal is time limited (e.g. by an accumulation of risks over time, or by a need to initiate reproduction within a certain time). 'Costs' and 'benefits' are always organism and situation specific.

The costs and benefits associated with particular patches in a landscape are also not indelibly fixed. Patch quality changes not only with density (as Fretwell and Lucas envisioned), but with seasonal changes in resources, breeding phenology, changes in temperature or moisture regimes, variations in numbers of predators or competitors, and a host of other factors. These changes produce fluctuations in the cost–benefit surface. As the relative quality of different portions of a mosaic changes over time, the optimal pathway will also change.

While it may be useful to think of landscapes in such cost–benefit terms, thinking of dispersal pathways through such a landscape in optimization terms is probably less useful. One problem, of course, is that the theoretical optimum may not be attainable. In a cost–benefit landscape surface, some peaks may not be accessible because they are isolated by deep valleys: high-benefit patches are surrounded by high-cost landscape components (but see Gavrilets (1997) for a solution to this problem). There may also be multiple ways to disperse optimally among locations in a landscape; for an example dealing with long-distance migration, see Farmer and Wiens (1998). An array of dispersal pathways through a landscape may be equally feasible, producing variation among individuals in net dispersal distances (Fig. 7.1) and in the conditions encountered during dispersal. And, of course, there is no reason to suppose that the movement pathways that organisms actually follow bear a close relationship to the optimal pathways, regardless of what theory says.

Clearly, a consideration of landscape structure complicates our view of how individuals disperse from place to place. Whether one considers landscapes strictly in terms of their spatial configuration and physical properties or in terms of the functional characteristics that affect the costs and benefits experienced by dispersing individuals, it is apparent that the simple assumption of linear or random movement between places is not likely to hold. This conclusion leads to two questions: (1) How do real landscapes affect movement and dispersal? and (2) How do changes in landscapes, such as those produced by human activities, affect the dispersal process and its consequences?

Movement in landscapes

One reason for the widespread acceptance of the view of dispersal represented in Fig. 7.2A, and the neglect of the spatial details exemplified in Fig. 7.2C, is the difficulty of charting actual movement pathways in real landscapes. Measuring the costs and benefits associated with locations in a landscape is likely to remain a formidable challenge for some time, given the variety of factors contributing to costs and benefits and the multiple scales of temporal variation in these factors. And although tools such as radio or satellite tracking, global positioning systems, and geographical information systems (GISs) provide the means to couple movement pathways with the physical structure of landscapes, relatively little work has been done in this area. This is partly because our awareness of landscape structure as a definable object of study in ecology is relatively recent (Hobbs 1994; Wiens 1995a), partly because studies of dispersal have customarily ignored the movement component, and partly because the size of land-scapes creates logistical difficulties in documenting movements through direct obser-vation or mark–recapture methods.

One way to circumvent at least some of these difficulties is through the use of experimental model systems (EMSs) that represent scaled-down versions of broad-scale landscapes (Ims *et al.* 1993; Wiens *et al.* 1993; Barrett and Peles 1999; Bowne *et al.* 1999; With *et al.* 1999). By adopting such an approach, one can focus on 'landscapes' at a scale of metres or hectares rather than square kilometres, and thereby on small organisms for which this scale of landscape structure is relevant. The advantages of such systems are that experiments can be properly designed and readily implemented, and sample sizes can be large enough to provide adequate statistical power, but the disadvantage is that it is not always obvious how or whether the results can be applied to other systems, particularly at broader scales.

My colleagues and I have used EMSs consisting of insects in simple grassland 'microlandscapes' to assess how landscape structure affects individual movements. In one series of studies, we documented the movement pathways of three species of *Eleodes* beetles (Coleoptera: Tenebrionidae) in natural mosaics that differed in both the coverage and the spatial patterns of patches of bare ground, grass, cacti, and small shrubs (Wiens and Milne 1989; Crist *et al.* 1992). Movement tracks differed among species, but the variations among microlandscape mosaics were even greater (Fig. 7.3). Net squared displacement (the straight-line distance moved per unit time) varied significantly, both among species and among mosaic types. Fractal dimension, a measure of the tortuosity of movement pathways, did not differ among species or landscapes, but did depart significantly from that generated by straight-line or random movements (Crist *et al.* 1992). Clearly, movement through these mosaics was neither linear nor random, and there were strong effects of the microlandscape on the probabilities that individuals would move certain distances over time.

We also conducted experiments to consider more explicitly the effects of landscape pattern on beetle movements. Following the framework of percolation theory (Gardner *et al.* 1989; With and King 1997), we recorded the movements of beetles in 25-m^2

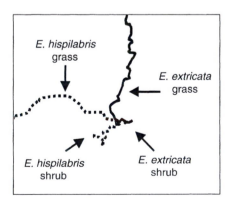

Figure 7.3. Typical movement pathways of two species of *Eleodes* beetles in 25-m^2 plots of shrub- or grass-dominated microlandscapes in Colorado shortgrass steppe. The pathways are 100 time steps of 5-s intervals (or until the beetle left the plot). (After Crist *et al.* 1992.)

bare-sand arenas in which small patches of grass sod were arrayed randomly to create mosaics with 0–80% grass cover (see With *et al.* (1999) for a similar experiment with crickets). In such percolating networks, the grass patches coalesce to form larger patches as coverage increases, and there is a threshold coverage value (approximately 60%, in simple versions of the theory) at which grass patches become linked across the grass-sand mosaic. Beetles moving through these experimental arenas also exhibited a strong threshold in movement patterns, but not at 60% grass cover (Wiens *et al.* 1995). Rather, there were no significant differences in most pathway measures when grass cover varied from 20% to 80%, but pathways were clearly different when no grass was present. The displacement rate, for example, was nearly five times greater in the unvegetated arena than when grass was present. Beetles moved rapidly and more or less linearly in the featureless microlandscape. As individual beetles spent more time in the grass patches, their movement rates declined, but not as a simple linear function of the amount of grass present. Instead, beetles moved more slowly in grass patches when only 20% of the plot was vegetated than when more grass was present. This seemingly counterintuitive result may reflect the influence of patch boundaries, which a beetle would encounter more frequently in the more finely patterned mosaic. In fact, it may be that the degree of lacunarity or 'gappiness' in the landscape influenced beetle movements more than did the coverage or pattern of grass patches *per se* (With and King 1999). In any case, it was again apparent that both the composition and the structural configuration of these microlandscapes strongly affected beetle movements.

These beetles do not appear to occupy individual territories or to exhibit local site fidelity. Consequently, the landscape-dependent variations in short-term movements may contribute to differences in the dispersal of individuals, and to variations in the abundance of individuals among patches over the landscape as a whole. With and Crist (1995) explored such effects by combining the empirical results of With's EMS studies of grasshopper movements in the same grassland microlandscapes with a spatially explicit individual-based simulation model. Cells in the model array represented patches of habitat that differed in their internal heterogeneity and in their resistance to movement by grasshoppers. The probability that an individual would leave a cell of a given type was a function of its movement rate within that type, and the movements of individuals across the mosaic of cell types were therefore determined by the landscape arrangement and by the movement-based transition probabilities among cells. Beginning from a random distribution of grasshoppers across the landscape, this interaction between landscape structure and movement characteristics could, over time, lead to nonrandom population distributions. To evaluate such interactions, With and Crist considered two species of grasshopper that differed in movement rates in the habitat types. A small species, *Psoloessa delicatula*, exhibited reduced movement rates in relatively homogeneous cell types. Because this habitat was relatively rare, individuals moved rapidly through most of the remaining landscape and the population continued to display a random distribution pattern over the landscape as a whole. A larger species, *Xanthippus corallipes*, moved more slowly in habitat types that collectively comprised 35% of the landscape. As a consequence, its population became highly aggregated in the cells in which movement was reduced. At the end of the model simulations, *Xanthippus* individuals were found in only 2% of the landscape, whereas *Psoloessa* was present in 67% of the cells.

It is difficult to extrapolate broadly to other taxa or landscapes from these EMS studies of insects in grasslands. Nonetheless, two conclusions do have broad implications for thinking about dispersal. First, it is clear that movements vary not only among species but also among landscapes, and that the details of landscape composition and configuration can affect the probability that an individual will move from one place to another. Second, because movement patterns differ among components of a landscape in ways that depend upon idiosyncrasies of the species, the spatial distribution of a population that results from individual movements may not map closely on to the physical distribution of habitat patches in the landscape. This lack of concordance may compromise attempts to derive population distributions from remote sensing of landscape patterns alone.

Another conclusion emerges less explicitly from these EMS studies. As any naturalist knows, organisms differ in how they respond to different habitats in a landscape. The grasshoppers that With studied, for example, differ in their responses to habitats that differ in vegetation composition and heterogeneity. But organisms also differ in the *scales* on which they perceive landscape structure. Thus, because the grasshoppers move through a mosaic at different rates, they experience the landscape at different scales (With 1994). Both the resolution with which they perceive patch structure (grain) and the area through which they move over a given time interval (extent) differ. In a similar fashion, the three beetles we studied differ in body size and, correspondingly, in movement rates (Wiens *et al.* 1995), and they are therefore likely to perceive the patch structure of their landscapes at different scales. Generally speaking, we should expect the grain–extent window, which determines the scales over which an organism perceives landscape structure, to increase with increasing mobility (Kolasa and Rollo 1991; Wiens 1992; see Lavorel *et al.* (1995) for a similar argument for plants). Landscape structure is also scale dependent, so a rapidly moving or widely dispersing organism may not only experience landscape structure at a broader scale than a slower, short-distance disperser, but may also encounter a different structure due to the scale dependency of the landscape patterns themselves. Differences in how individuals move among elements in a landscape, or in how costs and benefits are distributed among landscape patches for organisms that differ in ecology or life history, can create substantial variation in the relationship between landscape structure, the scaling of individual responses, and the movements that ultimately result in dispersal.

Landscape change and dispersal

This multi-scale tapestry of landscape patterns is set against a backdrop of time. Landscapes exhibit temporal dynamics. The contours of costs and benefits that define a landscape to an organism at one time change on multiple scales. Daily cycles in predator activity can alter predation risk, reproductive activity or resource phenology can change on weekly or seasonal scales, and yearly differences in climate can affect patch quality. Studies of Townsend's ground squirrels (*Spermophilus townsendii*) in Idaho shrubsteppe, for example, demonstrated that habitat quality (judged by adult survival) was greatest in perennial bunchgrass patches during years with normal precipitation, but decreased markedly during a drought, when survival was greater in habitats dominated by sagebrush (*Artemisia tridentata*) (van Horne *et al.* 1997). What were peaks in the cost–benefit surface became valleys, and *vice versa*.

These temporal dynamics involve changes in the quality, but not the physical structure, of a landscape. Landscapes also change in composition and structure over time, altering the properties of the mosaic through which an individual must move during dispersal. Conditions in the home or natal patch may affect the probability that dispersal will be initiated. Landscapes undergo ecological succession, for example, and change can be rapid in early successional stages. Organisms occupying such successional habitats ('fugitive species', 'weeds', or 'ruderals') often display suites of adaptations to cope with the ephemeral conditions in local patches (Grime 1979; Bazzaz 1996). Traits that affect the propensity to disperse or the means of dispersal are important components of these adaptive strategies. Early successional plants often produce large quantities of wind-dispersed seeds that travel over long distances, for example, and insects occupying variable environments employ a wide array of behaviours to move from one place to another (Drake and Gatehouse 1995). Some birds that breed in early stages of grassland succession, such as Dickcissels (*Spiza americana*), engage in 'distant flight', in which males leave their territories and fly several kilometres across the landscape, presumably to locate patches of habitat that will be suitable for occupancy in subsequent years (Zimmerman 1971; Wiens and Johnston 1977).

Exactly how changing conditions in local habitats may affect dispersal, either proximately or evolutionarily, will depend not only on the local patches, but on the patterns and dynamics of the surrounding landscape and the scales on which these patterns and dynamics are expressed. The adaptiveness of long-distance dispersal in a variable environment may depend strongly on the size of area affected. Temporal changes in arid landscapes, for example, often affect hundreds or thousands of square kilometres in much the same way, and many organisms occupying these environments undertake long migrations or nomadic movements in response to changing conditions (Davies 1984; Wiens 1991; Dingle 1996). Natural landscape change in more mesic environments, on the other hand, generally occurs on scales of square metres to hectares or a few square kilometres. The magnitude of environmental change in local patches is likely to be less severe, and there is a reasonable probability that suitable conditions may be found elsewhere within the local landscape. Under such conditions, limited dispersal or 'bet-hedging' dispersal polymorphisms (Roff and Fairbairn, chapter 13) may be more appropriate strategies. There is a feedback, from the scale of environmental variability, through landscape structure and its dynamics, to the decision of individuals about whether and how to disperse and, ultimately, to the evolution of dispersal strategies. Different landscapes impose different kinds of filters on this feedback process.

Increasingly, however, landscape change is produced not by natural events such as succession, disturbance, or droughts, but by human actions. Growing human populations and industrialization have contributed to changes in atmospheric carbon dioxide and ozone levels that may affect global and regional climates. Perhaps even more profound are the effects that population growth and climate change may have on land-use practices, which in turn alter landscape structure on many scales (Meyer and Turner 1994; Dale 1997). Agricultural practices convert wetlands and grasslands to monocultures, homogenizing landscape structure over broad areas and creating vast troughs in the cost–benefit profiles of landscapes for many species. Urban and ex-urban development replace natural communities with assemblages dominated by exotics and invading species, and the increased levels of human activity may affect predation risks, not only in the developed areas but also in adjacent parts of the landscape (Andrén 1992). Forestry practices create landscape patterns that differ in scale and intensity from those produced by natural disturbances such as fire or wind throw (Angelstam 1992).

While all of these land-use changes may influence movement patterns and therefore alter the probabilities that dispersing individuals will reach suitable habitat patches, habitat fragmentation has received the most attention from ecologists and conservation biologists, perhaps because the spatial structure of fragments in a landscape provides such an obvious analogy to oceanic islands (MacArthur and Wilson 1967) or to the spatially subdivided population structure envisioned in metapopulation or source–sink models (Pulliam 1988; Hanski and Gilpin 1997). In such models, dispersal rates and distances determine the colonization rates of unoccupied but suitable patches and influence patch extinction rates, and thereby determine the fate of local populations and the probability of metapopulation persistence. Because the frequency of dispersers (and therefore the probability of reaching a patch) is assumed to decline monotonically with increasing distance from the dispersal source, dispersal–distance functions (Fig. 7.1) determine the probability that fragments isolated by a given distance from sources of immigrants will in fact be reached and colonized or 'rescued' from local extinction (Fig. 7.4). By increasing the distance between patches of suitable habitat, fragmentation may move many patches of suitable habitat into or beyond the tail of the dispersal–distance function, reducing dramatically the colonization probability of empty patches (Fig. 7.4A). If this occurs, individuals representing the tail of the dispersal–distance function may play a disproportionately important role in population dynamics and gene flow (Goldwasser *et al.* 1994; Kot *et al.* 1996), and selection will favour long-distance dispersal.

There is considerable empirical support for a strong distance effect on the likelihood of fragment occupancy or colonization. For example, brown kiwis (*Apteryx australis*) in New Zealand are restricted largely to remnant patches of forest and scrub that are isolated by large expanses of pasture and farmland. Kiwis are flightless and therefore move between fragments by walking. In one area, all fragments isolated by less than 80 m were occupied, more isolated fragments were used only if they were relatively large, and fragments isolated by more than 330 m were not occupied (Potter 1990). Other, similar, examples abound in the vast literature on fragmentation. In nearly all instances, however, the details of the matrix separating the fragments of interest are not considered or, as in the kiwi example, the matrix is thought to be relatively homogeneous. In fact, both the pattern and the process of fragmentation are more complex than suggested by such examples or by simple island–mainland or isolation-by-distance models (Wiens 1995b, 1997b).

The pattern is more complex because the 'matrix' is in fact a mosaic of landscape elements whose quality, configuration, and relations with one another can have important effects on individual movements. Different species (e.g. habitat specialists versus habitat generalists) may perceive such a mosaic in quite different ways. To the degree that movements are impeded or facilitated by patch boundaries, variations in costs and benefits among landscape elements, or the scales at which landscape

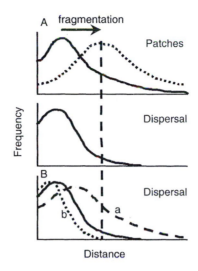

Figure 7.4. The relationship between the dispersal–distance function of a species and the frequency distribution of distances among patches of suitable habitat in a landscape. In (A), distances between patches in an undisturbed landscape (solid line) are small, and most patches are accessible to most dispersing individuals. The vertical dashed line indicates (arbitrarily) the distance beyond which approximately 95% of the population does not disperse. Fragmentation of this landscape shifts the frequency distribution of interpatch distances (dotted line, top graph), with the result that a greater proportion of the patches lie beyond the 95% dispersal demarcation and have a correspondingly low probability of colonization. With fragmentation, the individuals in the 5% tail of the dispersal distribution assume increasing demographic importance. Panel (B) illustrates the effects of shifting the dispersal–distance function, either (a) as a result of phenotypic plasticity in the population or of changes in the landscape mosaic that facilitate movement among the patches of suitable habitat, or (b) as a result of restrictions on movement due to behavioural changes or increased dispersal costs associated with landscape changes.

patterns are expressed, the dispersal–distance function may be constrained or expanded (Fig. 7.4B).

The process of fragmentation is more complex than portrayed by simple models because it incorporates two confounding effects (Angelstam 1992; Trzcinski *et al.* 1999). As an unbroken area of forest, for example, is altered by harvesting of timber, clearing for agriculture or development, or natural disturbances such as fires, forest habitat is lost and replaced by other landscape components. As habitat loss continues, a threshold is reached at which patches of forest are no longer interconnected; fragmentation has occurred (Andrén 1994; Fahrig 1997). The consequences of crossing this fragmentation threshold are sensitive to the form of the dispersal–distance function. If isolation of habitat patches is sufficient and the intervening mosaic does not facilitate dispersal, an increasing proportion of habitat patches will become unoccupied as local extinctions are not counterbalanced by patch colonizations. The level of structural habitat fragmentation at which this occurs, however, is determined by the functional isolation of the patches, which is a consequence of the dispersal behaviour and area requirements of the species (Pagel and Payne 1996; Andrén *et al.* 1997) *and* the characteristics of the mosaic in which the fragments are embedded. Thus, red squirrel (*Sciurus vulgaris*) densities in Sweden are affected by the loss of forest habitat but not by fragmentation, whereas elsewhere fragmentation has effects above and beyond those of habitat loss *per se*. Delin and Andrén (1999) attributed these differences to the shorter distance between fragments and the 'less hostile' landscape surroundings in the Swedish study area. Such observations raise the prospect that, by actively managing the entire landscape that contains the habitat fragments of interest, dispersal movements might be facilitated, lessening the actual effects of fragmentation from those predicted from a simple patch–matrix approach.

It is often assumed that the form of the dispersal–distance function of a species is a fixed trait; individuals may differ in the distances they disperse, but the population or species as a whole follows a particular dispersal pattern. Although dispersal distances are obviously constrained by morphology, physiology, and a variety of life history traits, several contributions to this volume also provide clear evidence of plasticity in dispersal–distance functions. Thus, the fungus beetles (*Bolitotherus cornutus*) studied by Kehler and Bondrup-Nielsen (1999) switch dispersal mode with increasing distance between suitable habitat patches, and European nuthatches (*Sitta europaea*) disperse over greater distances in highly fragmented landscapes than in more continuous forest (Matthysen *et al.* 1995; see also Verhulst *et al.* 1997). If the early experience of individuals with different habitat types leads to differing habitat preferences among individuals (Hildén 1965; Stamps, chapter 16), then between-phenotype variability in dispersal may occur, as individuals differ in their search images for suitable habitat during dispersal. Furthermore, if the cost–benefit surface of a landscape is dynamic over time, some degree of plasticity in dispersal behaviour should be selectively advantageous. Although the plasticity in dispersal distances may not be great, the tail of the dispersal–distance function may become increasingly important as landscapes change. Either proximate or evolutionary changes in the form of the dispersal–distance function may therefore have substantial effects on patch occupancy and metapopulation dynamics. Indeed, studies of damselflies and butterflies show that there may be evolutionary responses in flight morphology to landscape change (e.g. fragmentation) (Taylor and Merriam 1995; Thomas *et al.* 1998). Whether fragmentation should promote increased mobility, to enhance patch-colonization probability, or reduced mobility, due to increasing isolation of habitat patches (Fig. 7.4B), depends on the degree of resource or habitat specialization, dispersal abilities, and the scale on which a species perceives landscape structure. Thus, similar species occupying the same environment may differ markedly in their responses to landscape change and fragmentation, depending on how these traits are expressed in their populations. Differences in dispersal characteristics among species, coupled with environmental heterogeneity, may contribute to the distribution–abundance patterns of the species (Venier and Fahrig 1996), and thereby profoundly influence their population dynamics and evolutionary potential.

Concluding comments

It is because dispersal occurs in a heterogeneous environment that identifying the proximate or selective forces that shape a particular dispersal strategy is so difficult. During dispersal, individuals are exposed to an array of costs and benefits that constantly change in space and time, and different individuals are exposed to different combinations of costs and benefits depending on when, where, and how far they disperse. Clearly, the composition and structure of a landscape can affect dispersal in complex ways.

But how much of this complexity must be considered in order to understand dispersal? Must we deal with the details of individual movement pathways in individual landscapes at each of a multitude of scales in order to produce a mechanistically sound foundation for predicting the consequences of variations in dispersal? The answer, of course, is that we do not know. We do know that simple visualizations of the dispersal process (Fig. 7.2A) are generally wrong, and that being wrong *can* have important consequences. We need to know much more about how dispersal actually occurs in spatially complex landscape mosaics.

One route to understanding landscape effects on dispersal is through computer modelling. Spatially explicit, individual-based models, such as that of With and Crist (1995), have become popular tools for delving into the consequences of individual behaviour and distribution in spatially complex landscapes (DeAngelis and Gross 1992; Dunning *et al.* 1995; Schippers *et al.* 1996). Such modelling approaches have several clear advantages. They force one to develop explicit hypotheses, organize existing knowledge, and estimate values for unknown parameters. Because parameter values, movement algorithms, and landscape structure can be easily manipulated in the model structure, simulation can be used to explore the consequences of assumptions and the sensitivity of the outcomes to model parameters or functions. These analyses may indicate which parameters are likely to be important, and which relatively unimportant, for the question at hand, and how much complexity and detail must be considered to understand dispersal and its consequences.

There are also disadvantages and constraints to the application of spatially explicit, individual-based models, however (Wennergren *et al.* 1995; Schippers *et al.* 1996). As with any model, the results are largely a consequence of the model structure and assumptions, and a poorly structured or incomplete model will produce only poor or incomplete insights. The models are also data-hungry and may require quantification of many parameters to describe the structure of landscapes and the responses of animals to them. Spatially explicit dispersal models may be especially sensitive to errors in the estimation of dispersal rates and survival (Ruckelshaus *et al.* 1997, 1999; South 1999), both of which may be strongly influenced by the landscape mosaic that individuals encounter during dispersal.

To determine the estimation errors in dispersal parameters or to assess how dispersal patterns are affected by landscape structure, we need a renewed emphasis on empirical studies of the dispersal process itself. We need to know how organisms disperse under various conditions, how they cope (or fail to cope) with the risks they encounter along the way, and how the structure of landscapes at multiple scales influences their probability of success. As Ruckelshaus *et al.* (1999, p.1224) observed, 'no amount of clever modelling or detailed GIS habitat maps can circumvent our need for this natural history information'.

8

Dispersal, intraspecific competition, kin competition and kin facilitation: a review of the empirical evidence

Xavier Lambin, Jon Aars, and Stuart B. Piertney

Abstract

Theory suggests that intraspecific competition and kin competition are keys processes shaping the evolution of dispersal. Based on a review of the empirical literature we derived the following generalizations. The plasticity within taxa in the presence and direction of sex-biased dispersal revealed by recent empirical studies argues against a single factor accounting for interspecific patterns in dispersal. If they can be parameterized, models exploring the interactions between competitive interaction and inbreeding avoidance will provide rigorous predictions for different groups. Density-dependent dispersal (emigration) is common across most taxa considered, but data on the prevalence of effective dispersal in relation to the level of competition is generally lacking. For the limited set of species for which it is available, it appears that immigration is inhibited by crowding, leading to reductions in both immigration and the dispersal distance of successful dispersers. Despite kin competition being the cornerstone of theoretical arguments for the evolution of dispersal in stable environments, the empirical evidence of its prevalence is little more than anecdotal. A few exceptions restricted to vertebrates suggest that it is more prevalent between parents and offspring than between siblings. The rarity of studies in which the kin structure of populations is known precludes any inference about differences in the extent of kin competition in different habitats or taxa. In contrast, there is more evidence, primarily in vertebrates, that kin cooperation influences philopatry. Even in moderately social species, philopatry does create an advantage in acquiring key resources in the natal ground, such that the prospects of dispersers becoming established away from their natal ground may be greatly reduced. Where this is not the case, the inclusive fitness benefits of dispersal will be reduced if many non-related dispersers can usurp resources in the natal area.

Keywords: dispersal, kin competition, relatedness, facilitation, density dependence, local resource competition, local resource enhancement, sex ratio

Introduction

Intraspecific competition is a ubiquitous feature of animal and plant populations, and is best defined as the interaction between individuals brought about by a limiting resource, in which, for each, the birth and/or growth rate are depressed and/or the death rate increased by the other organisms (Begon *et al.* 1996). Competition may occur among a large number of mobile individuals widely dispersed in space (global competition) or among smaller groups of individuals inhabiting a limited area and exploiting local resources (local competition). When the prevailing pattern of dispersal results in relatives being aggregated in space and interacting primarily with one another, then local competition may become kin competition.

Using simple evolutionarily stable strategy (ESS) models, assuming an island model, Hamilton and May (1977) demonstrated that avoidance of kin competition could favour parents who enforce the departure of a fraction of their offspring even if dispersal entails a substantial survival cost. Hamilton and May's (1977) and subsequent models (reviewed by Johnson and Gaines 1990; Stenseth and Lidicker 1992) demonstrate that substantial dispersal can be selected for in a temporally stable environment because individuals dispersing from a habitat patch not only colonize empty sites, but also avoid competing with related individuals who do not disperse. Further predictions are developed by Gandon and Michalakis (chapter 11) and by Murren *et al.* (chapter 18).

The above models are primarily heuristic, as their assumptions, for instance of spatially structured populations in a stable environment, are rather restrictive and exclude other processes known to favour dispersal, such as the benefits of incest avoidance and escape from transient habitats. Furthermore, the cost of dispersal, which is the key variable, encompasses many ecological factors, including the cost of competition between established residents and prospective immigrants attempting to settle away from the natal site, and the loss by dispersers of the potential for cooperative interactions with their relatives.

Perrin and Goudet (chapter 9) show, theoretically, how inbreeding avoidance and kin competition may interact and lead to male- or female-biased dispersal according to the prevailing circumstance. In this chapter, we ask whether the available empirical evidence is consistent with the assumed role of kin competition in driving dispersal. We review the evidence that competition represents an ecological and evolutionary force that not only influences the fraction dispersing (emigrating) from the natal site, but also may prevent individuals from settling (immigrating), or cause them to disperse further than necessary to avoid close inbreeding. Thus competition also contributes to the cost of dispersal. In many species, there is a continuum of competitive and cooperative interactions among relatives influencing dispersal. Empirical evidence of kin competition associated with dispersal is relatively scant and appears to be more common between parents and offspring than between siblings. The evidence is restricted to a few case studies in lizards, birds, and mammals. There is more evidence that kin cooperation enables individuals to remain philopatric whereas it may decrease the ability of dispersers to become established away from their natal site. Thus cooperation among kin conveys an advantage to philopatric individuals in assuming occupancy of the natal site, hence increasing the cost of dispersal. Our review is not intended to be exhaustive and is unavoidably biased towards birds and mammals with which we are most familiar. Whenever possible, we use recently published studies to illustrate more general points.

Asymmetries in intrasexual competition and sex-biased dispersal

Since Greenwood's (1980) seminal paper describing patterns of sex-biased dispersal in birds and mammals, numerous studies have attempted to explain: (1) why there

Lambin et al.

commonly is a sexual dimorphism in dispersal, and (2) which sex actually disperses (reviewed in Wolff 1994; Clarke *et al.* 1997). Two potentially complementary evolutionary explanations have been proposed. First, that predominant dispersal by members of one sex can act to prevent close inbreeding. Second, that sexual asymmetries in the level of intrasexual competition for resources resulting from the prevailing mating system may lead to asymmetries in the costs and benefits of dispersal and philopatry for males and females (Greenwood 1980). Competition for territories among male birds is thought to be intense such that males benefit more from familiarity with the natal area than females. Female birds are argued to benefit from dispersing by the available choice of resources defended by different males. Conversely, in female-defence mating systems, which are prevalent in mammals, philopatric females benefit from familiarity with the natal ground, and males gain most in selecting areas with the largest number of defensible females. What ultimate mechanisms provide a general explanation for the presence and direction of the observed sex bias has been the focus of considerable debate. Moore and Ali (1984) considered the evidence for competition sufficiently compelling to state that 'the inbreeding hypothesis is both inadequate and unnecessary to explain general dispersal patterns'. Wolff (1994) favours the opposite point of view, and has argued that dispersal functions primarily to avoid close inbreeding, and that the sex that 'chooses' or has first access to the mating site for subsequent reproduction will prove the philopatric sex. He claimed 'that there is virtually no evidence that juvenile dispersal resulted from resource competition or parental aggression in mammals'. Both unifactorial hypotheses may provide satisfactory explanations for specific case studies but they do not account for the diversity of patterns observed. For instance, the evidence that, in the marsupial *Antechinus*, males disperse despite the death of their fathers directly after mating is clearly most consistent with a role of inbreeding avoidance (Cockburn *et al.* 1985). On the other hand, the lack of sex bias in dispersal in species such as song sparrows and nuthatches, in which males and females play a symmetrical role in territory establishment, is consistent with arguments involving competition alone (Matthysen 1987; Arcese 1989a). Furthermore, in song sparrows, *Melospiza melodia*, for which the evidence is available, dispersal does not prevent frequent close inbreeding with severe fitness consequences (Arcese 1989a; Keller 1998; Keller and Arcese 1998). Whereas some exceptions to the general pattern, such as bird species with sex-reversed roles in which the sex bias in dispersal is also reversed, have been interpreted as supporting unifactorial hypotheses, many others do not find any ready explanation. Some recent examples for birds and mammals with no sex bias or with reversed sex bias in dispersal include: badgers *Meles meles* (Woodroffe *et al.* 1993), red squirrels *Sciurus vulgaris* (Wauters *et al.* 1994), collared flycatchers *Ficedula albicollis* (Pärt 1991), garden shrews *Crocidura russula* (Favre *et al.* 1997), lesser kestrels *Falco naumanii* (Negro *et al.* 1997), American kestrels *Falco sparverius* (Miller and Smallwood 1997), and black brant *Branta bernicla* (Lindberg *et al.* 1998). Some explanations invoke often non-quantifiable species-specific features, hence reducing the generality and testability of the arguments. For instance, Wolff and Plissner (1998) invoked an unspecified benefit of familiarity with the natal ground in addition to inbreeding avoidance in order to explain the 'reversed' sex bias in dispersal in *Anatidae*. The authors of a long-term study documenting female-biased dispersal in the black brant (Lindberg *et al.* 1998) did not invoke inbreeding avoidance.

Thus, empirical studies of birds and mammals are consistent with the view that dispersal patterns have multiple causes and that dispersing individuals benefit from both increased access to unrelated mates and decreased intrasexual competition (e.g. Dobson 1982, 1985; Dobson and Jones 1985; Lambin 1994a). Attempts to build all-encompassing verbal models may be somewhat misguided. The lack of sex bias in philopatry in some species does not rule out the possibility that inbreeding avoidance has shaped their dispersal patterns, nor does it necessarily imply that competition is the sole factor involved. It remains difficult to partition the amount of variation accounted for by either mechanism, but the general framework presented in chapter 9 seems appropriate to make progress.

Intraspecific competition and dispersal

We find abundant evidence (below), from nearly all groups of organisms considered, that dispersal (emigration) rate increases with increasing competition for limiting resources. However, the sign of the relationship between immigration rate and the level of competition may be negative in some taxa where data are available, whereas it is not known for most other groups. It follows that the cost of dispersal is influenced by ecological circumstances including the prevailing level of competition. Members of some species with social structures appear able to reduce emigration rates when the prospects of successful dispersal are reduced by intense competition.

Wing-dimorphic insects such as planthoppers with environmental wing-form determination provide a good example of the selective pressures associated with dispersal and philopatry (Denno *et al.* 1989, 1991). There is a fitness trade-off between flight capability and reproduction for both females and males (Zera and Denno 1997) such that even successful dispersers experience a fecundity cost to dispersal (Roff and Fairbairn, chapter 13). However, it must be kept in mind that not all macropterous insects (the dispersive forms) actually disperse and little is known about the proportion doing so in the field. Nevertheless, of all the environmental factors known to affect wing form in planthoppers, population density is clearly the most influential (Denno *et al.* 1991). In most species, the production of dispersing forms is density dependent, associated with crowded conditions, and intensified by nutritionally inadequate host plants. According to Denno and Peterson (1995), density-dependent dispersal occurs at a similar frequency at intra-plant, inter-plant, and inter-habitat spatial scales. Thus, competition for resources clearly triggers dispersal. However, other factors correlated to habitat persistence and the need for males to locate females even in sparsely populated conditions result in sex- and habitat-specific response to density, highlighting the multiple influences on dispersal. Species and populations in persistent habitats are characterized by density-dependent macroptery in both sexes and much higher density thresholds above which dispersing forms are produced. Further evidence of density-dependent dispersal by flight for mono-morphic winged species, as well as dispersal by walking within and among plants, is found in the aphids (reviewed by Denno and Peterson 1995). An elegant set of experiments by Herzig (1995) demonstrated that emigration by the goldenrod beetle, *Trirhabda virgata*, is also density dependent. She manipulated densities of *Trirhabda* larvae early in the season to create crowded, heavily defoliated, host patches and uncrowded, lightly defoliated patches, and found that heavy host defoliation inflicted by high population densities induced long-distance emigration by flight. The emigration rates of beetles that had developed in crowded and uncrowded patches and were subsequently placed on lush, undefoliated plants as teneral adults did not differ.

Evidence from other groups of invertebrates and some marine organisms, although consistent with the view that intraspecific competition increases dispersal, makes it difficult to determine the exact nature of its influence on emigration and immigration. Bengtsson *et al.* (1994a), for instance, provided experimental evidence that population density and food availability both influence dispersal in a fungivore soil collembola. In the marine ascidian, *Clavelina molluccensis*, repeated mapping of settlement patterns suggested that settlement rate was related to density, and active spacing behaviour between individuals occurred as a response to water-borne cues, such that dispersal was clearly affected by intraspecific competition (Davis and Campbell 1996). However, small-scale movements resulting from competitive interactions may not influence effective (genetic) dispersal in species with external fertilization and/or planktonic larvae. For instance, individual limpets, *Siphonaria sirius*, undergo pushing contests close to algal food resources. Clear dominance ranks are established, and low-dominance individuals are forced to move to areas where desiccation may prove more of a problem (Iwasaki 1995). Such short-scale movements may have only small effects on the outcome of reproduction, given that limpets exhibit broadcast spawning with long planktonic larval longevity, but this is not necessarily so. Indeed, a few examples challenge the common assumption that the distribution of larvae of marine organisms with extended planktonic phases is dominated by physical forces, such as current patterns and tidal movement. For instance, Savage (1996) examined the density-dependent and density-independent relationships in the temporal dynamics of a number of benthic macro-invertebrates over a 27-year period. He concluded that density-dependent factors were involved in dispersal and changes in numbers in several cases. In the snow crab, *Chionoectes opilio*, Comeau *et al.* (1998) showed that the patchy distribution of individuals and the seasonal movements to shallow waters by adult individuals was density dependent. Such density-dependent adult dispersal has also been documented in the brittlestar, *Amphiura filiformis* (Rosenberg *et al.* 1997), and in the blue crab (Pile *et al.* 1996).

Demonstrations that dispersal by members of one sex functions to reduce the risk of inbreeding do not preclude an important contribution of dispersal. Even when the risk of incest is reduced by the dispersal of members of one sex, the options faced by members of the more philopatric sex can be constrained by competition for space or other resources in the vicinity of the natal ground. Dispersal is heavily male biased in many microtine rodents, such as in Townsend voles, *Microtus townsendii*. However, in this species, the proportion of females recruited near their natal area is negatively influenced within and between seasons by adult female density (Lambin 1994a). The limiting resource causing competition is often, but not always, a food resource. For instance, between-cohort variation in dispersal distance in the European kestrel, *Falco tinnunculus*, was related to prevailing food availability, with young birds more likely to disperse and settle further in years of low food availability (Adriaensen *et al.* 1998). Descriptive and experimental evidence from an intensively studied song sparrow population provides a further example of competition for food increasing dispersal. Both sexes played an equal role in territory defence in this species, and the proportion of juveniles recruited declined as the number of territorial adults in the autumn increased (Arcese 1989b). Food supplementation of selected territories greatly increased the rate of philopatry, demonstrating that competition for food resource was the proximate factor triggering dispersal at high density. A strongly contrasting result was obtained with female Belding ground squirrels, *Spermophilus beldingi* (Nunes *et al.* 1997). All males eventually disperse in this species whereas about 8% of females ordinarily leave their natal area (Holekamp 1984). Experimental food provisioning during pregnancy and lactation increased emigration by daughters of food-provisioned mothers to over 40%. Competition for food did not appear to influence female dispersal; instead behavioural data suggest that they reacted to the increased crowding in supplemented areas and to competition for non-food resources, such as space, where young can be defended against infanticidal conspecifics. The pattern of transfer between pine vole (*Microtus pinetorum*), social groups also suggests that competition for breeding position influences both emigration and immigration. Indeed, males and females immigrated to territories without same-sex conspecifics, and replacement females rapidly immigrated and filled experimentally created vacancies and started reproducing. Females did not immigrate into social groups unless they contained no females (Solomon *et al.* 1998).

In many vertebrate species with sex-biased dispersal, rates of dispersal and immigration of members of the less dispersive sex appear more influenced by density than that of members of the opposite sex. There is limited evidence that both sexes are equally affected by density in those species with no sex bias in dispersal. In the marsh tit, *Parus palustris*, for instance, dispersal distance increased with density in both sexes (Nilsson 1989), but males appeared to be severely affected by the density of already established individuals and had no time to choose flocks according to quality. Competition for establishment opportunities was lower in females and they could afford to be more selective in their choice of flock range, resulting in longer dispersal distances. A contrasting pattern may reflect similar asymmetries in the great tit, *Parus major*. The proportion of males changing habitats, natal dispersal distances, and the proportion of male immigrants among first-year recruits were reduced as local density increased (Delestrade *et al.* 1996). Thus, when competition increased, male immigrants found it more difficult to become established. No such effect was found among females. Great tits of both sexes that left their natal habitat settled successfully in habitats with low occupation rate, and presumably low quality.

Descriptive and experimental studies on the influence of population density on dispersal conducted primarily with small mammals have contributed to the debate on whether dispersal rates are density dependent or whether so-called 'pre-saturation dispersal' predominates (reviewed in Gaines and McClenaghan 1980; Stenseth and Lidicker 1992). Their seemingly contradictory outcomes reflect the fact that emigration and immigration show contrasting responses to density (see also Ims and Hjermann, chapter 14). For instance, experimental manipulation of density in common lizard, *Lacerta vivipara*, did not result in any increase in emigration, but immigration rates were negatively influenced by density (Massot *et al.* 1992). This is despite clear experimental evidence with juvenile common lizards that their dispersal tendencies, as indexed by their movement between connected small enclosures, increases with adult female density (Léna *et al.* 1998). Similarly, Krebs (1992) concluded that dispersal is positively density dependent in Townsend voles, but he used a measure of *per capita* dispersal rate based on the recovery rate of a population depleted by removal which overestimates immigration rate because competition with residents is relaxed. A very different conclusion was reached by Boonstra (1989) and Lambin (1994a,b), who considered only those individual voles who completed all three stages of dispersal, namely emigration, transience, and immigration (effective dispersal). They found effective dispersal was negatively density dependent, and that effective dispersers moved shorter distances at higher density. Evidence from an increasing number of species suggests that this pattern is widespread, with members of the less dispersive sex most affected (Jannett 1978; Waser and Jones 1983; Jones *et al.* 1988; Wolff *et al.* 1988; Woodroffe *et al.* 1993; Delestrade *et al.* 1996; Blackburn *et al.* 1998). A set of recent studies in semi-natural enclosures, where habitat patches are stocked at different densities with Tundra voles, *Microtus oeconomus*, provides a unique insight into the contrasting effects of density on emigration and immigration rates. Aars and Ims (1999, 2000) created habitat patches of contrasting density in large enclosures. Effective dispersal by adult females was strongly density dependent and strikingly effective in eliminating experimentally imposed differences in vole density between patches. In contrast, overall high density precluded immigration late in the breeding season, such that density differences frequently persisted between patches despite high emigration rates (Aars *et al.* 1999). In the above studies, emigrants that failed to establish themselves suffered high predation (Andreassen *et al.* 1998). Studies of dispersal in kangaroo rats, *Dipodomys spectabilis*, provide a more mechanistic explanation of how habitat saturation leads to reduced dispersal. Members of this species require ownership of a complex burrow system containing food reserves, in order to survive. In this species, dispersal tendency increased with density but effective dispersal was strongly depressed at high density (Jones 1986). This resulted in part from the fact that acquisition of the maternal burrow provided a survival advantage at high but not low density, and more so in females than in males. Thus females who dispersed at high density suffered high mortality rates and the dispersal distance of successful dispersers was therefore reduced at high density (Jones *et al.* 1988). The implication of reduced natal dispersal at high density on the resulting level of kin competition and an interpretation of the behavioural process that may explain reduced dispersal distance at high density are discussed below.

Whereas the high cost of dispersal at high density implies that dispersal in such circumstances provides little direct fitness benefit, it may nevertheless provide some indirect fitness benefits. Alternatively, dispersers may be heterogeneous and include both high-quality individuals with high prospect of successful dispersal and subordinates forced to leave their natal area by dominants. Dispersal by subordinate individuals is not necessarily more prevalent under saturated conditions (Hanski *et al.* 1991) and, in many species, departure from the natal group is typically not accompanied by aggression (e.g. Holekamp 1984; Pusey and Packer 1987; Brandt 1992). However, there is substantial evidence that the combination of aggression and spacing may be of primary importance in limiting immigration (reviewed in Brandt 1992).

In several polygynous mammals, successful male immigrants were larger than males that remained in their natal group, suggesting that competitive ability confers an advantage on prospective immigrants (e.g. Clutton-Brock *et al.* 1982; Le Boeuf and Reiter 1988; Anholt 1990). Immigrant male European badgers, *Meles meles*, were also larger in terms of head–body length, remained reproductively active until later in the autumn, and had higher testosterone titres than adult males that remained in their natal group (Woodroffe *et al.* 1993). On the evidence above, it is not possible to determine the exact nature of the contribution of differential emigration or immigration, and it remains unclear whether individuals with above-average competitive ability are more likely to attempt improving their reproductive opportunities through dispersal.

Avoidance of kin competition through dispersal

Models exploring the role of kin competition in the evolution of dispersal typically consider idealized populations and follow the island model, such that competition is, by definition, local, and takes place primarily among kin. Whereas this assumption is reasonable for species such as parasitoid wasps, and successfully predicts facultative sex ratios in response to mate competition, few studies have attempted to document dispersal patterns for species for which competition is clearly local, or even occurs among kin. In this section, we infer that competition is local when the decision to disperse depends upon an individual's environment rather than upon population level traits.

Competition among siblings for restricted opportunities in the natal territory has been inferred from direct behavioural interaction between dominant and subordinate siblings leading to the departure of the latter. Direct evidence that sibling competition leads to partial dispersal is found in gray jays (*Perisoreus canadensis*; Strickland 1991). Juvenile gray jays must be associated with a breeding pair during winter in order to acquire sufficient food for survival. Non-breeders are either philopatric offspring or juveniles that dispersed from their natal territory and successfully became attached to a non-related pair that failed to rear offspring. A single non-breeder remains associated with breeding pairs and actively enforces the departure of its sibling such that dispersers invariably originate from family groups that retain only a single philopatric juvenile. Dispersers are expelled by their siblings soon after independence and suffer a high mortality rate. Parents seemingly have no involvement in enforcing the departure of all but one of their offspring. A similar pattern prevails in Siberian jays, *Perisoreus infaustus*, and data on the lifetime reproductive success confirm that juveniles who delay dispersal breed more successfully and more often than juveniles who disperse early (Ekman *et al.* 1999).

The widespread relationship between litter size and dispersal rate provides more indirect evidence that competition between same-sex siblings may be an important contributing factor to dispersal in mammals. Female white-footed mice, *Peromyscus leucopus*, from small litters dispersed substantially less than females from median or large litters, but neither the absolute number of females in a litter nor number of previous litters produced by their mothers affected female dispersal (Jacquot and Vessey 1995). A similar relationship is found in the monogamous California mouse, *Peromyscus californicus*, as males from large litters and females with many sisters dispersed further (Ribble 1992). Similarly, in red foxes, *Vulpes vulpes*, living in suburban environments, small male cubs from large litters were more likely to disperse, and females that had dispersed tended to come from larger litters than did philopatric females (Harris and

Trewhella 1988). Thus, sibling competition leading to partial dispersal is not restricted to members of the philopatric sex in mammals.

Parthenogenetic reproduction is common in many aphid species and small number of clones may account for a large proportion of a population, even over large spatial scales (Hodgson and Godfray 1999). Thus kin competition could be a major force in such species. Unfortunately, we are not aware of any empirical study relating changes in the production of dispersers (alates) to variation in the fine-scale genetic structure of aphid populations, within or between host plants (but see Ozaki 1995). As seasonality, multiple host plants, and the need to escape transient habitats represent other strong evolutionary forces promoting dispersal, it is not possible to separate the contribution of kin competition from other forces in these species.

In many species occupying year-round territories, territory acquisition by juveniles is initiated shortly after independence, such that juveniles potentially compete with their parents and siblings for the natal territory. Territory ownership may be a prerequisite for survival, such as in red squirrels (Wauters *et al.* 1994) or red grouse, *Lagopus lagopus scoticus* (Watson 1985), or for early reproduction such as in many microtines (Bujalska 1973). The extent of competition between parents and offspring varies considerably. Instances of offspring aggressively expelling their parents have been documented in red grouse (Watson 1985) and song sparrows (Arcese 1987). In contrast, in at least four mammal species, parents relinquished part or all of their breeding territory to their offspring, and occasionally dispersed to a new breeding territory to avoid parent–offspring competition completely (reviewed in Lambin 1997). The requirement of exclusive resource ownership may prevent parents from retaining their offspring in the natal territory, but it would lead to a low expectation of survival for offspring forced to leave the parental territory. Accordingly, red squirrel (*Tamiasciurus hudsonicus*) mothers with late-born offspring, which have a lower prospect of acquiring a territory independently than early-born offspring, were more likely to abandon their territory than mothers of early-born juveniles (Price and Boutin 1993). Lambin (1997) critically reviewed the evidence for territory bequeathal through breeding dispersal in mammals and concluded that it is rare and restricted to those species for which ownership of a discrete resource requiring a long period to build (mound, midden, or burrow) is critical for survival. He found no evidence that breeding dispersal reduces parent–offspring competition in any of the 10 microtine species listed as engaging in this form of parental investment by Cockburn (1988, 1992). The ability of female microtines to share space with their female relatives or to delay dispersal and sexual maturation seemingly favours the tolerance of philopatric offspring by their parents (Jones 1986) over territory bequeathal.

A remarkable series of field and laboratory experiments with juvenile common lizards suggests that dispersal strategies are adjusted so as to minimize mother–daughter competition. There is no social interaction between parents and offspring in this species and no evidence that juveniles are forced to disperse. Juveniles disperse shortly after birth, but their tendency to disperse, as assayed by their willingness to transfer between pairs of small outdoor enclosures, is positively related to variables reflecting high life expectancy of mothers. These include maternal age (Ronce *et al.* 1998), body condition, parasitism rate (Sorci *et al.* 1994), and hormonal status during development (de Fraipont *et al.* 2000). Thus, philopatry appears to be more prevalent when daughters are likely to inherit the maternal home range and its resources, as implicitly predicted by Hamilton and May (1977) and formalized by Ronce *et al.* (1998). The above studies provide strongly suggestive evidence of a dispersal strategy conditional upon the level of kin competition; there are as yet no data quantifying how often non-dispersal of 10-day-old juveniles results in their occupancy of the natal site 2 years later when they reproduce for the first time. Indeed, the grain of lizard habitats is such that patches contain numerous unrelated competitors who could potentially usurp the maternal territory and reduce the realized inclusive fitness of dispersal.

Overall, the available evidence in favour of kin competition influencing dispersal is little more than anecdotal, and moreover is restricted to vertebrates. As more data on the genetic structure of populations become available, a clearer picture may emerge. Despite the paucity of empirical evidence for kin competition, individuals of most species must experience some level of competition with their immediate relatives. However, the contribution of kin and global competition is undetermined in most cases, and it is premature to determine whether habitat structure or life history traits can predict the relative prevalence.

Kin cooperation and dispersal

Models invoking kin competition as a driving force behind the evolution of dispersal normally assume that philopatric individuals compete on equal terms with individuals who leave their natal area. If philopatry created an advantage in acquiring key resources in the natal ground, for instance through cooperation with philopatric relatives, by comparison the prospects away from the natal ground would be greatly reduced. However, under kin cooperation, the magnitude of this advantage is not constant and depends on the proportion of the local population that is philopatric, resulting in a positive feedback on philopatry (Matthiopoulos *et al.* 1998), which may amplify any sex bias in dispersal (Perrin and Goudet, chapter 9).

In saturated environments, few opportunities for successful independent breeding are available, and the odds of effective dispersal are low. Thus, selection favours offspring with philopatric tendencies and parents who tolerate them (Jones *et al.* 1988). Delayed dispersal and habitat saturation have been implicated as important factors in the evolution of cooperative breeding in birds and mammals (Brown 1987), but there is also evidence of reduced dispersal and greater spatial association among kin at high density from non-social species (Jannett 1978; Jones *et al.* 1988; Wolff *et al.* 1988; Lambin and Krebs 1993; Lambin and Yoccoz 1998; MacColl 1998). In these instances, parents actively share spatial resources with their offspring and appear willing to pay the cost of increased depletion of limiting resources by their philopatric offspring. A unique study of dispersal and territory acquisition by MacColl (1998) in a highly saturated red grouse population revealed that many young males who settled in the close vicinity of their natal ground had previously attempted unsuccessfully to become established further afield. Thus, in this example, cocks first dispersed but had the option to return home where they were tolerated by their fathers. Very strong male philopatry and facilitation of philopatric recruitment by relatives are two other features of this species (MacColl 1998). Similar anecdotal accounts of philopatry following failed dispersal attempts are reported by Holekamp *et al.* (1993) and Nunes *et al.* (1997) for female spotted hyenas and Belding ground squirrels, respectively.

The cost to parents of tolerating a philopatric offspring in saturated conditions depends on the nature of the limiting resource. Female kangaroo rats require access to a burrow and its food reserves for survival, and Jones *et al.* (1988) argued that mothers who occasionally take over a satellite burrow protect themselves against the cost of tolerating their philopatric breeding daughters. On the other hand, if the resource held in territories is social space free of infanticidal individuals, this cost may be negligible to parents, whereas the benefit of philopatry by parental consent to offspring may be substantial. In some instances, referred to as local resource enhancement (LRE), the presence of philopatric offspring benefits the future survival and reproduction of parents. The position of each species on a continuum from competition for local resources through local resource enhancement by relatives through cooperation or helping has been shown to influence the production of offspring with different dispersal tendencies. Studies of sex allocations also suggest that parents respond to spatial and temporal variation in the levels of local resource competition (LRC) and enhancement. When local competition increases or local enhancement decreases, due to higher densities or degrading habitat, dispersal by members of the philopatric sex should increase and sex allocation should be biased against the dispersive sex. The most striking study of sex ratio allocation altering in this way is that of Seychelles warblers, *Acrocephalus sechellensis* (Komdeur *et al.* 1997). Primary sex ratios are finely adjusted according to local habitat quality and the number of helpers at the nest, hence the benefit of recruiting further helpers in this species. With no or one helper at the nest in good habitats, almost exclusively philopatric females are produced. When two or more helpers are present in good habitats, or when the territory is of low quality, mostly males are produced. A remarkable aspect of the study on the Seychelles warblers is that dispersal rates also respond to the level of LRC–LRE (Komdeur *et al.* 1997). When high-quality territories are vacant, both females and males will disperse, even from high-quality habitats with no helpers at the nest. As surrounding high-quality territories become saturated, females especially will tend to stay and help parents with no or one helper, rather than disperse to an available low-quality territory. In contrast, all juveniles disperse from low-quality territories in order to occupy another low-quality territory. Dispersal is therefore dependent both on the LRC–LRE level, and on the expected future success away from the natal territory. Sex allocation on the other hand depends only upon local conditions.

Another study showing sex ratio allocation within a species to be correlated to habitat variables is on spider monkeys. Chapman *et al.* (1989) found sex ratios to vary with habitat, in accordance with the LRC hypothesis. In this species, males are philopatric, and sex ratios were more male biased in good habitats than in less good habitats. In a study of enclosed populations of root voles, Aars *et al.* (1995) found that reproductive mothers occupying a patch alone produced more daughters than mothers having other reproducing females around. Males usually disperse in this species, whereas females are philopatric (Aars and Ims 2000). Similarly, Townsend vole females varied their population sex ratio temporally and produced female-biased litters in two springs of low density, whereas sex ratios were unbiased in a year with high spring density (Lambin 1994b). On a smaller temporal scale, a clan of spotted hyenas was shown to alter sex ratio allocation immediately after clan fission (Holekamp and Smale 1995). All males disperse in this species, and an excess of males was produced when clan size was large and intra-clan competition presumably high. After fission, an excess of philopatric females was produced.

Dispersal coalitions, whereby several individuals disperse as a group and join forces in order to increase their ability to compete with residents and to become established away from their natal area, are known of from several mammal and birds species. Male lions, *Panthera leo*, form such coalitions, usually including relatives, and this makes them more likely to take over prides, secure longer tenures in prides, and gain access to more females per male than do single males (Bertram 1975; Bygott *et al.* 1979; Packer *et al.* 1988). Similar coalitions exist in some primate species (Pusey and Packer 1987) and in many social carnivores (Frame and Frame 1976; Rood 1986; Doolan and MacDonald 1996). For instance, Woodroffe and MacDonald (1995) observed that in a high-density European badger population, where dispersal in females was slightly more common than in males, 13 of 14 females leaving their natal group formed coalitions with putative close relatives while dispersing, and in five of six cases, took over a group with a single resident female, who subsequently disappeared. The proportion of females that bred declined with group size in badgers, such that female dispersal reduced within-group competition for breeding status. Dispersal coalitions also made single females vulnerable to between-group competition. Cooperation with relatives during the process of dispersal results in relatives being spatially aggregated even after dispersal. Such an outcome was reported for Belding ground squirrels by Nunes *et al.* (1997), with six instances of sisters settling within metres of one another after dispersal.

Summary and conclusions

From our review of the empirical literature, several salient points emerge. The plasticity within taxa in the presence and direction of sex-biased dispersal revealed by new studies argues against a single factor accounting for interspecific patterns in dispersal. If they can be parameterized, models exploring the interactions between competitive interaction and inbreeding avoidance (Perrin and Goudet, chapter 9; Gandon and Michalakis, chapter 11) will provide rigorous predictions for different groups.

Density-dependent dispersal (emigration) is common across most taxa considered, but data on the prevalence of effective dispersal in relation to the level of competition are generally lacking. For the limited set of species for which data are available, it appears that immigration is inhibited by crowding, leading to reductions in both immigration and the dispersal distance of successful dispersers.

Despite kin competition being the cornerstone of theoretical arguments for the evolution of dispersal in stable environments, the empirical evidence for its prevalence is little more than anecdotal. A few exceptions restricted to vertebrates suggest that it is more prevalent between parents and offspring than between siblings. The rarity of studies in which the kin structure of populations is known precludes any inference about differences in the extent of kin competition in different habitats or taxa.

In contrast, there is more evidence, primarily in vertebrates, that kin cooperation influences philopatry. Even in moderately social species, philopatry does create an advantage in acquiring key resources in the natal ground, such that the prospects of dispersers becoming established away from their natal ground may be greatly reduced. Where this is not the case, the inclusive fitness benefits of dispersal will be reduced if many non-related dispersers can usurp resources in the natal area.

Acknowledgement

J. Aars was in receipt of a fellowship from the European Community.

9

Inbreeding, kinship, and the evolution of natal dispersal

Nicolas Perrin and Jérôme Goudet

Abstract

Using analytical tools from game theory, we investigate the relevance of a series of hypotheses concerning natal dispersal, focusing in particular on the interaction between inbreeding and kin competition, as well as on the components of mating and social systems that are likely to interfere with these phenomena. A null model of pure kin competition avoidance predicts a balanced equilibrium in which both sexes disperse equally. Inbreeding costs have the potential to destabilize this equilibrium, resulting in strongly sex-biased dispersal. This effect is mostly evident when the peculiarities of the mating system induce asymmetries in dispersal and/or inbreeding costs, or when kin cooperation counteracts kin competition. Inbreeding depression, however, is not the only possible cause for sex biases. The relevance of our results to empirical findings is discussed, and suggestions are made for further empirical or modelling work.

Keywords: competition, cooperation, inbreeding, kin selection, mating systems

Chapter aims and structure

The role of inbreeding avoidance in dispersal remains highly controversial. Some authors consider it to be central, and others irrelevant. Part of the problem stems from the fact that arguments usually remain purely verbal. Moreover, potential causes are too often considered as alternatives rather than as interacting forces. Inbreeding is likely to interfere with several other selective forces behind dispersal, and the whole story is too complex to be accounted for fully by verbal models. Rather than a review of empirical data, the present chapter is an attempt to bring models and arguments into a common mathematical framework. Our purpose is to provide a formalization of the interactions between inbreeding, kinship, and dispersal, in order to evaluate the importance of inbreeding relative to other selective forces.

The concept of inbreeding itself may be confusing, as it is often used in different contexts and with different meanings (e.g. Jacquard 1975; Templeton and Read 1994). Thus, the chapter starts with some definitions and basic empirical information. It then makes a brief excursion into the verbal arguments that have been invoked when attempting to link dispersal and inbreeding. The second part, which constitutes the core of the chapter, develops an evolutionary modelling approach. After introducing the analytical framework and assumptions, we investigate the relevance of a series of hypotheses concerning natal dispersal, focusing in particular on the interaction between inbreeding and kin competition, as well as on the components of mating and social systems that are likely to interfere with these phenomena.

Some definitions, and a few words about the context

What is inbreeding?

At the individual (behavioural) level, inbreeding designates a process, that of mating with a relative. Any offspring born from such a mating is likely to carry genes that are identical by descent, and is said to be inbred. In that sense, inbreeding is a universal and inescapable feature of finite sexual populations. Population biologists, however, usually refer to inbreeding in a relative sense, as may be quantified through Wright's fixation indices. If mating partners in a population are, on average, more related than expected by chance, then observed heterozygosity (H_i) will be lower than the Hardy–Weinberg expectation (H_p). Wright's (1921) *inbreeding coefficient*:

$$F_{ip} = \frac{H_p - H_i}{H_p} \tag{1.1}$$

measures this deficit in heterozygotes relative to random-mating expectation. A null value does not imply absence of mating among relatives, but only that mating partners are, on average, no more related than would be expected by chance. Therefore, in the following, the term 'inbreeding' will be used in its relative sense, as measuring the genetic similarity among mating partners *in excess of random-mating expectation* (or, equivalently, the probability that the two copies of a gene in any offspring are identical, *relative to random copies* from the population).

The most likely cause of inbreeding is population structure. If dispersal is low, individuals mate with neighbours, who are likely to be closer relatives than average individuals in the population. A convenient way to formalize spatial structure is to assume that populations consist of local groups of related individuals, genetically differentiated from other such groups. Genetic similarity among patch mates can be measured by their *co-ancestry*:

$$F_{gp} = \frac{H_p - H_g}{H_p} \tag{1.2}$$

building on the fact that, because some genetic variance occurs among groups, the variance within groups (H_g) is lower than total variance (H_p). The co-ancestry between two individuals measures the probability that their gametes carry identical alleles, relative to gametes taken randomly from the population. By this definition, co-ancestry among partners equals the inbreeding coefficient of their offspring, and co-ancestry with self is $(1 + F_{ip})/2$. Co-ancestry among patch mates as a proportion of self co-ancestry measures their *relatedness*:

$$r = \frac{2F_{gp}}{1 + F_{ip}} \tag{1.3}$$

If patterns of effective migration among groups are known, inbreeding and co-ancestry coefficients can be worked out by deriving the equilibrium conditions of a set of difference equations. This has been done in particular for situations that allow one to account for the peculiarities of breeding systems (sex-biased dispersal and degree of polygyny; Chesser 1991a,b; Sugg *et al.* 1996). A low dispersal results in significant differentiation among groups ($F_{gp} > 0$) but, if sex-biased, it also creates an excess of heterozygotes at the group level:

$$F_{ig} = \frac{H_g - H_i}{H_g} < 0 \tag{1.4}$$

F_{gp} and F_{ig} may actually diverge drastically in polygynous mating systems with male-biased dispersal (Dobson *et al.* 1997). However, as implied by equations (1.1), (1.2), and (1.4), these coefficients are linked by the constraint:

$$(1 - F_{ip}) = (1 - F_{ig})(1 - F_{gp}) \tag{1.5}$$

so that divergences may cancel out in such a way that inbreeding vanishes (i.e. $F_{ip} = 0$). Thus inbreeding (in its relative sense) is not a necessary consequence of population structure.

Why avoid inbreeding?

Inbred individuals often display phenotypic abnormalities, resulting in a loss of fitness through lower viability or fertility. Although *inbreeding depression* could be defined as the decline, with increasing homozygosity, in the mean phenotype of any trait (Lynch and Walsh 1998), we will refer to it as a decline in fitness. Consistent with our definition of inbreeding, the *cost of inbreeding* will be the fitness loss of an inbred mating *relative to* a random mating (Fig. 9.1).

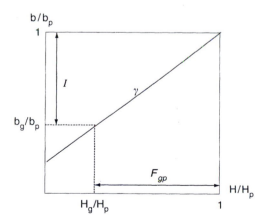

Figure 9.1. Relative fecundity of partners in a pair (b/b_p) as a function of their relative genetic dissimilarity (H/H_p). F_{gp} is the co-ancestry among partners stemming from the same patch, I the inbreeding cost of their mating, and γ the marginal decrease in relative fecundity with co-ancestry.

Inbreeding depression can occur only if alleles at a locus do not act additively. Dominance or over-dominance (heterozygote advantage) is needed. Little evidence exists for the impact on inbreeding depression of over-dominance, although it might be quite important (Lynch and Walsh 1998). In the dominance model, recessive alleles are assumed to be deleterious to various degrees. As inbreeding increases homozygosity (including that of recessive deleterious alleles), it decreases fitness. The number of recessive deleterious alleles in a population is referred to as the *genetic load*, and is quantified here as the marginal change in relative fitness with a change in relative homozygosity (Fig. 9.1).

This genetic load may actually be purged by recurrent inbreeding. If the recessive alleles are strongly deleterious, then inbred individuals carrying them are likely to die. After a few generations of inbreeding, the frequency of deleterious alleles in a group may have diminished drastically. This process, however, requires that groups must be able to cope with a high mortality rate for some generations, and that deleterious alleles are problematic enough to cause death (otherwise they might become fixed in small groups because of drift). It also implies some isolation, since newcomers are likely to bring new deleterious alleles. Social species living in closed groups might be good candidates (e.g. Reeve *et al.* 1990; Keane *et al.* 1996).

Inbreeding depression has been demonstrated repeatedly in mammals (including humans), other vertebrates, invertebrates, and plants (including, surprisingly, selfing species; Agren and Schemske 1993). It has been reported in the laboratory (e.g. Brewer *et al.* 1990; Keane 1990) and, more importantly, in the field (Chen 1993; Bensch *et al.* 1994; Jimenez *et al.* 1994; Keller *et al.* 1994; Madsen *et al.* 1996; Olsson *et al.* 1996; Keller 1998; Westemeier *et al.* 1998; reviewed by Lynch and Walsh 1998). Cases of strong inbreeding depression seem often to be linked to unusual situations such as recent habitat fragmentation (e.g. Madsen *et al.* 1996; Hitching and Beebee 1998) which isolates previously outbred populations and thereby exposes high genetic loads to selection.

A few cases showed no sign of inbreeding depression (e.g. Gibbs and Grant 1989; Reeve *et al.* 1990; Keane *et al.* 1996), which might be due to the purging of deleterious mutations. Whether such situations are common is difficult to estimate for obvious reasons of publication bias. Inbreeding might even be beneficial under some circumstances (see below).

Did dispersal evolve to avoid inbreeding?

Dispersal may intuitively appear as an obvious means to avoid inbreeding depression. The pervasive importance of dispersal in general, and of sex-biased dispersal in particular, has often been interpreted as the direct consequence of inbreeding avoidance (Johnson and Gaines 1990). Indeed, incestuous matings (defined as parent–offspring or sib–sib pairing) are rare in the field (normally less than 2% according to Ralls *et al.* 1986; Harvey and Ralls 1986). Such low levels clearly could not be achieved without dispersal (in interaction with kin recognition mechanisms when these exist).

This, however, does not imply that inbreeding avoidance is the ultimate cause of dispersal, which might evolve primarily as a response to other selective pressures such as non-equilibrium population dynamics, competition for resources, or competition for mates (Johnson and Gaines 1990). These and other causes may generate sufficient dispersal to prevent inbreeding anyway, in such a way that inbreeding avoidance itself plays only a marginal role.

Furthermore, the selective pressure induced by inbreeding may often not suffice to drive dispersal:

- Inbreeding might even be favoured, for at least two reasons. First, outbreeding may dismantle genetic co-adaptations built up locally through linked gene complexes (Shields 1982, 1983; Bateson 1983; Templeton 1986; Wiener and Feldman 1993). Second, inbred matings bring direct benefits to males (and inclusive benefits to females through increased reproductive output of related males), as long as they do not forfeit other breeding opportunities (Parker 1979, 1983; Waser *et al.* 1986).
- The costs of dispersal may outweigh those of inbreeding (e.g. Pärt 1996). The dispersal-induced mortality rate sometimes exceeds 50% (Johnson and Gaines 1990). High costs may force individuals to be philopatric, and the ensuing inbreeding history may largely purge deleterious mutations from populations, a process that would further reinforce philopatry.
- Other ways to avoid or limit inbreeding costs exist (Harvey and Ralls 1986; Blouin and Blouin 1988), including kin recognition (McGregor and Krebs 1982; Fletcher and Michener 1987; Keane 1990), promiscuity and multiple paternity (Brooker *et al.* 1990; Stockley *et al.* 1993), extra-group copulations (Sillero-Zubiri *et al.* 1996), and divorce (Kempenaers *et al.* 1998).

As a consequence, much debate has resulted over the exact role of inbreeding in the evolution of dispersal. Some authors consider inbreeding avoidance as a central cause (Bengtsson 1978; Packer 1979, 1985; Harvey and Ralls 1986; Pusey 1987; Wolff 1992; Wolff and Plissner 1998). Among the arguments presented are sex biases in dispersal and negative correlations between male and female dispersal (Pusey 1987). Others consider inbreeding to be only marginally important. Waser *et al.* (1986) argue that inbreeding depression should select for female-biased dispersal in polygynous mating systems, while the opposite is usually observed. Dobson (1982) concludes from his review that competition for mates is the primary reason for male-biased dispersal in mammals. Some authors even see inbreeding as totally irrelevant (Moore and Ali 1984), an extreme position that seems hardly tenable: in addition to clear-cut empirical evidence (e.g. Dobson 1979; Packer 1979, 1985; Cockburn *et al.* 1985; Clutton-Brock 1989; Wolff 1992), there are good logical arguments to expect inbreeding to play a significant role in dispersal. The question is not whether inbreeding affects dispersal or not, but how and by how much.

A modelling approach

Whatever the causes driving dispersal, optimal decisions are likely to depend on partners' behaviour, so that a game-theoretical approach is required. Furthermore, any consideration of the role of inbreeding has to incorporate kin selection arguments, simply because inbreeding itself cannot arise without kin structures. But its role must also be clearly differentiated from that of kin competition avoidance. For this reason our model has two starting points. The first considers a stable, structured population with no genetic load, in which local competition is the only cause affecting dispersal. The second starting point assumes a population free of any sort of competition (but not of genetic load), which will allow us to delineate the direct and indirect (kin-selected) effects of inbreeding. The interaction of both causes (a more general and realistic situation) will then be considered, before we investigate the possible influence of mating systems and social structures. However, several of the reasons why inbreeding avoidance may not drive dispersal (such as outbreeding depression, purging, or kin recognition) will not be investigated here.

Our model organism is diploid and annual, the sex ratio is even at birth, and dispersal precedes mating and is under offspring control. For analytical tractability we assume an infinite island model. Even though isolation by distance might constitute a more realistic framework, it would encumber analysis without adding much to the understanding of the basic principles delineated below.

Local competition without inbreeding

The idea behind kin selection is that individuals gain inclusive fitness benefits from enhancing the reproduction of relatives (Hamilton 1964). Hamilton and May (1977) showed that dispersal might be selected for in stable habitats and in the absence of inbreeding depression, even if it bears survival costs. This arises from the inclusive fitness benefits of avoiding competition with kin, not from direct fitness effects. Indeed, dispersal is clearly detrimental in terms of individual fitness, because it bears dispersal costs which, in stable habitats (notwithstanding demographic stochasticity), are not compensated for by a lower level of competition in the new habitat. But relatives will benefit from the release from competition, which may improve the disperser's inclusive fitness sufficiently for some level of dispersal to be favoured. Hamilton and May (1977) considered only one sex competing for a limiting resource, but we will follow here a formalization with two sexes, in order to allow for later inclusion of inbreeding and possible biases in sex-specific dispersal.

Let us consider patches of limited size (say N breeding opportunities). Let female i in patch j disperse with probability x_{ij}, and survive dispersal with probability s ($c = 1 - s$ is the mortality cost to dispersal). In her new patch, she will compete for N breeding opportunities among $Nb(1 - x + xs)$ unrelated females, where b is the average number of daughters per female, and x is the average dispersal rate in the population. With the complementary probability $1 - x_{ij}$, this female will stay in her patch, and compete there for N breeding opportunities among $Nb(1 - x_j + xs)$ females, from which $Nb(1 - x_j)$ are related (x_j is the average dispersal rate in patch j). In case of success, this female will produce b copies of her genes, so that her total fitness may be written:

$$W_x = x_{ij} \frac{sNb}{Nb(1 - x + xs)} + (1 - x_{ij}) \frac{Nb}{Nb(1 - x_j + xs)} \tag{2}$$

Selective pressures are derived using the direct fitness approach of Frank and Taylor (Taylor and Frank 1996; Frank 1997, 1998):

$$\frac{dw_\xi}{d\xi} = \sum_{z=x,y} c_z \left(\frac{\partial W_z}{\partial x_{ij}} + r_{xz} \frac{\partial W_z}{\partial x_j} \right) \tag{3}$$

where w_ξ is the fitness of an allele coding for female dispersal, and ξ the breeding value (expected phenotypic effect) of this allele. The sum is taken over all classes expressing, or affected by, the gene (here females and males). c_z measures the reproductive value of class z, and r_{xz} the relatedness of the focal female to individuals of class z born in patch j. Under our assumptions, males and females have identical reproductive value, and relatedness within classes equals that among classes. Coefficients c_z will therefore be omitted, and a single parameter r used to designate relatedness among offspring born in the same patch. Using equation (2) into (3) provides:

$$\frac{dw_\xi}{d\xi} = \frac{s}{1 - x + xs} - \frac{1}{1 - x_j + xs}\left(1 - r\frac{1 - x_{ij}}{1 - x_j + xs}\right) \tag{4}$$

The evolutionarily stable strategy (ESS) is found by setting this derivative to zero, while equating $x_{ij} = x_j = x$:

$$c = rk_x \tag{5a}$$

where $k_x \equiv (1 - x_j)/(1 - x_j + sx)$ is the probability that a breeding female is local. This simple result, first reached by Frank (1986) and Taylor (1988), has the intuitive meaning that, for an inner equilibrium, dispersal costs must meet kin competition costs. Similar reasoning for males (writing male dispersal as y) leads to:

$$c = rk_y \tag{5b}$$

where $k_y \equiv (1 - y_j)/(1 - y_j + sy)$ is the probability that a breeding male is local.

For an explicit solution to (5), one needs to express r as a function of dispersal rate and patch size. This is obtained by first writing down the difference equations for genetic variance H_i, H_g, and H_t, then substituting their equilibrium values into the equation for relatedness (1.3). The results are plotted in Fig. 9.2a in the x–y space, using Perrin and Mazalov's (1999) difference equations, which account for inbreeding depression under infinite island assumptions. (Finite island or stepping-stone models are expected to lower the evolutionarily stable dispersal probabilities by generating relatedness among neighbours.) The best-response curve of females (dashed line) is plotted as a function of male dispersal, and the best-response curve of males (plain line) as a function of female dispersal. These curves cross on the diagonal, because (5a) and (5b) are symmetrical. This joint equilibrium implies identical dispersal by both sexes, the amount of which depends on s and N. Note that at equilibrium the individual fitness of a philopatric individual exceeds that of a disperser. Because the male curve crosses the female one from above, the balanced equilibrium found here is a convergence stable strategy (CSS) (Taylor 1989; Motro 1994). Any random drift of the population away from this equilibrium creates a selective pressure on both sexes to return back to it. Convergence stability stems from the fact that an individual optimal decision depends mainly on what other individuals *of the same sex* are doing. (Here, dispersal is a means to avoid interactions with relatives of the same sex.) However, it should also be noted that, whereas individual fitness does not depend on dispersal by the other sex (eqn 2), the female best-response curve is a negative function of male dispersal, and *vice versa*. This interdependence stems from kin interactions: a low male dispersal increases r, which promotes female dispersal as a kin competition avoidance mechanism. As will now be

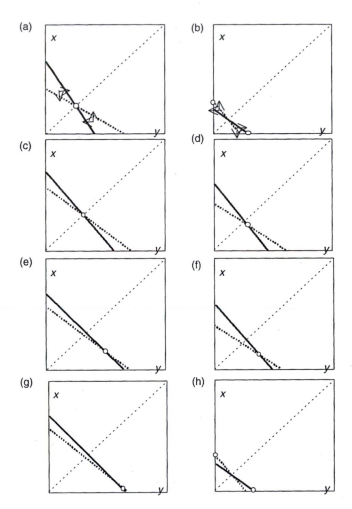

Figure 9.2. Evolutionarily stable dispersal patterns as functions of female (x) and male (y) dispersal. Plain lines plot male best response to female dispersal, and dashed lines plot female best response to male dispersal. Selection favours lower male dispersal to the right of plain lines, and higher dispersal to their left. Similarly, selection favours lower female dispersal above dashed lines, and higher dispersal below dashed lines (examples plotted on 2a and 2b). When male best response crosses the female one from above (e.g. 2a), the balanced equilibrium is a CSS (open circle). When it crosses it from below (e.g. 2b), the balanced equilibrium is unstable, and two border equilibria become CSSs. Parameter values were set to $N = 10$ and $c = 0.1$ in all simulations. (a) If kin competition is symmetrical and in the absence of genetic load, the balanced equilibrium is on the diagonal (open circle). (b) If inbreeding depression is the only selective pressure for dispersal ($\gamma = 1$) the balanced equilibrium is unstable. Two border CSSs coexist (open circles). (c) Combining kin competition and inbreeding depression destabilizes the inner equilibrium: the curves cross with a more acute angle than in (a). (d) A slight asymmetry in the benefits of philopatry (Greenwood's hypothesis; here $a_y = 1$, $a_x = 0.99$) induces a slight male bias in dispersal. (e) The male bias is much larger in the presence of genetic load ($\gamma = 1$), but other parameter values are as in (d). (f) A slight asymmetry in the benefits of kin cooperation (here $a_y = 1.0$ and $a_x = 0.98 + 0.02x$) induces a male-biased dispersal. (g) The male bias is much larger in the presence of inbreeding depression ($\gamma = 1$), but other parameter values are as in (f). (h) If both sexes benefit from kin cooperation, the balanced equilibrium may become unstable (here $\gamma = 1$, $a_x = 0.8 + 0.2x$ and $a_y = 0.8 + 0.2y$).

apparent appear, the sensitivity to dispersal by the other sex is one of the main points affected by inbreeding.

Inbreeding without competition

Inbreeding depression is unlikely to be the only selective pressure for dispersal, because inbreeding needs kin structures to arise, which, as soon as resources become limiting, create conditions for the evolution of kin competition avoidance. However, as inbreeding effects on dispersal are best understood when considered in isolation, we will assume for now unlimited breeding resources. This condition may be temporarily approximated under non-equilibrium dynamics, whereby extinction–colonization events generate important founder effects.

Let the fecundity of a pair be a negative function of co-ancestry among partners (Fig. 9.1). Female fecundity will be b_p if she mates with a male from another group, and b_g if her mate stems from the same patch. In absence of competition her total fitness is thus:

$$W_x = x_{ij}sb_p + (1 - x_{ij})(k_y b_g + (1 - k_y)b_p) \tag{6a}$$

while that of males is:

$$W_y = y_{ij}sb_p + (1 - y_{ij})(k_x b_g + (1 - k_x)b_p) \tag{6b}$$

The selective pressures on female and male dispersal are given by substituting equation (6) into (3). Setting these pressures to zero, while equating $x_{ij} = x_j = x$ and $y_{ij} = y_j = y$ provides the evolutionarily stable dispersal for females:

$$c = Ik_y + rI(1 - k_x)k_{yx} \tag{7a}$$

and for males:

$$c = Ik_x + rI(1 - k_y)k_{xy} \tag{7b}$$

where $I = (b_p - b_g)/b_p$ is the cost of inbreeding (Fig. 9.1), $k_{yx} \equiv (1 - y_j)/(1 - x_j + xs)$ is the number of local males per breeding female, and $k_{xy} \equiv (1 - x_j)/(1 - y_j + ys)$ the number of local females per breeding male. Equation (7) receives the intuitive interpretation that, at inner equilibrium, the marginal costs of dispersal match the marginal benefits, which are of two sorts. The first one (Ik_y for females) expresses the direct cost of an inbred mating (weighted by its probability), while the second ($rI(1 - k_x)k_{yx}$) quantifies the fact that a dispersing female decreases the risk of inbreeding for relative males (which adds to her own inclusive fitness).

For an explicit solution to equation (7), one needs to express I as a function of co-ancestry. For the purpose of illustration, we may assume a linear relationship, which has the merits of both simplicity and empirical support (e.g. Fig. 10.2 in Lynch and Walsh 1998): $I = \gamma F_{gp}$, where $\gamma = (db/dH)(H_p/b_p)$ measures genetic load (Fig. 9.1). Figure 9.2b plots female best response to male dispersal for $\gamma = 1$. Compared with Fig. 9.2a, the slope is much steeper, stemming from the fact that female inbreeding costs depend primarily on male behaviour (eqn 7a). If all males stay home ($y = 0$) then even low inbreeding costs may exceed dispersal costs. By contrast, if (all males disperse), even huge inbreeding costs will have no effect. Thus, contrasting with kin competition mechanisms, the optimal decision of a female depends mostly on dispersal by the other sex. (Here, dispersal is a means to avoid interactions with relatives of the opposite sex.)

Note, however, that there is a slight dependence on the behaviour of other females, mediated by their effect on co-ancestry: the more females disperse, the lower the average co-ancestry within groups, thus the lower the inbreeding costs. This dependence on relatives of the same sex makes possible inner equilibria (i.e. for some male dispersal values, female best response is intermediate between 0 and 1).

As equations (7a) and (7b) are symmetrical, both curves cross on the diagonal. However, because of the strong dependence on the other sex, they cross now in the opposite way (Fig. 9.2b). The male response curve crosses the female one from below. This implies that the balanced equilibrium at the crossing of the curves is not convergence stable. Any drift of one sex away from this equilibrium will select for a further drift apart of the other sex in the opposite direction, so that the whole system will move towards one of the two border CSSs. Border solutions mean that one sex will remain entirely philopatric. Which sex disperses eventually is purely random, and may depend on evolutionary history as well as genetic drift. This obviously leaves a significant role for phylogenetic inertia in deciding which sex disperses (details and further discussion in Perrin and Mazalov 1999). Note that, at equilibrium, fitness in the dispersing sex does not depend on dispersal decision (which is not true in the philopatric sex).

The main conclusion of this part is that inbreeding depression, were it allowed to act in isolation, would select for some dispersal, but by one sex only. The fact that dispersal usually occurs in both sexes, even when it is sex biased, adds empirical argument to the above *a priori* comment that inbreeding depression is unlikely to be the only reason for dispersal.

Combining kin competition and inbreeding depression

Under the combined effects of competition and inbreeding depression, female fitness becomes:

$$W_x = x_{ij}s\frac{Nb_p}{Nb(1-x+sx)} + (1-x_{ij})\frac{N(k_yb_g+(1-k_y)b_p)}{Nb(1-x_j+sx)} \tag{8a}$$

while male fitness is given by:

$$W_y = y_{ij}s\frac{Nb_p}{Nb(1-y+sy)} + (1-y_{ij})\frac{N(k_xb_g+(1-k_x)b_p)}{Nb(1-y_j+sy)} \tag{8b}$$

Substituting equation (8) into (3) provides:

$$\frac{dw_\xi}{d\xi} = \frac{sb_p}{b(1-x+xs)} - \frac{k_yb_g+(1-k_y)b_p}{b(1-x_j+xs)}$$
$$+ r\left(k_x\frac{k_yb_g+(1-k_y)b_p}{b(1-x_j+xs)} + (1-k_x)\frac{k_y(b_p-b_g)}{b(1-x_j+xs)}\right) \tag{9}$$

from which are calculated female evolutionarily stable dispersal:

$$c = rk_x(1-Ik_y) + Ik_y(1+r(1-k_x)) \tag{10a}$$

and male evolutionarily stable dispersal:

$$c = rk_y(1 - Ik_x) + Ik_x(1 + r(1 - k_y)) \tag{10b}$$

Or, in words, for a balanced equilibrium, marginal dispersal costs must match the sum of marginal benefits. A first benefit ($rk_x(1 - Ik_y)$ for females) can be attributed to kin competition avoidance (although it interacts partially with inbreeding), and the second ($Ik_y(1 + r(1 - k_x))$) to inbreeding (although it interacts partially with kinship).

Interactions are not symmetrical. The selective pressure stemming from kin competition avoidance is devalued by the effect of inbreeding, because dispersal then brings fewer benefits (the share of reproduction left behind for relatives is smaller). By contrast, inbreeding avoidance is strengthened by the inclusive fitness component, as already discussed.

This kin inbreeding–avoidance argument was first put forward by Bengtsson (1978). Starting from a situation in which all female offspring are philopatric, and all male offspring disperse, Bengtsson searched the conditions for a rare philopatric mutant male to succeed. As groups were assumed to consist of a single breeding female ($N = 1$), philopatric males would mate and reproduce with sisters. The model showed that male philopatry would not evolve unless

$$c > \frac{3I}{2} \tag{11}$$

a condition to which equation (10b) reduces under Bengtsson's assumptions of complete female philopatry ($k_x = 1$), complete male dispersal ($k_y = 0$) and a patch size reduced to $N = 1$ (so that offspring are full sibs and $r = 0.5$). This result has a straightforward interpretation: the higher the relatedness among patch mates, the lower the inbreeding costs needed to favour dispersal, because inbreeding depression affects both the male's own reproduction and that of his sisters.

Using an explicit genetic approach for an haploid organism in small patches ($N = 1$), Motro (1991, 1994) showed that, depending on the intensity of inbreeding depression, the male curve may cross the female one either from above or from below, resulting in either a stable or an unstable balanced ESS. In Fig. 9.2c, the balanced ESS is stable and, compared with Fig. 9.2a, shows only a slightly increased dispersal. (Effects are not simply additive, since dispersal rates induced by kin competition avoidance prevent inbreeding, and *vice versa*.) But the slopes of the best-response curves are steeper, because individual decisions are now more sensitive to behavioural decisions of patch mates from the other sex. Thus, even when inbreeding depression is not strong enough to make the inner ESS unstable, it contributes to destabilizing it, in the sense of making the response curves cross with a more acute angle (Fig. 9.2c). This matters in the case of sex asymmetry in costs or benefits, because it then induces much larger sex biases in dispersal, as will become apparent when the effects of mating systems and social structures are taken into account.

Mating systems

Greenwood (1980, 1983) suggested that interactions between inbreeding and dispersal were affected by mating systems. He first noticed that dispersal was often sex biased, the bias itself being assumed to stem from inbreeding avoidance: if one sex disperses, the other does not need to. Second, he observed that the identity of the dispersing sex was dependent on mating systems. Here, the amount of paternal investment is a central issue. At one end of the range lie polygynous or promiscuous species in which females invest (almost) all the time and energy necessary for reproduction, the male contribution being restricted to a few sperm cells and a few seconds of copulation time. Mammalian breeding systems often resemble this, owing to the physiological burden of pregnancy and lactation imposed on females. At the other end of the range lies the monogamous system of many bird species, in which nest feeding enlarges the scope for males to increase their fitness through paternal care. Males may then adopt a resource-defence strategy (as opposed to the female-defence strategy found in many mammals), and play a significant role in territory acquisition and defence. As a result of their restricted paternal investment, the males of polygynous or promiscuous species also enjoy a much larger potential reproductive rate than females, as opposed to monogamous species in which sex differences in potential reproductive rates may vanish.

Polygyny itself has an effect on dispersal. The fact that a few males are able to monopolize reproduction within breeding groups enhances gene correlations within these groups, which affects relatedness r as well as inbreeding depression I. This both increases dispersal and contributes to destabilization of the inner equilibrium, but it does not, *per se*, induce a bias (Perrin and Mazalov 1999). By contrast, the several correlates of mating systems delineated above have the potential to induce such sex biases.

The resource-competition hypothesis

Greenwood (1980, 1983) noticed that the philopatric sex is usually female in mammals and male in birds, which in both cases corresponds to the sex that benefits most from acquaintance with territory (i.e. the sex most involved in territory acquisition). A few exceptions in mammals and birds are linked to atypical mating systems (resource-defence systems in male mammals, or female defence in male birds), and thus corroborate the theory (e.g. Pusey 1987; Clarke *et al.* 1997; Wolff and Plissner 1998). Other exceptions seem best explained by inbreeding avoidance as well. Some cases of female-biased dispersal in mammals correspond to situations in which males are able to monopolize reproduction over a breeding group for a period longer than the maturation time of their daughters (Clutton-Brock 1989).

The benefits from philopatry assumed by Greenwood (acquaintance with natal territory) may arise in two ways: philopatric individuals may either make better use of local resources (which translates into higher fecundity) or have a greater chance of obtaining a breeding opportunity (territory or mate). As these alternatives are mathematically equivalent, we develop the second one only, assuming immigrants to suffer from a lower competitive ability. This may be expressed through a coefficient a_z that measures the relative weight of an immigrant of sex z when competing for a breeding opportunity. Female fitness may be written:

$$W_x = x_{ij}s\frac{a_xNb_p}{Nb(1-x+a_xsx)} + (1-x_{ij})\frac{N(K_yb_g+(1-K_y)b_p)}{Nb(1-x_j+a_xsx)} \tag{12a}$$

while male fitness is:

$$W_y = y_{ij}s \frac{a_y N b_p}{Nb(1 - y + a_y sy)} + (1 - y_{ij}) \frac{N(K_x b_g + (1 - K_x)b_p)}{Nb(1 - y_j + a_y sy)} \tag{12b}$$

where $K_y \equiv (1 - y_j)/(1 - y_j + a_y sy)$ is the proportion of breeding males that are of local origin and $K_x \equiv (1 - x_j)/(1 - x_j + a_x sx)$ the proportion of breeding females that are of local origin. Searching for ESSs as above provides the following best response for females:

$$C_x = rK_x(1 - IK_y) + IK_y(1 + r(1 - K_x)) \tag{13a}$$

and for males:

$$C_y = rK_y(1 - IK_x) + IK_x(1 + r(1 - K_y)) \tag{13b}$$

where $C_z = 1 - a_z s$ acts as a compound, sex-specific cost to dispersal, that affects the territorial sex more markedly.

Equations (13a) and (13b) become asymmetrical as soon as a_x and a_y differ, which induces a sex bias in dispersal. Figure 9.2d and 9.2e illustrates the case of a slight asymmetry in the benefits of philopatry for a female-defence system (i.e. $a_x < a_y$). Dispersal decreases in females and increases in males, but the effect is much stronger with inbreeding depression (Fig. 9.2e compared with 9.2d), first because co-ancestry induced by female philopatry promotes male dispersal through its effect on I (in addition to its effect on r), and second because the sensitivity to the other sex induced by inbreeding makes the response curves much steeper, thereby rendering the joint equilibrium much more sensitive to asymmetry (this illustrates the destabilizing effect of inbreeding depression discussed above).

Thus, Greenwood's argument indeed predicts a male-biased dispersal in female-defence systems, and a female-biased one in resource-defence systems. Synergistic interactions arise between inbreeding and resource competition, because inbreeding depression enhances the bias introduced by the weighting coefficient a_z. However, there is an important difference to note between this model and Greenwood's verbal argument. If, as in the latter, inbreeding (rather than kin competition) was the only ultimate reason for dispersal, then (13) would reduce to a system similar to that in equation (7), which would bring instability to the inner equilibrium and induce border CSSs. Asymmetry in the benefits of philopatry would result not in a lower dispersal rate by the territorial sex, but in a higher probability that the territorial sex would be the philopatric one (Perrin and Mazalov 1999).

Local mate versus local resource competition

The link between male-biased dispersal and female-defence strategy may also arise from different causes. In our formulation of the Hamilton–May model, the balanced CSSs were on the diagonal because both sexes were assumed to suffer identically from local competition. This might be true for a strictly monogamous species in which males and females would play an equivalent role in resource acquisition. But, in the female-defence systems of many mammals, sexes do not compete for the same items: females compete for resources (breeding sites) and thus display local resource competition (Clarke 1978), while males compete for females and thus suffer from local mate competition (Hamilton 1967). Furthermore, because of sex differences in potential reproductive rates, these items may not be equally limiting. As copulation normally takes only a few seconds, male fitness is likely always to be limited by female availability. By contrast, the process of transforming resources into nurturing of offspring is time consuming, so that female fitness may often be limited by her rate of processing resources, rather than resources themselves. As a result, local mate competition between males may often exceed local resource competition between females.

Perrin and Mazalov (2000) contrasted a situation like the one envisioned above (equal competition among males and females; eqns 8a and 8b) with a situation in which only males suffer from local mate competition (combination of equations 6a and 8b), corresponding to a female-defence system in an unsaturated environment. Female evolutionarily stable dispersal becomes:

$$c = k_y(I - r + rI) \tag{14}$$

whereas that for males remains identical to equation (11b). Kin competition thus provides a strong incentive for males to disperse (avoidance of local mate competition) which is not counterbalanced by local resource competition among females. This situation induces a strong sex bias in evolutionarily stable dispersal rates. In the precise situation investigated by Perrin and Mazalov, female philopatry was actually complete, partly because the bias was further amplified by inbreeding depression, and partly because resources were assumed to be not at all limiting. (Partially limiting resources are expected to have the potential to induce a balanced equilibrium, a point that deserves further investigation.) This analysis points to the balance between local resource and local mate competition as a central factor in deciding which sex disperses. Similar arguments were developed in the context of sex ratio theory, according to which kin competition is avoided by producing less of the sex that suffers more from local competition (Taylor and Bulmer 1980; Taylor 1981; Bulmer 1986).

Conclusions from this model rejoin those of the preceding paragraph in that, while inbreeding depression is not required to induce a sex bias in dispersal, it will enhance it. They also meet empirical patterns, as reviewed in mammals by Dobson (1982): while polygynous and promiscuous species display a male-biased dispersal, no such bias occurs in monogamous species. From his review, Dobson concluded that competition for mates was more likely to be responsible for the general mammalian dispersal pattern than either competition for resources or inbreeding, the latter acting mostly through synergistic interactions.

Asymmetry in inbreeding costs

The difference among sexes in potential reproductive rate also has the consequence that sexes may differ with respect to inbreeding costs. The argument was first made by Parker (1979, 1983) that inbreeding is most costly for the sex that contributes more to the parental investment. In polygynous or promiscuous systems, because of their low levels of paternal investment, males do not forfeit other mating opportunities by siring a relative's offspring. They therefore benefit directly from an inbred mating, which simply adds offspring to those stemming from other matings rather than replacing them. In females, by contrast, an inbred mating entirely forfeits one breeding opportunity, so that inbreeding depression bears large direct costs that will not be matched by inclusive fitness benefits (enhancement of related males' reproductive output). Inbreeding is therefore less costly for males in a female-defence system.

Building on this argument, Waser *et al.* (1986) suggested that females should be the dispersing sex in polygynous systems, because their threshold for accepting inbreeding is lower. Since this opposes what has been observed (the male is usually the dispersing sex in such systems), Waser *et al.* (1986) concluded that inbreeding plays only a marginal role (if any) in dispersal. However, as noted by Perrin and Mazalov (1999) this argument may actually be reversed if females have the ability to choose their partner according to his dispersal status (kin recognition).

Mate choice and sexual selection on dispersal

Indeed, because of their large commitment to reproductive effort, females should evolve choosiness. In particular, they should refuse copulation with relatives and prefer immigrant males. Female preference might be formalized by attributing a larger weight to immigrant males than to local males in their competition for local females. Setting $a_y > 1$ in equations (12b) and (13b) would obviously constitute a further incentive for males to disperse, and thereby induce a stronger male bias in the dispersal of polygynous or promiscuous species.

Interestingly, if males also evolve some form of choosiness in female-defence systems (which may arise if one mating at least partly forfeits other mating opportunities), then they should prefer philopatric females under Greenwood's resource competition hypothesis, at least in the case where females benefit from philopatry through a better use of resources. Indeed, philopatric females in this case have a higher fecundity, so that males have more to gain from mating with them. Thus, sexual selection should in any case reinforce the trend predicted under Greenwood's resource competition hypothesis, and enhance the link between male dispersal and female-defence systems.

Social structures

The two components of kin selection considered up to now (kin competition avoidance and kin inbreeding avoidance) favour dispersal. A third, potentially important, effect lies in the benefits of cooperation and reciprocal altruism, which, opposing the two other components, may provide strong benefits to philopatry. Social structures based on kin interactions are widespread among birds and mammals. How do they interact with the selective forces for dispersal investigated here?

In their simplest form, benefits may arise simply because territory owners are less aggressive towards related neighbours (e.g. Watson *et al.* 1994; Koprowski 1996). Local settlement is made easier because of acquaintance, not with territory, but with kin. Owners may even share part of their territory, if the inclusive benefits of allowing relatives to reproduce exceed the costs of sharing. More evolved forms of cooperation appear in social groups, such as helping at the nest (e.g. Emlen and Wrege 1991; Powell and Fried 1992; Komdeur 1996; Dickinson and Akre 1998), or cooperation among related adults (e.g. Mappes *et al.* 1995; Lambin and Yoccoz 1998). Kin cooperation may help in acquiring not only a territory, but possibly also social status or direct access to reproduction, as observed in the females of several monkey species, and in the males of hamadryas baboons or chimpanzees (Packer 1979; Pusey 1980). Kin cooperation has been shown to enhance fitness: 'there is evidence from various species that philopatric females have higher reproductive success than females that have left their natal area or Group' (Pusey 1987, p. 298).

On the one hand, kin cooperation thus constitutes an important selective pressure for philopatry. On the other hand, inbreeding depression opposes this pressure. Could inbreeding be sufficiently detrimental to impede the evolution of social structures? Not necessarily, as a sex-biased dispersal may suffice to prevent inbreeding, and meanwhile allow significant kin structures to arise (Chesser and Ryman 1986; Chesser 1991a,b). A male-biased dispersal, often associated in mammals with polygyny, allows the building of fairly high relatedness values ($r > 0.3$) without any noticeable inbreeding (i.e. $F_{ip} = 0$), as shown for instance in black-tailed prairie dogs (Chesser et al. 1993; Sugg et al. 1996; Dobson et al. 1997). It is worth noting in this context that 'particularly striking examples of sex differences in natal dispersal occur in species which live in permanent social groups' (Pusey 1987, p. 295). Why should this be so?

The difference between selective pressure in Greenwood's resource competition hypothesis and that stemming from kin cooperation lies in the fact that, in the latter, the benefits of philopatry are not constant, but depend on the proportion of local individuals among other settlers. This induces a positive feedback cycle in philopatry: the more local settlers among patch mates, the higher the benefits from settling locally. This can be formalized by making the coefficient a_x weighting immigrants in equation (12a) an increasing function of x_j, rather than a constant. Thus, when x_j is very small (i.e. most locally born stay), immigrants have a relatively low probability of settling successfully (because their local competitors help one another), while as x_j tends to unity (all locally born disperse), immigrants have the same weight in competition as any locally born. Applying the direct fitness approach (eqn 3) to equations (12a) and (12b), while considering a_x a function of x_j rather than a constant, provides the best-response curves for females:

$$C_x = r\beta_x K_x(1 - IK_y) + IK_y(1 + r(1 - \beta_x K_x)) \tag{15a}$$

and for males:

$$C_y = r\beta_y K_y(1 - IK_x) + IK_x(1 + r(1 - \beta_y K_y)) \tag{15b}$$

which are identical to equations (13a) and (13b), except that patch mates *of the same sex* (K_z) are now weighted by a coefficient $\beta_z = 1 - sz(da_z/dz)$ (N. Perrin and C.R. Lehmann, unpublished manuscript).

As this coefficient is smaller than unity (and all the smaller given that a_z increases strongly with z), it weakens markedly the dispersal pressure stemming from kin competition avoidance (because it affects the whole kin competition term), and enhances slightly the dispersal pressure from inbreeding avoidance (only its kin-selected component is affected). The net effect is a dispersal pressure that is both of lower intensity and more dependent on the other sex's strategy. Together with inbreeding depression, kin cooperation has the potential strongly to destabilize the balanced equilibrium.

The effect of inbreeding depression, in the case of a weak benefit from kin cooperation among females, can be seen by comparing Fig. 9.2f with Fig. 9.2g. A strong sex bias is observed in Fig. 9.2g, even though kin cooperation advantages to the philopatric sex are very low. The destabilizing effect of within-sex cooperation also appears when both sexes benefit from it. Added to inbreeding depression, it can make the inner ESS unstable, as plotted in Fig. 9.2h. Two border CSSs coexist, implying that one sex remains entirely philopatric, while the other disperses, even though both sexes benefit equally from cooperation.

In the example plotted in Fig. 9.2h, the 26% equilibrium dispersal by one sex results in about 10% co-ancestry and 20% relatedness within breeding groups, corresponding in our model to a 10% inbreeding depression. This last value may actually be much larger, provided kin cooperation brings enough benefits. Assuming, for example $a_x = x_j$ and $a_y = y_j$, the 8% equilibrium dispersal by one sex boosts relatedness within breeding groups to 77%, corresponding in our model to a very significant inbreeding cost of 62%. This extreme value shows that strong inbreeding does not stem necessarily from high mortality costs of dispersal (fixed to 10% in our simulations): the benefits of kin cooperation may induce individuals to withstand extreme inbreeding costs. This also means that selective pressure stemming from inbreeding avoidance may exceed that from kin competition avoidance and become the most relevant force driving dispersal in social species, constituting a very significant incentive for those few individuals that disperse (even though actual dispersal rates remain limited). It is worth noting that all studies on non-human primates reviewed by Johnson and Gaines (1990) offer inbreeding avoidance as an explanation for the evolution of dispersal.

Social species studied also often evolved kin recognition mechanisms (e.g. Hoogland 1982; Harvey and Ralls 1986; Keane 1990; Potts *et al.* 1991), which may allow behavioural incest avoidance (i.e. mate choice based on co-ancestry). If breeding groups are so small that incest cannot be avoided through mate choice, then even a moderate and occasional dispersal might suffice to lower inbreeding significantly, provided it is condition dependent (i.e. dependent on co-ancestry with local potential mates). This is usually the case in troop transfer among social mammals (e.g. Packer 1979; Clutton-Brock 1989; review in Hoogland 1995). The way in which kin recognition may co-evolve with dispersal deserves further theoretical formalization.

Another point deserving investigation is the dynamics of genetic load. (Load was assumed to be constant in our simulations.) It is unlikely that inbreeding costs as high as 60% will remain for long in a breeding group: recurrent inbreeding will lead to the purging of deleterious mutations, resulting in the progressive reinforcement of philopatry. There are several examples of social mammals showing no sign of inbreeding avoidance or inbreeding depression (e.g. Reeve *et al.* 1990; Keane *et al.* 1996), which is best explained by a history of strong inbreeding and purging of deleterious mutations.

Conclusions and perspectives

The formalization presented here, like any other model, is an approximation of the exact processes involved. In addition to the simplifying assumptions made above, the point should be made that the approach is deterministic (which may be important when inbreeding occurs in small patches), and that some approximations were used (e.g. the dependence of b on dispersal and inbreeding patterns of the previous generation was not taken into account in eqn 8). Similarly, the shape of the relationship between inbreeding depression and co-ancestry matters. Some authors found that the balanced ESS might be unstable when both kin competition and inbreeding depression interact (Motro 1991; Gandon 1999, Gandon and Michalakis, chapter 11), which presumably stems from different assumptions. (Motro assumes inbreeding depression is independent of co-ancestry, and Gandon assumes an exponential negative function of co-ancestry.)

The usefulness of our model lies in the fact that it provides a unifying approach, allowing us to account simultaneously for inbreeding and kin competition in the evolution of dispersal, making explicit in particular the ways in which these two forces interact. It also sheds light on the ways in which mating systems and social structures interfere with these selective forces. Our main conclusions can be summarized as follows:

(1) Were inbreeding the only reason for dispersal, then a strong effect on dispersal would arise, but be restricted to one sex only (the balanced equilibrium is unstable). This situation is, however, unlikely for logical reasons (inbreeding implies kin structures, and thereby also kin competition as soon as resources are limiting) as well as empirical reasons (usually both sexes disperse, even when dispersal is biased). This implies that inbreeding cannot be the only ultimate cause of dispersal when both sexes disperse (even at very different rates), opposing implicit or explicit assumptions of many verbal models (e.g. Greenwood 1980, 1983).

(2) In absence of inbreeding depression, kin competition favours a balanced dispersal. The inner equilibrium is continuously stable because dispersal evolves in this case as a way to avoid patch mates of the same sex.

(3) Combining inbreeding depression and kin competition has little influence on dispersal rates. More importantly, the stability of the inner equilibrium is weakened, because dispersal now becomes a way of avoiding patch mates of the opposite sex.

(4) This destabilizing effect is apparent in the case of sex asymmetries in selective forces, as may arise within certain mating systems. Several arguments converge to predict a male-biased dispersal in female-defence systems, and a female-biased dispersal in resource-defence systems. This may stem from sex biases in (a) benefits of territory, (b) local competition, (c) inbreeding costs, and/or (d) sexual selection.

(5) Within-sex kin cooperation also has the potential to induce asymmetries in dispersal. Because of self-reinforcing benefits of philopatry, this pressure further destabilizes the balanced equilibrium, and should thus induce a lower, but more sex-biased, dispersal. Inbreeding avoidance might constitute the most important selective pressure for dispersal in social species.

(6) Inbreeding depression is thus not required to produce sex biases in dispersal, even though it may boost such biases. Similarly, negative (interspecific) correlations between male and female dispersal do not imply inbreeding avoidance either (Pusey 1987). A strong philopatry by one sex (induced, for example, by a lower local

competition) enhances relatedness within groups, which then favours more dispersal in the other sex as a kin competition avoidance mechanism.

These conclusions are consistent with empirical patterns, since male-biased dispersal is associated with female-defence and polygynous mating systems, while female-biased dispersal occurs mainly in resource-defence and monogamous systems (Greenwood 1980, 1983; Dobson 1982; Clarke *et al.* 1997). Also, stronger sex biases appear in social species (Pusey 1987), and inbreeding avoidance seems to play a more significant role (Johnson and Gaines 1990), even though the average dispersal rate is lower. However, a general problem appearing recurrently throughout our approach is that the predictions arising from inbreeding depression-avoidance arguments differ only quantitatively (as opposed to qualitatively) from a null model of pure kin competition avoidance. This situation is bound to make extremely difficult any field test of the importance of inbreeding depression in moulding dispersal patterns.

Empirical investigations might try to focus on situations in which competition for resources vanishes, as may happen when prey populations are heavily controlled by predation, or when extinction–recolonization dynamics create strong founder effects. Unfortunately, these situations are also likely to impose other selective forces on dispersal (Ronce *et al.*, chapter 24; Weisser, chapter 12b). Further modelling work is necessary to delineate the interaction of these forces with inbreeding. Non-equilibrium dynamics offers a particularly interesting case. Founder effects create inbreeding and, reciprocally, inbreeding depression contributes to extinction rate and metapopulation dynamics (e.g. Saccheri *et al.* 1998). The dynamics of genetic load would have to be taken into account, as well as other selective forces stemming from extinction risk and the benefits of dispersing towards less crowded patches.

Other potentially fruitful perspectives include the study of conditional dispersal. The dependence of dispersal on the presence of parents or relatives of each sex may bring useful information, since inbreeding avoidance fosters dispersal as a means to avoid opposite-sex relatives, while kin competition fosters dispersal in order to avoid relatives of the same sex. The importance of kin selection (relative to direct inbreeding costs) might be inferred from individual fitness differences between residents and dispersers. Dispersal distance may also differ among males and females, and this may tell something about the relative impact of inbreeding and kin competition, although this avenue would obviously require further formalization. The same is true of the ideas about sexual selection mentioned above. Female choosiness is likely to depend on the balance between inbreeding risk and the competitive disadvantage of immigrants, as well as on her own status as an immigrant or philopatric. Finally, the co-evolution of kin recognition and dispersal as alternative ways to avoid inbreeding, as well as their interaction, deserve proper treatment.

Acknowledgements

We want to thank warmly J. Clobert, S. Dobson, S. Gandon, and particularly O. Ronce for thoughtful comments and editorial improvements on a first draft of this chapter.

Appendix: List of parameters

Genetic measures are taken among offspring before dispersal. Sex-specific parameters are listed for females only, but the same notation also applies to males, with relevant changes.

H_i, H_g, H_p	Observed heterozygosity and Hardy–Weinberg expectations at the group and population level, respectively.
F_{ip}, F_{gp}	Inbreeding and co-ancestry among patch mates.
$r \equiv 2F_{gp}/(1 + F_{ip})$	Relatedness among patch mates.
b_g, b_p	Fecundity of pair, as a function of partner's genetic similarity. Without a subscript, b is the average fecundity within a patch.
$\gamma = (\mathrm{d}b/\mathrm{d}H)(H_p/b_p)$	Marginal increase in relative fecundity with relative heterozygosity, a measure of genetic load.
$I \equiv (b_p - b_g)/b_p = \gamma F_{gp}$	Cost of breeding with a patch mate, assumed to be proportional to co-ancestry.
N	Patch size (number of breeding females).
x_{ij}	Dispersal probability of female i in patch j. x_j is the average for the patch j and x is the average for the population.
W_x	Fitness of a female; a function of its dispersal probability.
W_ξ	Fitness of an allele coding for female dispersal; a function of its breeding value (expected phenotypic effect) ξ.
$c = 1 - s$	Mortality cost of dispersal.
a_x	Relative competitive weight of an immigrant female. $C_x = 1 - sa_x$ combines both the mortality and competitive cost of dispersal.
$K_x = (1 - x_j)/(1 - x_j + a_x sx)$	Probability that a female breeding in patch j is of local origin.

10

Inbreeding versus outbreeding in captive and wild populations of naked mole-rats

M. Justin O'Riain and Stanton Braude

Abstract

In this chapter we review the evidence for inbreeding versus outbreeding as the principal mating strategy in both captive and wild colonies of naked mole-rats. The naked mole-rat is a cooperatively breeding mammal occurring in large colonies with non-breeding individuals working on behalf of a fecund minority. Evidence from preliminary genetic studies and laboratory observations suggests that these rodents routinely inbreed, with new colonies forming through fission of the parent colony. While occasional inbreeding is especially deleterious in normally outbred species, constant ecological pressure selecting for inbreeding becomes progressively less costly. Theoretical predictions, however, suggest that close inbreeding cannot continue indefinitely since the accumulation of genes with mild deleterious effects and the loss of the ability to track changing environments ultimately selects for dispersal. Recent laboratory and field studies provide evidence in support of the above prediction with the finding that naked mole-rat colonies do in fact harbour potential dispersers and that outbreeding in the field is far more common than previously suspected. Low levels of genetic variation in previous molecular genetic studies may be explained in part by a recent common ancestor of the population, in conjunction with population viscosity, typical of fossorial rodents. Naked mole-rats illustrate the importance of understanding the different F statistics, their biological meanings, and the methodologies used to estimate them. The dispersers from naked mole-rat colonies provide a good example of phenotypic plasticity and are consistent with the theory that dispersal patterns have multiple causes and that dispersing individuals benefit from both increased access to unrelated mates and decreased intrasexual competition.

Keywords: cooperative breeding, inbreeding, dispersal morph

Introduction

Mole-rats within the family Bathyergidae exhibit a range in social organization from strictly solitary and aggressive to highly social, cooperatively breeding species (Jarvis *et al.* 1994). However, only the naked mole-rat (*Heterocephalus glaber*), the most social of these species, will engage in consanguineous matings in the laboratory following the death or removal of a dominant breeder. Close inbreeding (parent–offspring and sib–sib matings) have frequently been observed in captive colonies. Genetic studies (Reeve *et al.* 1990) of wild populations reported the highest known coefficient of inbreeding ($F = 0.62$) yet recorded among free-living mammals. Early field data from Brett (1986, 1991) suggested that new colonies form through fission from the parent colony, and Lacey and Sherman (1997) have since documented an attempted fission event within a captive colony.

Close inbreeding is known in a variety of mammalian species, for example the yellow-bellied marmot, *Marmota flaviventris* (Schwartz and Armitage 1980; Armitage 1986), meadow vole, *Microtus pennsylvannicus* (Selander 1970), pika, *Ochotona princeps* (Smith and Ivins 1984), rock hyrax, *Procavia johnstoni* (Hoeck 1982), dwarf mongoose, *Helogale parvula* (Keane *et al.* 1996), in which the high energetic costs and risks of death during dispersal result in moderate to high degrees of inbreeding. Inbreeding is usually associated with lowered fitness, such as inbreeding depression and susceptibility to pathogens (Ralls *et al.* 1986), and it is generally suggested that inbreeding behaviour is strongly selected against in most species (Dobzhansky 1970; Bengtsson 1978). Indeed, in all the above-mentioned examples of inbreeding, dispersal and inbreeding avoidance mechanisms (e.g. kin recognition) do exist and promote some degree of outbreeding (Smith 1993). In addition, Bartz (1979) modelled cycles of close inbreeding punctuated with a dispersal phase and obligate outbreeding in termites and predicted that this would also promote the evolution of eusociality. However, the functional significance of this model remains controversial (see Myles and Nutting 1978).

In naked mole-rats, the apparent lack of incest avoidance (Jarvis 1991; Jarvis *et al.* 1994), high inbreeding coefficients (Reeve *et al.* 1990), the marked xenophobic response of colony members to foreign conspecifics (Lacey and Sherman 1991; O'Riain and Jarvis 1997), and the paucity of small nascent colonies in the wild (Brett 1991) have suggested that they are obligate inbreeders and that new colonies are formed only by colony fissioning (Brett 1991). Foreign conspecifics have thus been regarded as extraneous for the purposes of reproduction. Inbreeding in naked mole-rats has been attributed largely to the prohibitive costs associated with dispersal and outbreeding, and to the limited survival probabilities of small nascent colonies in the wild (Jarvis 1985; Lovegrove and Wissel 1988; Brett 1991; Jarvis *et al.* 1994). The enormous probability of death during dispersal or early colony foundation makes kin cooperation and inbreeding a relatively secure fitness option and may induce individuals to withstand extreme inbreeding costs. While occasional inbreeding is especially deleterious in normally outbred species, owing to the exposure of deleterious homozygotes to selection, constant ecological pressure selecting for inbreeding may result in the purging of the genetic load after only a few generations of inbreeding (Perrin and Goudet, chapter 9). Foreign conspecifics pose the threat of introducing new deleterious alleles and are typically not welcomed within groups (O'Riain and Jarvis 1997). Although inbreeding depression has not been studied systematically in naked mole-rats, they do routinely inbreed in the laboratory with no obvious deleterious consequences, whereas populations with a history of outbreed-ing exhibit strong inbreeding depression when initially forced to inbreed (Madsen *et al.* 1996).

Recent discoveries in the laboratories at Cape Town of a dispersal morph amongst captive-bred individuals (O'Riain *et al.* 1996), and at Ann Arbor of mole-rats showing a preference for unrelated mates (Ciszek 2000), in addition to recent fieldwork which demonstrates that nascent colonies comprising unrelated individuals are relatively common (Braude 2000), suggest that dispersal and outbreeding may have been over-looked in this species. Here we assess the available evidence for inbreeding and outbreeding from recent laboratory and field studies and interpret the findings within theoretical frameworks developed in other chapters of this book.

Dispersal rates

Of the 48 captive colonies at the University of Cape Town, six contained individuals with strong dispersal tendencies (O'Riain *et al.* 1996). All of these colonies were large (more than 40 individuals) and well established (more than 4 years since inception, with a minimum of 10 successfully recruited litters). While there were also large colonies without obvious dispersers, there were no small colonies with potential dispersers, suggesting a lower limit to colony size for the emergence of dispersers. Examples of density-dependent dispersal, and in particular their effects on social and demographic factors in mammals, have frequently been documented (Gaines and McClenaghan 1980; Pusey 1987; Wolff 1997). Similarly, of 24 wild colonies that were captured in more than 1 year, only six produced dispersers. These colonies were all large (more than 40 individuals).

Unlike the relatively small laboratory colonies of naked mole-rats (approximately 20 to 40 animals) which experience little turnover of individuals (Jarvis 1991), wild colonies are typically larger (up to 290 animals) and have high rates of both recruitment and loss. Wild colonies lost between 12% and 91% of their workers per year (mean $46 \pm 23\%$, $n = 42$ cases). Although it is not possible to determine what proportion of those disappearances were due to death, the fates of at least 21 animals are known because they dispersed and either formed new colonies or, in one case, migrated into a neighbouring colony (Table 10.1).

Table 10.1. Long-term life history studies of wild colonies of naked mole-rats reveal the frequency of occurrence of newly formed colonies (nascent colonies) by dispersers. The table includes data on the sex, colony of origin, and distance travelled by dispersers

Colony	Date captured	Members: Sex (i.d.)	Year last captured in natal colony	Distance travelled[†]
P	17.7.88	F (unmarked)		
		M (unmarked)		
MM88	23.7.88	F (no. 483)	1987	220 m
MM91	26.5.91	F (unmarked)		
	5.7.91			
CC	14.7.91	F (no. 2109)	1990	180 m
		M (no. 2730)	1990	180 m
		M (no. 2732)	1990	180 m
		M (unmarked)		
		7 pups (3F, 4M)		
CCC	7.6.92	F (unmarked)		
		M (no. 3570)	1991	120 m
		1 pup (M)		

Table 10.1.   *(cont'd)*

Colony	Date captured	Members: Sex (i.d.)	Year last captured in natal colony	Distance travelled[†]
HHH	9.6.92	F (unmarked)		
		M (unmarked)		
		M (no. 3171)	1991	280 m
		5 pups (2F, 3M)*		
II	11.6.92	F (no. 3180)	1991	< 5 m
		M (unmarked)		
		3 pups (2F, 1M)		
OO	10.7.93	F (unmarked)		
		F (unmarked)		
		F (unmarked)		
MM95B	17.6.95	F (no. 2627)	1990	2.4 km
	25.6.95			
MM95A	25.6.95	F (unmarked)		
MMA	20.5.97	F (no. 6324)	1996	150 m
	24.5.97			
MMB	24.5.97	F (no. 6354)	1996	100 m
MMC	25.5.97	F (no. 7275)	1996	200 m
MMD	25.5.97	F (no. 6367)	1996	280 m
VV	26.5.97	F (no. 6275)	1996	150 m
		M (no. 6283)	1996	150 m
		M (no. 6253)	1996	150 m
WW	28.5.97	F (unmarked)		
		M (no. 6214)	1994	2.25 km
XX	29.5.97	F (unmarked)		
		M (no. 6361)	1996	160 m
MME	29.5.97	M (no. 6550)	1995	20 m
MMF	4.6.97	F (6331)	1996	200 m
YY	15.5.98	F (unmarked)		
		M (unmarked)		
		M (no. 1855)	1988	670 m
		14 pups (6F, 8M)		
SS	28.5.99	F (no. 7070)	1998	400 m
		M (7305)	1998	500 m

*All five pups born in captivity (1F, 1M died immediately). [†]Distance from natal colony trapping location to location where disperser was retrapped.

Nascent colonies

Twenty-one nascent colonies were discovered in the field between 1988 and 1999 (Table 10.1). Seventeen of these were founded by marked animals from known colonies, and four were founded by unmarked naked mole-rats, presumably from colonies far from the main trapping transect of the study area. Of the nascent colonies with marked animals, six contained males and females from different natal colonies along with their young, two contained breeding pairs from different natal colonies but without offspring, one contained two males and a female from the same natal colony but with no pups, and eight contained only a single mole-rat who had not yet been joined by a mate (Table 10.1). Most of the nascent colonies that had already bred were captured late in the season, and most often in years with good rainfall. It is possible that the other dispersers found mates and/or bred after the annual research census ended.

These small, nascent colonies (with four or fewer adults) account for 37% of the 54 colonies in this study. Despite the frequency with which they arise, only one of the nascent colonies has been trapped in more than 1 year. This is not surprising when we consider the generally low success of dispersing rodents or moles (Chepko-Sade and Halpin 1987; Anderson 1989; Gorman and Stone 1990). However, the one retrapped new colony was founded by male no. 1855, who dispersed from his natal colony between 1988 and 1989 and was recaptured in 1998 and 1999 as the breeding male in his new colony (YY) (Table 10.1). In 1998 he had the pale, thin, elongate morphology typical of well established, old breeding males. The time of his dispersal is known because he came from a well studied colony that had been recaptured in 1989, 1990, 1993, 1994, and 1998. The fact that no. 1855 was not recaptured for 10 years suggests that we will discover additional cases of successful dispersal as the monitoring of this population continues.

Although dispersal typically exposes individuals to an increased probability of death (Bekoff 1977; Swingland 1983), successful dispersers may accrue sufficient reproductive benefits to ensure that selection for dispersers is maintained within the population (Stenseth 1983). The moderate frequency of dispersers within both captive (O'Riain *et al.* 1996) and wild (Brett 1991; Braude 2000) naked mole-rat populations may well reflect these opposing selective pressures. The fossorial habitat and high foraging risks of unproductive foraging in naked mole-rats (Jarvis 1985; Lovegrove and Wissel 1988; Lovegrove 1991; Jarvis *et al.* 1994) bias selection strongly in favour of philopatry.

Sex bias in dispersal tendencies

Individuals with strong dispersal tendencies in captive colonies were predominantly males (18 : 1, males : females), whereas successful dispersal of marked individuals in the field showed a more balanced sex ratio (10 : 14, males : females). While there is no clear relationship between dispersal tendencies and successful dispersers, it is important to interpret these findings with reference to the substantial literature that exists on sex-biased dispersal in vertebrates.

In most vertebrate species, dispersal occurs in both sexes but is often biased towards males in mammals and females in birds (Greenwood 1980; Dobson 1982). Sex-biased dispersal is commonly interpreted as an adaptation to avoid inbreeding (Johnson and Gaines 1990), but it is unlikely to function as the sole reason for dispersal (Pusey 1987). Dispersal patterns ultimately depend on the relative costs and benefits of dispersal and philopatry to each sex (Pusey 1987). Females are likely to suffer higher costs of inbreeding because of their larger investment in offspring. If dispersal is purely to avoid inbreeding, then females in most mammals should be the dispersing sex, a suggestion that is not supported by available data (Gaines and McClenaghan 1980; Greenwood 1983).

Pusey (1987) and Perrin and Goudet (chapter 9) have suggested that other assymetries that exist between the sexes might further our understanding of sex-biased dispersal.

Most mammals are polygynous, and male parental contribution is low relative to that of females. Male reproduction is typically limited by access to mates, whereas females are limited by food and/or safe breeding sites. Females may thus benefit from philopatry because of familiarity with the territory, whereas males may benefit by dispersing to areas with a greater number of potential mates. This reasoning has been summarized by Greenwood and Harvey (1982), who suggested that a male bias in dispersal is usually associated with a mate-defence mating system, and a female bias with a resource-defence mating system. Furthermore, Greenwood (1983) suggested that, in mammals, high maternal investment has predisposed females to a greater incidence of cooperative behaviours that may in turn facilitate philopatry.

Naked mole-rats of both sexes are strongly philopatric, and the mating system is essentially monogamous but with some evidence for polyandry (Faulkes *et al.* 1997). Non-breeders of both sexes contribute equally to all cooperative behaviours (Faulkes *et al.* 1991; Jarvis *et al.* 1991; Lacey and Sherman 1991). Burrow systems represent critical resources that are defended fiercely by all group members against foreign conspecifics, but only females contest breeding vacancies aggressively within the natal territory. Kin competition between females for territory inheritance thus emerges as the most salient difference between the sexes that might influence dispersal rates. Differences in the degree of competitive and cooperative interactions among relatives have important influences on dispersal, and kin competition has emerged as the cornerstone of theoretical arguments for the evolution of dispersal in stable environments (Lambin *et al.*, chapter 8). While there is no evidence that breeding females evict offspring from the natal burrow system, their death typically results in a rapid escalation in aggression amongst their female offspring. It is possible that overt kin competition would result in dispersal of the losers, which in this case would be females. Kin competition and subsequent dispersal by females may result in selection for dispersal, in the absence of inbreeding avoidance, as predicted by Hamilton and May (1977). This might explain why dispersal in captive colonies was seldom by females. All the colonies in O'Riain *et al.*'s (1996) study were socially stable, and no reproductive conflicts were evident at the time of the study.

Male–male competition for reproductive tenure is not observed in naked mole-rats, and individuals with dispersal tendencies were not subjected to heightened parental or sibling aggression despite their general lack of contribution to cooperative tasks. Male dispersers do not appear to be forced to leave their natal colony (see Christian 1970; Moore and Ali 1984; Waser 1985), but do so of their own volition. Male dispersal does, however, appear to be condition dependent (O'Riain *et al.* 1996; Ims and Hjermann, chapter 14) and thus driven by ontogenetic changes in individuals irrespective of reproductive vacancies in the natal territory. Potential male dispersers did show strong kin discrimination behaviour in mate choice experiments and showed a significant preference for unrelated females, avoiding reproducing with their in-oestrus mothers. Ciszek (2000) also found a strong preference for unrelated mates in her mate choice experiments. These results suggest that dispersing males actively avoid inbreeding and that dispersal may function primarily as an outbreeding mechanism.

Dispersal in naked mole-rats may thus be heterogeneous and include high-quality specialized individuals with the prospect of successful dispersal in addition to subordinates forced to leave by dominants. Theoretical predictions by Perrin and Goudet (chapter 9) demonstrate that, if inbreeding depression is the sole selective pressure for dispersal, then only one sex is predicted to disperse. The results presented here, that both sexes disperse, reinforce arguments that inbreeding avoidance is unlikely to be the only reason for dispersal in naked mole-rats. The difference in the sex ratio of dispersers in the laboratory and field suggests that males and females may use different social and environmental cues for dispersal. The cues that females rely on may not have been present or strong enough in the laboratory setting. Attempts to repeat O'Riain *et al.*'s experiments with captive colonies in which there is reproductive conflict amongst females may help us discover those cues.

Behavioural and morphological profile of dispersers

Environmental conditions may trigger dispersal directly as an immediate behavioural response, or indirectly by first altering the internal body condition that subsequently triggers dispersal. Indirectly induced dispersal involves persistent changes in behaviour, physiology, and morphology, and is thus more costly than direct behavioural responses (Ims and Hjermann, chapter 14). A preliminary investigation into possible behavioural, morphological, and physiological differences that may exist between dispersing and non-dispersing individuals within the colony context has been carried out for captive individuals (O'Riain *et al.* 1996). It was clearly evident from detailed behavioural observations of captive colonies that dispersers within colonies formed a behaviourally distinct group that participated little in cooperative maintenance tasks, but displayed a heightened frequency of both locomotory and feeding activities (O'Riain *et al.* 1996). Braude (2000) compared the morphology of wild dispersing versus non-dispersing individuals, but the subterranean niche of mole-rats precluded the recording of behavioural data.

Potential dispersers in laboratory colonies had higher basal levels of plasma luteinizing hormone than non-dispersers, suggesting that they are physiologically primed for reproduction before leaving their natal burrow (O'Riain *et al.* 1996). Although few studies have directly measured the relative proportion of fat in dispersing versus non-dispersing vertebrates, it is frequently assumed that the significantly heavier body mass of dispersers reflects the amount of energy stored as fat (e.g. Holekamp 1984). Potential dispersers in captive naked mole-rat colonies had significantly higher percentages of total body fat than same-aged non-dispersers (Fig. 10.1). Animals with large energy reserves may better endure the transience stage from leaving home to establishing their own territory, especially where this is energetically costly. Thus the higher percentage of fat in dispersers most probably serves as a nutritional safeguard against starvation during dispersal and colony establishment. Growth trajectories (O'Riain and Jarvis 1998) of dispersers show a growth surge which is probably indicative of the onset of fat deposition and may serve as a means of predicting which individuals within colonies are potential dispersers. Braude (2000) was able to compare the morphology of successful dispersers with similarly sized workers by examining the trapping records from the year before they

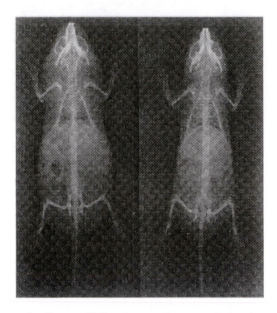

Figure 10.1. X-ray images of a disperser (left) and a non-disperser naked mole-rat of the same age and sex provide evidence for a distinct dispersal morph. Dispersers were skeletally similar to non-dispersers but had significantly greater deposits of subcutaneous fat. This fat reserve may serve as an important nutritional safeguard to offset the high costs associated with dispersal for a fossorial species. Reprinted from *Nature* (O'Riain *et al.* 1996).

dispersed. The relative fatness of dispersers and similarly sized workers was ranked in the nine colonies from which dispersers were known. The class of wild dispersers was significantly fatter than similar-length non-dispersing workers in these colonies.

Outbreeding and mate choice in dispersers

A problem that inevitably faces all dispersers is acquiring a mate or being accepted by a foreign colony and integrating with another social hierarchy. Naked mole-rats are known to be highly xenophobic (Lacey and Sherman 1991; O'Riain and Jarvis 1997); discrimination against foreign individuals and recognition of colony members appears to be achieved through the presence or absence of familiar colony odours (O'Riain and Jarvis 1997). However, when given a choice, potential dispersers in the laboratory chose to associate with foreigners significantly more frequently than colony members. In addition, dispersers showed a consistent preference for foreign breeding females over foreign non-breeding females. Interestingly, foreign breeding females showed no aggression to dispersers (or to foreign, reproductively active males) when paired after the choice experiment. Indeed, they readily consorted with these males, engaging in mutual naso-anal nuzzling, although no copulatory acts were observed. This was in contrast to non-breeding females, who responded aggressively and attempted to flee their advances. In addition, Ciszek (2000) has found strong preference for non-relatives in long-term mate choice experiments.

Table 10.2. Long-term life history studies of wild colonies of naked mole-rats reveal the frequency of occurrence of dispersal between established colonies. The table includes data on the sex, colony of origin, and distance travelled by dispersers

Animal no. (sex)	Moved from colony (in year)	Moved to colony	Last year captured in natal colony	Distance travelled[†]
2035 (M)	X	DD	1989	360 m[†]
3148 (F)	HH	B	1991	100 m[†]
3531 (F)	B	KK	1991	1.2 km
3712 (M)	AA	KK	1991	2.6 km
3595 (F)	C	B	1992	200 m[†]
4133 (F)	KK	EE	1992	650 m[†]
5093 (M)	D	SS	1994	1.4 km
6699 (M)	KK	QQ	1995	700 m
7184 (M)	HH	SS	1996	1.1 km[†]
7194 (F)	HH	B	1996	100 m[†]

[†]Adjacent colonies.

In three different cases documented in the laboratory, foreign males not only were accepted by a colony, but became the reproductive males. O'Riain *et al.* (1996) suggested that disperser males would attempt to join established colonies. In the field there are also 10 cases of animals dispersing between colonies (Table 10.2). These animals are of both sexes, and have moved between adjacent colonies or travelled longer distances, thus having a significant impact on gene flow. Of these animals at least one (male no. 5093) has achieved reproductive status 1 year after dispersing into the new colony.

Nascent colonies appear to be started by one or a few animals who then attract animals who have dispersed from other colonies. Nine females and one male were each found alone in small isolated burrow systems, apparently waiting for mates to join them (Table 10.1). A number of these females were vaginally perforate (reproductively active) at the time they were captured. Several were retrapped within a few days or weeks, confirming that they were alone. However, none was found to have attracted a mate between the first and subsequent trappings.

If solitary dispersers frequently start new colonies alone, how then do they attract mates? Fossorial rodents are typically sensitive to seismic vibrations (Nevo and Reig 1990) and often communicate by thumping with their hindlimbs on the floor of the burrow (Narins *et al.* 1997). Neither captive nor wild naked mole-rats have been observed to thump (Lacey *et al.* 1991; Pepper *et al.* 1991), but they are nevertheless very sensitive to seismic signals. It is possible that naked mole-rats can hear the burrowing activities of neighbouring colonies in the wild and dispersers could thus use this as as cue for locating unrelated conspecifics. Burrowing necessitates the animal gnawing with incisors at very hard soil faces, which generates considerable seismic vibrations. Indeed, it is possible for humans to hear underground digging activity by mole-rats from distances as great as 20 metres. Seismic vibrations might therefore serve as an important cue of the proximity of a foreign colony to a dispersing mole-rat. It is further apparent from field studies that naked mole-rats frequently open holes on to the surface along old animal tracks or roads, through which they eject debris excavated from their burrow systems (Braude 1991). Dispersers might be able to find other colonies or lone mole-rats by following these tracks and then cueing in on odours that emanate from the debris ejected by a foreign colony. O'Riain and Jarvis (1997) have shown that naked mole-rats are highly successful at discriminating their own versus foreign colony odours, and that dispersers show a significant preference for foreign odours. D. Ciszek (personal communication) has noted that captive animals often expel debris from their toilet chambers and suggests that lone females could advertize their presence to passing males by kicking up faeces and urine-soaked soil.

Regardless of the mechanism for finding a mate, only one of the nascent colonies with more than one animal contained exclusively marked adults from the same natal colony (colony VV, Table 10.1), while seven contained a mixture of marked resident and unmarked immigrant adults, and one contained marked breeders from different natal colonies (Table 10.1). Thus, naked mole-rat dispersers are likely to be outbreeding in at least 89% of the cases where we have information about the origins of breeders. It is also possible that the female of colony VV was avoiding mating with her brothers and was waiting for a non-related male to join them, as may have happened in colonies CC and HHH (Table 10.1). These data provide the first evidence for dispersal coalitions in naked mole-rats, whereby a few individuals disperse from their natal colony and join forces in order to increase their ability to become established away from their natal territory. Similar patterns of dispersal have been observed for other social mammals, for example lions (Bertram 1978; Packer and Pusey 1993) and suricates (Clutton-Brock *et al.* 1999).

Genetic variation in naked mole-rat populations

When successful, dispersers should contribute significantly to gene flow. Why, then, have various genetic studies (Faulkes *et al.* 1990; Reeve *et al.* 1990; Honeycutt *et al.* 1991) found that there is little genetic variation in naked mole-rats? The main reason is probably that a disproportionate number of the animals in those analyses was collected from the same small geographical area that has been isolated from other populations by a river (the Athi), which has served as a natural barrier to dispersal. Thus, despite the contention that localized wild populations of naked mole-rats have exhibited the highest known coefficient of inbreeding (Reeve *et al.* 1990), it is highly possible that this value was exaggerated by the geographical isolation of the population studied, and that its members are the descendants of a recent founder effect.

In fact, all of these molecular studies independently diagnosed the possibility of a relatively recent founder event that gave rise to this well studied population of naked mole-rats. Specifically, all of the naked mole-rats examined by Faulkes *et al.* (1990) came from this area, and they concluded that the low variation in the genes sampled is due to a population bottleneck. They also warned that any quantitative interpretation of their data would be invalid without data from complete families. Reeve *et al.* (1990) also noted the lower genetic variation in this area compared with the colony they sampled from north of the river and in contact with larger populations. They suggested that this is best explained by a population bottleneck and recent common ancestry.

Faulkes *et al.* (1997) obtained naked mole-rats from six locations north of the river, in addition to five colonies south of the river. They found less genetic variation in the colonies south of the river compared with other locations in their sample. Their analysis of mitochondrial loci showed that the population sampled south of the river must have a recent common maternal ancestor.

Low local genetic variation has also been noted north of the Athi river, but Faulkes *et al.* (1997) concluded that the low overall genetic variation found in the species is due to isolation by distance and limited gene flow. This is exactly the pattern found in other fossorial rodents (Lidicker and Patton 1987).

Confusion also exists over the term 'inbreeding', which has been used to describe an array of very different biological phenomena (Jacquard 1975; Templeton and Read 1994). Reeve *et al.* (1990) suggest that the genetic structure of the Mtito Andei population is equivalent to a population in which inbreeding has occurred in 80% or more of matings. They suggest that inbreeding is assumed to result from brother–sister or parent–offspring mating. However, their F statistic is an estimate of 'pedigree inbreeding' rather than the 'system-of-mating inbreeding' (cf. Templeton and Read 1994). These two terms are measures of very different biological processes. The F_p estimated by Reeve *et al.* (1990, p. 2499) 'is the probability that two alleles randomly drawn from an individual are identical by descent'. System-of-mating inbreeding measures deviation from Hardy–Weinberg equilibrium in a population due to violation of the assumption of random mating. It is important to note that system-of-mating inbreeding is a population parameter, not a characteristic of individuals. Pedigree inbreeding is a characteristic of individuals of known pedigree and can vary among individuals in a population. Pedigree inbreeding cannot be known without pedigree information, does not require inbreeding as the system of mating, and has a range of $0 \leq F_p \leq 1$, whereas system-of-mating inbreeding varies from $-1 \leq F_{ST} \leq 1$.

These two parameters can differ greatly, especially for a population with a recent founder event. For example, Templeton and Read (1994) found that inbreeding as a measure of the mating system in captive Speke's Gazelles was -0.291, while the pedigree inbreeding for the same population was 0.1490. Even when the mating system is avoidance of inbreeding ($F_{ST} < 0$), the pedigree inbreeding may be high if the effective population size is small. Although naked mole-rats may live in colonies of up to 300 animals, the number of breeding individuals, and thus the effective population size, is far smaller (Reeve *et al.* 1990). Thus the pedigree inbreeding tells us little about the system of mating for naked mole-rats. Although it is rarely cited, Reeve *et al.* (1990) have concluded that the genetic homogeneity in their sample is best explained by recent common ancestry due to a viscous population structure. This would result in a high F_p, but not necessarily a high F_{ST}. Furthermore, they point out that outbreeding must occur some of the time. This is in complete agreement with our results from both captive and wild colonies.

It is our suggestion that, despite the existence of dispersers within the population, the enormous costs and risks associated with dispersal limit the successful establishment of new colonies from pairs or small groups of individuals. Of all the nascent colonies found by Braude in North Kenya, only one survived for more than 1 year. More favourable environmental conditions associated with increased rainfall may temporarily improve the success of dispersers and the establishment of small nascent colonies. However, for the most part, the harsh ecological conditions are likely to favour philopatry and kin cooperation, with limited outbreeding opportunities and consequent reduced costs of inbreeding thus effectively restricting a large component of mate choice to those within each colony (Jarvis *et al.* 1994). In addition, any matings between members of neighbouring colonies are likely also to represent a degree of inbreeding. Ultimately, the low vagility of fossorial rodent populations leads to increases in the genetic homogeneity of local demes (Shields 1982), which may further explain the high inbreeding coefficients recorded for naked mole-rats within specific geographical localities. Gene flow into a colony with a small effective population size may produce a founder effect, thus dramatically increasing heterogeneity and maintaining selection for dispersal in a viscous population.

Social considerations of inbreeding and outbreeding

Philopatry and inbreeding may be adaptive, due to higher average survivorship of juveniles (through access to parental resources), accessibility to potential mates (siblings and other close relatives), and the maintenance of locally adapted gene complexes (Shields 1987). However, while there is no evidence to suggest that naked mole-rats suffer from the potentially deleterious effects of inbreeding, close inbreeding cannot continue indefinitely (Chesser and Ryman 1986). Although in the medium term (e.g. after 50 generations), chronic local inbreeding may be advantageous in diploid animals (because of reduced genetic load), it is likely to become detrimental for a variety of reasons. These include an accumulation of genes with mild deleterious effects (Werren 1993) and loss of the ability to adapt to changing environments (Maynard Smith 1978b). Indeed, Tyson (1984) defined the parameter D_n, which is the genetic bias for desertion, or antisocial behaviour, in the nth generation of inbreeders. After this point is reached, there would be selection against altruistic alleles that favour cooperation within genetically closed groups and selection for selfish dispersers.

Inbreeding may remain a necessary trade-off between the high somatic fitness costs associated with dispersal for a poikilothermic subterranean mammal and the limited opportunities for reproduction and outbreeding available at the natal site. More favourable environmental conditions (e.g. an increase in rainfall) may serve to promote both the frequency and survival of those dispersers in nascent colonies. Different combinations of morphological, physiological, and behavioural characteristics, which at a proximate level result in certain individuals showing dispersal tendencies, may serve ultimately to ensure optimal levels of outbreeding within localized populations.

11

Multiple causes of the evolution of dispersal

Sylvain Gandon and Yannis Michalakis

Abstract

Dispersal evolves under the action of opposing forces. Some factors, like the cost of dispersal, select against dispersal. Several other factors, however, select for dispersal. In this chapter we focus on three main factors: (1) the temporal variability of the environment will often select for dispersal; (2) dispersal may also be adaptive if it reduces competition between relatives; and (3) dispersal may represent a way to escape the cost of inbreeding. This simple qualitative description has heuristic value but is not sufficient to understand the evolution of dispersal when the effects of several factors are considered simultaneously. First, when several factors are acting, some of their interactions may complicate our simplistic qualitative views. Second, quantitative predictions are needed to evaluate the relative importance of a given factor. In this chapter we present the results of a model of the evolution of dispersal that incorporates the effects of: the cost of dispersal, the extinction rate of populations, the average coefficient of relatedness among individuals, and the cost of inbreeding. This formalization yields analytical expressions or numerical solutions for the evolutionarily stable (ES) dispersal rate. These results are very useful to make both qualitative and quantitative predictions. First, some qualitative non-intuitive results do indeed emerge from the interactions between the effects of several factors. For example, in some cases, the ES dispersal rate can increase with higher costs of dispersal. Second, an analysis of the sensibility of the ES dispersal rates allows us to make a hierarchy of the importance of the effects of the different factors. In particular, population extinctions have a higher impact on the evolution of dispersal than kin competition or the cost of inbreeding. The need for models that incorporate the effects of multiple causes of dispersal is discussed.

Keywords: dispersal, metapopulation, kin competition, extinction, inbreeding depression

Introduction

Dispersal is a central trait in ecological and evolutionary processes. It affects both population dynamics (movement of individuals) and population genetics (gene flow) of species. Reciprocally, the dynamics and the genetics of species populations are also likely to act on the evolution of dispersal. In this chapter we focus on the multiplicity of factors that may affect the evolution of dispersal, and which unavoidably leads to some level of complexity. As a first approximation, the effects of these multiple factors may be described by a simple qualitative (cost–benefit) argument: the evolution of dispersal results from a balance between opposing forces. On one hand, some factors may select against dispersal. For example, dispersers may incur a cost, due to either increased mortality during the dispersal phase or disadvantages during the settling period in the novel habitat. Moreover, in a spatially heterogeneous environment, dispersal will incur an extra cost because it will generally lead to bad environments (Balkau and Feldman 1973; Hastings 1983; Holt 1985). All these factors are often pooled and denoted by a single parameter: the cost of dispersal. On the other hand, several factors may select for dispersal. In this chapter we focus on three main factors. (1) Dispersal may be adaptive if it reduces competition between relatives (Hamilton and May 1977; Motro 1982a,b; Frank 1986; Taylor 1988). In this case, dispersal can be viewed as an altruistic trait: the cost of dispersal is balanced by the benefit, due to the reduced competition, experienced by related individuals. (2) The temporal variability of the environment also selects for dispersal. It is worth pointing out the difference between local temporal variance and global temporal variance (Venable and Brown 1988). Local temporal variance characterizes the temporal variation in demographic success in a single patch. For example, this variation may generate some between-populations variation in density, and dispersal may evolve in order to avoid crowding. In particular, the local extinctions of populations (an extreme case of local temporal variability) favour dispersal, because only dispersed offspring will be able to recolonize empty sites (van Valen 1971; Comins *et al.* 1980; Olivieri *et al.* 1995; Gandon and Michalakis 1999). Global temporal variance refers to the variation in the demographic success of all populations. For example, the proportion of extinct populations may vary from one year to the next. Such variation selects for risk-reducing (bet-hedging) strategies such as dispersal (Venable and Brown 1988). Indeed, dispersal allows one to escape the risk of low success in a particular site. (3) Finally, dispersal may represent an efficient way to avoid inbreeding depression (Bengtsson 1978; May 1979; Shields 1982; Waser *et al.* 1986; Motro 1991; Perrin and Mazalov 1999, 2000; Gandon 1999; Perrin and Goudet, chapter 9).

This simple qualitative description has heuristic value but is not sufficient to understand the evolution of dispersal under the joint and simultaneous influence of several factors, first, because interactions between the various factors may complicate our simplistic qualitative view of the evolution of dispersal, and second, because some quantitative predictions are needed in order to evaluate the relative importance of each factor on the evolution of dispersal. It is thus necessary explicitly to formalize the interactions of several factors.

In this chapter we analyse the effects of multiple factors on the evolution of juvenile dispersal. For the sake of simplicity, we focus on the case where dispersal is unconditional (i.e. does not depend on the state of the individual or on the state of the habitat) in a semelparous species (i.e. adults always die after reproduction). Under these assumptions, we present a model of the evolution of dispersal that incorporates, step by step, the effects of: the cost of dispersal (c), the extinction rate (e), the average coefficient of relatedness among individuals (R), and the cost of inbreeding (δ). This formalization is based on the direct fitness approach developed by Taylor and Frank (1996) and Frank (1998). It provides analytical expressions of the evolutionarily stable (ES) dispersal rates which allow us to make both qualitative and quantitative predictions. First, some qualitative non-intuitive results emerge from the interactions between the effects of several factors. For example, in some cases, the ES dispersal rate can increase with higher costs of dispersal. Second, quantitative results regarding the ES dispersal rates allow us to point out a hierarchy of the importance of the effects of the different factors. This analysis provides an alternative to the classical, and often misleading, cost–benefit view of the evolution of dispersal. The implications of these analyses for empirical approaches to the study of the evolution of dispersal are discussed.

From simple to complex models of the evolution of dispersal

General life cycle

The following life cycle will be assumed throughout this chapter. (1) the habitat is filled with an infinite number of populations. Each population contains an equal number, $2N$ (N males and N females), of diploid individuals. (2) Populations can go extinct with a probability e, and their habitats are immediately recolonized. Note that, since we assume an infinite number of populations, there is no global temporal variance (Venable and Brown 1988). Indeed, the proportion of extinct populations will always be equal to e. Therefore, extinctions induce only local temporal variation. This variation selects for dispersal through the recolonization of empty sites. (3) Females produce a very large number of offspring with a 1:1 sex ratio. The expected number of offspring for each female is the same. (4) The adults die. (5) Male and female offspring leave their natal population with a probability d (island model of dispersal). Dispersal is assumed to be under the control of the offspring and determined by a single locus. (6) Dispersing individuals (male or female) incur a cost of dispersal, c, during the dispersal phase or during the settling period. (7) Mating takes place randomly between juveniles (dispersers or philopatric) and the breeding pairs compete for the N breeding sites in the population. (8) The cost of inbreeding is paid during this competition phase: the breeding pairs formed by two philopatric individuals have a lower chance $(1 - \delta)$ than any other pair of settling successfully in the population. This all-or-nothing response is different from the classical definitions of inbreeding depression (e.g. a cost on the fecundity of inbred females). However, this definition greatly simplifies the algebra, since all the different breeding pairs have the same expected number of offspring (assumption 3). The way we model the cost of inbreeding is assumed to be an approximation of the effect of inbreeding depression. The difference between the cost of inbreeding and inbreeding depression will be discussed later in this chapter (9). After competition, each population contains N mating pairs (N males and N females).

Under these assumptions, we derive the ES dispersal rate using the 'direct fitness' approach developed by Taylor and Frank (1996). In the following, we will not detail the derivation of the results; these derivations are fully exposed in the papers we refer to. Instead, we prefer to focus on the results themseleves and discuss the effects, and the interactions, of four factors on the ES dispersal rates: (1) the cost of dispersal; (2) the level of kin competition; (3) the probability of extinction; and (4) the cost of inbreeding.

Simple models

We review the main results of simple models, when only one of the three factors selecting for dispersal is acting.

Kin competition

First, we assume that kin competition is the unique cause of dispersal evolution (i.e. no extinctions and no cost of inbreeding). Frank (1986) formalized the effect of kin competition on the evolution of dispersal in a very general way. He obtained the following expression for the ES dispersal rate:

$$d^* = \frac{R - c}{R - c^2} \tag{1}$$

Note the generality of this expression. Several factors act only through their effects on relatedness. For example, when each population supports only one breeding pair (i.e. $N = 1$), and when dispersal is under maternal control, $R = 1$, which leads to the classical result of Hamilton and May (1977): $d^* = 1/(1 + c)$. Taylor (1988) further showed that expression (1) holds under a variety of assumptions (i.e. different ploidy level, different life cycle, maternal or offspring control of dispersal). However, it is important to note that relatedness, R, is not a fixed parameter but a dynamical variable that depends on several factors including dispersal itself. Indeed, relatedness can be expressed in a general way as a function of the probabilities of identity in state between homologous genes in randomly chosen individuals (F. Rousset and S. Billard, unpublished manuscript). If we further assume that the locus that determines the dispersal behaviour has an infinite number of alleles, relatedness reduces to: $R = f_{XY}/f_{XX}$ (Michod and Hamilton 1980), where f_{XY} is the coefficient of consanguinity between juveniles X and Y, and f_{XX} is the inbreeding coefficient (Crow and Kimura 1970, p. 68). More precisely, f_{XY} and f_{XX} are the probabilities of identity between homologous genes, first, in two randomly chosen juveniles (X and Y) in the same population before dispersal, and second, in the same individual (X). These different coefficients depend on dispersal. For example, when there is no dispersal, the probability of identity between two juveniles is equal to 1. At the other extreme, when all juveniles disperse, the probability of identity between juveniles before dispersal is equal to $1/4N$. An 'unwound' (Frank 1998, p. 120) solution for the ES dispersal rate needs to incorporate the co-evolution between dispersal and relatedness, which yields (Taylor 1988):

$$d^* = \frac{H + 1 - 4Nc}{H + 1 - 4Nc_2} \quad \text{with} \quad H = \sqrt{1 + 8N(2N - 1)c^2} \tag{2}$$

This expression indicates that larger population sizes always select for lower dispersal rates.

Extinction of populations

Let us now assume that there is no kin competition. This may represent the limit case in which population sizes are very large and, consequently, individuals are not related (i.e. $R \to 0$ as $N \to \infty$). Under this simplifying assumption, we analyse how population extinctions affect the evolution of dipersal. Comins *et al.* (1980) showed that the ES dispersal rate depends in the following way on the extinction rate:

$$d^* = \frac{e}{1 - (1 - c)(1 - e)} \tag{3}$$

This expression collapses to van Valen's (1971) result, $d^* = e$, when we further assume that $c = 1$.

Cost of inbreeding

Finally, we consider the case where there is no kin competition or population extinction (i.e. $R = 0$ and $e = 0$). Gandon (1999) showed that, when only inbreeding depression acts, the ES dispersal rate is:

$$d^* = \frac{\delta - c}{\delta - c^2} \qquad (4)$$

Note the analogy between this expression and expression (1). This analysis has heuristic value, but it is unreasonable to consider the effect of the cost of inbreeding under the assumption that individuals are unrelated (because only consanguineous mating pairs pay the cost of inbreeding). As for relatedness, one could argue that the cost of inbreeding depression is not a fixed parameter but a dynamical variable that depends on inbreeding coefficient (the coefficient of consanguinity between parents). As shown by Gandon (1999) and, with slightly different assumptions, by Perrin and Mazalov (2000) and Perrin and Goudet (chapter 9), the effects of kin competition and the cost of inbreeding are intimately linked and need to be taken simultaneously into account. We therefore discuss this interaction in a subsequent section.

All these three different cases quantify the effects of c, R, e, and δ. Not surprisingly, R, e, and δ select for a higher dispersal rate, and c always selects for lower ES dispersal rates. Moreover, the above expressions of the ES dispersal rates – expressions (1) to (4) – allow us to quantify the relative importance of each of these factors. R and δ have exactly the same effect on d^*. However, population extinctions have a larger effect since, for example, $d^* > 0$ when $e > 0$, whatever the cost of dispersal. This is not the case for R and δ since some dispersal rate is selected for only if $R > c$ or $\delta > c$.

2.3. Multiple causes of dispersal

In the following we present more complex results, for the cases where more than one factor selects for dispersal. As kin competition always occurs in a structured environment, we only present models that combine the effects of kin competition with one other factor. We first discuss the case where there are extinctions but no cost of inbreeding ($\delta = 0$). In a second step we study the case of stable habitats ($e = 0$) where there is some cost of inbreeding. We finally discuss briefly the model in which all three factors are acting together.

Kin competition and extinctions

The effects of kin competition and extinction on dispersal evolution were first studied by Comins *et al.* (1980). Gandon and Michalakis (1999) studied a very similar model and derived an analytical expression for the ES dispersal rate:

$$d^* = \frac{A - \sqrt{A_2 - 4eB}}{2B} \qquad (5)$$

with

$$A = c + e^2(1 - c) + e - R(1 - e)$$
$$B = (c + e(1 - c))^2 - R(1 - e)$$

The above expression generalizes (1) and (3). The qualitative effects of each factor on the ES dispersal rate are similar (i.e. d^* increases with R and e). It is interesting to note, however, that higher costs of dispersal do not always select for lower dispersal rates. In particular, when some extinctions occur and when relatedness is very high, higher costs of dispersal select for higher dispersal. This effect has already been noted by Comins *et al.* (1980), but Gandon and Michalakis (1999) showed that it could be explained by a kin selection argument. Indeed, a higher cost of dispersal increases the probability of competing with a philopatric individual (because the immigration rate decreases) and thus increases the level of kin competition. When extinctions occur, the recolonization of empty sites offers an opportunity to avoid the increased cost of kin competition imposed by the higher cost of dispersal. This opportunity exists as long as colonizers are not themselves highly related and selects for a higher dispersal rate with higher cost of dispersal when some extinctions occur. This effect illustrates the interaction that may emerge from a combination of different factors acting on the evolution of dispersal (here, between kin competition, the cost of dispersal, and population extinctions).

Equation (5) was obtained assuming relatedness as a fixed parameter (Fig. 11.1). Gandon and Michalakis (1999) studied the case where relatedness is a dynamical variable (for a haploid organism) depending on all the other parameters of the model

(a) *R* is a fixed parameter:

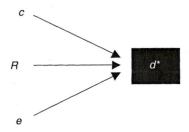

(b) *R* is a dynamical variable:

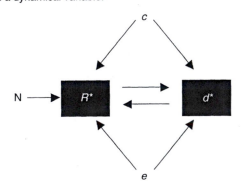

Figure 11.1.   Direct and indirect effects of various parameters on the evolution of dispersal. When relatedness is used as a fixed parameter (a), we study the direct effects of parameters on the evolution of dispersal. When relatedness is used as a dynamical variable (b), we can study both the direct and the indirect effects of parameters on the evolution of dispersal.

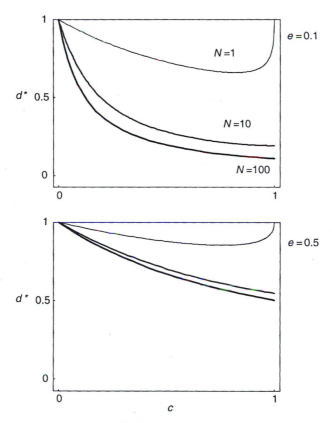

Figure 11.2. Numerical solutions of the ES dispersal rate (d^*) versus the cost of dispersal for asexual organisms for different sizes of the population: $N = 1$ (light), $N = 10$ (medium), and $N = 100$ (bold). On the upper graph $e = 0.1$, and on the lower graph $e = 0.5$.

(dispersal, cost of dispersal, extinction rate, population size). In this case there is no simple analytical expression, but the results can be studied numerically. Some results are presented in Fig. 11.2, which shows that most qualitative results hold. In particular, higher extinction rates always select for higher dispersal rates.

Kin competition and the cost of inbreeding

The evolution of dispersal that includes kin competition and inbreeding depression, for haploid organisms when there is only one breeding pair ($N = 1$) per population, was first studied by Motro (1991, 1994). Gandon (1999) generalized Motro's model for diploid individuals and any population size. Unfortunately, there is no simple analytical expression of d^* as a function of c, δ, and R (but see Gandon (1999) for analytical expressions in some limit cases). The effect of each factor can be studied through numerical calculation of the ES dispersal rates. Higher R and δ values always select for higher ES dispersal rates. However, higher costs of dispersal may select for higher dispersal when both R and δ are large. Again, this counter-intuitive effect results from the interaction between the different factors involved in the evolution of dispersal. Higher costs of dispersal decrease the immigration rate and, consequently, increase the risks of both mating (cost of inbreeding) and competing with related individuals.

In a second step, it is important to consider the case where R and δ are dynamical variables (Fig. 11.3). Here, the cost of inbreeding, δ, depends on the genetic load, Δ (the number of lethal equivalents), and on the coefficient of consanguinity between mates, g, in the following way: $\delta = 1 - e^{-\Delta g}$ (Gandon 1999). It is difficult to make intuitive predictions concerning the evolution of dispersal because relatedness, consanguinity, and the cost of inbreeding are intimately linked. A higher coefficient of consanguinity between mates, g, increases the cost of inbreeding because more related mates pay higher costs of inbreeding (Crow and Kimura 1970). Reciprocally, a higher cost of inbreeding decreases both relatedness, R, and consanguinity between mates, g. Indeed, the cost of

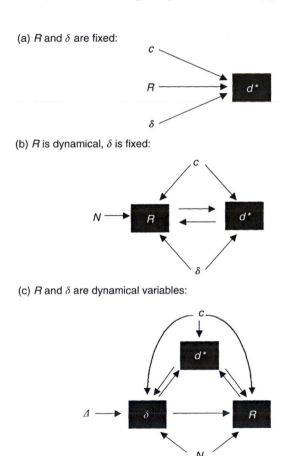

(a) R and δ are fixed:

(b) R is dynamical, δ is fixed:

(c) R and δ are dynamical variables:

Figure 11.3. Direct and indirect effects of various parameters on the evolutionarily stable dispersal rate, d^*. In (a), relatedness, R, and inbreeding depression, δ, are used as fixed parameters. In (b), relatedness is a dynamical variable and co-evolves with dispersal. In (c), both relatedness and inbreeding depression are dynamical variables that depend on various parameters (c, cost of dispersal; N, population size; Δ, number of lethal equivalents).

inbreeding increases the probability that immigrants successfully reproduce. Therefore, a higher cost of inbreeding increases the effective gene flow between populations and, consequently, decreases relatedness and consanguinity. The quantitative analysis of this model is presented in Gandon (1999). If relatedness is used as a dynamical variable, the above qualitative results generally hold (Fig. 11.4). However, note that when R and δ are dynamical variables, the ES dispersal rate always decreases with higher costs of dispersal (see Fig. 11.4).

The above approach would naturally lead to the full model where kin competition, extinctions, and the cost of inbreeding are all acting simultaneously on the evolution of

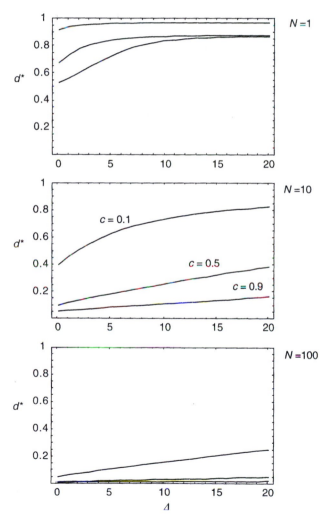

Figure 11.4. Evolutionarily stable dispersal rate, d^*, versus the number of lethal equivalents, Δ, when both relatedness, R, and inbreeding depression, δ, are used as dynamical variables. In (a), $N = 1$, in (b), $N = 10$, and in (c), $N = 100$. For each panel, three different values of the cost of dispersal are considered ($c = 0.1, 0.5,$ and 0.9).

dispersal. Several complications emerge, however, in the formalization of such a model. In particular, since relatedness within populations changes with the age of the population (i.e. time since foundation; Whitlock and McCauley 1990; Gandon and Michalakis 1999), the cost of inbreeding would also vary with the age of the population. However, the relevant measure of the cost of inbreeding, contrary to that of relatedness, is not a simple average over the different populations. A solution of the full model thus becomes analytically intractable and even numerically difficult. A simulation project, which still remains to be done, seems to be a possible alternative.

Discussion

What is the ultimate cause of dispersal? There is no single answer to this question (Dobson and Jones 1985). In this chapter, we focus on the effects of the three main causes of the evolution of dispersal: (1) avoidance of kin competition, (2) population extinctions, and (3) avoidance of the cost of inbreeding. Using a general model we discuss different cases when one or two of these factors are acting. This formalization is useful in several ways.

First, it provides qualitative results which allow us to test the validity of verbal arguments. For example, we found that higher relatedness, higher extinction rates, and higher costs of inbreeding always select for dispersal. This qualitative analysis also revealed some counter-intuitive results, especially when several factors interact. For example, when some extinctions occur and when relatedness and the cost of dispersal are high, the ES dispersal rate may increase with higher costs of dispersal. This effect had already been noticed by Comins *et al.* (1980), but the model we present above allowed us to explain this effect by a kin selection argument. It is worth pointing out, however, that, apart from this, the results we present revealed no qualitative interactions between the different factors.

Second, this analysis leads to quantitative predictions which allow us to study the relative importance of each factor in the evolution of dispersal. For example, we showed that relatedness and the cost of inbreeding affect the ES dispersal rate in similar ways: compare equations (1) and (4). The extinction rate acts differently and more strongly on the evolution of dispersal.

Finally this formalization helps to understand the processes involved. In particular, it highlights the fact that some factors are covarying (e.g. R and δ) and that variation in one of them may feed back through the effects of the others (Fig. 11.3). These general models may thus be very useful because they help to expand our intuitive understanding of the different processes involved in the evolution of dispersal.

However, the generality of the models we used comes at the cost of reduced realism. These models cannot lead to any testable quantitative predictions on the ES dispersal rate of specific biological models even though they can be used in comparative analyses. Our aim in this chapter has been to present a general framework which may then be adjusted to particular biological systems after relaxing some (or many) simplifying assumptions. In the following, we present the main assumptions which would need to be relaxed in more realistic models. Although this enumeration might be laborious, we think it is necessary since, as we saw in previous examples, even simple interactions among factors revealed unexpected results (e.g. the increase of ES dispersal rate with higher cost of dispersal).

In some cases, relaxing such assumptions might not qualitatively affect the results. For instance, most models using the kin selection formalism make the unrealistic assumptions, for mathematical convenience, of an infinite number of populations and an island mode of migration. It turns out that these assumptions only weakly affect the ES dispersal rates and do so only when the cost of dispersal and the size of the populations are small (Gandon and Rousset, 1999). Another assumption, common to most models of kin selection, is the absence of age structure (i.e. semalparous life cycle). This simplified life cycle differs greatly from that of many biological examples, but, again, this simplifying assumption affects only weakly the results regarding the ES dispersal rates in the absence of local extinctions (O. Ronce, S. Gandon, and F. Rousset, unpublished manuscript).

In other cases, however, relaxing the assumptions might potentially have a large effect. In the above model, we noted that the cost of inbreeding has been formalized in a simplistic way in order to avoid some mathematical complications. Indeed, we assumed that the cost of inbreeding acts before reproduction (inbred mating pairs are less likely to settle in the population) and that all females have the same number of offspring. This is very convenient because there is no variance in the number of offspring produced in the different types of populations (a type is defined by the proportion of inbred mating pairs). Relaxing this hypothesis would affect both the calculation of relatedness and the derivation of the ES dispersal rate, but it would model the cost of inbreeding in a more realistic way (i.e. closer to inbreeding depression). Another challenge would be to assume that the genetic load is a dynamical variable depending on several factors including dispersal itself. This would be analogous to the work on the co-evolution between selfing rates and inbreeding depression (Lande and Schemske 1985; Holsinger 1988, 1991; Charlesworth *et al.* 1990; Uyenoyama *et al.* 1993).

An analysis of the full model where kin competition, the cost of inbreeding and the extinction rate are acting together on the evolution of dispersal remains to be done. Indeed, the complex interaction between these three factors may reveal some unexpected interactions.

Another gap in the above model is the assumption of homogeneity of the populations. (Note, however, that when some extinctions occur, some populations are empty, and that relatedness varies with the age of the population, but only the averaged relatedness matters.) Classically, spatial variation is known to select against dispersal (Balkau and Feldman 1973; Hastings 1983; Holt 1985) but the interaction of spatial variation with other factors has been studied only in the case of temporal variability (McPeek and Holt 1992). What happens when kin selection and spatial variation are acting simultaneously?

We emphasize the importance of introducing some between-population variation (in quality and/or in size). This variation is a necessary condition for the evolution of plastic behaviours. Therefore, in such models, it would be particularly interesting to analyse the evolution of dispersal strategies that could be conditional on the state of the individual (the sex (Pusey 1987; Perrin and Mazalov 1999, 2000; Gandon 1999; Perrin and Goudet, chapter 9), or the age (Ronce *et al.* 1998)) or on the state of the population (age, density, etc.).

The effect of the demography has largely been neglected in the present model. Indeed, since we assume that each female can produce a very large (infinite) number of offspring, the carrying capacity of each population is assumed to be reached in only one generation. Therefore, even when extinctions and recolonization occur, there is no variation in the size of the different populations. This variation is known greatly to affect the evolution of dispersal and other traits (Olivieri *et al.* 1995; Ronce and Olivieri 1997; Ronce *et al.* 2000). In particular, these effects of the demography can interact with other factors in a very non-intuitive way (e.g. population extinctions can select against dispersal; O. Ronce, F. Perret, and I. Olivieri, unpublished manuscript). However, the interaction between demography and kin selection has not yet been studied. Undoubtedly, the link between demography and kin selection is of paramount importance for the understanding of the evolution of dispersal and, more generally, for the understanding of kin selection processes.

All the above modifications of our simplified approach would surely lead to a better understanding of the evolution of dispersal. However, such models are often highly complex because many factors are dynamical variables (e.g. R and δ). Even extinction rates can be a dynamical variable if the probability of extinction depends on some property of the populations, such as level of inbreeding (Saccheri *et al.* 1998) or population size (Hanski 1985; Gyllenberg and Hanski 1992). Such dynamical variations are often neglected in verbal arguments (where different factors are assumed to be fixed quantities). The formalization adopted in this chapter is a warning against a simplistic view of the evolution of dispersal. A change in a parameter may affect the evolution of dispersal through very different direct and indirect ways (e.g. the cost of dispersal). However, when models become more realistic, it becomes more and more difficult to estimate the relative importance of each factor in the evolution of dispersal since all the factors may affect one another.

We believe, however, that general and simple models of the type discussed here may be very helpful to understand the processes in action, and that they may actually lead to some broad-scale empirical predictions. Indeed, it is possible to compare the dispersal rates across different biological systems, where parameters such as population size, the cost of dispersal, or extinction rates are known, and see whether the observed patterns follow qualitative predictions. Our models have, furthermore, shown that some synthetic variables, such as relatedness, R, and inbreeding depression, δ, could be highly useful in such an approach.

Finally, even though in this chapter we have focused on the evolution of dispersal strategies, we are obviously convinced that dispersal may in turn affect the evolution of other traits. Reciprocally, these traits may also influence the evolution of dispersal. An obvious example, elaborated further by Ronce *et al.* (chapter 24), is the potential co-evolution between dispersal and selfing rates. Indeed, both traits may affect relatedness and the cost of inbreeding. A related line of inquiry would be to study the co-evolution between different ways developed by species to avoid the cost of inbreeding. Dispersal is a very efficient way to avoid inbreeding, but some kin recognition (Pusey and Wolf 1996) would allow avoidance of both inbreeding and the cost of dispersal. In plants, a way to avoid kin competition is delayed germination (i.e. dormancy) (Ellner 1986, 1987; Venable and Brown 1988; Nilsson *et al.* 1994; Hyatt and Evans 1998; Kobayashi and Yamamura 2000). How would dormancy co-evolve with dispersal? In a more general way, how does dispersal co-evolve with 'substitutable' traits (Venable and Brown 1988)? We are convinced that future theoretical and empirical studies investigating the evolution and occurrence of correlations between such traits will greatly improve our understanding of the evolution, and potentially the co-evolution, of such fundamental traits.

Acknowledgements

We thank J. Clobert, N. Perrin, O. Ronce and two anonymous reviewers for useful comments.

12

Parasitism and predation as causes of dispersal

Wolfgang W. Weisser, Karen D. McCoy, and Thierry Boulinier

In the past 30 years, much theoretical and empirical work has been devoted to exploring the relationship between intraspecific interactions and dispersal. As a consequence, kin competition, inbreeding, and demography have been shown to be major factors in the evolution of dispersal (see Gandon and Michalakis, chapter 11; Ronce *et al.*, chapter 24; Perrin and Goudet, chapter 9). In contrast, the effects of interspecific interactions have received considerably less attention. Although interspecific interactions such as predation, mutualism, parasitism, and competition find their way into studies of dispersal, for example as part of the environment to which an individual disperses (see Stamps, chapter 16; Danchin *et al.*, chapter 17 on habitat selection), to our knowledge no systematic attempt has been made to investigate the relationship between interspecific interactions and dispersal.

In the following two chapters, we discuss how parasitism and predation might affect the evolution of dispersal, and review the role that dispersal could play in these interspecific interactions. Emphasis differs between the chapters and reflects the fact that research efforts have not been devoted equally to different types of systems. In chapter 12a on parasitism, Boulinier *et al.* consider what is known about the impact of parasites on host dispersal decisions. Special reference is made to the intricate relationships between dispersal and transmission in host–parasite systems and, as a consequence, the association between parasite life cycles and opportunities for parasite dispersal is explored. In the subsequent chapter on predation by Weisser (chapter 12b), the influence of predation on dispersal is discussed, both for plant–herbivore and animal predator–prey systems, this chapter focusing on top–down effects.

The differences between the chapters point to the importance of detailed investigations into dispersal in predator–prey and host–parasite relationships, and our focus on only these two types of interactions further indicates the need for research on other types of interspecific interactions as causes of dispersal. Knowing whether the evolution of dispersal for one species is at least partly the result of interspecific interactions or, even more importantly, whether it is the result of a co-evolutionary process between dispersal types, is a necessary step in understanding the role of species dispersal on community stability and evolution (see van Baalen and Hochberg, chapter 21, and Mouquet *et al.*, chapter 22). In our reviews, we not only show that predation and parasitism interact with the process of dispersal in a number of interesting and unexpected ways, but also identify areas of research where progress can be made.

12a

Dispersal and parasitism

Thierry Boulinier, Karen D. McCoy, and Gabriele Sorci

Abstract

Host and parasite dispersal are among the most important factors affecting the evolutionary ecology of host–parasite interactions. Early epidemiological models have recognized the role of spatial processes in host–parasite interactions, and recent theoretical approaches have shown that host and parasite dispersal are key factors for the evolution of parasite virulence and host local adaptation. Conversely, from an empirical point of view, very little is known on the reciprocal role of parasite and host dispersal on the ecology and evolution of the association. Most knowledge on the role of dispersal in host–parasite interactions involves systems where there is a strong spatial subdivision of host and parasite populations. In particular, studies on colonial bird–ectoparasite systems have shown that differential levels of local infestation by parasites may affect dispersal among groups of hosts. In this chapter we review the studies that have investigated the connection between parasitism and host dispersal. We also focus on the interaction between parasite life cycle and host dispersal, and on how this might, in turn, affect parasite dispersal. Finally, we suggest that host and parasite dispersal should be considered to be interconnected traits, and that both drive the evolution of host–parasite interactions.

Keywords: co-evolution, epidemiology, habitat selection, immune defence, life cycle, meta-population, transmission

Introduction

Host–parasite associations involve long-term interactions between species that provide resources (hosts) and species that exploit these resources (parasites) (Combes 1995). Thus, parasites usually have fitness costs for their hosts, and the mechanisms that hosts have developed to avoid or resist parasitism will be related directly to the costs incurred. One of the mechanisms that hosts may use to lower the fitness consequences of parasitism is dispersal: individuals can leave areas where the risk of parasitism is high and attempt to settle on parasite-free habitat patches. Several studies have discussed the pathogenic effects of parasites on individual hosts, and a few have discussed the implications of this effect on host populations. However, there is a surprising lack of studies that have directly examined host dispersal patterns in relation to parasitism and have investigated the question of whether parasites affect host dispersal decisions.

It is very difficult (or even impossible) to discuss the effect of parasitism on host dispersal without considering differences among parasitic life cycles, host and parasite dispersion patterns, and the impacts parasites have on their hosts. In most instances, parasites will use their hosts to various degrees for dispersal; thus host dispersal will not always eliminate parasitism. Ectoparasites, for instance, can generally cover only short distances unless they are on their hosts. Similarly, parasites with indirect life cycles may have many more opportunities for dispersal when they move from the intermediate to the final host. This interdependence of host and parasite dispersal sets up a dynamic interaction where co-evolutionary outcomes will change depending on the pattern of the interaction in space and time and on the relative abilities for evolutionary change to occur.

In this chapter we review a few examples of the impact that parasites can have on host dispersal decisions. We then discuss the relationship between parasitic life cycles and their implications for dispersal of parasites among host populations. Finally, we discuss the dynamic nature of host–parasite interactions in an ecological and co-evolutionary framework. For all three aspects, we attempt to give a few brief illustrative examples involving animal–parasite systems. We end with a discussion of suggestions for future work in this area of research.

Parasites as a cause of host dispersal

Parasites are among the many factors that can affect the quality of habitat patches at a given point in time. The local presence of host-seeking parasites or of infested hosts with transmissible parasite stages may be a cause of host dispersal because (1) parasitism is usually associated with a negative impact on host fitness, and (2) it is likely to affect the temporal consistency of patch quality. Early epidemiological models have recognized the role of dispersal processes in host–parasite interactions (Bolker *et al.* 1995), and recent theoretical approaches have shown that host and parasite dispersal can be key factors affecting the evolution of parasite virulence and local adaptation (Gandon *et al.* 1996). Others have also pointed out that dispersal of hosts and parasites among subpopulations can potentially influence metapopulation dynamics (e.g., May and Southwood 1990). In a conservation context, there are concerns that corridors between favourable habitat patches may increase the dispersal of parasites and the extinction probability of sub-divided populations of endangered host species (Hess 1994). From an empirical point of view, however, very little has been published on the role of parasitism on host dispersal. In particular, this topic is barely addressed in several recent books and reviews about behaviour, ecology, and evolution of host–parasite systems (e.g., Barnard and Behnke 1990; Toft *et al.* 1991; Grenfell and Dobson 1995; Clayton and Moore 1997). This lack could be due, in part, to the practical difficulties associated with empirical studies of dispersal in host–parasite systems. Interesting examples exist of hosts possibly avoiding parasites by migrating or moving temporarily out of infested areas (e.g., Freeland 1980; Folstad *et al.* 1991), but such host movements do not represent dispersal *per se*. Most knowledge on the role of dispersal in host–parasite interactions involves systems where the spatial grouping of hosts is associated with relatively high and predictable local levels of parasite infestation (Rothschild and Clay 1952). In particular, studies on group-living vertebrates have provided insight on the importance of parasitism as a cause of host dispersal.

'Nest ectoparasites' and dispersal in colonial birds

In several colonial bird species, ectoparasites have been reported as a potential cause for colony abandonment (e.g., Feare 1976; King *et al.* 1977; Duffy 1983; Loye and Carroll 1991) and have been shown experimentally to have deleterious effects on the local reproductive success of their hosts (Moss and Camin 1970; Brown and Brown 1986, 1996; Møller 1987; Chapman and George 1991). In most of these cases, the parasites involved are haematophagous ectoparasites that spend a large part of their life cycle in the substrate of the breeding habitat of their hosts. They stay on the host for either a series of short blood meals (e.g., fleas, argasid ticks) or for one relatively long blood meal per developmental stage that lasts several days (e.g., ixodid ticks). These parasites may have a direct effect on the physiology and behaviour of their hosts through the injection of toxins, physical damage of tissues, and blood spoliation. They can also be the vectors of micropathogens, such as arboviruses and bacteria (e.g., Chastel 1988). For these reasons, ectoparasites may affect host reproductive success and survival, but also emigration rates. In this context, Brown and Brown (1992) specifically tested whether nestlings with higher parasite loads had a higher probability of dispersing from their natal colony than nestlings from non-infested nests. In their system, which involves the cliff swallow (*Hirundo pyrrhonota*) and the swallow bug (*Oeciacus vicarius*) they found that, as predicted, individuals coming from infested nests had a lower probability of recruiting in their natal colony than individuals that fledged from nests with no sign of these parasites. However, as this study was not experimental and did not involve the manipulation of local levels of infestation, one would still need an experimental test to ascertain the causal nature of the association observed. Strong deleterious effects of swallow bugs on cliff swallow chicks have nevertheless been shown by experimental removal of parasites by fumigation (Brown and Brown 1996).

Relationships between local levels of parasite infestation and host dispersal have also been investigated in a system involving the kittiwake (*Rissa tridactyla*) and the tick *Ixodes uriae*. The kittiwake is a pelagic seabird that breeds on marine cliffs in dense colonies numbering tens to tens of thousands of pairs. Its breeding season lasts from March to August at the Cap Sizun colony in Brittany, France, where a survey of a small population has permitted the collection of detailed demographic, behavioural, and epidemiological data over more than 12 years (Danchin and Monnat 1992). In particular, reliable information is available on kittiwake natal and breeding dispersal in relation to the local levels of breeding cliff infestation by the seabird tick *Ixodes uriae*. Observational and experimental evidence of a negative impact of tick infestation on host reproductive success and local demography has been provided (Boulinier and Danchin 1996; Danchin *et al.* 1998a; T. Boulinier *et al.*, unpublished manuscript). In addition, parasites are heterogeneously distributed among nests (Boulinier *et al.* 1996b), and local levels of infestation show positive temporal autocorrelations for time lags of up to 3 years, but not longer.

Using the detailed information collected on this system, T. Boulinier *et al.* (unpublished data) tested whether kittiwake natal and breeding dispersal were related to local levels of tick infestation. It was predicted that breeding dispersal, but not natal dispersal, should be positively associated with local infestation, as mean age at first reproduction in kittiwakes is around 4 years. In accordance with the predictions, there was no association between natal dispersal and level of infestation of the birth cliff (Fig. 12a.1a, $n = 803$ recruits), whereas there was a negative association between the proportion of breeding philopatric individuals and local parasite infestation the year before (Fig. 12a.1b, $n = 156$) (T. Boulinier *et al.*, unpublished results). Further, adding individual reproductive success as an independent variable in the statistical model indicated an additive effect of individual reproductive success and local level of infestation: unsuccessful birds in heavily parasitized cliffs showed a lower probability to be site faithful than unsuccessful birds from less parasitized cliffs (Fig. 12a.1b). Although these results are observational, they nevertheless suggest that parasitism may play an important role in the predictability of local patch quality, and thus on the habitat selection decisions of hosts. Parasitism may determine host dispersal either by a direct effect on exposed individuals, or through an indirect effect on a potential cue used to assess habitat quality (Boulinier and Lemel 1996; see also Stamps, chapter 16, and Danchin *et al.*, chapter 17).

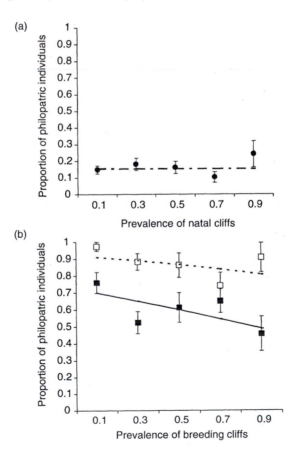

Figure 12a.1. Natal (a) and breeding (b) philopatry of kittiwakes in relation to prevalence of tick infestation. Breeding philopatry significantly decreases with tick prevalence but also with nest failure (b). The proportions of site-faithful individuals are reported (±s.e.) for the different classes of prevalence. Fitted lines are from logistic regression models. ●, first-time breeders; □, successful breeders; ■, failed breeders.

Parasites as a cause of host dispersal in other systems

Although we found few studies addressing the question of a potential effect of parasite infestation on host dispersal, some work has been carried out on certain taxa, with mixed results. In the common lizard (*Lacerta vivipara*), Sorci *et al.* (1994) found that the mite infestation level of pregnant females was related to the proportion of dispersing off-spring. Male offspring tended to disperse more when their mother harboured mites. Conversely, female offspring tended to stay in the maternal environment, but had higher growth rates and higher reproductive investment, when reproducing for the first time compared with females from uninfested mothers (Sorci and Clobert 1995). This result suggests that mothers could transfer information on the environment to offspring as a way to respond to spatiotemporal variation in habitat quality. In the yellow-bellied marmot (*Marmota flaviventris*), a trend for higher ectoparasite infestation of dispersing individuals has also been reported (van Vuren 1996). The author did not interpret this pattern as a strategy to avoid parasitism, but, instead, as evidence that individuals in poor condition are those that decide to disperse. Recently, a study reported patterns of local recruitment of great tit (*Parus major*) fledglings in relation to their sex, the size of their natal brood, and experimental infestation by hen fleas (*Ceratophyllus gallinae*) (Heeb *et al.* 1999). The authors proposed that the shorter dispersal distance of male great tits that grew in flea-infested nests could be a possible way for them to be better adapted or to show greater tolerance to local parasites. The only example we could find involving an endoparasite was part of an experimental study investigating the effect of a nematode (*Howardula aoronymphium*) on the survival of mycophagous *Drosophila*. Jaenike *et al.* (1995) reported no effect of the level of parasite infestation on dispersal of *Drosophila putrida* and *Drosophila neotestacea* under natural conditions, even though parasite-induced host mortality was high.

Thus, despite some evidence that hosts may disperse in response to parasites, little is known on the general effects of parasitism on host dispersal. For instance, we could find no study that combined detailed data on spatiotemporal patterns of parasitism risk and host response in terms of dispersal following an experimental manipulation of infestation levels. It is possible that an ultimate response to parasitism in terms of dispersal could be hidden in related and correlated proximate cues. For instance, when individuals disperse from overcrowded areas, this could be indirectly associated with reducing the risk of parasitic infection. Similarly, if individuals use an integrative cue to assess local habitat quality, such as the reproductive success of conspecifics, then parasites may also be involved as an ultimate factor.

Dispersal of parasites in relation to their hosts

The impact that parasites have on the decisions of their hosts will depend on the life history of the parasites (i.e., what stage in the life cycle and the degree of negative effect that this stage has on the host) and on their dispersion and abundance in the host's environment (e.g., the aggregated nature of some parasites can affect patterns of exposure of their hosts and thus host decisions to defend or flee). This impact will, therefore, be intimately associated with the ability of parasites to reach hosts, and thus to parasite transmission and dispersal.

Dispersal and transmission

There is a close relationship between the transmission of parasites from one host to the next and parasite dispersal. This makes the two difficult to define, particularly because of the ambiguity and debate about what constitutes a parasite population (see Bush *et al.* 1997). In most cases, a parasite population will be all individuals of all life phases in a given spatial location at a certain time. In relation to this, we define transmission as the passage of a parasite from a source of infection to a host, regardless of the point in its life cycle (Watt *et al.* 1995), and parasite dispersal as the movement of a parasite from its natal or former breeding location to a new spatial location where it undergoes repro-duction, regardless of which individual is the host or whether reproduction is sexual or asexual. In this sense, the two activities can occur in isolation (e.g., transmission without dispersal – movement from intermediate host to final host at a given spatial location; dispersal without transmission – a single host releasing parasite eggs in different loca-tions); in most cases, however, dispersal and transmission will be closely linked and can be considered in tandem (Fig. 12a.2).

Opportunities for parasite dispersal

Parasitic life styles can be divided into two main categories: (1) direct (parasite completes its life cycle in a single host) and (2) indirect (parasite uses at least two different hosts to complete its life cycle). Movements from host to host can be either passive (without additional energy expenditure) or active (parasite expends energy to seek a new host). Different strategies will have different implications for the dispersal potential of para-sites and thus for their probability to colonize host patches.

   In most cases, parasites with direct life cycles will disperse only if their host disperses. For example, a study that examined the genetic structure of five nematode parasites from three different host species, using mitochondrial DNA (mtDNA) sequence data, found that the degree of structure of parasite populations was dependent on host species (Blouin *et al.* 1995). The four nematodes of domestic livestock, *Ostertagia ostertagi* and *Haemonchus placei* from cattle, and *H. contortus* and *Teladorsagia circumcincta* from sheep, showed weak genetic structure consistent with high gene flow among populations at large spatial scales. Conversely, the nematode *Mazamastrongylus odocoilei* of white-tailed deer (*Odocoileus virginianus*) showed much more population subdivision with isolation by distance. Furthermore, this structure matched patterns of host mtDNA structure. Thus, the frequent movement of livestock over long distances greatly influ-enced parasite dispersal opportunities and, in turn, parasite population characteristics.

   In parasites that can exploit more than one host species (i.e., generalists), the opportunity for dispersal may change depending on the host species they infect. For example, the seabird tick, *Ixodes uriae*, has been found to infect up to 52 different host species. Dispersal in this tick is thought to occur near the end of the host breeding season when young individuals and failed breeders prospect among colonies (Danchin 1992). As different host species have different movement patterns at this period, there may be different possibilities for ticks to be dispersed among locations. For example, kittiwake fledglings remain at and around the nest, and can visit different colony sites during the period when they take their first flights. In common guillemots, *Uria aalge*, and other

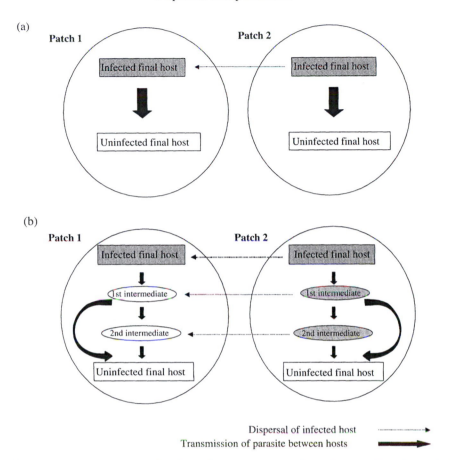

Figure 12a.2. Potential opportunities for dispersal and transmission of parasites with direct (a) and indirect (b) life cycles. Free-living stages of parasites in the environment could provide additional opportunities for dispersal to occur.

alcid species, however, chicks jump directly to sea at the point of fledgling with no movement within and among colonies. Thus, ticks that parasitize kittiwakes should have higher dispersal than those that parasitize guillemots. As direct measures of dispersal in this system are logistically impossible, a population genetic survey of tick populations from different host species could be used to examine this prediction (McCoy *et al.* 1999).

Indirect parasitic life cycles have much more variation in patterns of host use and opportunities for dispersal. For example, all digeneans use a gastropod intermediate host during part of their life cycle, and many have a second intermediate host that is preyed upon by the final host (Soulsby 1986). There have been surprisingly few studies that examined dispersal patterns of parasites with complex life cycles in natural populations. We found a single study that addressed such a question by comparing the population genetic structure of *Fascioloides magna*, a fluke of white-tailed deer, with that of its host (Mulvey *et al.* 1991). The authors found no isolation by distance among parasite populations at regional scales and no congruence between patterns of host and parasite population genetic structure. These results were attributed to both the large population size of the parasite, due to asexual reproduction in the intermediate host, and to the mobility of the final host. There are many other such systems where we could imagine that opportunities for parasite dispersal would increase with increasing life-cycle complexity.

Parasites can also directly alter the activity behaviour of their hosts. There are many examples of parasites that change the behaviour of their hosts either as an adaptation that increases the probability of transmission (for reviews, see Holmes and Bethel 1972; Dobson 1988; Combes 1991) or as a side effect of parasitic infection on host defences (Poulin 1995). In a review of 114 studies that compared host behaviour, Poulin (1994) found that parasites (nematodes, acathocephalans, and cestodes) moderately, but significantly, affected host activity and habitat choice. Thus, this kind of behavioural alteration may be a viable strategy that could favour dispersal for parasites with several types of transmission scenarios. A fascinating, but still unexplored, possibility would be that parasites induce host dispersal to facilitate their own dispersal.

It is clear that the opportunities for parasite dispersal will alter the dispersion of parasites in the environment, and thus will be an important factor influencing the impact parasites have both on their individual hosts and on host populations. This ability of parasites to re-infect or newly colonize hosts may, in turn, affect host dispersion patterns in the environment and host decisions to disperse to potentially parasite-free locations (Blower and Roughgarden 1989; Grosholtz 1993). Further, we expect that these decisions would be directly related to the life cycle of the parasite. For example, ectoparasites that can be left in the environment might be more likely to alter host dispersal decisions than endoparasites that hosts take with them.

Dispersal and host–parasite interactions: co-evolution and local adaptation

Because of the intimate relationship between parasites and their hosts, and because such systems are dynamic in space and time, it is necessary to consider factors relating to both members of the interaction when we want to study the role that parasites have played in the evolution of host dispersal. The direct consequences of host infection will be related to how virulent the parasite is, and this will depend, in part, on whether hosts are locally adapted to their parasite populations. Virulence has become a central issue of modern evolutionary biology, and recent theoretical works have emphasized the role that spatial structure and dispersal can play in its evolution (e.g., Gandon *et al.* 1996; Boots and Sasaki 1999). These studies have shown, notably, that the relative rate of dispersal of host and parasite will affect their abilities to be locally adapted, and that virulence will be higher if pathogens have the opportunity to infect distant hosts, rather than if their dispersal is restricted spatially. Empirical evidence now exists of relative dispersal rates affecting local adaptation in an animal–parasite system. Oppliger *et al.* (1999) showed that blood parasites (haemogregarine genus) were able to perform better on allopatric hosts (canarian lizard, *Gallotia galloti*) compared with sympatric ones, suggesting that, owing to restricted dispersal, parasites were locally maladapted to their hosts. In general, plant–pathogen systems have provided some of the best examples of metapopulation studies where both numerical and genetic dynamics have been investigated empirically (e.g., Jarosz and Burdon 1991). Such systems are thus good candidates for investigating patterns of local adaptation (Thrall and Burdon 1997; Kaltz and Shykoff 1998).

In animal–parasite systems, hosts can respond to local parasite infestation by dispersing, but also through different forms of resistance, which can be genetically based and/or environmentally induced. Little evidence exists for a genetic basis of differences in susceptibility to parasites in natural populations of animals (Sorci *et al.* 1997), and two of the scant examples come from studies carried out on bird–ectoparasite systems (Møller 1990; Boulinier *et al.* 1997). In another system involving an ectoparasite and a bird (hen fleas (*Ceratophyllus gallinae*) and great tits (*Parus major*)), some elements suggesting the existence of induced defence have been shown (Heeb *et al.* 1998). This opens the question of the potential existence of conditional responses to the risk of parasitism, with some individuals dispersing and others resisting. Theoretical examination of the importance of conditional response to parasitism has been neglected. Considering the insight that has been gained by introducing dispersal into theoretical approaches, it would be interesting to investigate the evolutionary stability and ecological effects of dispersal strategies conditioned by local parasite infestation.

Co-evolving interactions: an example of brood parasites and their hosts

As mentioned in the previous section, hosts might face the following dilemma: to leave a patch where risk of parasitism is high, potentially paying the cost of dispersal (e.g., an individual can never be sure that elsewhere it will find a safer patch), or to stay in the risky patch and resist. The costs and benefits associated with each of these strategies are likely to determine the outcome of this evolutionary game. Vertebrates have a very complex and sophisticated weapon against parasites: the immune system (Roitt *et al.* 1996). However, in spite of the diversity of immune responses, experimental work has shown that host resistance to some parasite species can increase susceptibility to other species or even other strains of the same parasite (Sorci *et al.* 1997). If this were a general pattern, it might prove extremely difficult to predict the optimal strategy hosts should adopt: whether to stay in a patch and invest in immune defence or to disperse.

One group of parasites, however, elicits very specific host responses which are unlikely to trade-off against resistance to other parasites. These are brood parasitic birds. Brood parasites lay their eggs in the nest of another species which incubates and takes care of the parasitic egg and chick (Rothstein 1990). Parasitized hosts have lower reproductive success than non-parasitized individuals; this should therefore select for some sort of host resistance. Egg discrimination is thought to be the most effective host response to brood parasites. If a host can recognize and reject the parasitic egg, then it should avoid part of the cost of parasitism. This type of resistance is obviously very specific and does not seem to have other functions than to prevent brood parasitism. However, egg discrimination cannot be achieved without costs; hosts may make mistakes and either eject or damage one or more of their own eggs when ejecting parasitic eggs (Davies *et al.* 1996). Finally, abundance of brood parasites varies greatly among areas (e.g., Soler *et al.* 1998), and therefore hosts face the dilemma of staying in a patch where the risk of being parasitized is high or of leaving that area. This sets up a similar scenario to the one described above involving classical parasites. However, there is a major difference: here, dispersing hosts are likely to be as resistant to the local brood parasites as to the allopatric ones, because egg mimicry should be maximal for sympatric parasites.

A recent study (Martinez *et al.* 1999) has investigated the population structure of a brood parasite, the great spotted cuckoo (*Clamator glandarius*), and its main host in Europe, the magpie (*Pica pica*). Populations of both species appear to be genetically differentiated, but gene flow is higher among magpie populations in sympatry with the cuckoo than among allopatric populations. Magpie dispersal is also higher than cuckoo dispersal. Could these findings be the result of the interaction between the cuckoo and the magpie? On the one hand, susceptible magpies would certainly benefit from leaving patches where cuckoos were abundant; on the other hand, as resistance arises in para-sitized populations, it should spread very rapidly, and as resistant hosts are those with the highest reproductive success they are also likely to produce more dispersing off-spring. Obviously, this has several consequences for the maintenance of both parasite and host populations. Dispersal of resistant genotypes to new populations would speed up the spread of resistance; however, dispersal of susceptible hosts to parasitized populations would provide further opportunities for the parasite to persist at the local scale. From a metapopulation perspective, naive host populations exploited by cuckoos could be considered sinks, whereas non-parasitized magpie populations would be con-sidered sources. Dispersal from sources to sinks might maintain both cuckoo and magpie populations at the regional scale (May and Robinson 1985). However, as resistance appears and increases in frequency in parasitized populations, cuckoos might be forced to leave the patch to find more susceptible host populations. Given that resistance is costly in the absence of parasitism (Marchetti 1992), it might decrease after parasite dispersal, and render the patch appropriate again for parasite colonization. Such processes may eventually end up in cycles of parasitism and host resistance, driven by reciprocal host and parasite dispersal. Interestingly, this could also account for the coexistence of susceptible and resistant host genotypes.

Conclusion

Overall, it appears that, despite its potential importance in host–parasite systems, relatively few studies have focused on dispersal *per se*. Parasites may represent a neglected factor of habitat quality which may condition host dispersal. It is important to determine how often parasites affect host dispersal, and whether host decisions to dis-perse are based on direct evidence of parasites themselves or integrative cues. For example, we could predict that more 'visible' parasites, such as some ectoparasites, could represent direct cues for hosts while less 'visible' parasites, such as some helminths, could induce host dispersal only if associated with external or indirect cues.

Clearly, parasite distribution and abundance in the environment can alter the deci-sions hosts make with regard to whether to stay and resist or to disperse from habitat patches. Likewise, host dispersal decisions should also depend on the dispersal strategies of their parasites. With the increased use of molecular techniques (Nadler 1995), population genetic studies should provide complementary information for testing pre-dictions linking parasite and host dispersal to parasite life cycles and host ecology.

For example, if parasites disperse with their host (e.g., helminths), parasite-conditioned dispersal could be non-advantageous for the host, whereas if hosts can leave their parasites in the habitat patch (e.g., some ectoparasites), parasite-conditioned dispersal could evolve. Conversely, as some parasites can manipulate host behaviour to favour transmission, and as parasite transmission and dispersal are intricately linked, parasites could actually modify host dispersal behaviour to increase their own dispersal. The development of theoretical models could provide insight into such scenarios. Further, models could also be used to investigate the combined effects of intraspecific and interspecific interactions on the evolution of dispersal. For instance, what are the costs and benefits of dispersal to reduce parasite transmission to kin? If this behaviour has evolved, we would expect to find it only in species where individual hosts form highly related groups and for parasites with intermediate levels of transmission. In addition to theoretical approaches, we require studies examining the dynamical distribution of genotypes and phenotypes of hosts and parasites, and specific experiments to address the causal nature of observed relationships.

With more empirical elements on the reciprocal effects of parasitism on host dispersal in contrasting host–parasite systems, specific models including conditional dispersal strategies could be refined to investigate the importance of such processes on the ecology and evolution of host–parasite interactions. This is an important issue as dispersal in host–parasite systems is likely to play a prominent role in the evolution of virulence or resistance, local host–parasite adaptation, the use of corridors for the conservation of subdivided populations, and other aspects of the ecology of infectious diseases in natural populations.

Acknowledgements

We thank J. Clobert, C. Combes, A.P. Møller, W. Weisser, and an anonymous reviewer for comments on this chapter. T.B. thanks E. Danchin and J.-Y. Monnat for their important contribution to the study on ectoparasitism and dispersal in kittiwakes. This work benefited from support by the CNRS, the Ministère de l'Environnement, the French Polar Institute (IFRTP), and NSERC, Canada (to K.D.M.).

12b

The effects of predation on dispersal

Wolfgang W. Weisser

Abstract

Predation interacts with the dispersal process in a number of ways. In this chapter I discuss the evolutionary consequences of predation for dispersal in plant–herbivore and animal predator–prey systems. Predation is shown to have three main effects on the dispersal process. First, predation influences whether an individual leaves its natal population, either because predators or herbivores transport the individual to a new habitat, or because dispersal serves as an adaptation to escape from predation risks. Second, predation influences to which habitat an organism disperses, either because the predator or herbivore transports the dispersing propagules, or because the disperser selects a new population based on the risk of predation in the habitat. Third, predation influences the survival of dispersing individuals. Predation is an evolutionary force that may (1) select for dispersal, (2) select against dispersal, or (3) select for a conditional dispersal strategy. While some of the evolutionary consequences of predation on dispersal have received extensive coverage in the literature, other effects have scarcely been investigated. Some areas that merit future research are identified.

Keywords: predator–prey, plant–herbivore, emigration, immigration, transit, dispersal

Introduction

Most organisms are preyed upon by a natural enemy, and most disperse in one way or another. In this chapter, I explore how predation by herbivores and carnivores influences the dispersal of plants and animals, respectively. For reasons of space limitations, I concentrate on top-down effects and do not discuss the effects that plants or prey have on dispersal of their predators. Some bottom-up effects such as resource-dependent habitat selection are discussed in the chapters by Stamps and Danchin *et al.* (chapters 16 and 17; see also Rosenzweig 1991). As a further restriction, I do not consider interactions that may occur between two different food chains. For example, many seeds disperse by adhering to the fur of passing animals (Sorensen 1986), which may be predators (e.g. shrews searching for invertebrate prey) but which do not eat the plants or seeds that they transport. Finally, I ignore the fact that in many diploid organisms genes disperse not only by being carried within a zygote, but also in the gamete stage (e.g. pollen dispersal of angiosperms or free-swimming sperm in some marine organisms).

For this review, I define *dispersal* as the process of individuals leaving their natal population and, if successful, arriving in or founding another population. The places where populations live are referred to as *habitats*. Thus, I do not consider repeated movements such as diurnal vertical migration of zooplankton, or any other movement where the prime objective is different from leaving the natal population (e.g. foraging movements of animals). Successful dispersal requires both surviving the journey to a new habitat and breeding there. The dispersal process can therefore be subdivided into the successive phases (as in Danchin *et al.*, chapter 17): emigration, transit (or transience; Ims and Hjermann, chapter 14), and immigration, starting with the commitment of an individual to dispersal, and ending with its establishment in a new habitat.

In the following sections, I discuss how predation affects the different phases of the dispersal process in plant–herbivore and predator–prey systems. The consequences of predation for dispersal can be proximate or ultimate, and it is important to distinguish between these two types of effects. In this chapter I concentrate on the ultimate (i.e. evolutionary) consequences of predation for dispersal. Thus, when a particular effect of predation on dispersal is discussed, I ask how this effect influences the evolution of dispersal rate. Proximate consequences of predation for dispersal, i.e. the immediate changes in physiology and behaviour of prey caused by predation, are not the subject of this review. I illustrate each effect with examples. Because herbivores are often defined as animals that eat living plant tissue (Crawley 1983), my examples not only involve true predators or grazers, but also include organisms such as aphids that functionally are parasites. In choosing these examples I do not claim to be exhaustive. After reviewing the effects of predation on the dispersal process, I investigate how predation influences the trade-off between dispersal and other life history traits. In the concluding section, I summarize the effects of predation on the evolution of dispersal rate and identify areas of research that merit further work.

The effects of predation on the different phases of dispersal

Emigration

Emigration is completed when an individual has left its natal population. In most organisms, both genetic and environmental factors influence whether a particular individual leaves a population (Roff and Fairbairn, chapter 13; Ims and Hjermann, chapter 14; and Dufty and Belthoff; chapter 15). Importantly, there may be a time lag between the point at which an individual commits to emigration and its actual departure from the population (see Ims and Hjermann, chapter 14). For example, if dispersal requires special morphological adaptations, these structures have to be developed during the ontogeny of the individual. One such example is wing-dimorphic insects, where only the dispersing individuals (the macropterous morph) develop a fully functional wing apparatus (Zera and Denno 1997; Dixon 1998). In the case of a time lag, there is a period of time during which a prospective disperser will still be in the natal population, and during which it may suffer from a different risk of predation than the non-dispersers. Predation may therefore interact with emigration in two ways: first, predation, or the risk of being preyed upon, may be one of the reasons why an individual commits to emigration. Second, in the case of time lags, predation may cause mortality risks for the individuals already committed to dispersal. It seems, therefore, useful to discuss separately the effects of predation on:

(1) the development of dispersal ability in an individual
(2) the survival of a specialized disperser until the age at which it leaves its natal population
(3) the actual departure of a specialized disperser from the natal population.

Predation and the development of a dispersal ability

Mathematical models on the evolution of dispersal mostly assume unconditional dispersal (i.e. a purely genetic basis for the determination of dispersers). In such models, the risk of population extinction generally selects for an increase in dispersal rate, i.e. an increase in the frequency of alleles associated with emigration (Johnson and Gaines 1990; Dieckmann *et al.* 1999). Thus, where predation causes population extinction, models would predict that the fraction of offspring that disperse, or the allocation of resources into dispersing propagules, is higher in areas where populations suffer heavier rates of predation. In addition, predation may also affect the evolutionarily stable dispersal strategy when it does not lead to population extinction. Spatial and temporal variability in the vital rates of the population strongly affect the evolution of dispersal (Gandon and Michalakis, chapter 11), and predation may be a cause for such variability. Unfortunately, there are very few studies relating population growth rates or population extinctions to predation, so this hypothesis remains untested. In some wing-dimorphic insects, dispersal has a strong genetic component, and the proportion of winged morphs in a population may be used as an indirect measure of the population dispersal rate (Zera and Denno 1997). It has been demonstrated that the proportion of winged morphs in plant-hopper populations is correlated with habitat persistence (i.e. the average survival time of populations), but predation was not the main cause of population extinctions (Denno *et al.* 1996). Thus, although there are scenarios under which mathematical models would predict an effect of predation on the evolution of unconditional dispersal, no such effect has been reported.

In many organisms, dispersal depends not only on genetic but also on environmental factors (i.e. conditional dispersal; see Ims and Hjermann, chapter 14). In such cases, predation or the presence of a predator or herbivore may be one of the environmental factors involved in the determination of dispersers. For organisms with a specialized morphology for dispersal, a plastic response to predation would imply that more dispersal morphs are produced as a direct response to a predation event. This dispersal morph could then escape to colonize new habitats. Such a conditional strategy has recently been shown for pea aphids (*Acyrthosiphon pisum*), which responded to the presence of predatory ladybirds by increasing the proportion of their offspring that developed into winged morphs (Weisser *et al.* 1999). It is not known whether predation has similar effects in other species with dispersal polymorphisms. In the majority of species, however, dispersal is conditional, and there are no obvious morphological differences between the dispersing and the non-dispersing individuals. Any member of the population may, in principle, emigrate from the natal population as a response to predation, or to the presence of a predator. In such cases, emigration is immediate, and individuals disperse to reduce the mortality risks for themselves or for their offspring. The largest number of examples of an effect of predation on emigration falls into this category. For instance, dispersal of spider mites from plants has been shown to be triggered by the presence of predatory mites (Bernstein 1984; Janssen *et al.* 1998). Other examples come from birds, where predators are a major cause of egg or nestling loss. In many species, nest predation induces females to leave their territories to search for new habitats elsewhere (Sonerud *et al.* 1988; Jackson *et al.* 1989).

It is striking that all the above examples come from animal predator–prey systems. In plant–herbivore systems it is extremely difficult to quantify dispersal rates or investment into dispersal, because seeds represent both investment into reproduction and investment into dispersal. Disentangling these two components has proven problematic. As a result of these difficulties, research on the evolution of dispersal in plants has concentrated on heteromorphic seeds where the relative proportion of long-distance to short-distance dispersers can be used as a measure of investment in dispersal (Olivieri and Gouyon 1997). While predation is generally seen as a cause of seed dispersal (Willson 1992), to my knowledge no study has linked rates of dispersal in plant populations to the risk of population extinction, or to the variability in population growth rates caused by predation.

In summary, predation has been shown to induce the emigration of individuals from populations. However, most of the examples concern a conditional response of animals to immediate risks of predation, in species without a specialized dispersal morph. Little is known about the relationship between population growth rates, population extinction, predation, and the evolution of dispersal rates, in particular for plant populations. Thus, there is a need for comparative studies in plant–herbivore and predator–prey systems.

Predation on specialized dispersal morphs

Examples of preferential predation on dispersal propagules are restricted to plant–herbivore systems. In many plants, a large proportion of flowers and seeds is lost as a result of predation before the seeds are dispersed. Pre-emigration seed predation, mostly by insect herbivores, has been the subject of many reviews (Crawley 1992). Plants have developed a number of adaptations to reduce the risk of seed predation, such as mass fruiting or chemical defences (Crawley 1983, 1992). Nevertheless, pre-emigration seed predation is common and an important determinant of plant communities (Louda 1989). For seed dispersed by food-hoarding animals, some seed predation is a prerequisite for successful emigration (Price and Jenkins 1986; Vander Wall 1990). In contrast to the wealth of studies in plant–herbivore systems, little is known about pre-emigration predation on dispersers in animal predator–prey systems (Murren *et al.*, chapter 18). In wing-dimorphic insects, the macropterous morph often has a longer developmental time than the non-dispersing morph (Zera and Denno 1997), but whether this translates into a higher mortality risk due to predation has not been investigated. Thus, at least in plants, individuals committed to emigration are affected by predation, and a number of adaptations can be observed that are designed to increase propagule survival and the chances of successful emigration.

Predation and the departure of specialized dispersal morphs

After a dispersal propagule has completed its development, it has to leave the natal population. In plants dispersed by frugivores or food-hoarding animals, an encounter with a herbivore is required for emigration. In such systems, the fraction of propagules that disperse will depend on the level of frugivory or food-hoarding. The effects of herbivores on seed or fruit dispersal have been reviewed extensively (Howe and Smallwood 1982; Howe 1986; Fleming 1991; Jordano 1992; Stiles 1992; Chambers and MacMahon 1994). In contrast, very little is known about the effects of predators on inducing emigration of specialized dispersal morphs in animals. Although it is known that not all individuals with morphological adaptations for dispersal leave the natal population (Shaw 1970b), it remains unclear whether predators can trigger emigration in these individuals. The differences between plants and animals may in part be due to an intrinsic difference between herbivores and carnivores: carnivores are often true predators that kill their prey, while this is less often the case for herbivores. Thus, in animals, specialized dispersal morphs are likely to be killed rather than transported by predator attack. Nevertheless, it is conceivable that the actual departure of specialized dispersal morphs in animals may also be influenced by predator attack or the risk thereof.

To summarize, dispersal of seeds by frugivores or seed dispersers is the only case in which predation affects the departure of specialized dispersal propagules. For animals with a specialized dispersal morphology, no such effect has been recorded.

Transit

Quantitative estimates of predator-related mortality during dispersal are difficult to obtain because of the often small size of seeds or dispersing animals. The existing evidence suggests, however, that predation is one of the major causes of mortality for dispersing individuals. In plants, dispersal propagules that have left the natal population but have not yet reached a safe site (i.e. the seed bank) are subject to discovery and consumption, often by generalist or specialist vertebrates (Chambers and MacMahon 1994). Seed mortality rate can be very high, with consequences for the structure of plant communities (Louda 1989). In animals, travelling individuals also suffer from mortality risks due to predation. A number of examples can serve to illustrate the scale of these risks. In *Latrodectus revivensis*, a widow spider of the Negev desert that constructs web in shrubs, travelling between plants was associated with a 40% chance of death during a move (Lubin *et al.* 1993). In the colonial ascidian *Diplosoma similis*, between 41% and 47% of the larvae released die as a result of planktonic and benthic sources of mortality before settlement (Stoner 1990). In marine bivalves, the juvenile mortality rate often exceeds 90%, with predation being one of the main causes of death (Gosselin and Qian 1997). Theory makes very clear predictions about the effects of travel mortality risks on dispersal. Models of the evolution of dispersal predict that emigration rates decrease if the costs of dispersal increase (Dieckmann *et al.* 1999; Johnson and Gaines 1990). Similarly, optimal foraging models predict that travel mortality leads to longer residence times in habitat patches (Houston and McNamara 1986). Empirical evidence for these effects is, however, very limited, as there are few comparative studies on dispersal rates in natural populations (see above). In contrast, adaptations that reduce the predation risks associated with dispersal are commonly observed, and include the production of a high number of dispersal propagules, or chemical defence against predators (Lindquist and Hay 1996).

Thus, during transit, predation is a major cause of the loss of dispersal propagules. Although quantitative data are scarce, for many animals, and possibly also some plants, transit is likely to be the phase in the dispersal process where the risks of predation are greatest.

Immigration

Immigration consists of two processes: the selection of a suitable habitat, and establishment in that habitat. Predation may therefore influence to which habitat an individual disperses, and it may also influence the establishment of immigrants in the new habitat. I will discuss these effects in the following two sections.

Predation and habitat selection by dispersers

Animals often select habitats based on suitability (Danchin *et al.*, chapter 17; Stamps, chapter 16). One important component of suitability is the predation risk in this habitat. For a variety of taxa, it has been shown that dispersers select habitats with lower risks of predation. For example, small passerine birds in Finnish farmlands avoid nesting close to the breeding grounds of the European kestrel, *Falco tinnunculus* (Suhonen *et al.* 1994). In the pelagic dipteran, *Culiseta longiareolata*, females avoid laying eggs in waters with the predatory backswimmer, *Notonecta maculata* (Blaustein 1998, 1999). Larval ascidians also preferably settle on predator-free sites (Young 1989). Other examples appear counter-intuitive at first sight. Curlews (*Numenius arquata*) prefer to nest *close* to the nests of kestrels (Norrdahl *et al.* 1995). By comparing predation rates of curlew nests at different distances from kestrel nests, Norrdahl *et al.* (1995) found that the survival rate was highest close to these nests, because nest predation by kestrels was less than that due to other predators, which were deterred by the presence of the kestrels.

A different way in which predation influences where dispersers go is when the dispersing propagules are transported by the predator or herbivore. In seeds or fruit dispersed by animals, the destination of the dispersal propagules is determined by the predator or frugivore. Frugivorous birds deposit seeds at places where they defaecate (Kollmann and Pirl 1995). Food-hoarding animals often initiate new plant populations by failing to retrieve stored seeds from their caches (Vander Wall 1990). Although seeds dispersed by animals are sometimes thought to have no control over their final destination, plants may actually achieve targeted dispersal by influencing what type of animal transports the seed (Hanzawa *et al.* 1988; Bazzaz 1991). Given the high rates of post-dispersal seed predation (see below), one would also expect targeted dispersal of seeds to habitats with few or no natural enemies. With respect to parasites, such an effect has been described: bellbirds (*Procnias tricarunculata*) transport seeds of the tropical tree *Ocotea endresiana* to sites where seedling mortality from fungal attack is significantly reduced (Wenny and Levey 1998).

In some cases, animals are also dispersed by their predators. Vander Wall (1990) lists a number of instances where food-hoarding animals store prey items that then escape their predators. For example, crested tits store sawfly (*Lophyrus* sp.) cocoons unharmed, and individuals not eaten might escape to breed at the new sites. Similarly, earthworms stored by moles may escape after regeneration of segments injured by the moles to make them immobile. In these examples, predators influence where the dispersers go, but the significance of this dispersal is controversial (Vander Wall 1990).

In summary, predation affects the destiny of dispersers. In animals, habitat selection by the dispersing individuals is often based on predation risks associated with a habitat. In plants, seeds are often transported by a herbivore which determines where the seeds will go.

Predation and the establishment of immigrants in a new habitat

Successful dispersal requires both arrival in a new population and successful reproduction. Only when reproduction is successful does the immigrant contribute to the dynamics of the resident population. In plants, post-dispersal seed predation is common, and mainly due to generalist and specialist vertebrates (Crawley 1992; Chambers and MacMahon 1994). Even when a seed is in the seed bank, it might not be safe from predators that search for buried seeds. The greatest effect of herbivores on plant mortality is, however, at the seedling stage (Crawley 1983). Darwin (1859) already recognized the importance of herbivores at this stage of plant development when he described his observations on the survival of marked seedlings (less than 20%). In animals, immigrants often have a higher mortality risk than residents (Tinkle *et al.* 1993), but little is known about the sources of this additional mortality. A long-term study on the Columbian ground squirrel (*Spermophilus columbianus*) in Alberta showed that immigrants occupy a more peripheral position in colonies than residents, and have lower reproductive success, partly because their offspring are more likely to die from predation (Wiggett and Boag 1993). Thus, in both plants and animals, predation may have a major impact on the establishment of immigrants, although the effects on animals remain to be investigated in more detail.

Predation and the trade-off between dispersal and other life history traits

In most organisms, dispersal is an energy-intensive process. Resources are needed, both to build the structures necessary for dispersal (e.g. a seed pappus or wings) and to power the organism during transit. Because resources are limited, organisms have to trade investment into dispersal against investment into other life history traits. In wing-dimorphic insects, for example, investment into wings often results in winged morphs having a lower fecundity than unwinged morphs (Zera and Denno 1997). Predation may influence this trade-off between dispersal and other life history traits. For example, life history theory predicts that high juvenile mortality, for example due to selective predation on smaller individuals, leads to lower reproductive effort (Stearns 1992; Charlesworth 1994), which is supported by a number of examples from natural populations (Reznick *et al.* 1990). However, when predation is local, so that parents could escape predation by dispersing, natural selection might favour a higher investment into dispersal. This in turn would reduce investment in reproduction and may change other life history traits as well. Thus, rather than locally adapting to a higher predation risk, another possibility might be to escape from predation by changing allocation patterns and investing more in dispersal. Conversely, there are cases (e.g. in stream fish; Reznick *et al.* 1990) where escape from predation is not possible, so that selection can act only on life history traits other than dispersal. Similar scenarios can be thought of when predation is on adults. Thus, one can predict that predation affects the evolution of life history traits such as reproductive effort and age at maturity, not only directly but also via its effect on investment into dispersal. Up to now, life history theory has largely neglected dispersal as an evolving trait, and models of the evolution of dispersal rate only start to incorporate other life history traits into their analysis (see Gandon and Michalakis, chapter 11). Theoretical work that analyses the joint evolution of dispersal and other life history traits is likely to make useful contributions to our understanding of variation in life history patterns among organisms.

Conclusions

The survey shows that the evolutionary implications of predation for dispersal may vary depending on which phase in the dispersal process is affected. Predation is an evolutionary force that may:

(1) select for dispersal (e.g. to escape predation risk)
(2) select against dispersal (e.g. when predators increase the cost of dispersal), or
(3) select for a conditional dispersal strategy (e.g. conditional on the state of the habitat of origin or the state of the new habitat).

Thus, predation may have conflicting consequences for the evolution of dispersal. The study of the evolutionary consequences of predation therefore requires a formal analysis that extends beyond the existing models. Theoretical studies are needed that explore the links between the different effects of predation on the dispersal process and the evolution of dispersal.

Theoretical studies are also needed to explore co-evolutionary relationships between predator and prey dispersal rates. For example, the dispersal rate of prey is not only influenced by predation and the risk thereof, but may also be affected by the dispersal rate of the predator itself (M. Egas *et al.*, unpublished manuscript). If the dispersal abilities of the predator exceed those of the prey, emigration from habitats attacked by predators may not improve the situation of prey. In such a situation, it is conceivable that the evolutionarily stable prey dispersal rate decreases, which in turn affects the evolution of the predator dispersal rate. Thus, co-evolutionary processes may occur whereby dispersal rates of predator and prey evolve in response to each other (see also van Baalen and Hochberg, chapter 21; Mouquet *et al.*, chapter 22). Because dispersal rates also affect the stability properties of predator–prey models (Weisser and Hassell 1996), co-evolutionary models should also investigate the interaction between population dynamics and the evolution of dispersal rate in predator–prey systems.

Finally, the survey shows that empirical evidence is still rudimentary for several of the effects of predation on dispersal. What is needed is a quantification of predation-related mortality for all phases of the dispersal process, in both plants and animals. Similarly, comparative studies of plant–herbivore or predator–prey systems that link predation rates to population growth and to investment into dispersal may unravel strong effects of predation on the evolution of dispersal rate.

Weisser

Acknowledgements

I thank Thierry Boulinier, Jean Clobert, Dieter Ebert, Tad Kawecki, Tom Little, Karen McCoy, and Gabriele Sorci for critically reading previous versions of the manuscript, and Mark Westoby for suggestions. Comments from three anonymous referees are gratefully acknowledged. This study was supported by grant no. 31-053852.98 of the Swiss Nationalfond, by the Roche Research Foundation, and by the Novartis Stiftung.

Part III

Mechanisms of dispersal: genetically based dispersal, condition-dependent dispersal, and dispersal cues

13

The genetic basis of dispersal and migration, and its consequences for the evolution of correlated traits

Derek A. Roff and Daphne J. Fairbairn

Abstract

'Dispersal and migration' cannot be considered a single trait: it comprises physiological, morphological, behavioural, and life history components, all contributing to the probability that an organism will move from one location to another. Furthermore, traits that form the suite of 'migration' characters may themselves influence other fitness-related traits. Here we address three questions. (1) Is there genetic variation for migratory traits? (2) Are migratory traits genetically correlated with one another and/or other components of fitness? (3) Is it important to understand the genetic architecture of migratory and associated traits to predict the evolution of such traits? For the first question we present evidence that, in general, there is considerable additive genetic variation for migratory traits. Data to address the second question are fewer but indicate that genetic correlations between migratory components and other traits are probably the norm. Using the sand cricket as a model organism we show how quantitative genetics can be used to predict direct and correlated responses of traits directly associated or genetically correlated with migration.

Keywords: heritability, genetic correlation, correlated response, threshold trait, selection, trade-offs, quantitative genetics, fitness, liability, dimorphism

Introduction

Migration, here defined as the movement of an organism from one location to another either permanently (also called dispersal) or on a seasonal cycle, is a ubiquitous part of the natural world: all organisms at some point in their life cycle have a migratory phase, whether or not the individual actually migrates. Nevertheless, the considerable variation in the habitats in which organisms live necessitates different degrees of migration, with some organisms barely moving each generation, and all individuals in some other species moving considerable distances (Dingle 1996). This enormous range in migration distances and rates is *prima facie* evidence that migration strategies have evolved. For evolution to have occurred there must be genetic variation in the traits involved. Thus, the above 'casual' observations of the diversity of migration types suggests that genetic variation for migration and its components is probably widespread.

The purpose of this chapter is threefold. First, we ask: 'Is there genetic variation for migration and its components?'. Finding the answer to be in the affirmative leads to the second question: 'Are migratory traits genetically correlated with one another and with other components of fitness?'. If the answer to the latter were 'no' then migratory strategies would be free to evolve to the optimal combination. However, genetic correlations either between the migratory components themselves or with other traits could severely constrain, if not the final combination of traits, the evolutionary trajectory (in the sense of both time to equilibrium and the sequence of character states passed through). Finally, we draw together data from our work on the genetic basis of migratory components in the cricket species *Gryllus firmus* to illustrate how the genetic architecture of migration and fitness components is fundamental to an understanding of the evolution of migration.

Is there genetic variation for migratory traits?

Variation in migration can be analysed from two perspectives. First, it can be viewed as a threshold trait in which an individual may be classified as a migrant or a non-migrant. Such a dichotomous classification does not imply a simple mendelian inheritance: according to the threshold model there is some continuously distributed trait termed the 'liability' and a threshold of sensitivity such that individuals on one side of the threshold will be migrants and individuals on the other will be non-migrants. It is important to note that this model makes no statement about the suite of traits correlated with the trait(s) used to classify an individual as a migrant or non-migrant, i.e. the traditional model of a dichotomous suite of traits does not follow from this perspective.

The simplest case in which the above dichotomous classification might be applied is that in which there are morphological changes that prevent migration in some individuals. For example, in paedomorphosis in salamanders or wing dimorphism in insects there is a morph that clearly cannot disperse to the same extent as the other morph (thus the neotenic form of the salamander cannot leave its natal pond, and the flightless insect morph is restricted, generally, to migration by walking, which is clearly more limiting than flight). In these cases we can call one morph the 'non-migrant' but it is strictly correct to call the other morph only a 'potential migrant'. Having the morphological capability of migration does not mean that this capability will be exercised, a point to which we shall return later.

Where migrants are classified into two groups the threshold model can be used to calculate the heritability of the trait. In the threshold model we are not calculating the heritability of the dimorphism *per se* but the heritability of the liability, assuming that the threshold of sensitivity is itself fixed (although it is equally plausible to suppose that there is genetic variation in both the liability and the threshold, and even a genetic correlation between them).

An alternative approach is to view migration as a continuous trait as measured, for example, by duration of directed movement. In this case we calculate the heritability of the migratory trait directly (transforming the data, if required, to fulfil the assumption of normality). In some cases the migratory trait may be continuous and an arbitrary cut-off used to define the two forms. The heritability of the trait is then estimated using the threshold model. If the distribution of the trait is normal, the heritability so calculated is the heritability of the original continuous trait; if it is not normal then it is the heritability of the original trait transformed according to some undefined function such that it is normal. Thus there is nothing to be gained by dichotomizing continuous data; indeed the loss of information will increase the standard error of the heritability.

So far we have not explicitly defined what is meant by 'migrant' and 'non-migrant'. We suggest that, in fact, such a simple distinction may be very misleading: to understand the evolution of migration we must understand the evolution of the components of migration. The actual act of migration is only one component; the traits involved can be divided into physiological, morphological, behavioural, and life historical. As an example, consider the components involved in a wing-dimorphic insect. First, there are a series of physiological components that determine the development of wings, flight muscles, and associated body changes (e.g. changes in the shape of the thorax). These physiological processes thus lead to particular morphological differences between individuals that either restrict or permit migration by flight. Having the capacity to migrate does not necessarily mean that the individual will migrate: to do so involves a behavioural component. Finally, migration may be confined to a narrow ontogenetic window because of either behavioural or physiological changes that switch the morph from a migratory phase to a reproductive phase (the so-called oogenesis flight syndrome).

Interspecific comparisons of plants have shown that seed size and shape play important roles in migration (Pijl 1982; Murray 1986). Some plant species possess dimorphic seeds, which appear to act as short- and long-distance dispersers (Payne and Maun 1981; Morse and Schmitt 1985; Telenius and Torstensson 1989; Zhang 1995). In milkweed (*Asclepias syriaca*), dispersal distance decreases with seed mass, the further travelling, lighter seeds showing lower germination, survivorship, and seedling mass (Morse and Schmitt 1985). Variation among clones in the foregoing characters hints at genetic variation in dispersal and correlated traits, but maternal effects cannot be ruled out (Morse and Schmitt 1985). However, there is abundant evidence that seed size is typically genetically variable (see Roff 1992, Table 10.5) and hence it is likely that dispersal distance is also genetically variable within plant species.

Most analyses on genetic variation in dispersal traits have been on animals. In all the studies that we have examined there is evidence for additive genetic variance in migratory traits (Table 13.1). Morphological components that can be used to define migrants (body size, wing length, or wing morph), behavioural measures of migration (propensity to initiate migration, duration of migration), and physiological traits associated with the ability to disperse (enzymes associated directly or indirectly with locomotion) typically have heritabilities greater than 0.30 (Table 13.1). Such heritabilities permit rapid response to selection, as has been observed in a number of studies (Table 13.2).

Are migratory traits genetically correlated with one another and with other components?

Establishing the existence of a genetic correlation between traits is considerably more difficult than demonstrating additive genetic variance for a single trait. It is, therefore,

Table 13.1. A review of heritability estimates of migratory traits

Species		Trait		h^2	SE	Reference
Morphology						
Cricket	Gryllus firmus	Wing morph		0.65	0.08	Roff (1986)
Cricket	Allonemobius socius	Wing morph		0.62	0.15	Mousseau and Roff (1989)
Cricket	Gryllus rubens	Wing morph		0.98	0.16	Roff and Fairbairn (1991)
Cricket	Gryllus pennsylvanicus	Wing morph		0.74	0.19	Roff and Simons (1997)
Cricket	Dianemobius fascipes	Wing morph		0.30	0.04	Masaki and Seno (1990)
Plant-hopper	Laodelphax striatellus	Wing morph		0.32	ng	Mori and Nakasuji (1990)
Salmon	Oncorhynchus tshawytscha	Body morph		0.32	0.14	Heath et al. (1994)
Behaviour						
Mite	Tetranychus urtica	Dispersal behaviour		0.28	ng	Li and Margolies (1994)
Fruitfly	Drosophila melanogaster	Anemotaxis		0.03	0.01	Johnston (1982)
Fruitfly	Drosophila melanogaster	Locomotor behaviour		0.12	0.04	Van Dijken and Scharloo (1979)
Fruitfly	Drosophila melanogaster	Activity		0.51	0.1	Connolly (1966)
Fruitfly	Drosophila mojavensis	Host acceptance		0.18	ng	Lofdahl (1986)
Fruitfly	Drosophila aldrichi	Flight duration		0.21	0.10	Gu and Barker (1995)
Fruitfly	Drosophila buzzatii	Flight duration		0.28	0.10	Gu and Barker (1995)
Caterpillar	Choristoneura rosaceanna	Settling behaviour		0.73	0.17	Carriere and Roitberg (1995)
Polychaete	Spirobis borealis	Settling behaviour		0.27	0.1	Mackay and Doyle (1978)
Blackcap	Sylvia atricapilla	Migratory activity		0.45	0.08	Berthold and Pulido (1994)
Robin	Erithacus rubecula	Migratory activity		0.52	ng	Biebach (1983)
Milkweed bug	Lygaeus kalmii	Flight duration		0.31	0.06	Caldwell and Hegmann (1969)
Lepidopteran	Heliothis armigera	Flight duration		0.39	0.08	Colvin and Gatehouse (1993)
Lepidopteran	Epiphyas postvittana	Flight capacity		0.43	0.02	Gu and Danthanarayana (1992)
Lepidopteran	Spodoptera exempta	Flight duration		0.40	ng[a]	Gatehouse (1986)
Lepidopteran	Cydia pomonella	Flight distance		0.57	ng[a]	Schumacher et al. (1997a)
Physiology						
Fruitfly	Drosophila melanogaster	12 metabolic enzymes		0.40	0.15—0.59[b]	Clark (1990)
Fruitfly	Drosophila melanogaster	Power output in flight		0.33[c]	ng	Curtsinger and Laurie-Ahlberg (1981)
Cricket	Gryllus firmus	Flight muscle histolysis		0.51	0.09	Roff (1994a)

ng, not given; [a] significantly different from zero ($P < 0.05$); [b] range in estimates; [c] broad sense h.

Table 13.2. A review of selection studies on migratory traits. In all cases a response to selection was observed

Species	Trait selected	Reference
Tribolium castaneum	Migration rate	See references in Dingle (1984)
Tribolium confusum	Migration rate	See references in Dingle (1984)
Oncopeltus fasciatus	Flight propensity	Palmer and Dingle (1989)
Oncopeltus fasciatus	Wing length	Palmer and Dingle (1986)
Sylvia atricapilla	Migratory activity	Berthold (1988)
Gryllus firmus	Wing morph	Roff (1990a)

not surprising that there are fewer studies of such correlations. Selection in *Oncopeltus fasciatus* for either flight propensity or wing length produced a correlated response in the unselected trait (Palmer and Dingle 1986, 1989). Sib analysis in the cricket *Allonemobius socius* demonstrated a highly significant genetic correlation between the shape of the thorax (which contains the flight muscles) and wing morph (Roff and Bradford 1998). The most detailed examination of the genetic correlation structure between migratory components has been carried out in our laboratories using the sand cricket, *Gryllus firmus* (Fig. 13.1). We have shown that all migratory components investigated (juvenile hormone esterase activity, muscle histolysis, and flight propensity) are genetically correlated, and thus selection for changes in migration rate will result in a suite of changes in physiology, behaviour, and morphology.

Not only are the components of migration genetically correlated, but these components are themselves genetically correlated with other fitness-related traits. Three such

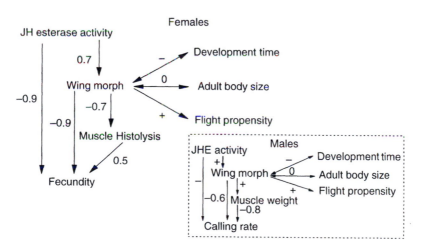

Figure 13.1. Genetic correlations between migratory and related traits in the cricket, *Gryllus firmus*. Arrows indicate the hypothesized direction of influence (e.g. wing morph is determined in part by the level of juvenile hormone (JH) esterase activity). Double arrows indicate that the direction of influence is uncertain. Wing morph is defined as a 0, 1 variable with $0 = SW$ and $1 = LW$; thus a positive correlation between a trait X and wing morph means that the value of X increases with the proportion LW. Note that muscle histolysis and muscle weight measure the same phenomenon but one is the inverse of the other. Data sources: Fairbairn and Roff (1990); Roff (1990b, 1994a); Roff *et al.* (1997, 1999); Crnokrak and Roff (1998).

Table 13.3. Genetic correlations between a migratory trait and another trait

Second trait	Migratory trait	Species	r_G	Reference
Body size	Migration rate	*Tribolium* (two spp.)	Positive	See references in Dingle (1984)
Body size	Flight propensity	*Oncopeltus fasciatus*	Positive	Palmer and Dingle (1986, 1989)
Body size	Wing length	*Oncopeltus fasciatus*	Positive	Palmer and Dingle (1986, 1989)
Body size	Wing morph[a]	*Gryllus firmus*	Zero	Roff (1990b)
Body size	Wing morph	*Allonemobius socius*	Positive	Roff and Bradford (1998)
Body size	Flight distance	*Cydia pomonella*	Positive	Schumacher *et al.* (1997a)
Development time	Migration rate	*Tribolium* (two spp.)	Negative	See references in Dingle (1984)
Development time	Wing length	*Oncopeltus fasciatus*	Negative	Palmer and Dingle (1986)
Development time	Wing morph	*Gryllus firmus*	Negative	Roff (1995)
Early fecundity	Migration rate	*Tribolium castaneum, T. confusum*	Negative	See references in Dingle (1984)
Early fecundity	Migration rate	*Tribolium castaneum*	Positive	See references in Dingle (1984)
Early fecundity	Flight propensity	*Oncopeltus fasciatus*	Positive	Palmer and Dingle (1986, 1989)
Early fecundity	Wing length	*Oncopeltus fasciatus*	Positive	Palmer and Dingle (1986, 1989)
Early fecundity	Wing morph	*Gryllus firmus, Allonemobius socius*	Negative	Roff (1995); Roff and Bradford (1996)

[a]Wing morphs are coded as 0 = SW and 1 = LW. Therefore, selection for increased migration will produce a correlated response in the second trait that is in the direction designated by the sign of the correlation (e.g. selection for increased macroptery leads to a decrease in development time within the population as a whole and within each morph).

traits that have been investigated in *Tribolium, Oncopeltus, Cydia* and two cricket species (*Gryllus firmus, Allonemobius socius*) are body size, development time and fecundity (Table 13.3). There is evidently no general pattern: body size may be positively correlated with migration or show no relationship; the genetic correlation with development time is negative in all cases shown in the table, but phenotypic correlations suggest that in taxa other than the Orthoptera, development time is either uncorrelated or positively correlated with wing morph in wing-dimorphic species (Roff 1995); early fecundity appears to be negatively correlated with wing morph in wing-dimorphic insects (Roff and Fairbairn 1991), but is positively correlated in *Oncopeltus*, and the sign of the correlation varies with study in *Tribolium casteneum* (Table 13.3).

Predicting the correlated response of fecundity to selection on migratory potential

Selection on the proportion of macropterous *Gryllus firmus* produced an asymmetrical correlated response in fecundity, with a significant decrease in the fecundity of the long-winged (LW) morph when its proportion was increased, but no significant correlated response in the short-winged (SW) morph when its proportion was increased (Fig. 13.2). On the other hand, selection for increased or decreased fecundity produced a roughly symmetrical correlated response in the wing morph (technically the liability; Roff *et al.* 1999). We can account for this difference in correlated response by reference to the underlying mechanism determining fecundity in the two morphs. Fecundity is assumed to be influenced in two ways by pleiotropic effects of the liability genes: one effect is due directly to the liability genes, and corresponds to the usual genetic correlation between traits. A second effect is due to the 'competition' for resources between the flight muscles

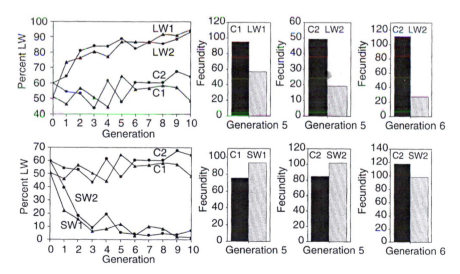

Figure 13.2. Response to selection on proportion of macropterous *Gryllus firmus* and the correlated response in fecundity during the first week after final eclosion. LWi = the ith line selected for increased macroptery; SWi = the ith line selected for decreased macroptery; Ci = the ith control line.

and the gonads. In the latter case the pleiotropic effects of the liability genes are produced via the threshold response, because the effect will be felt almost solely in the macropterous females, micropterous females having no or very small muscles (Roff 1989; Fairbairn and Roff 1990; Zera and Tanaka 1996; Zera *et al.* 1997). For simplicity of discussion we shall assume that the two pleiotropic effects are uncorrelated. Fecundity can now be decomposed into two components:

$$F = X - cA \tag{1}$$

where X is the fecundity in the absence of flight muscles, which may or may not be correlated with the liability, A is the fecundity lost from allocation to flight muscle maintenance, and c is a constant. Assuming, without loss of generality, that macropterous females are those that lie above the threshold (large liabilities), the genetic correlation between the liability and X is negative, whereas the genetic correlation between the liability and A is positive (i.e. large liabilities lead to increased commitment to maintenance of flight muscles and hence to reduced fecundity).

Now consider selection on the incidence of macroptery: to calculate the correlated response of fecundity we proceed as follows:

(1) The liability is assumed to be normally distributed with some initial mean, μ_{X0} and unit variance (Roff 1994a). We assume that the initial proportion that is macropterous is 0.5, as observed in the selection experiment (Roff 1990a). Without loss of generality we set $\mu_{L0} = 0$, from which it follows that the threshold, T, equals zero. Now suppose selection changes the incidence macroptery from 0.5 (our baseline proportion) to p. The direct response on the underlying (liability) scale, R_L, is given by the solution to the equation:

$$p = \int_0^\infty \phi(x)\,dx$$

where $\phi(x)$ is the standard normal deviate. Because the initial mean liability is 0, R_L is the mean liability, μ_L, when the incidence of macroptery is p.

(2) Calculate the new mean value of X, μ_X:

$$\mu_X = \mu_{X0} + r_{GLX}\frac{h_X^2}{h_L^2}\frac{\sigma_X}{\sigma_L}R_L \tag{2}$$

where μ_{X0} is the mean value of X in the initial population, r_{GLX} is the genetic correlation between X and the liability (L), h_X^2 is the heritability of X, h_L^2 is the heritability of the liability (wing morph), σ_X and σ_L are the phenotypic standard deviations of X and the liability, respectively.

(3) Calculate the mean value of X within morphs, which for macropterous (LW) females is given by:

$$X_{LW} = \left(\frac{\int_{-\infty}^\infty \int_{f(a)}^\infty \left(r_{GLX}a\sqrt{0.5h_X^2} + r_{ELX}b\sqrt{1 - 0.5h_X^2}\right)\phi(b)\phi(a)\,db\,da}{\int_{-\infty}^\infty \int_{f(a)}^\infty \phi(b)\phi(a)\,db\,da}\right)\sigma_X + \mu_X \tag{3}$$

where

$$f(a) = \frac{T - \mu_L - a\sqrt{0.5h_L^2}}{\sqrt{1 - 0.5h_L^2}}$$

a and b are standard random normal deviates, and r_{ELX} is the environmental correlation between the liability and X. In the present analysis we set the genetic (r_{GLX}) and phenotypic (r_{PLX}) correlations and computed the environmental correlation from:

$$r_{ELX} = \frac{r_{PLX} - 0.5r_{GLX}\sqrt{h_X^2 h_L^2}}{\sqrt{[1 - (h_X^2/2)][1 - (h_L^2/2)]}}$$

The mean value of X within the micropterous morph (X_{SW}) is determined by setting the limits of integration of the second integral from $-\infty$ to $f(a)$.

(4) Using the above formulae, compute the mean value of A, μ_A. The allocation must always be positive, which we ensured by assuming, arbitrarily, that at $p = 0.5$, $\mu_A = 4$ with a phenotypic standard deviation of 1 (hence the probability of a value less than zero was negligible). Both morphs inherit an allocation value but it is expressed only in the LW females.

(5) Calculate the mean fecundities of the two morphs and the population:

$$F_{LW} = X_{LW} - cA_{LW}$$
$$F_{SW} = X_{SW}$$
$$F_{POP} = pF_{LW} + (1 - p)F_{SW}$$

The above model requires the estimation of 13 parameters. Using data from two experiments, we estimated some parameters and assigned values to others to give the fecundities (eggs laid + fully developed eggs in the ovaries) expected at a proportion macropterous of 50% (Table 13.4). We then used the above quantitative genetic model to predict the correlated change in fecundity when the macropterous proportion was changed by selection to 10% and 90%. As predicted, there was a negligible change in the fecundity of the SW females but the mean fecundity of the LW females and the mean population fecundity decreased with the proportion that was macropterous (Fig. 13.3). The predicted change in LW fecundity was actually less than observed (approximately 40% observed compared with 11% predicted; Fig. 13.2; and Roff 1994a), which suggests a non-linear relationship rather than the linear one assumed in equation (1). (A non-linear function complicates the algebra considerably but can be analysed using numerical methods. Analysis of other geographical populations has shown that the relationship is indeed non-linear.) The foregoing model also qualitatively predicts the correlated response of macroptery when selection acts directly on fecundity (Roff *et al.* 1999).

While the above model mimics the observed behaviour with the particular set of parameter values, alternative responses are possible with other combinations. We have examined a wide variety of parameter combinations and present here two combinations that illustrate the potential complexity of the evolutionary response. These combinations differ from the set appropriate for *G. firmus*, primarily in a large correlation between the

Table 13.4. Parameter estimates for the quantitative genetic model predicting the correlated response of fecundity to selection on wing morph in *Gryllus firmus*

Parameter	Estimate	Source
h_L^2 (liability)	0.5	Roff *et al.* (1997)
h_X^2	0.2^a	Roff *et al.* (1997)
μ_{X0}	400^b	Roff *et al.* (1997)
ϕ_L	1	By definition
ϕ_X	122^b	Roff *et al.* (1997)
r_{GLX}	-0.1^c	—
r_{PLX}	-0.1^c	—
h_A^2	0.5^d	Roff (1994a)
$\mu_{A0},\ \phi_A$	$5, 1^e$	—
r_{GLA}	0.8^f	Roff *et al.* (1997)
r_{PLA}	0.4^g	Roff *et al.* (1997)
c	30^e	—

[a]Approximated by the heritability of fecundity using only the SW females, the fecundity of which, according to the model, is not affected by allocation to flight muscle. [b]Approximated by the fecundity of SW females. [c]No estimate available. We have assumed that this is small relative to the correlation resulting from the 'competition' between fecundity and wing muscles. [d]No direct estimate is available. We have used the heritability of muscle histolysis as an approximate estimate. [e]No estimates available. Values selected to produce approximately the observed fecundity of LW females when the proportion of macropterous females is 20%. [f]No direct estimate available. An approximate estimate is the genetic correlation between fecundity and liability (-0.9; Roff *et al.* 1997); the value used has been reduced slightly (the change in sign simply reflects the scaling relationship), because the foregoing genetic correlation results from the combined effects of X and A, the latter assumed to be more important. [g]Same rationale as above.

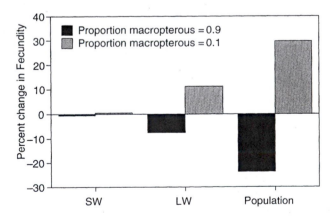

Figure 13.3. Predicted correlated response in fecundity to selection on the proportion of macropterous females using parameter values derived from experiments with the sand cricket, *Gryllus firmus*.

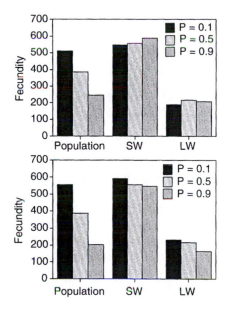

Figure 13.4. Changes in fecundity with selection on the proportion of macropterous females (P), where the initial proportion is 0.5 (middle bars). Parameter values used are: $h_L^2 = 0.5$, $h_X^2 = 0.5$ (lower panel), 0.1 (upper panel), $h_A^2 = 0.5$, $r_{GLX} = -0.6$, $r_{PLX} = -0.7$, $r_{GLA} = 0.6$, $r_{PLA} = 0.7$, $c = 50$, $\mu_{X0} = 500$.

liability and the fecundity component X, and in the first case by a relatively large heritability of X. In both combinations, the population mean fecundity declines with macroptery (Fig. 13.4). In the first case (lower panel, Fig. 13.4), fecundity within both morphs decreases as the proportion of macropterous females increases. However, a change in the heritability of X (from 0.5 to 0.1) changes the pattern observed within morphs (upper panel, Fig. 13.4). The cause of these changes resides in the non-linearity of equation (3). In the SW females the pattern is reversed from the previous case, with fecundity increasing as the proportion of macropterous females increases. A change in the macropterous proportion from 0.5 to 0.1, or from 0.5 to 0.9 produces a correlated decrease in the fecundity of LW females in both cases.

The above results indicate that evolutionary change in the traits genetically correlated with migration may be complex, and serve as notice that predictions of such change require genetic analysis. However, we should not approach such analyses as if the organism were a black box: rather, we should attempt to reconcile the evolutionary responses with a model that incorporates not only the genetic architecture but also a biological interpretation of the interactions. Indeed, we suggest that it is only by an appreciation of the physiological basis of the trade-off between wing morphology and fecundity that the model we have presented above could be derived.

Conclusions

Typically, individuals in a population are divided into two groups, migrants and non-migrants (or dispersers and non-dispersers), to which each has an assumed associated suite of life history, morphological, behavioural, and physiological traits. This dichotomization may be convenient, but it may also be grossly misleading, as it implies a very particular and unlikely genetic architecture underlying the traits. In reality there is a continuum between migrants and non-migrants. Indeed, it is generally more realistic to speak of a probability distribution of migration rather than two categories. Even when there appears to be two types in the population, as in the case of wing-dimorphic insects, the division is not abrupt, for the presence of the capability of migration does not mean that this capability will be exercised.

Whenever appropriate investigations have been made, additive genetic variation has been found in traits that are concerned directly with migration. Thus, as would be suspected from the huge variety of species exhibiting migratory behaviour, migration capability and propensity are traits that can be readily moulded by natural selection. Equally important is the finding that many traits that contribute directly to the fitness of an organism, such as fecundity, are genetically correlated with migratory traits. The pattern of correlations indicates trade-offs both between migration capability and migration *per se*; thus, for example, fecundity of migrants may be reduced, both by virtue of the morphological requirements of migration and dispersal (e.g. flight capability in insects and seed size in plants) and by the direct costs of migration (e.g. energetic costs of movement).

As a consequence of the additive genetic variance and covariance between traits, selection on migratory propensity will generate indirect selection on other traits, and hence changes in migratory patterns will be accompanied by changes in several other traits, such as the age at maturity and the pattern of reproduction. Because of non-linear relationships between traits, the direction of change in traits may not always be obvious and may require the specification of a quantitative genetic model. The theoretical basis for such models is now available but at present we have few analyses that have combined quantitative genetics, life history, behaviour, and physiology. The challenges in such a project are enormous but, as illustrated by the above analysis in the sand cricket, it is only with such an approach that the complexity of evolutionary processes can be dissected and properly understood.

14

Condition-dependent dispersal

Rolf A. Ims and Dag Ø. Hjermann

Abstract

We review conditions found to induce dispersal in animals. Condition-dependent dispersal is due to both environmental conditions (typically habitat and food quality, population density, local demographic and social structure) and internal conditions (typically fat reserves, body size, and competitive ability). The pathway between an environmental condition causing dispersal and the response (i.e. dispersal) can be either direct (immediate behavioural response) or indirect (mediated by changed internal condition, i.e. the phenotype). Indirect pathways can be extended in the environmental domain by the action of indirect cues, and in the internal domain by the action of maternal effects. Alternative pathways can incur different costs, either in terms of physiological and behavioural trade-offs when phenotypic traits aiding dispersal develop, or when there are time lags between the environmental cue and the response. Long time lags due to complex pathways can result in a temporally mismatched dispersal if the environmental condition exhibits low temporal predictability or stability. In particular, maternally mediated dispersal is not expected to evolve when there is low temporal autocorrelation in environmental conditions, because such a pathway imposes a considerable time lag. We also point out the consequences of spatial environmental autocorrelation and use density-dependent dispersal as an example. Depending on the spatial scale of the covariance of population density fluctuations (i.e. population synchrony) relative to the maximum ranges of exploration and dispersal of the focal species, different forms of density dependence may result. We argue that the lack of an explicit consideration of the spatial scale of important environmental conditions relative to organism's dispersal or exploration capacity underlie many seemingly inconsistent results concerning condition-dependent dispersal in the literature. Also, the fact that the same condition may have different effects on different dispersal stages (dispersal stage dependence), and that several conditions or cues operate simultaneously and interactively, complicates the study of condition-dependent dispersal. There is a strong need for future studies to take a multifactorial approach on appropriate spatial and temporal scales.

Keywords: phenotypic plasticity, maternal effects, environmental cues, internal condition, density dependence, environmental autocorrelation

Introduction

In a spatiotemporally varying environment it is critical for animals to be able to match their own internal condition (i.e. phenotype) against prevailing environmental conditions in terms of fitness prospects. When the outcome of this matching indicates low fitness prospects in a given environment, the individual may choose to seek better opportunities elsewhere by means of dispersal. This chapter concerns such *condition-dependent dispersal*. The literature on condition-dependent dispersal is overwhelming, and it is not our task to provide a comprehensive review of this literature here. We rather emphasize some issues that are of general importance for interpreting the causes and consequences of condition-dependent dispersal and which we think have not received sufficient attention. In particular, it appears that there are many inconsistencies in the literature concerning how dispersal is conditioned. We provide some suggestions for why such inconsistencies have arisen and how they may be solved. Empirical examples will be drawn exclusively from the animal kingdom, mainly from terrestrial environments.

Different types of condition dependence

Internal and environmental condition dependence

We find it appropriate to distinguish between two main classes of conditions that make dispersal more or less likely: (1) those characterizing the individual organism, hereafter termed *internal conditions*, and (2) those characterizing the external environment, hereafter termed *environmental conditions*.

Concerning environmentally conditioned dispersal, most (if not all) aspects of the environment (both biotic or abiotic) have been reported to affect dispersal propensity in animals. However, some environmental conditions seem to be more general and more pervasive than others, and can be related to the main hypotheses of ultimate causes of dispersal (several chapters in this book). Most studies of environmentally conditioned dispersal have focused on habitat quality, population density, social or demographic structure, and photoperiod or season. We will not deal here with such factors as photoperiod. Photoperiod indicates with 100% certainty that the seasons are changing at the regional scale and thus functions more as a signal than a cue. Typically, organisms meet such temporally long-lasting and regionally synchronized environmental changes by escaping in time (e.g. diapause), rather than by escaping in space by means of dispersal (Southwood 1977) (although migratory movements evolve under such circumstances). The effects of density and/or food limitation are well known for a number of insects, especially such sap-feeding insects as aphids and plant-hoppers (reviewed by Dixon 1985; Denno *et al.* 1994; Denno and Peterson 1995), but also for a range of other invertebrates such as collembolans, grasshoppers, true bugs (Heteroptera), moths, beetles, and flies (e.g. Johnson 1969; Bengtsson *et al.* 1994a; Denno and Peterson 1995; Fonseca and Hart 1996; Scholz *et al.* 1997; Applebaum and Heifetz 1999). Study of habitat stability has been emphasized in studies of invertebrates (Southwood 1977; Solbreck 1978), sometimes in conjunction with population density. Population density can be expected to have a smaller influence when the habitat is temporary, which has been confirmed by interspecific and intraspecific comparisons of plant-hoppers (Denno *et al.* 1991). Studies of density-dependent dispersal, and in particular the effect of social and demographic factors (including genetic relatedness), have focused mainly on mammals and birds, often in relation to hypotheses on the evolution of dispersal (e.g. Gaines and McClenaghan 1980; Pusey 1987; Anderson 1989; Johnson and Gaines 1990; Cockburn 1992; Krebs 1992; Wolff 1997; Lambin *et al.*, chapter 8; Murren *et al.*, chapter 18). However, local demographic structure may affect invertebrates as well. For instance, emigration of males of the red milkweed beetle (*Tetraopes tetraophtalamus*) increases dramatically when the sex ratio is male biased (Lawrence 1987). Finally, interspecific interaction may induce dispersal in parasite–host and predator–prey systems (see Boulinier *et al.*, chapter 12a).

By internal conditions, we refer to temporally persistent and measurable behavioural, morphological, and physiological traits (i.e. the phenotype) that have been modified by the organism's present and past environment (including prenatal life), rather than differences produced by genotype. Common internal conditions correlated with dispersal propensity are body size and energy reserves. Typically, such conditions also correlate with fecundity, mating success, and survival probability. For example, McNeil *et al.* (1995) reported increased propensity to fly among the offspring of females of the spruce budworm (*Choristoneura fumiferana*) when the females had mated with poor-quality males (fed on limited, poor foliage). Similarly, fat male Belding's ground squirrels (*Spermophilus beldingi*) dispersed earlier than lean males, and winter survival (during hibernation) increased with the amount of fat reserves (Nunes *et al.* 1997). Competitive ability may also influence whether an individual disperses or not, although few experimenters have set out to test this (see Lambin *et al.*, chapter 8). Behavioural phenotype, developed at an early ontogenetic stage, may determine whether an individual will disperse at a later stage (Bekoff 1977; Holekamp 1986). For instance, young female grey-sided voles (*Clethrionomys rufocanus*) which showed an asocial behaviour type in arena tests against an unfamiliar female were the most likely to disperse from their natal site (Ims 1990).

Our definitions of internal and environmental conditions are somewhat related to W.E. Howard's (1960) original concepts: *innate dispersal* and *environmental dispersal.* Howard defined environmental dispersal as dispersal induced by specific conditions of the prevailing environment (environmental cues). Our definition of environmentally conditioned dispersal is equivalent. In contrast, our view of the innate (or internal) component of dispersal differs from Howard's view. Howard used innate dispersal to denote what he thought were rigid predispositions programming certain animals to disperse regardless of the prevailing environmental conditions. Hence, innate dispersal was inherited, inflexible, and not necessarily adaptive at the level of the individual. Actually, Howard (1960) felt a need to invoke group selectionistic arguments to explain its evolution.

Direct and indirect pathways

According to Howard's (1960) definition, the origins and functions of dispersal behaviour differed fundamentally between innate dispersal and environmental dispersal. In contrast, we will emphasize that internal predispositions for dispersal can be both flexible and adaptive, and, moreover, that environmental and internal determinants of dispersal often seem to be different mechanisms that may trigger dispersal based on the same environmental condition and the same ultimate (evolutionary) cause. Thus, an environmental condition may trigger dispersal directly as an immediate behavioural response, or indirectly by first altering the internal body condition which in turn triggers dispersal (Fig. 14.1A). We will refer to these two alternatives as *direct* and *indirect pathways*, respectively.

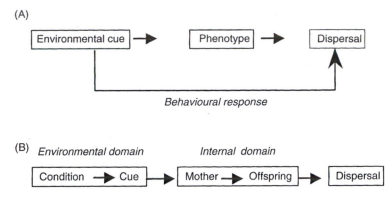

Figure 14.1. Alternative causal pathways between an environmental cue and the consequent dispersal response. (A) Direct pathways following from an immediate behavioural response, and indirect pathway mediated through a changed phenotype. (B) Extended, indirect pathways, due to several links in the environmental and internal domains.

For instance, it is often adaptive to react to a local deteriorating food resource by dispersing. In some species (or in some situations), low food quality may induce dispersal behaviour without any change in the body condition (except for the necessary nervous reactions in the brain). In this case the dispersal event has a mechanism similar to any other movement event used in foraging activities (Ims 1995). In other animals or situations, dispersal will not be induced until the deteriorated resource level has led to changes in the animal's internal condition. For example, low body growth rate at the nestling stage and resulting low juvenile body mass delayed natal dispersal in young crested tits (*Parus cristatus*) (Lens and Dhondt 1994). Similarly, rapid body fat accumulation in female Belding's ground squirrels enhanced both the timing and rate of natal dispersal (Nunes *et al.* 1997). In many insect species, it has been found that the food quality experienced during development influences the proportion of adults that disperse (or develop into dispersing morphs). Thus, food quality apparently leads to a change in the phenotype, usually assumed to involve a change in the hormonal state (e.g. the level of juvenile hormone). This particular hormonal state then stimulates dispersal (Dufty and Belthoff, chapter 15), and in many (but not all) species it simultaneously delays reproduction (the so-called 'oogenesis–flight syndrome'; see also Roff and Fairbairn, chapter 13).

When the presence or absence of certain compounds in the food induces dispersal, this does not necessarily mean that these compounds are advantageous or disadvantageous. They may simply act as cues indicating increasing population density, before the quantity or quality of food becomes a limiting factor. In this case, the causal pathway is extended in the environmental domain (Fig. 14.1B). In other situations (perhaps most commonly), dispersal may be induced only when the food quality *per se* becomes low (whether the change in the food is caused by high densities or not).

Environmental conditions (i.e. population density or habitat characteristics) often provide good predictions of average dispersal rates at population level. However, there is usually substantial variation in dispersal propensity among the individuals within populations. The reason for this may be genetic differences, different internal conditions, or because each individual experiences different environmental conditions. Moreover, we can expect that these factors interact with one another, rather than affecting dispersal propensity independently. For example, the effect of food quality or quantity on dispersal propensity depends on sex in Belding's ground squirrels (Nunes *et al.* 1997). Moreover, there are many examples demonstrating that density-dependent dispersal interacts with sex and social status (for a review see Brandt 1992). Different genotypes sometimes vary in their reaction norm towards environmental conditions. For instance, low temperatures induce dispersal in temperate, but not in Puerto Rican, populations of the milkweed bug (*Oncopeltus fasciatus*) (Dingle 1994). For this species, Groeters and Dingle (1987) also found variation between families from the same population.

The dispersal propensity of individuals is sometimes controlled by the conditions experienced by the mother. Such maternally induced dispersal adds another link in the 'internal domain' of the causal pathway between the environmental cue and the dispersal response (Fig. 14.1B). Maternal effects on dispersal propensity are especially well known from aphids. Aphid mothers fed on poor food, or kept under crowded conditions, give birth to winged daughters (Dixon 1985). Even 'grandmaternal' effects have been reported for aphids (MacKay and Wellington 1977). Outside aphids, relatively few species have been reported to have maternal effects on dispersal traits (but see Diss *et al.* (1996) for more examples in insects, and Massot and Clobert (1995), Léna *et al.* (1998), Ronce *et al.* (1998), de Fraipont *et al.* (2000) for the particularly well studied common lizard *Lacerta vivipara*). Maternal effects have been reported for many other traits in a great variety of taxa (reviewed by Mousseau and Dingle 1991; Rossiter 1996). We suspect that future studies will demonstrate that maternally induced dispersal is a fairly common phenomenon.

Costs and effects relative to pathways

Costs at the individual level

Dispersal is generally thought to be costly in terms of other fitness-related traits, such as survival (reviewed by Rankin and Burchsted 1992; Johannesen and Andreassen 1998; Aars *et al.* 1999). We focus here on two types of costs, and how they are influenced by the pathway between the environmental cue and the dispersal event. First, because indirectly induced dispersal (by definition) involves persistent changes in behaviour, morphology, and physiology, this pathway can be expected to be more costly than when dispersal is induced as a direct behavioural response. This is because physiological trade-offs may be imposed. For instance, in insect species with flight polymorphism, the fairly significant amount of energy used to develop and maintain wings and flight muscles is not available for reproductive investment. Although there are exceptions, flight capability is associated with decreased female fecundity for a wide variety of species (Zera and Denno 1997, and references therein). For a few species, winged males have been shown to have decreased mating success.

The second type of cost is associated with the length of time between when the environmental cue elicits the chain reaction to when dispersal takes place. The longer the time lag between cue and dispersal (see Fig. 14.2), the larger is the risk that the ecological

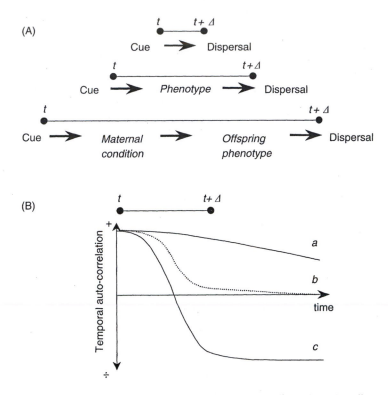

Figure 14.2.   Time lags imposed by alternative causal pathways for condition-dependent dispersal, relative to environmental stability or predictability. (A) Increasing time lags due to increasing number of links in the causal chain. (B) Time lag of the dispersal response in relation to the temporal predictability or stability of the environmental condition inducing dispersal, depicted as temporal autocorrelation functions. Function *a*: Long-term stability and predictability. Function *b*: No long-term stability or predictability. Function *c*: No long-term stability, but high predictability.

condition ultimately triggering the response may have changed by the time the response is made (Gatehouse and Zhang 1995). This would render the dispersal response maladaptive.

Effects at the population level

Positive density-dependent dispersal will generally have a stabilizing effect on populations. However, if there is a time delay from the cue (high density) to the actual dispersal event, population oscillations of a large amplitude are likely to result (as would any other process with time-delayed negative feedback; Maynard Smith 1978a). Time lags may be increased by a chain of reactions in the environmental or the internal domains (Fig. 14.1B), for example if the animal uses host plant condition as a cue for density, as most aphids and plant-hoppers do (Denno and Peterson 1995), or if population density affects maternal condition (Ims 1990).

Dispersal rates have also been reported to be negatively density dependent, so that a smaller fraction of individuals disperses at higher densities (Wolff 1997; Lambin *et al.*, chapter 8). Furthermore, those that emigrate may be expected to move relatively short distances (McCarty 1997). This has been reported in mammals (e.g. Lambin 1994a) as well as in birds (e.g. Delestrade *et al.* 1996; Wicklund 1996). Although positive density dependence may be the most common pattern in insects (see review by Denno and Peterson (1995) for sap-feeding insects), negative density-dependent emigration (and positive density-dependent immigration) was reported in a large-scale field experiment on the butterfly *Melitaea cinxia* (Kuusaari *et al.* 1996). Furthermore, Diss *et al.* (1996) reported that, when genetic effects had been accounted for, daughters of the gypsy moth (*Lymantria dispar*) had a shorter prefeeding stage (i.e. a shorter 'dispersal window') when their mothers fed on leaves with higher levels of tannin (a result of herbivore damage). It has been proposed that this and other maternal effects are responsible for the outbreak pattern frequently observed in North American forest lepidoptera (Rossiter 1991; Ginzburg and Taneyhill 1994; but see also Harrison 1995; Myers *et al.* 1998).

Predicting pathways of condition-dependent dispersal

Direct versus indirect pathways

Given the alternative pathways by which a given environmental cue can induce a response in terms of dispersal (Figs 14.1 and 14.2A), one may wonder why organisms often seem to adopt indirect pathways that involve several steps and long time lags. Of course, whether the pathway is indirect or direct will not matter if the environmental condition shows temporal stability or predictability that encompasses the time lag imposed by the actual pathway. On the other hand, we should not expect to find pathways imposing time lags when the lag exceeds the characteristic temporal scale of an important environmental condition (Fig. 14.2B). In other words, erratically changing conditions require direct and rapid responses, and gradual changes do not. Examples of rapid and unpredictable changes, which would be expected to prompt dispersal as an immediate response, include a lack of availability of receptive mates, invasion of natural enemies, and extreme climate events, while predictable changes in the environment include factors such as season and (sometimes) resource availability.

Whether to adopt direct or indirect causal pathways of condition-dependent dispersal may be optional in some species, especially in those invariably expressing one mobile, monomorphic phenotype (the typical case in vertebrates). Other species may not have any alternative but a pathway that imposes a significant time lag, due to developmental constraints. For example, a pathway that includes an intermediate link is obviously necessary when certain phenotypic traits that facilitate dispersal, such as wings in insects, must be developed. The length of the time lag between the moment the animal perceives the environmental cue and when the dispersal event can take place depends on the development time of the dispersal-aiding trait. For instance, in insect species with both flight-capable and flightless morphs, the insect must necessarily 'decide' whether it will develop a flight apparatus at some point before the adult stage (when the body form is fixed). This development time appears to be constrained by higher-level phylogeny. The general pattern is that aphids make this decision just before birth or during the early juvenile stages, plant-hoppers during the middle instars, whereas many cricket species can wait until the last instar to decide whether to develop into the long-winged morph or not (Zera and Denno 1997). It is likely that choosing the developmental pathway early in life will lessen the fecundity cost of the flight apparatus. However, generation times also differ between taxa: aphids have six to seven generations per year (or more), while crickets have only one or two. Generally, we can expect that the selection pressure for postponing the decision is higher in animals with a long generation time.

Of course, an adult insect that has developed a flight apparatus can decide to stay if it turns out that conditions have not deteriorated as 'expected' when it decided to develop wings. For instance, wingless mothers of the aphid *Aphis fabae* react to crowding by producing winged offspring, but the offspring will disperse before reproduction only if it experiences crowding itself. Uncrowded offspring will reproduce without dispersing at all, and offspring crowded only early in life will reproduce both where it was born and where it arrives after dispersal (Shaw 1970a). This usually has a cost in terms of lower fecundity.

Some species have the ability to histolyse their wing muscles and use the energy for reproduction (e.g. some aphids, water striders, and crickets; see references in Zera and Denno 1997); this probably reduces the cost of 'changing one's mind'. Some species, for example bark beetles, go a step further: they histolyse their wing muscles when they establish breeding galleries, but males are able to regenerate the wing muscles completely in 5 days if breeding fails (Robertson 1998). However, in many experiments, density manipulations appeared to have no effect (e.g. Englund 1993; Morse 1993; Fadamiro *et al.* 1996; Midtgaard 1999). One reason for this may be that the insects are sensitive to density only when it is experienced for a prolonged period, or in another stage; indeed, in one of the experiments by Fadamiro *et al.* (1996), it was found that the rearing density affected dispersal, although the density in the flight experiment did not. Below, we discuss other sources of seemingly inconsistent results concerning density-dependent dispersal.

The shortest and seemingly simplest pathway for environmentally conditioned dispersal includes only a direct behavioural response (see Fig. 14.1). Behavioural flexibility is widely demonstrated, even in what have been considered to be relatively 'hard-wired' taxa (Thornhill and Alcock 1983), and it is often taken for granted that there is more room for flexibility in behavioural than in morphological or physiological traits. However, there are also limits to behavioural flexibility. The fact that certain behavioural disperser phenotypes seem to originate during early ontogenetic stages (Holekamp 1986; Dufty and Belthoff, chapter 15) indicates that behavioural flexibility at later stages may be somehow constrained. It may also be questioned whether behavioural processing is efficient and reliable in every situation. Perhaps, when confronted with cryptic environmental conditions or the simultaneous action of multiple dispersal cues, the best way of processing and integrating complex or diffuse information is through a gradual change in the internal condition (e.g. physiology).

When should maternally induced dispersal be expected?

The existence of pathways in which maternal effects constitute a link in the causal chain between an environmental condition and dispersal is intriguing, especially as such pathways impose long time lags (Fig. 14.2). Still, the adaptive significance of maternally induced dispersal generally has received few explicit considerations in the literature (but see Sorci *et al.* 1994; Massot and Clobert 1995; Léna *et al.* 1998; Ronce *et al.* 1998; de Fraipont *et al.* 2000). The long time lags that may be induced by maternal intervention accentuate the importance of temporal stability or predictability of the causal environmental condition. Thus, one should not expect environmental cues to be mediated through maternal effects, unless they are very predictable in time. Concerning the question about the role of developmental constraints, we may expect maternal factors to be involved when phenotypic specialization for dispersal must take place at an early ontogenetic stage, and when the offspring themselves at that stage are not capable of perceiving cues from the external environment (for example, when the offspring are still inside their mother). In this sense, a maternal link in the pathway of condition-dependent dispersal conveys information about the environment not yet accessible to the offspring. This inter-generation transfer of information may benefit both generations. However, it also opens the possibility for mothers to control the dispersal propensity of their offspring, which may set the stage for offspring–parent conflict games when the optimal reaction norm differs between the two parties (Murren *et al.*, chapter 18). How often, and under which conditions, maternally induced dispersal involves conflicting interests between mother and offspring has yet to be explored, both theoretically and empirically.

Sources of inconsistent condition-dependent dispersal responses

Different empirical studies examining the effect of the same dispersal-inducing condition often come to different conclusions. For example, better internal condition (measured as the amount of energy reserves) is sometimes found to increase dispersal propensity, while it in other cases seems to decrease it (for discussions see, for example, Massot and Clobert 1995; Nunes and Holekamp 1996). Similarly, there has been considerable controversy about the presence and signs of density-dependent dispersal (Gaines and McClenaghan 1980; Stenseth 1983; Anderson 1989; Johnson and Gaines 1990; Krebs 1992; Wolff 1997). Of course, there may be many reasons why there are inconsistencies among empirical studies in ecology, perhaps especially so among studies of dispersal where both conceptual and methodological issues are far from settled (see Stenseth and Lidicker 1992; Ims and Yoccoz 1997). Here, we will set the methodological and definitional problems aside, and focus on four biologically non-trivial reasons why inconsistencies between studies may emerge. We believe that these issues have not been dealt adequately with in the literature on condition-dependent dispersal.

Stage-dependent dispersal

The process of dispersal consists of three interdependent stages: emigration from a site, transience, and immigration on to a new site. A condition influencing dispersal may exert its influence at any one of the three stages either to promote or to inhibit dispersal (Fig. 14.3). In this way the same condition may act to reduce dispersal propensity at one stage (e.g. immigration), while it may increase it at another stage (e.g. emigration). The realized dispersal response (i.e. the one we will observe) will, owing to such *stage-dependent dispersal*, be shaped by the combined actions of a condition during the entire

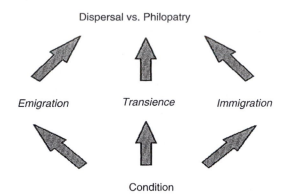

Figure 14.3. Stage-dependent dispersal. A condition may exert different influences on the three stages of dispersal (emigration, transience, and immigration). Dispersal or philopatry will result from the joint response.

dispersal process. A condition's relative impact on the different dispersal stages will be crucial in determining whether an individual disperses or not and, eventually, how far it will disperse.

The significance of stage-dependent conditional dispersal can be illustrated in the cases of density-dependent and internal condition-dependent dispersal. High local population density will generally tend to promote emigration, owing to intensified local competition. On the other hand, high density may act to inhibit movements in the transience stage (and thus dispersal) if residents interact aggressively with the moving individual. Such aggressive interactions are assumed to be responsible for the reduced dispersal distances that sometimes are observed at high population densities (e.g. Lambin 1994a; Wiklund 1996). Also, high density in putative recipient habitats may act to inhibit immigration. The mode of interaction or competition between individuals is crucial for the outcome of this dispersal stage-dependent action of density. Competitive inhibition of dispersal is likely to be most pronounced in territorial or strongly socially structured populations, where resident animals may form social fences (see Hestbeck 1982) against transients and immigrants. Dispersal may be assumed to be more 'free' in populations where individuals do not interfere directly with one another (e.g. scramble competition for resources). Moreover, there will be little or no effect of population density during the transience stage in very patchy habitats, where the matrix between the patches is devoid of conspecifics. Finally, conspecifics are not just competitors, but also potential mates and sources of information (Danchin *et al.*, chapter 17), and conspecific attraction can increase rate of emigration into recipient habitats, at least when densities are low to moderate (Stamps, chapter 16).

Considering dispersal stage-dependent effects of body fat reserves, we may expect that large fat reserves indicate high home-site quality, which should then cue for philopatry. On the other hand, individuals with large energy reserves may better endure the transience stage, especially when it is energetically costly (Nunes and Holekamp 1996; O'Riain and Braude, chapter 10). Consequently, individuals in good condition may be more prone to disperse, and to do so over longer distances, than individuals in poorer condition. Finally, individuals in good condition may be most capable of acquiring resources or space when they attempt to settle in a new habitat. Thus, the resultant effect of fat reserves on dispersal propensity will depend on the particular characteristics of the species and the external ecological circumstances under which it is found.

Spatial scale dependence

Dispersal may be viewed as a way by which organisms can escape unfavourable environmental conditions by means of spatial displacement. Consequently, a condition for dispersal to be at least a potentially successful escape mechanism is that the spatial scale of the unfavourable condition should not exceed the maximum dispersal (i.e. displacement) capacity of the organism. This kind of spatial scale dependence can be considered explicitly by plotting the spatial covariance of the dispersal-triggering condition (Fig. 14.4). The covariance function, which depicts the spatial predictability of an environmental condition (for example, habitat quality or population density) in space, should be compared with two other scaled entities, namely the maximum dispersal and exploratory ranges of the focal organism (Fig. 14.4). With respect to the exploratory range, it has been shown that some animals actively explore their surroundings before deciding to disperse (Bondrup-Nielsen 1985; Larsen and Boutin 1994; Price *et al.* 1994). The scale of exploration may be the same or smaller than the actual dispersal distance. The need for exploration before dispersal can be expected to be largest when the spatial predictability of the environment is low (i.e. covariance function *b* in Fig. 14.4).

The presence of spatial autocorrelation in ecological factors has received increasing interest from ecologists (Koenig 1999). In particular, there has been much focus on estimating autocorrelation (i.e. synchronization) in population density fluctuations, at scales both smaller (Sutcliffe *et al.* 1996; Ims and Andreassen 1999) and larger than the dispersal range of species (Bjørnstad *et al.* 1999). Few empirical studies of condition-dependent dispersal have taken into account considerations about spatial constraints on animal movement capacity in conjunction with environmental auto-covariance, to any significant degree, for example as illustrated in Fig. 14.4 (see Ray and Hastings (1996) for general consideration on the effect of scale on detection of density dependence).

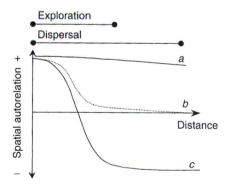

Figure 14.4. Spatial scaling considerations of importance to condition-dependent dispersal. The spatial autocorrelation functions depict the spatial predictability of an environmental condition some distance from a given site. The spatial predictability of the environment should be related to two spatial characteristics of the organism: its maximum exploration and maximum dispersal range. See main text for further explanation.

We suspect the confusion about the signs of density-dependent dispersal, which is especially pronounced in the literature on small mammals (e.g. Stenseth 1983; Krebs 1992), has its origin in a lack of explicit analysis of spatial covariance and of synchronization of density fluctuation at a spatial scale relevant to dispersal processes. We can illustrate this by considering the following hypothetical example.

Assume that we are to test for density dependence in dispersal rate by regressing a time series of emigration rate estimates obtained from a local census plot against the contemporal density estimates from the same plot (e.g. Krebs 1992; Lambin *et al.*, chapter 8). The census plot is typically much smaller than the exploration–dispersal scale of the animal in question (Koenig *et al.* 1996). The general population within the dispersal range from the census plot may exhibit density variation both in time and space. Then, when spatial population synchrony is consistently larger than the dispersal range (i.e. covariance function *a* in Fig. 14.4), dispersal rates can be expected to be generally low and density independent, because density and the intensity of competition will be spatially homogeneous. On the other hand, consistently asynchronous density fluctuations (i.e. covariance function *c* or *b*) would be expected to yield positive density-dependent dispersal (i.e. emigration rate correlates positively with density on the census plot), because the individuals may find space with lower density within their dispersal range. Finally, a more complex spatiotemporal density fluctuation scenario with spatial covariance patterns alternating between function *a* (Fig. 14.4) at high local density and function *c* (or *b*) when local density is lower, will give rise to a negative slope of the dispersal rate/density regression (negative density-dependence dispersal rate). This last scenario may fit cyclically fluctuating populations experiencing habitat saturation and 'frustrated dispersal' at peak densities, and a more patchy distribution and 'presaturation dispersal' at low densities (Lidicker 1975; Bondrup-Nielsen and Ims 1988).

The relationship between density-dependent dispersal and spatiotemporal population dynamics may be more complicated than depicted in the examples above. One potentially complicating aspect is that dispersal can also determine the degree of population synchrony (Bjørnstad *et al.* 1999). In that case there will be an interaction between population dynamics as a cause, and as a consequence, of dispersal. Furthermore, stage-dependent dispersal (see above), habitat heterogeneity (Stamps, chapter 16), and social organization (Wolff 1997) may mould the relationship between population density and dispersal propensity. Only new studies addressing these issues at appropriate temporal and spatial scales (see Ray and Hastings 1996), and with study designs that are able to tease the various factors apart, will be able to determine the causes and consequences of different forms of density-dependent dispersal.

Multiple condition dependence: confounding and interaction

Dispersal can have multiple causation (Fig. 14.5), and this is obviously a possible source of inconsistent results among empirical studies examining the effect of single conditions. When several conditions act in concert on dispersal propensity, both interaction and confounding among variables may intrude on the results in unknown ways if only a single variable is considered at one time. Multifactorial approaches to the study of conditional dispersal are therefore warranted. Different approaches may be adopted depending on the particular study system and the kind of data to be analysed. In an

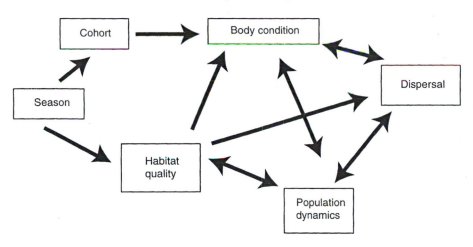

Figure 14.5. Inter-relations that may cause confounded and interactive effects between some common dispersal-related conditions.

observational setting, different *a priori* causal pathways may be evaluated using path analysis (Mitchell 1993). Path-analytical approaches may help to elucidate the relative effects of simultaneously acting predictor variables on dispersal propensity (as the response variable), and whether the predictors are likely to operate in direct or indirect ways. We are not aware of studies in which path analyses have been used to analyse condition-dependent dispersal. Factorial experiments have just started to be used to examine condition-dependent dispersal (e.g. Léna *et al.* 1998), and this will certainly be the most powerful approach to demonstrate multiple causation and interactive effects. Although there are still few studies that have explicitly tested for interactions between dispersal-triggering cues or conditions by applying appropriate study designs and statistical analyses, there is now sufficient evidence available to suggest that interactive effects between dispersal-triggering conditions may be common and powerful. Internal conditions can interact with one another; a few examples are: interaction between mated status (mated or unmated) and age (the noctuid *Helicoverpa armigera*, Armes and Cooter 1991; the tortricid *Cydia pomonella*, Schumacher *et al.* 1997b) or mated status and body size (e.g. the diamondback moth, Shirai 1995). External factors such as density and habitat quality may interact; for example, in the English grain aphid, habitat quality has a larger effect when density is high (Dixon 1985), while the opposite was found to be true in the collembolan *Onychiurus armatus* (Bengtsson *et al.* 1994a). Finally, internal conditions may interact with external conditions in complex and subtle ways, for instance in aphids the complex interaction between the age of the grandmother at reproduction and response to crowding, reported by MacKay and Wellington (1977), and body condition interacting with female density, in the common lizard (Léna *et al.* 1998).

Conclusions and perspectives

Very few researchers have tried to make quantitative predictions about the expected dispersal rate under different conditions. One of the few is Ozaki (1995), who tested a modified version of Hamilton and May's (1977) model on the migration rates of gall-forming aphids (*Adelges japonicus*), and found a good fit between the data and the model's predictions. We have in this chapter suggested some crucial aspects that need to enter such models for them to give reliable prediction, even with regard to qualitative aspects of dispersal (e.g. the sign of density dependence). In particular, it is important to consider the fact that dispersal often is a multi-cause phenomenon that is conditioned by different factors operating at different ontogenetic stages during the organism's life cycle, and during different stages in the dispersal processes (i.e. dispersal-stage dependence). The simultaneous action of multiple conditions opens the possibility for interactive effects, which need to be considered in predictive models, and especially in appropriately designed factorial experiments.

A great challenge to the studies of causes and consequences of condition-dependent dispersal is to examine this phenomenon from the perspective of spatial and temporal environmental variability (Danchin *et al.*, chapter 17). The idea that the predictability of influential environmental factors in time and space is crucial for the evolution of dispersal is quite old; however, it has been expressed mainly in hypothetical terms (e.g. Gadgil 1971; Southwood 1977; Solbreck 1978). Currently, the statistical tools for estimating the spatiotemporal auto-covariance patterns in ecological conditions are well developed and widely available (Bjørnstad *et al.* 1999). What has yet to be done is to combine such pattern-oriented approaches with the relevant biological aspects of the focal species, in particular constraints and flexibility in causal pathways leading to dispersal, as well as the capacity for organisms to explore their surroundings. The adaptive value of dispersal hinges on whether organisms are able to track their environment when crucial conditions are changing in space and time. Ecologists need to explore condition-dependent dispersal, both to understand dispersal as a fascinating biological phenomenon, and to predict the fates of the many organisms that are now challenged by human-induced environmental degradation.

15

Proximate mechanisms of natal dispersal: the role of body condition and hormones

Alfred M. Dufty, Jr. and James R. Belthoff

Abstract

We examine proximate issues related to the initiation of natal dispersal in birds and mammals, and focus on the interplay among hormone secretions and body condition in determining dispersal by individuals. In at least some mammals, early exposure to testosterone in males affects their subsequent dispersal, the timing of which is related to acquisition of a threshold level of body mass and body fat. Because in birds both sexes generally disperse, it is less likely that a gonadal hormone associated more with one sex than with the other plays a major role in that process. Indeed, the adrenal hormone corticosterone has been implicated in the timing of avian dispersal. Increases in corticosterone concentration near the time of dispersal come about either through endogenous or external mechanisms, and it is likely that these changes in hormone secretion are affected by attainment of body condition sufficient to disperse. Other substances, such as leptin, affect the hypothalamo-pituitary–adrenal axis, locomotor activity, and/or feeding behaviour, and may play a role in vertebrate dispersal.

Keywords: corticosterone, gonadal steroids, body condition, maternal effects, leptin, physiology

Introduction

As with any behavioural pattern, questions about natal dispersal can be addressed on proximate and/or ultimate levels (Mayr 1961; Tinbergen 1963; see Holekamp and Sherman 1989, for a discussion of proximate/ultimate issues and dispersal). Considerable attention has focused on factors that affect the patterns, sex biases, and demographic, genetic, or population forces affecting natal dispersal (Stenseth and Lidicker 1992). However, far fewer studies have addressed the underlying endocrine and physiological factors influencing natal dispersal in vertebrates. That is, questions regarding the proximate mechanisms of this behaviour pattern remain largely unanswered. For example, what are the hormonal influences on natal dispersal of young animals? What physiological processes are involved? Finally, what is the effect of physical condition on natal dispersal? Our objective in this brief chapter is to bring to light both endocrine effects and other physiological factors affecting natal dispersal and to highlight but a few avenues of future research that might bear fruit. We concentrate on mechanisms in mammals and birds. However, we readily acknowledge the growing literature examining physiological aspects of locomotor activity and dispersal in other taxa (e.g., Boyd 1991; Lowry and

Moore 1991; Clobert *et al.* 1994b; Lowry *et al.* 1996; McCormick 1998), including invertebrates. For example, it has become well known that in insects (for reviews, see Dingle 1996; Dingle and Winchell 1997; Zera and Denno 1997) juvenile hormone is a primary hormone influencing movement ('migration' in the insect literature) and dispersal polymorphism, as well as reproduction. This hormone is synthesized in the corpora allata, which are paired neurosecretory organs located along a short nerve pathway behind the brain. In many species, exogenous application of juvenile hormone increases flight behaviour in tethered flight situations, and inhibition of the pathway that produces juvenile hormone has the opposite effect, i.e., suppression of long-duration migratory flight (Rankin *et al.* 1986; Rankin 1989, 1991). Moreover, in crickets (*Gryllus* spp.), long- and short-winged forms reflect higher and lower propensity for movement, respectively, and degradation of juvenile hormone by the enzyme juvenile hormone esterase permits metamorphosis into a macropter (Zera and Tiebel 1989; Zera *et al.* 1989; Zera and Denno 1997). There is no comparable hormone in birds and mammals, nor is there the degree of understanding of the proximate bases for movement in these vertebrate taxa. However, while the study of hormonal mechanisms affecting dispersal movements in birds and mammals is in its infancy, it nonetheless offers much promise, as we summarize below.

Body condition and natal dispersal

Because natal dispersal is potentially energetically expensive, animals that disperse before they achieve the necessary energy reserves almost certainly increase the already high risks associated with these movements. Dispersers are vulnerable to predation and may encounter aggression from conspecifics as the former attempt to establish themselves in new territories or groups (Holekamp and Smale 1998; Nunes *et al.* 1998). Thus, acquisition of a sufficiently robust physical condition prior to natal dispersal probably plays a critical role in determining the timing and, perhaps, other aspects of natal dispersal by young animals. Indeed, natural selection could favour animals that delay departure from natal areas until their energy reserves are adequate to support the potential demands of the dispersal movement (Nunes and Holekamp 1996; Belthoff and Dufty 1998; Nunes *et al.* 1998).

Whether the proximate stimulus for natal dispersal is endogenous, exogenous, or a combination of both, an organism's metabolic and behavioural response to a particular stimulus may depend, at least in part, on its physical condition at the time the stimulus is applied (Ketterson and King 1977). In birds, the normal postabsorptive overnight-fasted state does not involve tapping into protein reserves for fuel because most glucose production initially occurs via glycogenolysis (Riesenfeld *et al.* 1981). However, birds have limited glycogen stores (Blem 1976) and, as these decline during food restriction, lipids, the most energy-dense fuel source (Johnston 1970; Cherel *et al.* 1988), are mobilized to provide an excellent supply of energy (Ramenofsky 1990). Birds have a large capacity to store lipids, and their lipid depots vary daily and seasonally (Blem 1976, 1990). With prolonged food restriction, however, or with food restriction in birds in poor body condition, gluconeogenesis is enhanced by amino acid metabolism, largely from muscle protein (Siegel 1980; Cherel *et al.* 1988) and is stimulated by increased glucocorticoid secretion (Veiga *et al.* 1978). Furthermore, the extent to which muscle protein is mobilized during fasting is directly related to body fat content at the onset of food restriction and to the size of the energy deficit. Lean individuals lose significantly more muscle protein than fat individuals (Cherel *et al.* 1988). Thus, in relation to natal dispersal, this suggests that juveniles dispersing with adequate fat depots would be able to avoid mobilizing muscle protein and would more likely survive natal dispersal than individuals in poor body condition. The latter probably must delay natal dispersal until their condition improves, through either increased foraging in the natal area or provisioning by adults.

Body condition and the timing of natal dispersal

Birds

Variation in the timing and distance of natal dispersal may be related to differential ability of young to acquire food in natal areas and to compete for dispersal areas in which to settle (Nilsson and Smith 1985, 1988; Verhulst *et al.* 1997; Ellsworth and Belthoff 1999). In nonmigratory species of birds, for example, there are benefits of early natal dispersal that include increased access to high-quality territories or favourable habitats that are limited in number and are acquired on a first-come, first-served basis (Drent 1984; Lens and Dhondt 1994).

A juvenile's ability to disperse early probably depends on its ability to acquire sufficient food during the postfledging period which, in turn, is dependent upon its competitive ability and relative social status (Nilsson and Smith 1985, 1988). Thus, the ability to out-compete siblings or conspecifics for access to food resources may be the means by which individuals can accumulate sufficient energy reserves to disperse early when the pay-off for doing so is sufficiently high. For example, early dispersing marsh tits (*Parus palustris*) are heavier than individuals that disperse later, and marsh tits in poorer physical condition forage longer in the natal area than birds in good condition (Nilsson and Smith 1985). As a result, juveniles initially in poor body condition disperse later than juveniles in good condition. In Spanish imperial eagles (*Aquila adalberti*), early dispersers have lower blood urea and uric acid levels than later dispersers (Ferrer 1992, 1993). High plasma values of nitrogenous wastes could indicate increased amino acid catabolism (Mori and George 1978; Le Maho *et al.* 1981) and may reflect individuals in poor condition that are tapping into protein (muscle) reserves. Finally, dominant juvenile western screech-owls (*Otus kennicottii*) disperse before subordinate siblings (Ellsworth and Belthoff 1999). If high dominance status is a predictor of good body condition, then the owl results are consistent with the idea that individuals in better body condition disperse earlier. In contrast, when benefits of philopatry outweigh those of natal dispersal, and the potential for competition exists, dominant individuals appear to disperse later or less frequently than more subordinate siblings (Ekman and Askenmo 1984; De Laet 1985; Arcese 1989a; Strickland 1991; Gese *et al.* 1996).

The response of birds to changing environmental parameters, such as food availability, is at least partially dependent on initial body condition and also varies with age (Witter and Swaddle 1997). In spontaneously fasting penguins, increased locomotor activity associated with refeeding is related to body condition (Robin *et al.* 1998).

This suggests a physiological mechanism to monitor body mass which, once a threshold mass is achieved, triggers feeding activity. A similar age-dependent mechanism could be involved in stimulating natal dispersal behaviour in young birds, in a manner akin to that proposed for mammals (below).

Mammals

In Belding's ground squirrels (*Spermophilus beldingi*), there is a relationship between body mass and natal dispersal by juvenile males. This relationship, combined with hormonal data, enabled Holekamp (1984, 1986) to suggest that natal dispersal by young males is controlled by an ontogenetic switch, which is triggered by attainment of a particular body mass or body composition. Supplemental feeding increased the mass of provisioned young and advanced the onset of dispersal relative to controls (Nunes and Holekamp 1996). Because body mass at natal dispersal was fairly constant among juvenile males regardless of the age at which they dispersed, it appears that a threshold body mass may exist (Nunes and Holekamp 1996) above which natal dispersal is stimulated. However, body mass alone cannot explain the timing of natal dispersal because some heavier-than-average males delay dispersal (Holekamp 1984, 1986). Instead, it appears that fat as a proportion of body mass strongly influences the timing of natal dispersal in young male ground squirrels (Nunes and Holekamp 1996). Nunes *et al.* (1998) suggest that the threshold percentage of body fat that triggers natal dispersal increases with the approach of hibernation, which itself requires considerable fat stores. These authors suggest that a circannual timing mechanism in juvenile males interacts with a physiological mechanism that monitors fat reserves to regulate the timing of dispersal behaviour. Natal dispersal is inhibited when body fat is below a seasonally appropriate threshold level. The fat threshold is initially low when the juvenile active season begins early in the calendar year, and available energy can be used to support natal dispersal. However, the fat threshold increases rapidly when the juvenile active season begins later, delaying natal dispersal. Late in the juvenile active season, dispersal is inhibited regardless of body fat content. The plasticity of natal dispersal behaviour results in a trade-off between energy for natal dispersal and for hibernation in response to changes in body mass and fat content. This may enable juveniles to coordinate their physical and physiological condition to fine-tune the timing of natal dispersal.

Hormones and natal dispersal

It is well established that endocrine secretions facilitate the expression of behaviours in a wide range of taxa (see Nelson 1994). In some cases, a particular hormonal milieu and its associated behaviour patterns occur repeatedly at predictable times in the annual cycle, such as the seasonal changes in gonadal steroid secretions associated with reproduction (Wingfield and Farner 1978; Bronson 1989). At other times, the relationships are unpredictable, such as when severe storms disrupt breeding (Wingfield and Ramenofsky 1997). Still others may occur only once in the lifetime of an individual, such as during sexual differentiation (Phoenix *et al.* 1959). It is likely that this last category also includes hormonal changes associated with the natal dispersal of juveniles.

Mammals

Natal dispersal in some mammals occurs in conjunction with increased sex hormone secretions (Bronson 1989). This temporal concurrence suggests that sex hormones may be involved in triggering natal dispersal. Because natal dispersal in mammals is frequently male biased, investigations into proximate endocrine mechanisms underlying mammalian natal dispersal have focused on androgens, the principal sex hormones in males (Pusey 1987).

Androgens have profound consequences for the development of male phenotypes through both prenatal and postnatal processes. Prenatal exposure to testosterone organizes gender-typical neural circuits, whereas postnatal exposure activates this circuitry, resulting in sexually differentiated behaviours (Phoenix *et al.* 1959). Because the female phenotype is the 'default' phenotype in mammals, masculine differentiation of the brain requires perinatal exposure to testosterone. For example, in rats, there is an increase in testosterone on day 18 of gestation and a second pulse shortly after birth, which establish male-like behavioural patterns in adulthood (Weisz and Ward 1980; Baum *et al.* 1988).

Female embryos positioned next to male siblings during gestation are exposed to the androgens produced by the latter and can be partially masculinized (vom Saal 1984). In grey-sided voles (*Clethrionomys rufocanus*), which exhibit male-biased natal dispersal, females from litters with a high proportion of males are more likely to disperse (Ims 1989, 1990) and to begin dispersing sooner (Andreassen and Ims 1990) than females from other litters. Such findings suggest that prenatal exposure of females to testosterone predisposes them to disperse. Similar findings have been reported in female (Zielinski *et al.* 1992) and male (Drickamer 1996) house mice (*Mus musculus*). However, this phenomenon is not seen in females of two species of *Microtus* (Bondrup-Nielsen 1992; Lambin 1994c), nor does neonatal exposure to testosterone affect natal dispersal behaviour of female *M. pennsylvanicus* (Nichols and Bondrup-Nielsen 1995). Interestingly, female spotted hyenas (*Crocuta crocuta*), which normally are masculinized by prenatal exposure to androgens (Glickman *et al.* 1992; Licht *et al.* 1992), do not disperse like males of this species (Frank 1986). Clearly, more work is needed in this area to clarify the extent to which prenatal exposure to sex-steroid hormones affects subsequent dispersal activity.

Testosterone levels often are increased at the time of puberty in mammals, and there are correlational data that link the onset of sexual maturity with natal dispersal. Therefore, testosterone may have an activational role in natal dispersal. For example, dispersing male European badgers (*Meles meles*) have higher testosterone levels than males remaining in natal groups (Woodroffe *et al.* 1993). Additionally, the few male naked mole-rats (*Heterocephalus glaber*) that appear to be dispersers exhibit higher levels of luteinizing hormone than non-dispersers (O'Riain *et al.* 1996). Similarly, late-winter social activity in pine martens (*Martes martes*) is characterized by raised testosterone levels and natal dispersal of juveniles (Audy 1976; Helldin and Lindström 1995). Despite these correlational data, for most species it remains to be determined whether increased testosterone levels stimulate natal dispersal or are a consequence of it. In spotted hyenas, increased testosterone concentration appears to result from natal dispersal behaviour (Holekamp and Smale 1998).

Holekamp and colleagues have taken one of the most comprehensive experimental approaches to teasing apart organizational and activational effects of testosterone on natal dispersal behaviour (e.g., Holekamp *et al.* 1984; Holekamp and Sherman 1989). Neonatal females treated with testosterone exhibit male-like natal dispersal patterns as juveniles, supporting an organizational role for testosterone in ground squirrels. However, castration of juvenile ground squirrels of either sex has little influence on subsequent natal dispersal or lack thereof, which indicates the absence of activational effects. If androgen changes associated with puberty do not trigger natal dispersal in male ground squirrels, then what does? Body condition has been implicated, and Holekamp and colleagues suggest that the effect of body condition on natal dispersal can be modified, depending on how close the animals are to the beginning of hibernation (Nunes *et al.* 1998; see above).

Birds

Because the environmental and evolutionary factors underlying natal dispersal behaviour may differ among taxa, it should not be surprising that proximate mechanisms also may differ. For example, natal dispersal in birds is female biased (Greenwood 1980), although both sexes usually disperse. While this does not preclude an organizational effect of gonadal steroid hormones on avian natal dispersal, it does suggest that, if proximate endocrine factors are involved, then they should not be found predominately in only one sex.

As in mammals, sex-steroid hormones appear to organize sexual differentiation in birds (reviewed by Ottinger and Abdelnabi 1997). But, in contrast to mammals, males appear to be the 'default' genotype in birds, and ovarian secretions feminize many sexual characteristics. The mechanism is complex, for while some functions are feminized by oestrogens, others are masculinized by androgens (Adkins-Regan 1987). Furthermore, the ratio of oestrogens to androgens, rather than absolute hormone levels, may affect some aspects of sexual differentiation (Ottinger and Abdelnabi 1997).

Clearly, avian embryos are influenced by their own hormonal secretions, but, because they develop oviparously and, thus, independently from one another, they are not affected by hormonal secretions of their developing siblings, as occurs *in utero* with mammalian litters (vom Saal 1984). In general, it has been assumed that opportunities for external factors to influence avian embryonic development are minimal. For example, within an egg, catecholamine and amino acid levels of embryos respond to changing environmental conditions experienced during incubation (Epple *et al.* 1997). But only recently has it been determined that the hormonal milieu of developing embryos varies, both within and between clutches, and that some of this variation is of maternal origin. Schwabl (1993) found that maternal testosterone is deposited in the yolk during avian egg formation. The concentration of testosterone within eggs of either sex varies with laying order and with stage of the breeding season. Eggs laid late in a clutch have higher concentrations of testosterone than those laid early, and clutches laid earlier in the breeding season have a higher concentration of testosterone than later-laid clutches (Schwabl 1996a). Furthermore, social interactions (probably aggression) and increased breeding population density stimulate adult females to deposit increased amounts of testosterone in eggs (Schwabl 1997a,b), which enhances nestling development, stimulates food begging, and increases aggressive behaviour in the young (Schwabl 1996b, 1997a). Adkins-Regan *et al.* (1995) have demonstrated that females also can modify the level of oestradiol in the yolk of their eggs, and that yolk oestradiol affects sexual differentiation of the resulting offspring.

The extent to which yolk steroids organize the brains of avian male and female embryos is unknown, as is the extent to which natal dispersal behaviour is linked to maternal steroids in yolk. But this represents a mechanism whereby females could exert subtle organizational control over the phenotypes of their offspring. For example, maternal modification of the level of yolk steroids, or their ratio, could enhance the competitive abilities of young in the local environment. It could provide the physiological basis for regulating the level of natal dispersal, enhancing departure from an overcrowded or resource-poor environment, reducing natal dispersal from favourable habitats, or establishing sex-specific sensitivities to stimuli that activate natal dispersal. That is, the physical condition of the mother could affect the prehatching environment of the young in ways that affect the probability, timing, and/or extent of natal dispersal behaviour. Similarly, in mammals, the condition of pregnant females is known to affect fetal hormone levels, sexual differentiation, and subsequent juvenile and adult behaviour patterns (Ward 1972; Ward and Stehm 1991; Henry *et al.* 1996). However, effects of the embryonic environment on natal dispersal behaviour in mammals, as with birds, is not well understood and deserves considerably more attention.

In terms of proximate mechanisms of natal dispersal, many temperate-zone birds disperse in late summer or autumn, when gonads are small and nonfunctional; this suggests that sex steroids are not involved in an activational sense. Conversely, corticosterone, a nongonadal steroid hormone produced by the adrenal glands of both sexes, has been implicated in natal dispersal (Belthoff and Dufty 1995, 1998; Silverin 1997). Glucocorticoids help to regulate the availability of energy resources within the body (for a review see Dallman *et al.* 1993). In birds, corticosterone affects overall body composition (Wingfield and Silverin 1986; Bray 1993) and influences specific behavioural patterns. For example, corticosterone stimulates feeding (Nagra *et al.* 1963) and locomotor activity (Astheimer *et al.* 1992). Similarly, unpredictable environmental events that stimulate corticosterone secretion may be responsible for facultative movements of birds from breeding territories and wintering areas when conditions deteriorate (Wingfield and Ramenofsky 1997).

We suggest that corticosterone influences avian species-typical natal dispersal behaviour (Belthoff and Dufty 1995, 1998). In captive juvenile western screech-owls, corticosterone levels exhibit a natural increase in conjunction with increased locomotor activity, and the latter correlates with the timing of natal dispersal behaviour of free-living juvenile owls. Thus, increased corticosterone levels may be part of the triggering mechanism that stimulates natal dispersal movement in juveniles. As we discuss below, the precise timing of this corticosterone increase may be related to the energy state of the individual. Furthermore, there may be a limited ontogenetic period during which increased levels of glucocorticoids will stimulate natal dispersal. The extent of this period may be determined by changes in external cues (e.g., photoperiod) and/or internal factors (e.g., body condition). However, we hypothesize that, once the birds leave this stage, increased levels of corticosterone no longer affect natal dispersal, although they may induce temporary movements (Wingfield and Ramenofsky 1997). Interestingly,

Silverin (1997) demonstrated that corticosterone implants stimulate natal dispersal in juvenile willow tits (*Parus montanus*) during winter flock formation, a time when birds normally leave their natal territories. However, similar implants given at the same time to adults, or to juveniles after winter flocks have stabilized, have no effect.

Glucocorticoids and natal dispersal: is it stress?

It is important to note that a relationship between changes in glucocorticoid secretion and natal dispersal does not make the latter a stress-related phenomenon. This may, at first, seem counter-intuitive, given that the most widely recognized role of adrenal glucocorticoid secretion is in stressful situations that threaten the survival and/or reproductive fitness of an individual. Indeed, chronic high levels of glucocorticoid secretion deplete protein reserves, interfere with the responsiveness of the immune system, inhibit reproduction, and retard growth rate (see reviews by Johnson *et al.* 1992, McEwen 1998). Clearly such effects would diminish the prospects for survival of a dispersing juvenile. However, studies on the physiological consequences of chronically high levels of corticosterone typically use captive animals that cannot escape from the stressor and, thus, cannot avoid the negative effects of corticosterone. In contrast, rising corticosterone levels in free-living organisms orchestrate a suite of acute physiological and behavioural responses that help them to survive threats with minimal deleterious effects (Wingfield *et al.* 1998). That is, the response occurs quickly, and the stressful event is resolved before corticosterone titres reach the level at which they do damage.

Corticosterone also exhibits diurnal and seasonal changes that are unrelated to stress. For example, daily corticosterone levels peak at or just before the onset of the active period (i.e., dawn in diurnal species, dusk in nocturnal species; Boissin *et al.* 1969; Dusseau and Meier 1971; Morimoto *et al.* 1977; Dufty and Belthoff 1998). In temperate breeding birds, corticosterone concentration is increased during the breeding season relative to the levels found during the post-breeding period (e.g., Wingfield and Farner 1978; Dufty and Wingfield 1986) and is also maintained at increased levels during migration (Holberton *et al.* 1996).

The mechanism whereby seasonal increase in corticosterone level occurs in the absence of the harmful effects described above is unknown, but it may have to do with the population of corticosterone receptors in the brain and elsewhere. Two intracellular corticosterone receptor types are known, one (type I receptors) with a high affinity for the hormone, and the other (type II receptors) with a low affinity (de Kloet and Ruel 1987; de Kloet *et al.* 1987; Evans and Arriza 1989). Type I receptors are thought to mediate the diurnal and seasonal responses to levels of corticosterone that are involved in regulating the use of energy resources, while type II receptors respond to higher levels of corticosterone. A third, as yet unidentified, receptor with an even lower affinity for corticosterone may exist (Orchinik 1998) and may elicit responses to the highest titres of corticosterone associated with stressful stimuli. Membrane glucocorticoid receptors, reported for amphibians (Orchinik *et al.* 1991) and hypothesized for birds (Breuner *et al.* 1998), may mediate rapid behavioural responses to corticosterone. Changes in the number or relative proportion of these receptors could result in different daily, seasonal, or life-stage responses to a given level of corticosterone.

Relating this to natal dispersal suggests that the interactions among internal and/or external factors that produce changes in corticosterone secretion reflect normal developmental changes within a specific life-history stage of young birds and do not produce the harmful physiological effects associated with chronic application of a stressor. This is not to say that external, unpredictable disturbances that elicit an unpredictable increase in glucocorticoid secretion do not affect the timing of natal dispersal. Such disturbances, termed 'labile modifying factors' by Wingfield *et al.* (1997), may include such things as reduction in food availability, inclement weather, or an increase in predation pressure, and would be expected to alter the timing and/or amplitude of the corticosterone response and, consequently, to alter the timing of natal dispersal. Another phenomenon, similar in outcome to natal dispersal, is facultative dispersal (or irruptive behaviour). Facultative dispersal occurs when adults respond to labile modifying factors with increased corticosterone secretion and movement (either temporarily or permanently) to a new location away from the environmental perturbation (Wingfield *et al.* 1998). However, whereas natal dispersal of juveniles will occur in the absence of labile modifying factors, facultative dispersal is a direct consequence of their presence. Therefore, the mechanisms that underlie the two phenomena may have fundamental differences.

Natal dispersal and other types of movement

Natal dispersal and migration

It is important to distinguish natal dispersal from other types of movement, most notably migration. For, despite some obvious behavioural similarities, the underlying mechanisms appear to differ between the two. Indeed, even within the accepted framework of migration, different migratory behaviours have different physiological underpinnings. For example, in birds, many of which migrate seasonally, the physiology of vernal and autumnal migratory movements is very different. Spring migration of north temperate birds requires the presence of intact gonads for normal hyperphagia and migratory restlessness to occur, while the absence of gonads has no effect on the development of autumnal migration (reviewed in Wingfield *et al.* 1990). Natal dispersal cannot be under the same hormonal control as vernal migration because the gonads of juvenile birds are undeveloped, and their hormonal profile differs from that of spring migrants (Williams *et al.* 1987; Wingfield *et al.* 1990).

Because so little is known of the hormonal changes that occur during natal dispersal in migratory avian species, a detailed comparison with the endocrinology of autumnal migration (which is itself poorly understood) is not possible. Birds hatch in a photorefractory state, do not become photosensitive until they experience a number of short day-lengths, and do not undergo gonadal growth until they encounter the vernal increase in day-length (McNaughton *et al.* 1992). Thus, dispersing juveniles, like autumnal migrants, have regressed gonads, and the overall hormonal physiology of juveniles and photorefractory adults may, indeed, be similar (Williams *et al.* 1987). However, adults, but not juveniles, undergo a full photo-induced gonadal cycle, complete with exposure to hypothalamic, pituitary, and gonadal hormones, before the autumn.

Such experiential differences result in different endocrine responses in juveniles than in adults in the autumn (McNaughton *et al.* 1995). Other aspects of the endocrine system also differ between dispersing juveniles and migrants. For example, Silverin *et al.* (1989) and Silverin (1997) found that the adrenocortical response to handling stress differs between juvenile and adult willow tits during the time when young willow tits seek to join winter flocks. Although this species does not migrate, this finding further highlights the fact that endocrine responses of juveniles and adults can differ at a given time of year. Furthermore, dispersal movements are rarely as persistent or as unidirectional as migratory movements (e.g., Belthoff and Ritchison 1989; Holekamp and Sherman 1989), thus violating two of the five behavioural criteria proposed by Dingle (1996) to define migration. Thus, we do not believe that natal dispersal and migration represent variants of the same phenomenon; however, additional work clearly is called for.

Natal dispersal and ranging

The term 'dispersal' has been applied to a wide variety of phenomena and suffers from the fact that, historically, it has meant different things to different investigators. Dingle (1996) points out that dictionary definitions of dispersal refer to population-level effects (i.e., the scattering of individuals within a population), which relegate dispersal to a distance measurement that omits the behavioural component associated with these movements. Consequently, Dingle (1996, p. 36) suggests that dispersal be removed from the lexicon to describe movement, that it be restricted to a description 'of the *process* leading to a distributional outcome of movement' (Dingle's emphasis), and that it be replaced by the term 'ranging'.

We agree with Dingle that dispersal is a process, involving distinct behavioural and physiological mechanisms. We also agree that the term 'dispersal' is too often misused, for it frequently is applied only to the end result of dispersal. That is, it is used to indicate the linear distance between an organism's starting and ending points (dispersion), rather than to indicate a process (dispersal). However, we disagree that the term 'dispersal' has lost its utility. Natal dispersal refers to events that occur within a distinct ontogenetic stage of development and that, once completed, do not occur again. On the other hand, ranging (Dingle 1996) encompasses not only the natal dispersal of juveniles, but also the movements of adults between breeding attempts, events that appear to have quite different ontogenetic, ecological, and physiological bases. Dispersal, as defined by Howard (1960) in the context of natal dispersal, neither explicitly nor implicitly suggests population-level events. Indeed, Howard specifically refers to the 'movement an *individual* makes' (our emphasis). Therefore, we believe that, by adhering carefully to the limits of the term, natal dispersal can have an important and unambiguous role in the investigation of animal movement.

Natal dispersal and other locomotor behaviour

Animals obviously engage in other types of locomotor activity besides natal dispersal, and some of these behaviours also have hormonal correlates. For example, glucocorticoid levels vary seasonally in voles (Seabloom *et al.* 1978), and corticosterone appears to play a role in exploratory behaviour (Veldhuis *et al.* 1982). Furthermore, testicular hormones increase open-field behaviour in male voles (Turner *et al.* 1980), and this behaviour varies with age and reproductive status (Turner *et al.* 1983). In birds, intact chicks treated with testosterone exhibit more persistent food searching behaviour than do control chicks (Andrew and Rogers 1972). It would be interesting to relate these other types of locomotor activity to an individual's propensity for natal dispersal, but such associations remain, for now, purely speculative.

Another piece of the puzzle?

Our discussion of birds focuses on the effect of corticosterone production on natal dispersal behaviour. However, we recognize that the relationship between corticosterone and natal dispersal need not be direct because body condition, feeding behaviour, and locomotor activity are affected by a host of interrelated regulatory substances that influence one another and the hypothalamo-pituitary–adrenal axis (HPAA). We highlight only one such substance, the newly discovered hormone leptin, which may well play a significant role in integrating body condition, the HPAA, and natal dispersal.

Leptin, a protein hormone discovered in 1994, is released by adipose cells (Zhang *et al.* 1994). Leptin suppresses food intake by inhibiting the production of neuropeptide Y (Ahima *et al.* 1996), a strong appetite stimulant. Leptin is released when individuals achieve a particular body condition as reflected by their level of fat deposition (reviewed by Caro *et al.* 1996). Physical activity and energy expenditure correlate with plasma leptin concentrations, and both are increased following the administration of exogenous leptin (Pelleymounter *et al.* 1995). Furthermore, fasting-induced anoestrus is reversed by exogenous leptin through leptin's effect of stimulating the oxidation of metabolic fuels (Schneider *et al.* 1998).

Leptin currently is under intense study in mammals. However, avian leptin has also been cloned (Taouis *et al.* 1997; but see Friedman-Einat *et al.* 1999), and leptin expression has been noted in chicken adipose and liver cells (Ashwell *et al.* 1999). Nonetheless, the role of leptin in regulating food intake in birds remains uncertain, for leptin has been shown to reduce food intake of chickens in one study (Raver *et al.* 1998) but not in another (Bungo *et al.* 1999). Indeed, the overall relationship between leptin and the HPAA and body condition is far from clear. It has been shown that leptin synthesis and release is stimulated by glucocorticoids in fed (but not fasted) individuals (Laferrère *et al.* 1998), and leptin acts both centrally (hypothalamus) and peripherally (adrenal glands) to inhibit glucocorticoid production (Pralong *et al.* 1998). Leptin levels fall during fasting, which allows glucocorticoid and adrenocorticotrophic hormone (ACTH) levels to rise, and exogenous leptin dampens this fasting-induced rise in glucocorticoids and ACTH concentration (Ahima *et al.* 1996). Leptin has also been shown to block stress-induced increases in glucocorticoids and ACTH levels by blocking the secretion of corticotrophin-releasing factor (CRF) by the hypothalamus (Heiman *et al.* 1997). However, other studies indicate that the feedback inhibition of leptin on glucocorticoid secretion disappears under some conditions, resulting in increased levels of both leptin and glucocorticoids (Bornstein *et al.* 1998; Roberts *et al.* 1999), and leptin itself has been shown to activate the HPAA (Malendowicz *et al.* 1998). Thus, while the relationship between leptin and the HPAA appears to be complex and variable, there is accumulating evidence to suggest that leptin may provide a link between body condition and glucocorticoid secretion (Bornstein 1997).

Interestingly, the pattern of leptin secretion differs between young and adult animals. Leptin levels increase markedly in young mice during their second week of life, and this increase is independent of fat mass and food intake (Rayner *et al.* 1997; Ahima *et al.* 1998). A decline to adult values of leptin occurs after weaning. Furthermore, the effect of leptin on behaviour can vary with body condition. For example, leptin treatment enhances sexual behaviour in female hamsters fed *ad libitum*, but inhibits sexual behaviour in females deprived of food (Wade *et al.* 1997). In addition, young animals given a high-fat diet had higher leptin levels and lower HPAA responsiveness during development than individuals fed a low-fat diet, but lower leptin levels and a more responsive HPAA upon reaching independence (Trottier *et al.* 1998). It is unknown what role, if any, leptin plays in signalling that a body condition has been attained that is capable of surviving the rigours of natal dispersal, or in stimulating concomitant changes in energy expenditure and locomotor activity. But its known effects on energy use and locomotor behaviour, the apparent diversity of its effects in organisms in different ontogenetic stages and body conditions, and its interactions with the HPAA suggest that research into the role of leptin in natal dispersal will be fruitful.

Leptin represents but one substance that affects feeding behaviour and bodyweight (for a more complete list, see Inui 1999). We believe that, as the effects of such factors on one another and on the HPAA become better understood, they will further clarify the proximate physiological mechanisms regulating natal dispersal.

Conclusions

The physiological mechanisms underlying natal dispersal behaviour are poorly understood. Much of the information used to construct the framework for our discussion is correlational, reflecting a lack of hard data relating physiological parameters and natal dispersal. Nonetheless, acquisition of a body condition sufficient to endure the rigours of natal dispersal appears important in mammals and birds, and is probably a critical prerequisite to natal dispersal across taxa. Indeed, studies of ground squirrels suggest that good body condition acts as an ontogenetic trigger to initiate natal dispersal behaviour. Leptin, a newly discovered hormone produced by adipose cells that signals the level of lipid reserves in mammals, warrants consideration in any model of natal dispersal. Studies in birds also suggest a relationship between locomotor activity, corticosterone secretion, and natal dispersal.

In mammals, where natal dispersal is male biased, androgens appear to activate natal dispersal in some species, but their role in organizing the mammalian brain to produce gender-specific natal dispersal patterns appears more widespread. Organizational effects of hormones on natal dispersal patterns have not been examined in birds. However, recent reports indicate that avian embryos contain steroid hormones from their mother, that the level of exposure is (at least in part) a function of environmental and seasonal effects, and that differential exposure to steroids has behavioural and developmental ramifications. Thus, as in mammals, female birds may affect embryonic development of their young in ways that influence their subsequent natal dispersal.

Acknowledgements

We are grateful to the National Science Foundation (award IBN-9509079), the Idaho State Board of Education (Specific Research Grant S95-042), Boise State University (Faculty Research Grants), and the Raptor Research Center at Boise State University for supporting our research on natal dispersal in screech-owls. We thank the National Science Foundation for travel support to attend the conference on dispersal in Roscoff (April 1999), from which this book emerged. We also thank J. Clobert and three anonymous reviewers whose comments helped us to improve this manuscript.

16

Habitat selection by dispersers: integrating proximate and ultimate approaches

Judy A. Stamps

Abstract

Behavioural research is shedding new light on the complex relationships between the proximate mechanisms involved in habitat selection and the selective pressures that may have contributed to the evolution of those mechanisms. Habitat selection by dispersers can be divided into three stages (search, settlement, and residency); recent studies suggest that the adaptive significance of behaviour at each of these stages may differ from the assumptions of traditional habitat selection theory. For instance, dispersers may benefit from the presence of conspecifics or heterospecifics while searching for, settling in, or living in new habitats, and individuals may prefer to settle in post-dispersal habitats similar to their pre-dispersal habitats, because this behaviour reduces the costs of detecting or assessing suitable habitats (habitat cueing) or because experience in a pre-dispersal habitat improves performance if an animal settles in the same type of habitat after dispersing (habitat training). Dispersers have evolved a variety of proximate behavioural mechanisms to reduce search and settlement costs in natural environments, but if they currently rely on these processes, species living in areas modified by human activities may not exhibit 'ideal' habitat selection behaviour. Insights from recent studies of habitat selection may help solve specific problems in conservation biology and, more generally, help biologists understand the intimate relationship between dispersal and habitat selection behaviour.

Keywords: habitat selection, proximate mechanisms, conspecific attraction, habitat imprinting

Introduction

This volume focuses on dispersal, a term derived from the Latin *dispersus*, meaning 'scattered abroad'. Students of dispersal tend to focus on the movement of individuals away from their point of origin to other regions of the landscape. However, one can look at the same process from the opposite perspective, i.e. one that emphasizes the 'aggregation' into a new habitat of individuals that originated from other areas in the landscape. This second perspective is the one adopted by students of habitat selection, who consider the processes by which individuals search for and select new habitats, and the consequences of those processes for the distribution, density, and fitness of individuals in different habitats (reviews in Hildén 1965; Morris 1991; Stamps 1994; Kramer *et al.* 1997). Habitat selection can occur at a number of different spatial scales (Lima and Zollner 1996); this chapter focuses on the intermediate spatial scales that are most relevant when individuals select unfamiliar habitats in which to establish new home ranges or territories (Fretwell and Lucas 1970; Morris 1987, 1992).

Proximate mechanisms in search of ultimate explanations: conspecific attraction and habitat imprinting

Since the goal of this chapter is to link proximate and ultimate approaches to habitat selection, we begin by considering two behavioural mechanisms that may play an important role in habitat selection in many different species: conspecific attraction and habitat imprinting. Conspecific attraction occurs when the presence of conspecifics in a patch of habitat increases the probability that other individuals will settle in that patch (Stamps 1988, 1991). Not surprisingly, conspecific attraction is common in species that humans consider to be colonial (e.g. Shields *et al.* 1988; Sarrazin *et al.* 1996; Danchin and Wagner 1997), but this behaviour has also been observed in many animals thought to be 'solitary' or 'territorial' (e.g. crabs, Weber and Epifanio 1996; reef fish, Booth 1992; terrestrial arthropods, Muller 1998; lizards, Stamps 1988; non-colonial birds, Muller *et al.* 1997; and benthic marine invertebrates belonging to at least eight different phyla, Pawlik 1992). Most studies of conspecific attraction have focused on animals settling in empty or nearly empty habitats, but a few workers have studied the effects of conspecifics on settlement over a wide range of densities, and found that at low densities residents increase the probability that newcomers will settle, whereas the reverse is true at high densities (e.g. marine invertebrates, Meadows and Campbell 1972; larval damselfish, *Dascyllus trimaculatus*, Schmitt and Holbrook 1996).

Habitat imprinting occurs when exposure to stimuli from a type of habitat early in life increases the probability that an individual will select an unfamiliar (hereafter 'new') habitat of the same type later in life (Hildén 1965; Klopfer and Ganzhorn 1985). As is the case with filial or sexual imprinting (review in Shettleworth 1998), habitat imprinting may be most likely to occur during a particular age or developmental stage (i.e. a sensitive period), and early experience and innate predispositions may interact in complex ways to influence later habitat preferences. Laboratory studies of birds and mammals have shown that individuals exposed to particular stimuli as juveniles are more likely to prefer new habitats containing the same stimuli when tested at older ages (e.g. Wecker 1963; Arvedlund and Nielsen 1996; Teuschl *et al.* 1998). Similarly, studies of parasitic hymenopterans indicate that chemical cues from both the host and the host's food plant affect habitat preferences after emergence (e.g. Honda and Walker 1996; Monge and Cortesero 1996), while in ants, exposing larvae to cues from nest plants increases their preference as adults for those nest plants (Djieto-Lordon and Dejean 1999). Thus far, evidence for habitat imprinting in the field is suggestive rather than conclusive (Hildén 1965; Temple 1978). For instance, Tordoff *et al.* (1998) recently showed that peregrine falcons (*Falco peregrinus*) that fledged from nests on cliffs were more likely to breed on cliffs than on buildings, and *vice versa*, but it was not clear that both types of habitat were equally available to all dispersers.

Conspecific attraction and habitat imprinting are only two of many behavioural mechanisms that may influence the habitat selection behaviour of a wide range of taxa. For instance, the presence of certain heterospecifics in an area may increase the probability that newcomers will settle in that area (heterospecific attraction; see Monkkonen *et al.* 1999). These and other general mechanisms of habitat selection have something else in common: each of them is consistent with more than one adaptive (ultimate) explanation. This last point can be appreciated most readily by shifting focus to the ultimate level, and considering factors that may affect the fitness of dispersing individuals during the three stages of habitat selection.

Three stages of habitat selection

When considering the fitness consequences of habitat selection behaviour, it is convenient to divide the habitat selection process into three stages: searching for a new habitat (search), settling in a new habitat (settlement), and living in a new habitat (residency). Frequently these processes occur in the temporal order indicated above and do not overlap, but this is not always the case. For instance, dispersers may begin searching for a new habitat while still residing in the old one, or may begin settling in a habitat, encounter aversive stimuli, then leave and begin searching for another one. However, it is still useful to differentiate among these processes, because each typically involves different types of behaviour, with different potential fitness consequences for dispersing individuals.

Instead of considering search, settlement, and residency in the temporal order that most dispersers exhibit them, we will consider them in reverse order. This is because many of the benefits of habitat selection are accrued during residency, while many of the costs of habitat selection are paid during search and settlement, and, generally speaking, economic approaches to behavioural questions are easier to understand if benefits are discussed before costs.

The residency period

Density, neighbourhood size, and fitness

A central premise of habitat selection theory is that an individual's fitness after settling into a new habitat varies as a function of the characteristics of that habitat, and as a function of the conspecifics and heterospecifics with whom the individual shares the habitat after settling. However, while there is general agreement that conspecifics, heterospecifics, and habitat quality affect individual fitness after settlement, new ideas are emerging about the functional relationships between these variables.

With respect to conspecifics, the familiar ideal free distribution (IFD) (Fretwell and Lucas 1970) assumes that conspecifics compete with one another for resources, as a consequence of which individual fitness declines monotonically as a function of the density of conspecifics within a patch (reviewed in Holt and Barfield, chapter 6). For the past 30 years, most theoreticians have continued to assume that individual fitness is negative density dependent after individuals have settled into new habitats (e.g. Tregenza 1995; Sutherland 1996; Holt and Barfield, chapter 6, but see Rodenhouse *et al.* 1997). However, earlier this century, Allee (1951) suggested another possible relationship between the density of conspecifics and fitness: a unimodal distribution in which fitness increases as a function of density at low to moderate densities, and then declines as a function of density at moderate to high densities (Allee effects). But, until recently, the implications of Allee effects for habitat selection theory have not been widely appreciated (Stephens and Sutherland 1999; C. Greene and J.A Stamps, unpublished findings).

One possible reason why Allee effects have been ignored is that behaviourists and ecologists prefer to study areas with high population densities when embarking on long-term studies of individual animals (e.g. Danchin *et al.* 1998a). As a result, most field data on the effects of conspecifics on fitness come from studies in which intraspecific competition for resources may really be of paramount importance, because factors that reduce population densities in other areas (e.g. predators, parasites, inclement weather, competing heterospecifics, intermittent food shortages) are unimportant or lacking in habitats that support the highest densities. One consequence of this understandable bias is that we lack information about the effects of conspecifics on fitness at the other end of the density spectrum, for example, habitats in which densities are maintained at medium to low levels by factors other than competition for resources, or empty habitats that are just beginning to be colonized by new arrivals.

In fact, there is growing evidence that even 'solitary' or 'territorial' animals may benefit from the presence of conspecific neighbours when living at low to moderate densities (reviews in Hildén 1965; Morse 1985; Stamps 1988; Stephens and Sutherland 1999). The list of potential benefits of neighbours is extensive, including: (1) increased efficiency in expelling intruders from territories (e.g. Eason and Stamps 1993; Meadows 1995), (2) reduced risk of predation (Smith 1986; Wisenden and Sargent 1997), and (3) improved access to mates or extra-pair partners (Levitan and Young 1995; Wagner 1997).

Although Allee effects are typically expressed as a function of population density, this practice is valid only if interactions with conspecifics occur at random. However, in many species, individuals direct their social behaviour towards particular conspecifics, as a result of which the *per capita* rate of social interactions is unrelated to population density (e.g. McGuire and Getz 1998). In some birds, for instance, territories appear to be uniformly distributed across a landscape, but individuals have divided themselves into neighbourhoods whose members share songs or calls that differ from the vocal signals produced by individuals living in adjacent neighbourhoods (e.g. Jenkins 1978; Payne and Payne 1997; Bell *et al.* 1998; Miyasato and Baker 1999).

The fact that social interactions are typically directed at particular conspecifics implies that Allee effects should be measured as a function of 'neighbourhood size' (defined as the number of individuals with whom an individual typically interacts) rather than as a function of population density (see also Addicott *et al.* 1987; Engstrom and Mikusinski 1998). Another reason for de-emphasizing population density when studying Allee effects is the mismatch between density and neighbourhood size that can occur when animals live in fragmented habitats. For instance, one pair of birds in an isolated 10-ha woodlot lives at the same population density as do 10 pairs in a 100-ha woodland, but neighbourhood size is 1 in the first case and 10 in the second (assuming that all 10 pairs interact socially with one another, which seems reasonable given the sizes of avian song neighbourhoods).

Traditionally, theoretical models of the effects of heterospecifics on individual fitness after settlement have also emphasized competitive interactions and negative density dependence (e.g. Rosenzweig 1981, 1985). However, there are many situations in which the members of one species may benefit from the presence of members of other species (Danielson 1991; Forsman *et al.* 1998; Monkkonen *et al.* 1999). For instance, some birds benefit by nesting in the proximity of other species whose members are proficient at repulsing predators from the neighbourhood (e.g. Veen 1977; Slagsvold 1980; Bogliani *et al.* 1999), or by selecting areas containing heterospecifics with whom they can form mixed species flocks (Monkkonen *et al.* 1996). Similarly, sessile invertebrates benefit by settling near heterospecifics in high intertidal rocky habitats, because neighbours buffer one another from thermal and desiccation stresses in these locations (Bertness and Leonard 1997). Given the recent interest in positive interactions in ecological communities, it would be surprising if the next generation of heterospecific habitat selection theory did not consider positive as well as negative effects of heterospecifics on one another.

Habitat quality and habitat training

A second key assumption of the IFD and its intellectual descendants is that habitats vary with respect to their intrinsic quality (termed 'basic suitability' in Fretwell and Lucas 1970), where intrinsic quality specifies the fitness an individual can expect after settling in a habitat, after controlling for the effects of conspecifics on fitness (Holt and Barfield, chapter 6). More important, current theory assumes that, if habitats differ with respect to environmental factors affecting fitness, the rank order of habitat quality is the same for every individual in the species. In other words, a 'high quality' habitat is better for every individual than is a 'low quality' habitat. This assumption holds even if fitness varies predictably among settlers within any given habitat, for example, as a result of individual differences in competitive ability (Parker and Sutherland 1986; Hugie and Grand 1998) or condition upon arrival (Ims and Hjermann, chapter 14), as long as every individual ranks different habitats in the same order with respect to expected fitness during the residency period.

A growing literature on behavioural development suggests that it may be time to re-evaluate the assumption that every individual in a species ranks habitats in the same order with respect to habitat quality. In particular, evidence from a variety of sources suggests that individuals may achieve higher reproductive success after settling if they choose a post-dispersal habitat that is similar to their pre-dispersal habitat, as a result of a process that can be termed 'habitat training'. Habitat training describes any situation in which experience in a particular type of habitat prior to dispersal improves performance in a new habitat of the same type after dispersal, as a result of modifications in behaviour, physiology, or morphology.

So far, the strongest proponents of habitat training have been conservation biologists faced with the practical problem of inducing animals to settle and live in unfamiliar habitats. Low success rates following the release of captive-reared birds and mammals into natural habitats have suggested the hypothesis that these animals acquire a variety of perceptual and motor skills in their natal habitat that affect their success in the wild (Beck *et al.* 1994; Shepherdson 1994). More important for our purposes, these workers assume that acquired skills are habitat-specific, for example, that foraging performance is best improved by providing captive individuals with the same type of prey they will encounter in the post-release habitat, or that predator avoidance is enhanced by exposing individuals to the same types of predators that they are likely to encounter after release (McLean 1996; Yoerg and Shier 1997). Recent studies indicate that habitat-specific training may improve post-release survival in a range of species, from mammals (black-footed ferrets, *Mustela nigripes*; Biggins *et al.* 1998) to marine fish (Olla *et al.* 1998).

Although basic scientists have lagged behind their applied colleagues when considering the implications of habitat training for habitat selection, they have provided a number of reasons why animals might benefit by choosing a post-dispersal habitat comparable to their pre-dispersal habitat. The behavioural literature indicates that experience during ontogeny can improve perceptual and motor skills in a variety of contexts, including foraging (e.g. Oretega-Reyes and Provenza 1993), anti-predator behaviour (Curio 1993), and social interactions (Nelson 1997; West *et al.* 1997). Similarly, exposure to particular pathogens during ontogeny can produce specific immunity to those pathogens, so if different types of pathogens are predictably associated with different types of habitat, a disperser who developed immunity to the pathogens in a pre-dispersal habitat might survive better if it selected the same type of habitat after dispersing (e.g. Boulinier *et al.*, chapter 12a). Experience during early development can affect a wide range of morphological and physiological systems and traits (review in Schlichting and Pigliucci 1998), implying that individuals who developed a morphology or physiology appropriate for a certain type of pre-dispersal habitat would perform better if they selected the same type of habitat after dispersing.

Settlement costs

Settlement costs include any factor with a negative impact on fitness that occurs during the 'settlement period'. In turn, the settlement period begins when an individual arrives at a new habitat, and ends when that individual either establishes a home range or territory in that habitat or does not become a resident in that habitat (e.g. because it dies, or leaves to search for another suitable habitat). When studying settlement costs, one must consider not only individuals that settled in the habitat, but also those that were unsuccessful (i.e. who paid settlement costs in a habitat but did not establish residency in that habitat).

Traditionally, theoreticians and empiricists have focused on positive density-dependent settlement costs, which occur as a consequence of competition and aggressive interactions among settlers (Fretwell and Lucas 1970; Ens *et al.* 1995; Sutherland 1996; Tobias 1997). Settlement costs as a consequence of competitive interactions were first described in the ideal despotic (dominance) habitat selection model of Fretwell and Lucas (1970) (see also Fretwell 1972), which assumed that earlier settlers behave aggressively to newcomers when the latter try to settle in occupied habitats. As a result, some latecomers absorb settlement costs but manage to settle, while others pay settlement costs but fail to settle. After these settlement costs are included in the fitness equation, the ideal despotic model predicts that latecomers will be less likely to settle in a habitat than is predicted by an IFD model with the same values of habitat quality and population density.

Of course, aggression is not the only behaviour that occurs during the settlement period. For instance, in many animals, exploration is a major component of the settlement process, as newcomers investigate the features of a new habitat, practise motor patterns that allow them to use that habitat safely and efficiently, and establish social relationships with the other individuals with whom they will share the neighbourhood (Stamps 1994, 1995; Stamps and Krishnan 1999). Since animals cannot simultaneously explore and exploit a habitat, time and energy invested in exploration and meeting conspecifics during the settlement period detracts from activities that would otherwise have been used to increase fitness in that habitat. If gathering information about an area and its inhabitants is an important component of the settlement process, then individuals might be able to use conspecifics to reduce some of these costs (see also Danchin *et al.*, chapter 17). In that case, settlement costs might be lower in the presence of conspecifics than in their absence – the opposite of the situation expected if settlement costs depend on competitive interactions among prospective settlers (C. Greene and J.A. Stamps, unpublished manuscript).

Although theoretical studies of the effects of settlement costs on habitat selection are still in their infancy, incorporating settlement costs into habitat selection models changes predictions about the optimal behaviour of settling individuals, compared with the behaviour expected in the absence of settlement costs (C. Greene and J.A. Stamps, unpublished manuscript). For instance, if settlement costs are reduced in the presence of conspecifics, conspecific attraction is predicted even if fitness is strictly negatively density dependent after individuals have settled in the habitat. Habitat selection models that incorporate settlement costs continue to assume that habitat selection is 'ideal', in Fretwell and Lucas, original sense that every individual adopts the behaviour that maximizes its fitness. However, these models yield different predictions from traditional models because they begin measuring the potential fitness consequences of habitat selection behaviour earlier in the process, when individuals first arrive at a habitat, as opposed to after they have settled in that habitat.

Searching for suitable habitats

Both the IFD and the related ideal despotic distribution assume that dispersers incur no costs detecting, travelling to, or assessing new habitats before attempting to settle in one of them. However, when habitat selection occurs at the spatial scales that are relevant to dispersal, the assumption that habitat selection is free is usually suspect (e.g. Ward 1987; Morris 1991; Danielson 1992). Detection of suitable habitat is a potential problem for any species that does not habitually live in large continuous tracts of suitable habitat (Zollner and Lima 1997). The process of searching for a suitable new habitat may involve mortality costs, as a result of predation, accident, exposure, starvation, etc. (e.g. Stoner 1990; Lubin *et al.* 1993), and the process of assessing and rejecting potential habitats may require time or energy that would otherwise have enhanced reproductive success in a new habitat (Ward 1987; Danielson 1992; Morris 1992).

Studies of search costs for dispersers are still rare (but see Morris 1987, 1992), although related literature implies that high search costs would encourage dispersers to accept low-quality habitats, even if higher-quality habitats exist in the same landscape. For instance, theoretical studies of host plant selection by phytophagous insects predict that if dispersal is time limited, optimal habitats are scarce, and a short time is available for searching an appreciable proportion of dispersers would accept host plants of inferior quality, even if higher-quality host plants were available elsewhere in the same region (Levins 1968; Ward 1987; Mayhew 1997). Similarly, studies of mate search, in which a mobile individual travels across the landscape before finding and accepting a single mate, indicate that if travel costs are high and high-quality mates are limited many searchers will accept a mate of relatively low quality (e.g. Real 1990; Reid and Stamps 1997).

More generally, the incorporation of search costs into habitat selection implies that dispersing individuals may face trade-offs between the costs involved in finding habitats of high quality versus the fitness benefits they can expect to enjoy after settling in new habitats (Morris 1992; Danchin *et al.*, chapter 17). All of the individuals within a population might resolve these trade-offs the same way, or different individuals within the same population might resolve them in different ways. For instance, some individuals might adopt high-risk behaviour, resulting in higher search costs, but also higher habitat quality if they are successful, while others might adopt low-risk behaviour, resulting in lower search costs, but lower habitat quality for successful dispersers. This pattern may occur in female turkeys, *Meleagris gallopavo*: the area over which females travel while searching for a nest site is inversely related to the quality of the nest sites (as reflected by nest survival) obtained by those females (Badyaev *et al.* 1996).

In summary, when estimating the fitness consequences of habitat selection for dispersers, it is important to consider all of the phases of the habitat selection process that apply to the species in question. If habitat selection is 'free' (no search or settlement costs), one can evaluate the fitness consequences of habitat selection by measuring the individual fitness of individuals after they have settled in a habitat (cf. the IFD). However, if search and settlement are costly, as is likely to be the case for habitat selection at the spatial scales relevant to dispersers, then one must begin measuring the fitness consequences of habitat selection as soon as those costs begin to accrue. If animals make physiological or behavioural preparations for dispersal while still living in the pre-dispersal habitat (e.g. habitat training; see also Ims and Hjermann, chapter 14), it may be necessary to begin measuring potential costs of habitat selection behaviour before dispersers even leave their original habitat.

Behaviour patterns that reduce costs while dispersers are searching for new habitats

Indirect cues of habitat quality

One of the most important components of the search phase of habitat selection is assessment, the process by which individuals evaluate the relative quality of prospective habitats, before accepting and settling in one of them. However, direct assessment of intrinsic habitat quality, conspecific density, heterospecifics, and other factors affecting fitness during the post-settlement period is not a simple proposition. Many different factors can affect fitness after an individual settles into a home range or territory, and the longer the period of tenure, the longer the list of factors that are likely to affect fitness. In addition, settlement decisions are frequently made before environmental factors that might affect fitness during the residency period are in evidence (Hildén 1965). For instance, larval barnacles select permanent attachment sites during high tides, but run the risk of selecting a site that will overheat or dry out during low tides on subsequent days or months (Lively and Raimondi 1987; Raimondi 1988).

Given the problems inherent in directly assessing important features of new habitats, it is not surprising that many dispersers take 'short-cuts', and use indirect cues when evaluating intrinsic habitat quality. The use of indirect cues for habitat selection has been documented in a wide range of species, ranging from endoparasites (Sukhdeo and Sukhdeo 1994) to vertebrates (e.g. Hildén 1965; Rangeley and Kramer 1998). A particularly striking example of indirect assessment occurs in larval barnacles, *Chthamalus anisopoma*, which have difficulty finding the appropriate zone of the intertidal in which to settle (Raimondi 1988). These larvae rely on a variety of indirect cues when selecting a habitat, one of which is the odour of a gastropod, *Acanthina angelica*, which preys on the members of this species (Raimondi 1988). Raimondi suggests that the odour of *A. angelica* enhances the probability of settlement in *C. anisopoma* because this gastropod confines its activities to, and thus leaves its odour trails in, the 'correct' zone of the intertidal. In this case, the cue in question is clearly important because it is correlated with another factor (location) that affects fitness after settlement, not because the cue is produced by a feature of the habitat that enhances individual fitness after settlement.

Although many of the indirect cues that are used in habitat selection by dispersers are species specific, this is not true of all of them. In particular, conspecific cueing, or the use of conspecifics to assess habitat quality, seems important in a wide range of species (Shields *et al.* 1988; Stamps 1988; Forbes and Kaiser 1994; Danchin and Wagner 1997; Danchin *et al.*, chapter 17). The idea here is that the presence, behaviour, type, etc. of conspecifics in a habitat provides newcomers with an indirect way to assess the intrinsic quality of that habitat. Variants of the conspecific cueing theme include colonial species, in which dispersers assess the reproductive success of residents in one year and then use this information when selecting a habitat the following year (Danchin *et al.*, chapter 17), and heterospecific cueing, in which newcomers use the presence of particular heterospecifics to assess the quality of new habitats (Monkkonen *et al.* 1990).

Habitat cueing

Habitat cueing can be defined as occurring when individuals use stimuli encountered in the pre-dispersal habitat to reduce the costs of searching for a suitable post-dispersal habitat. By learning stimuli associated with the pre-dispersal habitat, and then searching for a post-dispersal habitat containing those same stimuli, a disperser might be able to reduce its search costs, especially in situations in which suitable habitats are difficult to detect or assess. The process of habitat cueing would be expected to evolve in situations in which particular stimuli are reliably correlated with habitat quality at a temporal scale of several generations, in which case stimuli associated with the pre-dispersal habitat would be indicative of at least a minimum level of habitat quality, since a habitat containing those cues produced at least one successful individual (that disperser). As is true for any indirect cue used in habitat selection, the stimuli used for habitat cueing need not be produced by factors that directly affect fitness during the post-settlement period. For example, a mammalian disperser might prefer a habitat with odours similar to those in its natal habitat, not because those cues are produced by food items, predators, conspecifics, or other factors affecting fitness after settlement, but because those odours are easy to detect, and reliably indicate that a new habitat is suitable for long-term occupancy.

Information pooling among dispersers

Recent studies indicate that conspecifics pool information when assessing mates, predators, food locations, and other salient features within established home ranges or territories (Templeton and Giraldeau 1995; Prins 1996). Not surprisingly, animals are most likely to rely on information from conspecifics when distinctions between objects of interest are subtle and difficult to detect, and more likely to rely on their own judgement when differences between objects of interest are distinct and easily discerned (Dugatkin 1996a; Galef and Whiskin 1998; Witte and Ryan 1998).

Choosing a habitat in which to establish a long-term home range or territory is at least as important as choosing a mate or avoiding a predator, and, as was noted above, dispersing individuals often have difficulty assessing habitat quality on their own. We have already considered one form of sequential information pooling that may occur during habitat selection: conspecific (or heterospecific) cueing occurs when newcomers copy the choices made by earlier settlers in the same or another species. However, if dispersers are faced with an array of vacant habitats, conspecific or heterospecific cueing is not an option. In this situation, dispersers might pool information by simultaneously assessing different sites, combining information, and then settling in the area that most or all of them deem to be of the highest quality. A sophisticated version of this process has been described in honeybees, in which the choice of a new home site is resolved by pooling the evaluations of potential home sites provided by many different workers (Seeley and Buhrman 1999). Similar, albeit less sophisticated, information pooling by dispersers could occur when other species seek new home-sites. Information pooling predicts that individual dispersers will interact socially with other individuals (conspecifics and/or heterospecifics) with the same habitat preferences, and that the members of these social groups will end up settling in the same neighbourhood. In fact, many animals disperse in groups (e.g. Lessells *et al.* 1994), but this observation alone does not provide strong support for information pooling, as there are other ultimate reasons why dispersers might travel in groups (e.g. reduced search costs as a consequence of improved predator protection or food-finding *en route*, or reduced settlement costs in the new habitat, as a consequence of settling with familiar companions).

Discussion

Alternative ultimate explanations for behavioural phenomena

At the beginning of this chapter, I discussed two proximate behavioural processes that may play an important role in habitat selection by dispersers: conspecific attraction and habitat imprinting. By now, the careful reader will have noticed that there are several reasonable adaptive explanations for each of these phenomena. Observations of conspecific attraction (newcomers are attracted to areas containing previous settlers) are consistent with at least three ultimate hypotheses: (1) animals benefit from the presence of conspecifics after settling (Allee effects), (2) settlement costs are reduced in the presence of conspecifics, and (3) search costs are reduced if dispersers use conspecifics as an indirect cue of habitat quality (conspecific cueing). Similarly, habitat imprinting (a preference for a post-dispersal habitat similar to the pre-dispersal habitat) is consistent with at least two ultimate explanations: (1) habitat training – experience in the pre-dispersal habitat improves performance if an individual settles in a similar post-dispersal habitat, and (2) habitat cueing – cues associated with the pre-dispersal habitat reduce the costs of detecting and assessing a suitable post-dispersal habitat. These and other examples of alternative ultimate hypotheses for the same proximate mechanisms arise because habitat selection is a multi-stage process, and any given proximate behavioural process may have evolved because, in past environments, it reduced search costs, reduced settlement costs, and/or improved fitness after settlement.

Adaptive explanations for habitat selection mechanisms are not exclusive; for example, in a given species, conspecific attraction might have evolved because it allowed dispersers to locate suitable habitats more quickly and because dispersers benefited from the presence of conspecifics after settling. However, a given behavioural mechanism need not improve fitness at every stage of habitat selection, and, in fact, the best illustrations of the multi-stage nature of habitat selection come from species in which a given process seems beneficial at one stage, but not at another. For instance, conspecific attraction is very highly developed in bark beetles (e.g. Wood *et al.* 1986), but there is no indication that beetles benefit from the presence of conspecifics after settling in a new habitat. Indeed, a recent study of one species (*Ips pini*) showed a strong inverse exponential relationship between the lifetime reproductive success of both male and female beetles and the density of same-sex conspecifics on the same log (Robins and Reid 1997). Given the difficulties that bark beetle dispersers face in detecting newly fallen trees suitable for occupancy (especially in regions without widespread logging), and given the likelihood that many dispersers die before they find any suitable habitat, we might entertain another explanation for conspecific attraction in this species: that individual dispersers use cues from conspecifics to reduce the costs of searching for suitable post-dispersal habitats. Other species also exhibit strong conspecific attraction in conjunction with negative density-dependent relationships between fitness and density after settlement (e.g. larval damselfish, *Dascyllus trimactulatus*; Schmitt and Holbrook 1996, 1999). These examples hint that reductions in search costs as a result of conspecific cueing might sometimes be high enough to outweigh reductions in fitness that occur as a result of living with conspecifics after surviving settlers have settled in their new habitat.

Implications for conservation

Thus far, we have assumed that dispersers always behave in an 'ideal' fashion when selecting habitats, because this assumption allows us to predict the optimal behaviour for each individual as a function of experience in the pre-dispersal habitat, search costs, settlement costs, habitat quality, presence or density of conspecifics and heterospecifics, and so forth. However, if the proximate mechanisms involved in habitat selection are the product of evolution, the assumption that habitat selection is ideal is valid only if current selective regimes are comparable to past selective regimes. The problem, of course, is that as a result of human activity many species today live in habitats that are very different from those in which they lived in the recent past (e.g. Vitousek *et al.* 1997).

One consequence of widespread anthropogenic effects is that selective pressures that were responsible for the evolution of particular behavioural mechanisms may no longer exist (Lack 1933; Hildén 1965). For example, habitat selection behaviour that evolved in response to raptor predation is unlikely to yield any discernable benefits if studied in one of the many areas where large raptors have been eliminated as a result of human persecution (Forbes 1989; Regehr *et al.* 1998). Another consequence is that indirect cues that once reliably indicated habitat quality in past environments may no longer do so (Lack 1933; Hildén 1965). This problem is especially acute for species that rely on indirect cues which are 'innate', in the sense that they develop in the absence of any previous exposure to those stimuli (e.g. fish, Elliot *et al.* 1995; mammals, Wecker 1963; birds, Morton 1990).

If habitat selection is not ideal, then from an applied perspective it may be more important and productive to study the proximate behavioural mechanisms that affect habitat selection than to try to determine the adaptive significance of those mechanisms. For instance, if the members of a species exhibit such strong habitat imprinting that they refuse to settle in a habitat different from their natal habitat, it may not matter much whether this behaviour originally evolved to reduce search costs (habitat cueing) or to improve performance after settlement (habitat training). Indeed, ecological factors that originally favoured the evolution of habitat imprinting in this species may no longer exist, for example, if all available habitats now fall into two easily discernable classes, suitable reserves versus unsuitable disturbed areas, or if practice evading a particular predator is no longer relevant because that predator is extinct. If the members of a species exhibit strong habitat imprinting, from an applied perspective it might be better to try to provide individuals with pre-dispersal habitats as similar as possible to the post-dispersal habitats in which humans intend them to live, than to attempt to determine the reasons why habitat imprinting evolved in the first place.

Another relevant point for conservation is that, when considering any selective process involving positive social interactions with conspecifics (or heterospecifics), neighbourhood size may be more important than population density. Currently, most field workers assume that, if a fragment of habitat remains unoccupied, that fragment is of poor intrinsic quality or is located too far from a source population to attract dispersers. However, if individuals prefer to settle and live in neighbourhoods containing a minimum number of conspecifics, then a fragment that is too small to support a minimum neighbourhood size will remain unoccupied or underutilized even if it is of high quality and is visited by prospective settlers. Possible examples include pikas (*Ochotona pinceps*), in which patches able to support three territories remain unoccupied more often than predicted on the basis of habitat quality or distance to sources of dispersers (Smith 1974a, 1980), or nut-hatches (*Sitta europaea*) in which patches suitable for up to 10 territories are underutilized in comparison to predictions based on habitat quality and distance to inhabited patches (Bellamy *et al.* 1998).

Conclusions

One theme of this chapter is that we seem to be moving away from a view of habitat selection based entirely on competitive interactions, to one that also encompasses

positive interactions among conspecifics and heterospecifics. The phenomenon of conspecific attraction has been documented in a wide range of taxa, and heterospecific
attraction is beginning to attract attention from behavioural biologists and ecologists. In
addition, since dispersers often direct positive social interactions to particular individuals after settling, neighbourhood size may provide a better tool for estimating the
effects of conspecifics or heterospecifics on one another than population density, especially for animals living in fragmented habitats.

A second theme of this chapter is that it may be time to focus on the development of
habitat selection processes. In particular, if animals prefer post-dispersal habitats similar
to their pre-dispersal habitats (habitat imprinting), then habitat selection theory needs to
revisit the assumptions that all dispersers rank intrinsic habitat quality the same way, and
that every disperser prefers the same habitats. Habitat imprinting may also help to
explain puzzling anomalies in the empirical habitat selection literature. For instance,
Morris (1991) found that female white-footed mice, *Peromyscus leucopus*, born in edge
habitats settled and raised their young in edge habitats, even though *per capita* reproductive success was higher for females living in adjacent forested habitat, and even
though there was plenty of room for additional females in the forested habitat. This
result is inconsistent with traditional habitat selection theory, but it is compatible with
the hypothesis that female mice prefer to settle in the type of habitat in which they were
raised.

A final implication of this chapter is that the relationship between dispersal and
habitat selection is more intimate than previously imagined, because, at the proximate
level, processes affecting habitat selection begin as soon as (or even before) dispersers
leave their original habitat. On the one hand, understanding dispersal may be useful for
understanding habitat selection behaviour, because the factors that are considered to be
'costs of dispersal' broadly overlap with the factors that affect the costs of searching for
and settling in post-dispersal habitats. On the other hand, habitat selection may shed
light on behaviour observed in dispersers, such as travelling in groups, or situations in
which the same patch of empty habitat is accepted by some dispersers but rejected by
others. Instead of merely viewing habitat selection as the end point of dispersal, it may be
more profitable to view both as part of a developmental continuum, which begins early
in life before dispersal and which lasts well after the period that the disperser has settled
in its new habitat.

Acknowledgements

Special thanks to Alex Badyaev, Thierry Boulinier, Jean Clobert, Bruno Ens, Etienne
Danchin, Correigh Greene, and Dik Heg for their comments and suggestions on previous drafts of this chapter.

17

Public information and breeding habitat selection

Etienne Danchin, Dik Heg, and Blandine Doligez

Abstract

In most animal species, natal and breeding dispersal is bounded by departure and settlement decisions. Such decisions are based on the individual perception of habitat suitability, which varies with intrinsic habitat quality and social components. Individual perception of habitat suitability may be based on internal information (maternal effects) or on external cues revealing habitat suitability. Such cues can include: the measurement of key environmental factors (direct habitat assessment), the fact that the birthplace was favourable (philopatry), the presence (social attraction) or performance (public information) of conspecifics, or direct patch sampling (private information). We suggest that the use of private information (acquired by trial and error tactics) is unlikely in the context of breeding habitat selection. We also claim that, in temporally auto-correlated environments, public information (i.e. information on the performance of conspecifics) is likely to provide particularly valuable information for breeding patch decisions because it is easy to gather (i.e. at low cost) and reliable (conspecifics cannot afford to hide their performance). Hence, public information allows individuals to predict future fitness by integrating the effect of most factors influencing breeding performance. This is supported by a review of the evidence for the use of public information in a wide variety of taxa. Information on habitat suitability is gathered during a prospecting phase for which we provide a general classification according to the constraints resulting from the biology of the species. Animals are thus facing a trade-off between keeping on prospecting to enhance future reproduction and settling immediately as a breeder to enhance current reproduction. The nature of that trade-off varies according to the kind of prospecting strategy. A theoretical analysis strongly supports our claim for an important role of public information in dispersal decisions when the environment shows intermediate rates of temporal autocorrelation. Our model also leads us seriously to question the significance of the evidence in favour of social attraction. This implies that proximate mechanisms of habitat selection, as well as the pattern of temporal and spatial variation of the environment, are key elements if we are to understand the multiple ultimate functions as well as the consequences of dispersal.

Keywords: breeding habitat selection, cues, evolutionary stability, mate choice, natal and breeding dispersal, prospecting, public information, value of information

Introduction

In essence, dispersal (i.e. changing sites between two consecutive breeding events or between birth and first breeding) is bounded by two breeding habitat selection decisions: to depart from the current breeding patch (departure decision) and to settle in another

Table 17.1. Empirical and experimental evidence of the use of public information in breeding habitat selection

Species	Type	Cue	Effect on	Trend	Habitat variation	References
Mollusks						
Crepidula fornicata and plana	Ex	Qual	S	+	No	Review: Meadows and Campbell (1972) McGee and Targett (1989)
Crustaceans						
Balanus sp. barnacle cyprids	Ex	Chem, Qual	S, SpRec	+	No	Meadows and Campbell (1972)
Balanus amphitrite Elminius adelaidae Elminius adelaidae	Ex	Chem, Qual	S	+	No	Crisp (1974), Bayliss (1993)
Fish						Reviews: Gibson and Höglund (1992), Dugatkin and Höglund (1995), Brooks (1996, 1998)
Poecilia reticulata	Ex	Cop	MC	+	No	Dugatkin (1992, 1996a,b), Dugatkin and Godin (1992, 1993, 1998), Briggs *et al.* (1996)
Gasterosteus aculeatus	Ex	PRS	MC	+	No	Goldtschmidt *et al.* (1993), Patriquin-Meldrum and Godin (1998)
Betta splendens	Ex	PM, Qual	Ter	+	No	Oliveira *et al.* (1998)
Betta splendens	Ex	PM, Qual	MC	+	No	C. Doutrelant and P.K. McGregor (unpublished manuscript)
Oryzias latipes	Ex	Cop	MC	+	No	Grant and Green (1996)
Pomatoschistus minutus	Ex	PRS	MC	+	No	Forsgren *et al.* (1996)
Poecilia latipinna	Ex	Cop	MC	+	No	Schlupp *et al.* (1994), Schlupp and Ryan (1997)
Reptiles						
Anolis auratus	Ex	Pres, Qual, SpRec	S	+	No	Kiester (1979)
Birds						
Actitis macularia	Em	Pres, PRS	P, S	+	No	Reed and Oring (1992)
Bucephala clangula	Em	PRS	P, S	+	No	Dow and Fredga (1985), Zicus and Hennes (1991)
Tetrao tetrix	Em	Cop	MC	+	No	Höglund *et al.* (1990)

Species						Reference
Centrocerus urophasianus	Em	Cop	MC		No	Gibson et al. (1991)
Theristicus caudatus	Em	PRS	PD	+	No	Boulinier (1996)
Phalacrocorax carbo	Em	PRS	EF	+	S+	Schjørring et al. (1999, 2000); auto-correlation not provided
Gyps fulvus	Em	PRS	PD, S	+	S+	J. Lecomte et al. (unpublished manuscript); auto-correlation of PRS: 0.16 and 0.26
Rynchops niger	Em	PRS	PD	+	Dis	Burger (1982)
Rissa tridactyla	Em	PRS	P, D, S, PD, EF	+	S+	Monnat et al. (1990), Danchin et al. (1991, 1998a), Cadiou et al. (1994), Boulinier et al. (1996a), Cadiou (1999); autocorrelation of parasites: 0.54; auto-correlation of PRS: 0.22
Alca torda	Ex	PRS	P, S	+	No	Boulinier et al. (1999)
	Em	PM	MC, EPM	+	No	Wagner (1991)
Troglodytes aedon	Em	Pres, PRS	S	+	No	Muller et al. (1997)
Ficedula albicolis	Em	PRS	D, S, PD	+ in females	S+	Doligez et al. (1999); auto-correlation of PRS: 0.21
Turdus merula	Em	PRS	MC	+	S+	Desrochers and Magrath (1993); auto-correlation of PRS: 0.25
Panurus biarmicus	Em	PM	MC, EPM	+	No	Hoi and Hoi-Leitner (1997)
Poecile atricapillus	Em	PM, Qual	S	+	No	Ramsay et al. (1999)
Parus major	Ex	PM, Qual	EPM	+	No	Otter et al. (1999)
Luscinia megarhynchos	Ex	Qual, PM	Ter	+	No	Naguib and Todt (1997)
Petrochelidon pyrrhonota	Em	PRS	D, S, PD	+	S+	Brown et al. (2000); auto-correlation of PRS: 0.51
Progne subis	Em	PM	EPM	+	No	Morton et al. (1990), Wagner et al. (1996)
Dolichonyx oryzivorus	Em	PRS	D	+	No	Bollinger and Gavin (1989)
Agelaius poeniceus	Em	PRS	D	NS	No	Beletsky and Orians (1991)

Type of evidence: Em, empirical; Ex, experimental; Th, theoretical. Cue: Chem, chemical cues indicating the presence of conspecifics; Cop, mate choice copying; Gen, genes involved; ON, presence of old nests from the previous season early in the season; PM, information on potential mates or rivals (through extra-pair behaviour or eavesdropping); Pres, presence (includes group size and structure, tape lure, etc. = social attraction); PRS, patch reproductive success; Qual, different individuals shown to have contrasted impact. Effect on: D, departure decision; EF, effect of choice on future fitness; EPM, extra-pair mate; MC, mate choice; P, prospectors attracted; PD, effect on population dynamics; S, settlement decision (linked to recruitment); SpRec, species recognition; Ter, impact on territorial contests. Trend: +, significant positive effect; −, significant negative effect; NS, non-significant. Habitat variation – whether temporal auto-correlation was tested: Dis, discussed, not tested; No, not tested; S+, tested, and positive in the short term (at least 1 year at least); NS, tested, but non-significant; S−, tested and negative in the short term (at least 1 year). When significant, the parameter on which it was measured, as well as the level of auto-correlation between successive years, is provided in the reference column.

patch to breed (settlement). Animals may use any cue predictive of breeding success to make these decisions. Breeding habitat suitability may differ because of variation in intrinsic habitat quality as well as variation in the social components (potential mates, density dependence, kin competition). Because there are many commodities that are critical for breeding, animals are unlikely to assess all of them independently. Animals may use specific cues more parsimoniously, integrating the effects of various commodities simultaneously (Danchin and Wagner 1997). Because they share many needs, conspecifics are likely to provide such integrative cues. In particular, the success of conspecifics in various activities provides 'public information' on current environmental suitability.

The term public information was introduced by Valone (1989), in the context of foraging, to encompass readily available information about habitat suitability that can be extracted from the foraging success of other individuals in that patch (feeding success, food-associated calls, etc.; Valone 1996). Public information differs from 'private information' (or 'patch sample information'), that is, information gathered by individual patch sampling with trial and error tactics (Valone and Giraldeau 1993). More generally, public information encompasses any cue about the performance of conspecifics in a given activity (foraging, breeding, aggressive interactions, etc.). Mates being an important resource, of variable quality, for reproduction, organisms might observe the mate choice of conspecifics ('mate choice copying'; Pruett-Jones 1992; Andersson 1994; Dugatkin 1996a); or the fighting ability of potential mates ('eavesdropping'; McGregor and Dabelsteen 1996) as other sources of public information in breeding habitat choice (Nordell and Valone 1998). This shows the generality of that concept, which concerns not only social species but also territorial or solitary species.

In this paper, we focus on the role of public information, for social, territorial, and solitary individuals, in breeding habitat selection, although most of the reasoning and results could be applied to any other kind of site selection. By 'breeding habitat selection' we mean the departure and settlement decisions involved in natal and breeding dispersal. Published evidence of the use of public information in breeding habitat selection shows that public information is used in a wide variety of zoological groups representing contrasting life cycles (Table 17.1). Important issues are that of the value of the information and that of the behavioural patterns allowing animals to gather information about their environment. Costs and benefits of information gathering generate a trade-off between keeping on gathering more information (at the cost of skipping breeding) and breeding rapidly (at the risk of breeding in a bad patch). Because of the complexity of these processes, only a theoretical approach can allow us to determine what is the evolutionary stability of the use of public information in breeding habitat selection. We thus briefly report on a theoretical determination of the evolutionarily stable habitat selection strategy under various conditions of temporal environmental variability. It appears that public information is likely to be used under a wide range of environmental variation regimes.

The value of information

The value of information (its prediction of fitness) depends on the balance of the costs of gathering the information and the benefits that can be gained from its use.

What enhances the value of information?

By definition, to be informative, a cue should reliably predict fitness, which depends on several factors. (1) Environmental suitability should show some degree of temporal autocorrelation between information gathering and the subsequent decision (a condition inherent to any information gathering process). (2) There should be a covariation between the cue and environmental variation (e.g. there should be no time lag between the occurrence of environmental variation and the induced variation in the cue). (3) The size of the sample from which the cue can be assessed also affects the value of the information it conveys. (4) Competitors should not be able to change the cue to provide false information. (5) The cue should integrate the effect of several environmental factors to reveal global habitat suitability (i.e. it should be closely linked to fitness). (6) The cue should be measurable during a sufficiently long and easily determined time window (Boulinier *et al.* 1996a). (7) In the case of public information, the phenotype of the prospector should not be too different from that of the conspecifics it is observing. (8) Finally, to be valuable a cue must be easy to assess, which depends on the costs associated with information gathering (below).

What diminishes the value of information?

While organisms have a lot to gain from acquiring information on which to base their decisions about where to breed and with whom to mate, these benefits must outweigh the costs of acquiring information. We can think of three major types of costs of pro-specting: (1) direct costs due to sampling of the environment (e.g. reduced survival due to travel costs or due to predation and competition during sampling), (2) indirect costs due to the time lost sampling which might have been spent in reproduction instead, and (3) ulterior costs that arise after the choice as a consequence of the competition between all individuals making the same choice.

Evidence of costs associated with prospecting, notwithstanding actually measuring these costs, is scarce in the context of breeding site selection. Indirect evidence of direct costs comes from correlative studies (Smith 1984; Eden 1987; Clobert *et al.* 1988; Smith and Arcese 1989; Bélichon *et al.* 1996; Murren *et al.* chapter 18). For instance, juvenile magpies, *Pica pica*, following the 'disperser strategy' encounter relatively more breeding vacancies but have a lower annual survival rate compared with more sedentary juveniles (Eden 1987). However, that difference in survival may only be apparent, because individuals leaving the study area are considered to be dead (Clobert 1995). An example of ulterior costs is provided by the oyster-catchers (*Haematopus ostralegus*), where individuals compete for high- and low-quality breeding positions and spend several years queuing for vacancies (Ens 1992; Ens *et al.* 1995; Heg 1999). Individuals queuing for high-quality territories must compete, on average, for two more years than those competing for the low-quality territories. The expected Lifetime Reproductive Success, however, is similar for the two strategies because of the higher breeding success in the high-quality territories (Heg 1999). This shows that the value of information depends strongly on the choices of other individuals in the population, and in particular on the frequency of the various habitat selection strategies within a population.

The importance of public information in breeding habitat selection

To assess patch suitability, individuals may have internal information through maternal effects (in such cases they may not have to use other information because they have an internal measure of their environment), or may use external cues. External cues can involve: (1) the direct probing of the environment (by trial and error tactics, i.e. private information), (2) the consideration that the patch where they were born was favourable (i.e. philopatry), (3) the presence of conspecifics, (4) some key characteristics of a favourable environment (thereafter called direct habitat assessment), (5) public information, when it is readily available, or (6) any combination of these cues. Here, we develop only the cases involving some information gathering in relation to natal and breeding dispersal because the role of maternal effects in the development of the dispersal phenotype is detailed in other chapters (Ims and Hjermann, chapter 14; Dufty and Belthoff, chapter 15).

The importance of these different habitat selection strategies is likely to differ greatly between natal and breeding dispersal. Usually, factors influencing departure in natal dispersal are likely to be of a much more involved nature than in breeding dispersal. For instance, maternal effects are probably involved only in natal dispersal (at least as a major factor influencing decisions). Information about kin competition and risks of inbreeding probably plays a prominent role in natal dispersal (Perrin and Goudet, chapter 9; Ims and Hjermann, chapter 14), but much less so in breeding dispersal. Similarly, private information on reproductive success (i.e. experience) cannot be involved in natal dispersal, but is expected strongly to influence breeding dispersal. We thus expect the proximal mechanisms involved in natal and breeding dispersal to differ greatly.

The various potential cues for breeding habitat selection have advantages and disadvantages. Private information reveals the direct interaction between the phenotype and its environment, but it can be available only for breeding dispersal, and the number of feasible trials may remain very small. In breeding habitat selection, a trial and error tactic would mean settling to breed at random to test habitat quality, which would mean no habitat selection. Because of the length of such trials relative to an individual's lifetime, this would be prohibitively costly. Thus, contrary to other types of choices, individuals are not expected to use private information even in the breeding settlement phase of breeding dispersal. An exception may be when experienced individuals have an image of what a favourable patch is. Such individuals may not need other information and may decide where to breed according to habitat characteristics alone, which may allow them to pioneer empty patches. Philopatry accounts for the interaction of the phenotype with its environment at the time when the recruit was born, which may be misleading if the environment has changed in between. The presence of conspecifics (social attraction) may reveal habitat quality, but individuals may keep on settling in a highly occupied patch even though its suitability is deteriorating rapidly. Direct habitat assessment may be difficult to perform because individuals would need to assess all (or at least several) of the factors potentially affecting reproductive success before settling.

One particular characteristic of the environmental suitability is potential mate quality, and mate selection constitutes a particular form of habitat selection in which the mate is the main cue. Depending on the distribution of mates of various qualities, and the likelihood of encountering higher-quality mates and breeding successfully, we may expect individuals to accept a mate or continue searching (Janetos 1980; Johnstone 1997). In some instances, we might expect a correlation between patch and potential mate quality; hence, individuals might select mate and patch by cueing on one of them alone (Alatalo *et al.* 1986; Slagsvold 1986; Andersson 1994). Whether mate or patch cues predominate will depend on the speed and reliability of the information acquired from each of them, and on the relative importance of mate versus patch quality for reproductive success. Mate and habitat choice may also be totally separated, as in leks.

The last type of cue, that is public information, probably fulfils most of the characteristics of a high-value cue for breeding site or mate selection. Public information is highly reliable, because conspecifics cannot afford to bias their performance. As a component of fitness of conspecifics, public information gathered at the end of the previous breeding season integrates the effects of all factors acting on reproductive success, including density dependence, into a single proximate cue. This allows individuals to account for the choices of other individuals. For breeding patch selection, the most integrative public information is the reproductive success of conspecifics (patch reproductive success (PRS); see Switzer 1997). PRS may also provide information on both patch and mate quality (Wagner 1999).

In socially breeding species, the value of public information may be particularly high relative to the information conveyed by other cues because individuals can observe many conspecifics simultaneously, thus allowing social individuals to sample more patches. Socially breeding individuals are thus expected to use public information in dispersal decisions to a greater extent than solitary ones. However, solitary individuals may also gather public information by visiting neighbouring pairs to assess their performance, or through cues such as song rates or family size. Although solitary species are less prone to using public information, there is evidence for the use of public information in some solitary species (Table 17.1). The superiority of public over private information has been shown theoretically and experimentally (Valone 1989, 1991, 1993; Valone and Giraldeau 1993; review in Templeton and Giraldeau 1996). For instance, theoretical considerations have shown that foragers with access to public information should be able to assess the suitability of a resource more efficiently than those relying on private information alone (Clark and Mangel 1984). Such evidence regarding breeding habitat selection is sorely lacking (Boulinier and Danchin 1997). More generally, the advantages of public information suggest that it should be used in breeding patch selection (empirical evidence in Table 17.1). However, public information is unlikely to reveal inbreeding risks and kin competition directly, except if individuals have spent enough time in the patch to know who their close kin are.

Finally, as with any cue, public information allows individuals to predict their own fitness (see Templeton and Giraldeau 1996, for a full discussion in the foraging context) only if the environment is sufficiently predictable over time. This crucial assumption has rarely been tested in the context of habitat selection (Table 17.1): five estimations of the autocorrelation in the reproductive success of individuals breeding in a given patch (range 0.16–0.51), and only one study estimating the autocorrelation in a parameter determining intrinsic habitat quality (tick prevalence, 0.54). It is thus of great importance to describe the natural pattern of temporal and spatial variation of the environment if we are to understand breeding patch selection and the resulting dispersal patterns

(Boulinier and Danchin 1997; Hanski, chapter 20). We know from literature on learning that animals are capable not only of detecting environmental autocorrelation, but also of changing their behaviour in sophisticated ways to take advantage of it (Stephens 1991). Furthermore, we suspect that, because of its integrative quality, public information gathered at the end of the breeding season may better predict next-year fitness than any cue gathered early in the following breeding season in seasonal environments.

Behavioural patterns of information gathering

Prospecting is widespread in certain taxa, as in birds and marine invertebrates (Meadows and Campbell 1972; Doyle 1975; Ward 1987). It is rather surprising that, with the exception of marine invertebrates, prospecting has almost never been studied *per se* (review in Reed *et al.* 1999), even though its high incidence in some groups suggests that it plays an important function. In particular, prospectors may gather any kind of information on their environment that they can use to make settlement decisions. The possible behavioural patterns of information gathering that a species may adopt depend on the various biological parameters affecting the value of public information, such as the type of life cycle and the mobility of breeders, as well as many aspects of the breeding biology, such as breeding synchrony within and among patches. Behavioural patterns of prospecting can be classified into two main categories (in all cases, patch choice is determined by patch quality dynamics and the relative costs and benefits of gathering information during prospecting).

Category 1: Prospectors (semelpar species or organisms with a sessile adult phase, such as marine invertebrates) choose a patch for a lifetime, during a single prospecting phase.

Category 2: Prospectors of iteropar species may choose a different patch for every reproductive episode. Every choice involves a prospecting phase. There may be two different options, according to the capacity individuals have to prospect during reproduction. In category 2a, reproduction allows no further habitat sampling, and only private information is acquired during breeding (e.g. extremely isolated breeding organisms). In category 2b, reproduction allows further habitat sampling, at least after a breeding failure (e.g. some birds and mammals). In such a case, reproduction may allow habitat sampling only within the current patch (category 2b1) or at the larger scale of several patches (category 2b2). Individuals in categories 2a and 2b1 have to decide whether to take off from their original home range to venture through novel space in search of new patches encountered sequentially (sequential prospecting). Individuals in category 2b1 can make a series of forays from their current home base, and eventually settle in one of the areas visited during the forays (centrally-based prospecting).

The trade-off between prospecting and breeding

The relationship between departure and settlement decisions varies according to the prospecting tactic. Because of the costs and benefits of information gathering, animals face a trade-off between keeping on prospecting to enhance future reproduction or settling as a breeder to improve current reproduction. The nature of this trade-off differs according to the category of species relative to prospecting.

In sessile or semelpar species (prospecting category 1), individuals only perform natal dispersal for which they have no private information. Philopatry seems to be an efficient strategy in some fish, such as salmon, but it is probably impossible in marine inverte-brates (because of tidal streams), where settling individuals can rely only on the presence or performance of conspecifics or on direct habitat assessment. The time constraint is particularly obvious in marine invertebrates with a pelagic larval stage. In these taxa, larvae have a specific time-window during which they can fix on to the substratum and perform their metamorphosis into a sessile adult (Meadows and Campbell 1972). After that favourable time-window, unfixed individuals die. Such species are expected to use threshold tactics. Theoretically, in species with a constant instantaneous mortality rate during prospecting, an initial 'exploratory phase' may be adaptive if the abundance of optimal sites is unpredictable (Doyle 1975). In time-limited species, the length of the prospecting phase depends on the relative abundance and the difference in quality of good and bad patches (Ward 1987). Because of that large time constraint, direct habitat assessment may be impracticable, and individuals are likely to use the presence and/or performance of conspecifics as the major cues. Accordingly, experimental and obser-vational evidence that these species use the presence of conspecifics (Table 17.2) or public information (Table 17.1) has been accumulating for a long time (e.g. review in Meadows and Campbell 1972).

In category 2 species, both natal and breeding dispersal occurs, and individuals usually make the two major habitat selection decisions, to depart and to settle. The relative importance of public information in making these two decisions may differ greatly. In category 2b species, the departure decision has been shown to involve both private and public information (Danchin *et al.* 1998a). In patchy but temporally auto-correlated environments, the use of public information rather than private information to decide to depart from the current patch has been shown always to lead to higher breeding output over the lifespan in mobile species with several reproductive cycles (Boulinier and Danchin 1997). In such species, individuals staying in the same patch for several breeding episodes may also use public information to assess the trend in local suitability.

In centrally based prospectors, highly mobile individuals are able constantly to sample their environment during their various activities. They are expected to use simultaneous comparisons to select a new patch. Such comparison tactics may involve similar cues in both decisions, and the departure and settlement decisions are also related. Because public information is valuable, it is likely to be the main comparison cue used. Individ-uals may be able to limit the indirect costs of prospecting by keeping their ownership in their former patch if no better patch (or mate) is found. Hence, prospecting costs mainly involve travel direct costs.

In sequential prospecting, individuals decide to depart from their current breeding patch without any direct information about other potential breeding patches. Infor-mation on local suitability alone leads to departure. It may involve maternal effects (Ims and Hjermann, chapter 14). Information about the prospective patches is acquired sequentially, allowing sequential comparisons involving threshold tactics. Individuals are expected to disperse only when local suitability deteriorates below a certain threshold, which varies according to the probability of finding a better patch. We can

Table 17.2. Selected list of studies showing evidence for social attraction

Species	Type	Cue	Effect on	Trend	Habitat variation	References
Molluscs						
Ostera edulis	Ex	Chem	S, SpRec	+	No	Review: Meadows and Campbell (1972)
						Meadows and Campbell (1972)
Dreissena polymorpha	Ex	Pres	S, EF	+	No	Chase and Bailey (1996)
Polychaetes						
Spirorbis sp.	Ex	Chem	S, SpRec	+	No	Review: Meadows and Campbell (1972)
						Knight-Jones (1951), Meadows and Campbell (1972)
Sabellaria alveolata						
Coelenterates						
Aurelia aurita	Ex	Pres	S	±[a]	No	Review: Meadows and Campbell (1972)
						Gröndahl (1989)
Ascidians						
Botryllus schlosseri	Ex	Gen	S	+	No	Review: Meadows and Campbell (1972)
						Grosberg and Quinn (1986)
Reptiles						
Anolis aeneus	Ex	Pres	S	+	No	Stamps (1987, 1988)
Mammals						
Spermophilus colombianus	Em	Pres	S	+	No	Weddell (1986)
Dama dama	Ex	Pres	P	+	No	Clutton-Brock and McComb (1993),
						McComb and Clutton-Brock (1994)
Birds						
Diomedea immutabilis	Ex	Pres	MC	+	No	Podolsky (1990)
Phoenicopterus ruber	Ex	On	S 7	+	No	Rendon Martos and Johnson (1996)
Oceanodroma leucorhoa	Ex	Pres	S	+	No	Podolsky and Kress (1989)
Sterna antillarum	Ex	Pres	P, PM	+	No	Burger (1988)
Sterna sandvicensis	Ex	Pres	S	+	No	Veen (1977)
Fratercula arctica	Ex	Pres	S	+	No	Kress and Nettleship (1988)
Forest passerines	Ex	ON	S	NS	No	Erckmann et al. (1990), Yahner (1993)
Ficedula hypoleuca	Ex	Pres	S	+	No	Alatalo et al. (1982)
Hirundo rustica	Ex	ON	S	NS	No	A.P. Møller (personal communication)

Legend as for Table 17.1. [a]Established polyps predated planulae larvae (negative effect), but the rate of metamorphosis of planula larvae into polyps was increased by the presence of established polyps (positive effect).

expect that threshold to be lower in environments with low average patch quality than in environments with high average patch quality (adjustable threshold tactic; Reid and Stamps 1997). In sequential prospecting species, the time constraint is more compelling, and there are costs involving the risk of loosing breeding opportunities as well as direct costs. If an animal waits too long before deciding, it may find when it returns to settle in the site of its final choice that, meanwhile, another individual has taken it. The role of private and public information depends on the prospecting strategy. In species of category 2a, only private information is available before departure, while in category 2b1 private and public information on the local patch is available. In any case, the departure and settlement decisions are clearly separated in time, implying that the cues may be more different. Costs of sequential prospecting may be balanced by the benefits of better habitat choices if the environment is patchy and predictable, even if this implies losing one or several breeding opportunities until a favourable patch is found (Boulinier and Danchin 1997).

In summary, because of its potentially high information value, public information is likely to be involved in the departure and settlement decisions linked to natal and breeding dispersal. But it has been suggested that individual animals use public information as a prominent cue in those two decisions in only three species of birds (Table 17.1). However, the departure and settlement decisions were rarely studied simultaneously, and the role of public information has rarely been investigated. This may be of importance, since some ultimate functions of dispersal, such as avoidance of inbreeding or kin competition, influence departure (Clobert *et al.* 1994b; Perrin and Goudet, chapter 9), but much less so settlement.

Different phenotypes may face different trade-offs

Within a population, we may expect the relative importance of public and private information in dispersal decisions to vary among phenotypes according to their characteristics. In particular, young potential recruits have no information on the current average habitat suitability to make natal dispersal decisions. They may even have very little knowledge about the characteristics of a good breeding patch. They are thus likely to invest more in prospecting before settlement, and they may cue on public information more than experienced individuals (Muller *et al.* 1997). At the other extreme, pioneers may have search images, due to preference or imprinting (two mechanisms partly under genetic control), with which to select patches. Various phenotypes may use public information in different ways. For instance, if a group has had low breeding success due to high parasite infestation, the individual's decision of whether to disperse or not would be different depending on whether the individual is resistant to those parasites or not. In the collared flycatcher (*Ficedula albicolis*), the sexes do not use public information in the same way (Doligez *et al.* 1999). In that species, males are territorial, and they use PRS as an index of future territorial competition. Low-quality males thus tend to emigrate from high PRS woodlands, whereas high-quality males do not appear to be influenced by PRS. Contrary to the male, all females use PRS as indicating resources and emigrate from low PRS woodlands.

Although prospecting behaviour seems to be widespread in the animal kingdom, the balance of the costs and benefits of prospecting, and the value of private versus public information in dispersal decisions, are virtually unexplored. Moreover, the decision to continue prospecting might be influenced by the choice of other prospecting competitors and by the individuals already settled, particularly if the critical resource(s) cannot be shared (Nicholson 1954). This implies that it is crucial to account for competition between various phenotypes to determine which breeding habitat selection strategy is likely to be selected for. In the next section we report briefly on a theoretical study that aims to formalize the verbal arguments developed above in order to account for the competitive interactions between individuals using the various possible types of cues to select their breeding habitat, with the objective of determining their evolutionary stability in environments with different temporal autocorrelation.

Breeding patch selection and public information: evolutionary stability

Early models of the evolution of dispersal ignored habitat selection processes (Johnson and Gaines 1990): behavioural aspects were concentrated into a single migration parameter simulating a genetically fixed dispersal propensity among offspring. Accounting for conditional dispersal (i.e. dispersal probability depending on local conditions), however, would lead to an expectation of dispersal under almost all regimes of habitat variability, including a spatially varying and temporally constant environment (McPeek and Holt 1992; Lemel *et al.* 1997). These models underlined the importance of breeding habitat selection processes (i.e. departure and settlement decisions) in the evolution of dispersal (see Ronce *et al.*, chapter 24). These models parallel early behavioural models, which incorporated conditional strategies. Fretwell and Lucas (1970) and Fretwell (1972) proposed their classic ideal free distribution (IFD) model, which still provides a general framework for the study of the evolution of dispersal (Holt and Barfield, chapter 6; Stamps, chapter 16). Subsequent models relaxed the unrealistic assumption that every individual in the population has a perfect knowledge of the entire environment by allowing individuals to sample (i.e. learn) their environment before settling in a given patch (Clark and Mangel 1984; Stephens 1987; Bernstein *et al.* 1988; Valone 1989, 1993; Yoccoz *et al.* 1993). Although previous models have studied certain evolutionary implications of foraging strategies based on public information (Valone 1989, 1991, 1993), to our knowledge only a few studies have tried to analyse the evolutionary stability of breeding patch selection based on various types of information, in relation to the level of temporal autocorrelation in habitat quality (Switzer 1993; Forbes and Kaiser 1994; Boulinier and Danchin 1997).

In this section, we report briefly on a theoretical study of the evolutionary stability of various breeding patch selection strategies that will be published in detail elsewhere (B. Doligez *et al.*, unpublished manuscript). The model confronts the five types of strategies described on p. 246, in a two-patch habitat. Patches are homogeneous and characterized by their intrinsic quality (α_i), corresponding to the proportion of pairs that breed successfully. Intrinsic quality (α_i) varies independently in both patches with a temporal autocorrelation coefficient (r), which ranges from 0 (randomly varying environment) to 1 (constant environment). Population size is regulated by density dependence on survival during winter (all strategies are affected similarly), and on the number of offspring produced per successful pair (in proportion α_i) during breeding (according to patch-specific density, implying that strategies may be affected differently).

In all strategies, successful adults remain in their patch, and failed breeders and juveniles may disperse. The probability of dispersing depends on the habitat selection strategy. It was assumed that every individual can breed in the chosen patch. Two strategies are condition independent: (1) individuals of the '*random strategy*' have an equal probability of settling into each of the two patches (this corresponds to gathering private information); and (2) individuals of the '*philopatry strategy*' remain at their natal patch. The dispersal propensity of individuals of the three other strategies is condition dependent. Settlement in the patches relies on different cues: (3) individuals of the '*presence strategy*' use the relative densities of breeders in the two patches in the previous year as the cue to select a breeding patch in the current year; (4) individuals of the '*quality strategy*' use the relative intrinsic patch qualities of the two patches in the previous year (this corresponds to the above 'direct habitat assessment' strategy); and (5) individuals of the '*success strategy*' use the relative reproductive success of conspecifics (i.e. public information) of the two patches in the previous year. Animals using the three last strategies are assumed to have prospected the two patches at no cost in the previous year. However, they tend to aggregate in the best patch (highest α_i) and thus pay an ulterior cost through the effect of density dependence on the fecundity of successful breeders. Each strategy was confronted by the four other strategies separately, first as the mutant strategy, then as the resident strategy.

Results show that the strategy that outcompetes the others differs greatly, according to the temporal autocorrelation of patch quality (Fig. 17.1). When habitat quality is totally unpredictable, strategies tracking environmental variability cluster at last-year-best-patch, and, on average, individuals endure a lower fecundity because of density dependence. Hence, the random strategy is favoured, suggesting that plastic strategies

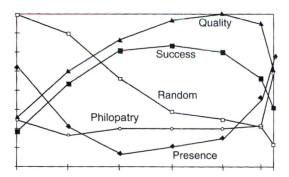

Figure 17.1. Scores obtained by the different breeding patch selection strategies (random, philopatry, presence, quality, and success) when confronted with one another according to the temporal autocorrelation coefficient of environmental variation, for a short-lived passerine. Y-axis scores correspond to the numerical importance of each strategy at the end of the confrontations involving this strategy, over at least 100 confrontations (each simulating different time series of the α_i parameters of intrinsic patch quality) for each autocorrelation coefficient. Open symbols: strategies of habitat choice independent of local conditions; black symbols: strategies conditional on local conditions (see text). Similar results were obtained for a long-lived life cycle. (Simulations were performed with the ULM software; Legendre and Clobert 1995.)

are not necessarily adaptive, even when they are not costly. When the environment is totally predictable, the philopatry strategy is expected. In the present model, however, because of the absence of information gathering costs, the quality and presence strategies perform equally well (Fig. 17.1). In situations of intermediate predictability, the selected strategy depends on the value of information, and the quality and success strategies are favoured. As such intermediate situations are likely to be the most common in nature, this underlines the importance of simulating environmental predictability to determine the evolutionarily stable breeding habitat selection strategy. The quality strategy corresponds to direct habitat assessment. It is an unrealistic strategy, involving individuals with perfect knowledge of their environment at time t, selecting their breeding patch at time $t + 1$. In practice, such intense information gathering would necessarily imply attending every breeding area for the entire breeding season to assess every factor potentially affecting fitness, hence skipping breeding. The success strategy may be based on the assessment of the relative densities of offspring in various patches (Desrochers and Magrath 1993; Boulinier and Danchin 1997). This is much more parsimonious than the quality strategy because animals assess only a single parameter to estimate patch suitability (Boulinier and Danchin 1997; Danchin and Wagner 1997; Danchin *et al.* 1998b; B. Doligez *et al.*, unpublished manuscript).

In conclusion, these simulations support the idea that breeding patch selection based on public information gathered in the previous year may be selected for under the natural ranges of environmental temporal autocorrelation. Our confrontations also show that the success strategy is capable of invading populations with the other strategies, meaning that this evolutionarily stable strategy (ESS) is reachable.

Discussion

In this chapter we suggest that, when available, public information (i.e. cues about the performance of conspecifics) is likely to provide valuable information for the two main dispersal decisions (departure and settlement). This is because of the close link between public information and fitness. This result is supported by the occurrence of prospecting-like behaviour in relation to breeding patch selection in a wide variety of animal taxa (Reed *et al.* 1999). Simulations support our claim that public information constitutes valuable information for breeding patch selection within the range of temporal auto-correlation that can be expected in nature (see Table 17.1): individuals using public information perform better than the others and this strategy invades and resists invasion by other strategies (Fig. 17.1).

Results also show that, except in temporally constant environments ($r = 1$), individuals using the presence of conspecifics as a cue to select breeding habitat perform poorly. This result may be expected for two related reasons. First, the lag between environmental variation and the induced variation of the presence of conspecifics is much longer than with other cues. Hence, individuals using the presence strategy track environmental variation only poorly. Second, the above simulations show that that strategy leads to the highest departure from the IFD, implying that individuals using that strategy pay a high indirect cost through competition. The poor performance of the presence strategy questions the abundant literature on social attraction (as in Stamps, chapter 16;

examples in Table 17.2). First, we suspect that, in many cases, the presence and the performance of conspecifics might have been confounded, so that researchers interpreted their results as revealing the presence of conspecifics, while in fact animals were attracted to some indices of performance. For instance, Veen (1977) used decoys of incubating sandwich terns (*Sterna sandvicensis*) to create a new breeding group in the vicinity of an active colony (Table 17.2). He interpreted those decoys as indicative of the presence of conspecifics, while attracted individuals may have been interested mainly by the high breeding success that those decoys revealed (100% of the pairs were incubating and thus breeding successfully). Similarly, in playback experiments, people may have used soundtracks of successfully breeding males (e.g. Alatalo *et al.* 1982; Podolsky and Kress 1989). To test this, one would need to contrast the attractive effect of different categories of individuals (successful versus failed breeders, high versus low-quality individuals; examples in invertebrates and vertebrates in Tables 17.1 and 17.2). For instance, although Podolsky (1990) found that decoys simulating courtship behaviour were more attractive to young laysan albatrosses (*Diomedea immutabilis*) than any other decoys, he interpreted the results of his experiment as evidence for social attraction (Table 17.2). But this could indicate potential mate performance. Similarly, Burger (1988) showed that least terns (*Sterna antillarum*) were more attracted to plots containing single birds together with paired terns rather than to plots containing only paired or solitary decoys (Table 17.1). Second, in some situations, such as under strong Allee effects, the presence of conspecifics may be a key component of habitat suitability; in such situations, fitness drops down at low density because of the lack of social interaction. The generality of such situations depends on the generality of Allee effects, a process that is still poorly documented (Courchamp *et al.* 1999; Stephens and Sutherland 1999). Third, the above theoretical results provide another interpretation for the evidence regarding social attraction: in many simulations, the presence strategy persisted at low frequency by 'parasitizing' the information from the other strategy. If this result is general, it may well be that some experimental evidence of the existence of social attraction in the wild was due to a small portion of the population using the presence of conspecifics as a cue to habitat selection while the bulk of the population used different cues.

A provocative way of looking at the relationship between breeding patch selection and dispersal is to view dispersal as a byproduct of breeding habitat selection: dispersal results from the animal departure and settlement decisions that generate it. Thus, we certainly need to take into account the mechanistic aspects of dispersal if we are to understand the link between proximate mechanisms and the multiple ultimate functions of dispersal (Ronce *et al.*, chapter 24). An important development would be to consider both the natural selection (habitat variability, interspecific and intraspecific competition with kin or non-kin) and sexual selection (mate choice, fertility insurance, inbreeding avoidance) dimensions of the proximate mechanisms of dispersal. The relative importance of these two groups of processes still remains to be clarified. Because the expected patch selection strategy differs according to the regime of environmental temporal variability, we also have to incorporate that crucial component into the models. The use of public information alone probably cannot account for ultimate functions of dispersal such as inbreeding and kin-competition avoidance, and phenotypic differences in patch selection strategies also have to be accounted for in order to understand animal dispersal and spatial distribution. More generally, breeding habitat selection has implications for several aspects of *evolutionary biology*, such as sympatric speciation (Rice and Salt 1988), and has obvious links with the study of learning, cognition, and memory (Stephens 1987, 1991). It has also been suggested that the use of public information in breeding patch selection may explain the evolution of certain forms of sociality (Wagner 1993, 1997; Boulinier and Danchin 1997; Danchin and Wagner 1997; Danchin *et al.* 1998a,b), and our simulations revealed large departures from the IFD with the success strategy. Thus, even elaborated strategies using highly valuable information do not succeed in ideally tracking environmental variability. This means that relaxing the assumption of the IFD, that animals have a perfect knowledge of the environment, has important consequences for animal distribution (Bernstein *et al.* 1991). This might account for the numerous observed deviations from the IFD.

In terms of *conservation*, the impact of the use of public information on population viability has to be studied in detail. Although the use of public information may constitute the ESS, its impact on aggregation may, as with social attraction (Smith and Peacock 1990; Ray *et al.* 1991; Reed and Dobson 1993; Lima and Zollner 1996), increase population extinction risks by concentrating individuals in fewer patches than under the IFD. However, these models of social attraction do not account for temporal environmental variability, and ignore the fact that relying on conspecifics should lead to the preferential use of patches that are currently the best. Furthermore, most models neglect the possible existence of phenotypes with different sensitivities to environmental cues; even a small proportion of individuals adopting a pioneer strategy might, for instance, counterbalance those viability effects by maintaining the population closer to the IFD (the simulations reported here often led to the coexistence of different strategies). Finally, individuals may use different habitat selection strategies under different conditions. For instance, the relative use of private and public information in foraging may vary according to the availability of public information (Templeton and Giraldeau 1995, 1996). From a practical point of view, conservationists may exploit this tendency of animals to rely on public information to improve the impact of management actions, for instance by luring animals with cues of apparently high local breeding success (examples in Danchin and Wagner 1997).

Acknowledgements

We thank Judy Stamps, Thomas Valone, Anders P. Møller, and Jean Clobert for constructive comments and suggestions on a previous version of this chapter. Françoise Saunier, Molly Brewton, and Monique Avnaim kindly helped to edit the text and references.

Part IV

Dispersal from the individual to the ecosystem level: individuals, populations, species, and communities

18

Dispersal, individual phenotype, and phenotypic plasticity

Courtney J. Murren, Romain Julliard,
Carl D. Schlichting, and Jean Clobert

Abstract

We examine the role of phenotypic plasticity (genetically controlled response to the environment) in dispersal decisions from two perspectives. Specifically, we examine the alternatives 'to stay' or 'to go' from the standpoint of the offspring and the maternal parent. For an offspring, phenotypic plasticity may play an important role in determining whether the decision is to remain philopatric or to disperse to a new location. From the maternal perspective, the issue is not the survival of any individual offspring, but the maximization of fitness through all philopatric or dispersing offspring. We use an evolutionarily stable strategy (ESS) model to examine resource allocation to dispersing and philopatric offspring in a variety of scenarios incorporating temporal and spatial heterogeneity. We find that kin interactions may be extremely important in determining the fitness of dispersing and philopatric offspring, and thus the optimal ESS dispersal rate. In sum, there may be multiple layers of plastic decisions that an organism must make when it weighs dispersal options. The frequency of particular environmental conditions and the reliability of information an organism can gather about its environment are both crucial for shaping the evolution of plasticity as it relates to dispersal.

Keywords: phenotypic plasticity, ESS dispersal model, philopatry, temporal heterogeneity, spatial heterogeneity, genotype by environment interactions, resource allocation, trade-offs

Introduction

It is a common observation that dispersing and philopatric individuals differ in their morphology and/or behaviour: presence or absence of wings in some insects, the pappus of a seed (Venable *et al.* 1998), degree of fatness in the naked mole-rat (O'Riain and Braude, chapter 10), and exploratory behaviour in the common lizard (Clobert *et al.* 1994b; general review in Swingland 1983). These and other less obvious anatomical, physiological, and life-history variations can lead to differential fitness between dispersing and philopatric individuals (Bélichon *et al.* 1996). Broadly speaking, such differences could have two causes: genetic polymorphism (see Roff and Fairbairn, chapter 13) or phenotypic plasticity, although phenotypic plasticity is itself a genetically controlled response to the environment. In the case of the choice between dispersing or remaining 'at home', adaptive phenotypic plasticity may be one way of avoiding the negative consequences of both the choice and the resultant effects of these two dispersal modes. In addition, costs associated with dispersal or philopatry may be overcome through phenotypic plasticity.

Individual fitness is based on numerous components summed across the life of an organism and includes various trade-offs, as well as any costs of dispersal and settlement. The type of cost will depend on whether the individual disperses or remains philopatric; for example, traits beneficial to the disperser may be costs for a philopatric individual and *vice versa*. The proportion of dispersing and philopatric offspring produced (defined here as the dispersal rate) is one clear target of selection. Selection may also operate to reduce costs associated with dispersal and philopatry, and may favour particular suites of morphological structures or life-history traits that are associated with either philopatry or dispersal.

In this contribution we examine the choice 'to stay' or 'to go' from two perspectives. First, we take an *offspring*-centred view of the role of phenotypic plasticity for an individual that remains philopatric *or* disperses either through choice or necessity. (We do not examine the role of phenotypic plasticity in dispersal itself, as that is the purview of two other contributions to this volume; see Stamps, chapter 16; and Ims and Hjermann, chapter 14). Beyond these broad patterns of phenotypic plasticity that enable individual organisms to stay or to go, we are also interested in addressing the specific case of how a mother should partition her resources to dispersing offspring or to philopatric offspring. Thus, our second focus is on a specific form of *maternal* phenotypic plasticity: the resource allocation to different types of offspring and the consequences of this choice. We examine an evolutionarily stable strategy (ESS) model of how fitness may be maximized when there is a trade-off in producing dispersing versus philopatric offspring.

The offspring's perspective

Phenotypic plasticity is a particularly important means of maintaining fitness within and across generations of both philopatric and dispersing offspring. We are interested in *two specific times* when plasticity can play an important role in dispersal evolution. First, it can operate as a set of responses that allow certain individuals to remain philopatric. Second, it can increase the phenotypic alternatives available to an individual, following dispersal. We examine the question of costs of dispersal versus philopatry from the perspective of the individual, and in particular with respect to methods by which individuals can ameliorate these costs via phenotypic plasticity. The consequences of the choice between philopatry and dispersal will tend to be most distinct between organisms that are mobile versus sessile as adults, and between actively versus passively dispersing organisms. Concomitantly, the types of plasticity that will be advantageous will vary, especially among these groups.

Philopatry by choice or necessity

When escape by dispersal is not possible, organisms rely on other mechanisms for maintaining fitness in the face of new environmental conditions. Phenotypic responses to altered environments can be in the form of physiological, morphological, developmental, or behavioural changes, or a combination of these changes (Schlichting and Pigliucci 1998). Sessile organisms, such as plants and many invertebrates, lack the ability to move to a better environment during their lifetime. For these organisms, phenotypic plasticity must be the solution to the challenges that dispersal cannot solve. Even in some mobile animals, biotic and abiotic environmental features may induce behavioural modifications in lieu of dispersal.

Sessile organisms

Organisms unable to 'choose' a new site, such as plants and sessile animals, may change their phenotype several times during their ontogeny in response to new environmental conditions. For plants, this may include responses such as the balance between seed dormancy and germination (Vleeschouwers *et al.* 1995), changes from sun to shade leaves (Ashton and Berlyn 1992; Ellsworth and Reich 1993), or changes in physiology of photosynthesis (Holbrook and Lund 1995). In most of these responses, integrated changes in physiology and morphology occur.

Seed dormancy and germination

The interaction between seed dormancy and seed germination is a good example of an integrated phenotypic response to the environment. Numerous environmental factors such as light (van der Schaar *et al.* 1997), light quality (van Hinsberg 1998), temperature, and even smoke (Keeley and Fotheringham 1998) have been shown to interact with hormones or scarification to induce germination (Pyler and Proseus 1996). Finnimore *et al.* (1998) found that temperature affected both seed dormancy and germination of wild oat (*Avena fatua*) at multiple points: (1) by maintaining the level of dormancy during seed development, (2) continuing dormancy when seeds were dry, (3) inducing secondary dormancy in previously non-dormant seeds, and (4) inducing germination when water was imbibed (Finnimore *et al.* 1998). They argued that sensing of the appropriate environment and timing for germination is an adaptive plastic response in wild oat. Some authors consider the dormant stages, such as seeds in seed banks and egg dormancy in copepods, to be plastic responses that allow for 'dispersal' through time (Evans and Cabin 1995; Husband and Barrett 1996). These 'sit and wait' phenotypes are one way of buffering against bad years, by adjusting the level of dormancy to the changing environment. Vleeschouwers *et al.* (1995) discuss the evolution of dormancy as an adaptation to highly variable environments.

Light

Light environments can vary quite dramatically within a forest canopy or an individual plant canopy. A plant in the understorey grows beneath an often-dense canopy, where light availability can be very low. At this life stage, leaves tend to be thick, and the photosynthetic apparatus saturates at low light levels: $200–500\,\mu\text{mol}^{-2}\,\text{s}^{-1}$ (Chazdon 1986). Additionally, these same plants may change leaf morphology, anatomy, and properties of photosynthesis when light and water environments are altered dramatically, for example after gap formation. Chazdon and Kaufmann (1993) planted two species of *Piper*, a tropical understorey shrub, across a light gradient in forest gaps. *Piper sanctifelicis*, which grows naturally in a variety of light environments in disturbed

habitats, showed both biochemical and anatomical responses to changes in light availability. *Piper arieianum*, a more shade-tolerant species, had a weak physiological
response to changes in the light environment, but a dramatic response in leaf thickness.
The authors concluded that differences in selective pressures have resulted in differential
adaptations of morphology, anatomy, and physiology of these two species. Changes in
rates of photosynthesis and transpiration, leaf thickness, stomatal density, and meso-
phyll thickness all constitute phenotypic responses that enable plants to persist and cope
with the new environmental regime of full sunlight conditions.

In four *Shorea* species growing in Sri Lanka, net photosynthesis, transpiration, and
stomatal conductance are tightly linked to morphological and anatomical changes which
respond to shifts in the light environment (Ashton and Berlyn 1992). The greatest
amplitude of phenotypic plasticity was measured in the species most commonly found in
high light environments. Plasticity can be detected not only in the rapid induction of the
photosynthetic apparatus of plants that obtain their light mainly through sunflecks
(Chazdon *et al.* 1996), but in the physiological pathways themselves. In *Clusia* species
(epiphytic trees), continuous net fixation of carbon dioxide is possible through Cras-
sulacean Acid Metabolism (CAM) to 3-carbon (C_3) switching over a 24-hour period.
This functional switch is highly dependent on external environmental conditions correlated with photosynthesis, such as water availability (Medina 1996). Throughout the
ontogeny of a plant, light environments can change drastically multiple times; plastic
responses are one way of maintaining growth in highly variable environments.

Mobile animals

Mobile animals 'decide' to stay or to go. Within a species, individuals may not disperse because of developmental or environmental states. Under these conditions, novel
behavioural strategies, such as helpers to reproductives (Brown and Vleck 1998) and
sneaker males (Fisher and Peterson 1987), may evolve. Other behavioural alternatives to
dispersal include, for example, changes in timing of reproduction (such as sex switching in reef fish; Warner 1985) and induced responses to predators (Komers 1997). As in
sessile organisms, the responses of mobile animals to the environment often integrate
behavioural, physiological, and morphological phenotypic changes.

Behavioural modification can occur when environmental conditions are poor, such
as mate scarcity or absence of suitable breeding sites. A delay in dispersal of offspring
creates a family composed of a breeding pair and 'helpers'. Emlen (1994) argued that
such helping behaviours increase survival in groups, increase fitness through indirect
fitness, and increase the probability of finding a better breeding site in the subsequent
year (which often is the natal breeding site). In the fairy wren, *Malurus cyaneus*,
helping behaviour occurs when neither mates nor suitable breeding sites are available
for 1-year-old juveniles to disperse into. Helpers' responsibilities include feeding of
nestlings, caring for fledglings, and assisting in nest and territory defence. In addition,
fairy wrens are punished for not helping (Mulder and Langmore 1993). Similar helping
behaviours associated with delayed dispersal can be found in a broad range of animals
including: golden lion tamarins (Baker *et al.* 1993), grey-capped social weavers
(Bennun 1994), Florida scrub jay (Woolfenden and Fitzpatrick 1990), and many canids
(Moehlman 1996).

Harsh environmental conditions of food shortage and other cues for the onset of winter can change behaviour, indirectly increasing fitness. Hibernation is typically a behavioural response selected in favour of saving energy under extreme environmental conditions (Kortner and Geiser 1998). Blumstein and Arnold (1998) found that social hibernation in golden marmots (*Marmota caudata aurea*) results in fatter individuals than does solo hibernation. Their data suggest that the probability of dispersal is reduced in harsher environments, and increases indirect fitness benefits.

Mating behaviours themselves can be plastic phenotypes. Subordinate males are often observed attempting copulations with females, while their mates are otherwise occupied with defence or feeding. Unpaired horseshoe crabs (*Limulus polyphemus*) hang around successfully coupled mating pairs, and 'sneak up' to attempt to fertilize the eggs. This alternative mating strategy appears to be directly related to the fact that these unpaired males have been found to have encrusting epibionts on their eyes, which impairs their ability to disperse to find a mate (Brockmann and Penn 1992). 'Sneaking' behaviour may often be a significant component to the mating strategy when non-relative helpers attend bird nestlings (Macgrath and Whittingham 1997).

A number of families of coral reef fish have species that change sex from male to female or from female to male (Warner 1985). Social factors (such as the presence or absence of smaller conspecifics) may induce or inhibit the growth of adult fish, determining a change to male phases in the saddleback wrasse, *Thalassoma duperrey* (Ross 1987). Sequential hermaphroditism increases overall lifetime fitness when sex change is size dependent.

Induced defences evolve in both mobile and sessile organisms when dispersal is not an option (Harvell 1990). Both biotic factors such as predators (Dodson 1989; Boersma *et al.* 1998; Yamada *et al.* 1998) and herbivores (Karban and Baldwin 1997), and abiotic factors such as increased density (Holopainen *et al.* 1997) and water turbidity (Nemtozov 1997), can be the environmental factors initiating induced defences. Carp are induced to a hiding mode with increased density of fish (Holopainen *et al.* 1997). Razorfish sense biotic and abiotic environmental changes which cause them to shrink their home ranges (Yamada *et al.* 1998). More directly, *Daphnia* detect the presence of kairomones of their fish predators and alter both morphology and behaviour (Boersma *et al.* 1998). Mussels at higher risk of attack have thicker shells and greater shell mass and are more strongly attached to substrates (Leonard *et al.* 1999). Harvell (1998) showed that the bryozoan *Membranipora* produced two types of spines, membranous or corner, depending on the concentration of a predator extract.

Dispersal by choice or necessity

Mobile, active dispersal

For individuals that can move and have specific habitat requirements, selection will favour the ability to make choices determining the ultimate site of relocation. Such organisms will tend to evolve systems that allow discrimination of environmental cues. Some form of behavioural plasticity will be favoured that involves information processing and decision making regarding the suitability of a site. The detection and processing of these cues will form the basis for niche or habitat selection (Maynard Smith 1962; Levins 1963; Stamps, chapter 16). Once successful dispersal to a new home site has been achieved, such organisms may revert to mechanisms of plastic response that are already in place (i.e., those utilized by non-dispersing conspecifics as well).

Dispersal into the unknown

Passively dispersed organisms do not have the capacity to choose a destination site directly. However, they will also be selected to evolve the ability to discern the features of the environment; these cues will be useful in establishing subsequent responses to the local conditions. The breadth of plastic responses that ultimately evolve for such organisms will depend on the likelihood of dispersal into the various different habitat types. This likelihood in turn will be determined by the *mode of dispersal* and the *frequency of that habitat* in the landscape. Mode of dispersal is important because it may influence the specificity of dispersal. Although dispersal mediated by the wind, for example, is typically quite random, some forms of passive dispersal can have relatively specific targets; water-dispersed seeds or larvae will generally end up in aquatic or marshy environments, and seeds dispersed by vertebrates may tend to be deposited in similar sites because of the habitat preferences of their carrier. The habitat frequency will determine the intensity of selection for particular forms of plastic responses. Common or predictable conditions will favour the evolution of plastic responses appropriate to them. Rarely encountered conditions will be unlikely to elicit appropriate responses because of the lack of concerted selection pressures.

Generalists

These rules of thumb for when to be plastic, however, do not satisfactorily explain the variation in the colonization success of different species across a range of environments. As has been debated for years, there are two endpoints for 'solving' the dilemma of adaptation to heterogeneous environments: (1) the production and maintenance of genetic variability for adaptation to subsets of the environmental range (specialists), and (2) an increase in individual niche breadth through phenotypic plasticity (generalists). The theory of when to be a generalist or specialist (e.g., Futuyma and Moreno 1988; Kawecki and Stearns 1993; Gilchrist 1995; van Tienderen 1997; Robinson and Wilson 1998), however, has far outstripped the empirical work (e.g., Sultan 1995; Mackenzie 1996; Condon and Steck 1997; Rebound and Bell 1997).

This theme has been especially well explored in experiments on plants by Bazzaz and colleagues (e.g., Zangerl and Bazzaz 1983; Garbutt *et al.* 1985; Bazzaz and Sultan 1987). Sultan has investigated these issues with four species of *Polygonum* that vary considerably in their distributions in the field (Sultan *et al.* 1998a). Glasshouse studies on the amounts and patterns of plastic responses found that the pair of species with the broadest ecological distributions (*P. persicaria* and *P. lapathifolium*) also maintained high photosynthetic activity across a series of high light treatments that varied in moisture and nutrient levels (Sultan *et al.* 1998b). When switched to low light, only *P. persicaria*, the species with the widest field distribution, maintained photosynthetic efficiency. These results implicate the importance of phenotypic plasticity in enabling the colonization of a variety of habitats following dispersal.

Parsons (1997, 1998) compared the plasticity of two species of marine gastropod, one with short-term mobile planktonic larvae and the other with direct development in benthic egg masses. Reciprocal translocation experiments and patterns of temporal variation in growth revealed that the more broadly dispersed species with the planktonic larvae was also the most plastic. The direct developer showed patterns of genetic differentiation among sites.

Dormancy

The causes and consequences of dormancy have already been covered as a possible adaptation for avoiding dispersal. Dormancy is also an option for those propagules that may be passively dispersed quite widely. Even if the new homesteads are potentially suitable, temporal environmental variability may make them quite inhospitable at the time the organism arrives. There are two modes of dealing with this type of uncertainty: either go dormant upon arrival, or, quite frequently, disperse as a dormant seed, spore, or fertilized egg. One consequence of examining both the maternal and offspring perspectives is that it becomes clear that offspring performance need not exactly reflect the expectation of the maternal plant.

The maternal perspective: allocating resources to philopatric and dispersing offspring

Maternal allocation patterns are a result of a complex set of interactions between the current philopatric environment and maternal resources. The choice to disperse is related to the fitness differential between an individual that remains philopatric or disperses. However, this offspring-level fitness is rarely an explicit parameter in existing theoretical models. Here, we attempt to clarify the relationship between dispersal rate and offspring fitness and examine the relationships between costs, fitness, and dispersal behaviour under several specific environmental cases. In particular, we examine how spatial and temporal heterogeneity in the environment influence fitness of both parents and offspring. Given possible differences in fitness of dispersing and philopatric individuals, how should dispersal evolve under changing environmental regimes? Evolution of dispersal has not been examined explicitly through this angle. This is because most models focus on an *individual* that may or may not disperse (offspring point of view), so that evolution of dispersal becomes a decision-making problem of the type 'to disperse or not to disperse'. The evolution of dispersal can be reformulated from the maternal point of view as a resource allocation problem that can change with different environmental cues. Parental fitness (W) can thus be written as a function of individual fitness of philopatric offspring (W_P), individual fitness of dispersing offspring (W_D), and dispersal rate d:

$$W = (1 - d) \times W_P + d \times W_D \tag{1}$$

ESS dispersal rate can be calculated from equation (1) using the method of Frank (1998). ESS dispersal rate d^* is found at:

$$\frac{\partial W}{\partial d}(d^*) = 0 \tag{2}$$

Here, we explore the properties of W_P and W_D that are required for dispersal to evolve. This formulation allows us to link the fitness of dispersing and philopatric individuals to the evolution of dispersal (see also Gandon and Michalakis, chapter 11; Perrin and Goudet, chapter 9).

Offspring fitness is independent of dispersal rate

Three cases can be considered: (i) $W_P = W_D$ (i.e., no selection for or against dispersal); (ii) $W_P > W_D$ and $d^* = 0$, (i.e., philopatry is the ESS); (iii) $W_P < W_D$ and $d^* = 1$, (i.e., dispersal is the ESS). The main factors leading to differences in fitness between dispersing and philopatric offspring are the costs of dispersal (e.g., increased risk of predation, delay in settlement, loss of familiarity with site and neighbours). In addition, spatial and temporal variation of fitness may affect philopatric and dispersing offspring differently. Indeed, when there is variation around the fitness expectancies, the realized fitness is lower than the predicted mean, with the reduction of fitness proportional to the variance around the mean (Gillespie 1977). If the fitness of one group of individuals is less variable than that of the other group, the former will have the advantage.

Temporal heterogeneity

Assume fitness at time t to be a random variable distributed normally with mean W and standard deviation σ. The mean fitness is then given by the geometric mean (Gillespie 1977) approximated by $W - \sigma/2$. If no cost of dispersal occurs, dispersing and philopatric offspring have the same fitness. There is thus no selection for or against dispersal.

Spatial heterogeneity

Assume fitness in site i to be a random variable distributed normally with mean W and standard deviation σ. The fitness of a dispersing individual is averaged over the n sites on which a dispersing individual is likely to settle. There is thus some variance around the fitness expectancy of a dispersing offspring. Using the approximation of Gillespie (1977), $W_D = W - \sigma/2n$. However, the fitness of a philopatric offspring is invariant and thus $W_P = W$. Hence, $W_P > W_D$ and $d^* = 0$ (i.e., no dispersal is the ESS – case ii).

Spatial and temporal heterogeneity

Assume fitness in site i at time t to be a random variable distributed normally, with mean W and standard deviation σ. For a philopatric offspring, variation of fitness is only temporal, thus $W_P = W - \sigma/2$. For a dispersing offspring, variation of fitness is also spatial, thus $W_D = W - \sigma/2n$. Hence, $W_P < W_D$, and $d^* = 1$; i.e., dispersal is the ESS (case iii). In this case, the production of dispersing offspring reduces the fitness variance for the maternal parent; it is a bet-hedging strategy.

The prediction that spatial variation will select against dispersal while spatiotemporal variation should favour dispersal is by no means new (see, for example, reviews in Johnson and Gaines 1990; McPeek and Holt 1992). The point of interest here is that when offspring fitness is independent of the actual dispersal rate (equation 1), one should observe either no dispersal (case ii) or no philopatry (case iii), unless costs of dispersal are exactly balanced by the benefits of reduced fitness variance (case i). This latter case is biologically rather unlikely. Hence, in order to obtain an ESS dispersal rate between 0 and 1, offspring fitness should depend on dispersal rate.

Offspring fitness is influenced by interaction between kin

Costs of interactions between kin may appear either through competition for resources (e.g., intrasexual competition) or through inbreeding. Kin interaction is more likely to occur between philopatric offspring, and, therefore, costs of kin interactions are likely to decrease with dispersal rate. If we assume W_P to be minimum for $d = 0$ ($W_{P,0}$) and maximum for $d = 1(W_{P,1})$ (corresponding to the theoretic fitness of a philopatric off-spring without kin), then, as a rough approximation, W_P will increase linearly with d:

$$W_P(d) = W_{P,0} + (W_{P,1} - W_{P,0}) \times d \qquad (3)$$

Solving equation (2) gives:

$$d^* = \frac{W_D + W_{P,1} - 2 \times W_{P,0}}{2 \times (W_{P,1} - W_{P,0})} \qquad (4)$$

ESS dispersal rate (d^*) can be evaluated for the limits of W_D. If $W_D \geq W_{P,1}$ (i.e. fitness of dispersing offspring is always greater than that of philopatric offspring), then $d^* = 1$. If $W_D = 0$ and $W_{P,0} \geq \frac{1}{2}W_{P,1}$, then $d^* = 0$. If $W_D = 0$ and $W_{P,0} < \frac{1}{2}W_{P,1}$, then $d^* > 0$. Hence, non-zero dispersal rate may evolve even though dispersing individuals all die. Thus, in the case of strong competition between kin, dispersal could theoretically evolve as a method of reducing kin competition.

Philopatric offspring fitness at the ESS is found by substituting equation (4) into equation (3): $W_P = \frac{1}{2} \times (W_D + W_{P,1}) = W_D + \frac{1}{2} \times (W_{P,1} - W_D)$. Hence, for $W_D < W_{P,1}$ (i.e., the condition for $d^* < 1$), $W_P > W_D$. In the case of kin interaction, at the ESS dispersal rate, the observed fitness of philopatric offspring is higher than that of dispersing offspring. The ESS dispersal rate calculated here maximizes the fitness of the mother (which coincides here with the success of a gene). However, unless individuals are clonal, it is clear that philopatric offspring achieve a higher fitness than dispersing offspring. Hence, in the case of kin competition, dispersal represents a potentially significant conflict between mother and offspring (see also Motro 1983).

Offspring fitness is influenced by frequency dependence at the population scale

When stochastic variation occurs between population patches (either demographic or environmental stochasticity), or when exchanges between two subpopulations are numerically unequal (McPeek and Holt 1992), some subpopulations may become overcrowded relative to others. Philopatric individuals of the less crowded subpopula-tions have a relatively high fitness compared with dispersing individuals. In addition, the fitness of both types of individuals depends on the actual dispersal rate in the population: philopatric individual fitness increases with increasing dispersal rate (i.e., as the sub-population size decreases). The fitness of philopatric and dispersing individuals may depend on their proportions in the population. Such frequency dependence will also occur in cases of asymmetrical competitive relationships among dispersing and philo-patric individuals, i.e., when competitive interactions between two philopatric individ-uals differ from those between two dispersing individuals or between individuals of each type. For example, philopatrics may be more territorial and aggressive than dispersers; the interaction between them may resemble the hawk–dove game. As in game theory, the fitness of each strategy depends on the frequency of the types of individuals in the population.

Typically, W_D may decrease with increasing proportions of dispersing individuals in the population, while the opposite may be the case for W_P. The proportion of dispersing individuals in the population is equal to or a function of the dispersal rate d^* (the resident strategy in the population). Ignoring kin interactions where d is a mutant strategy, equation (1) thus becomes:

$$W = (1 - d) \times W_P(d^*) + d \times W_D(d^*) \qquad (5)$$

ESS dispersal is then found at $W_P(d^*) = W_D(d^*)$. Thus, if fitness of philopatric and/ or dispersing offspring depends on the proportion of dispersing individuals in the population, ESS dispersal rate is such that philopatric and dispersing individuals have the same fitness.

Combination of factors

Even in cases of frequency dependence at the population scale as described above, kin interactions are likely to occur. Briefly, it seems that two cases can be considered. (i) The ESS dispersal rate that would be obtained from frequency dependence only is large to the extent that costs of kin interaction become negligible. This should lead to a situation where philopatric and dispersing individuals have the same fitness. This points out an interesting property of dispersal. Dispersal always reduces the costs of kin interaction, even though the ultimate factors determining the actual dispersal rate may not be related to such costs. (ii) There are still costs of kin interaction at the dispersal rate expected from frequency dependence, and the ESS dispersal rate is higher. Dispersing offspring should then have a lower fitness than philopatric ones.

Fitness is the combination of many components

The model predicts the outcome of dispersal evolution in relation to the relative fitness of dispersing and philopatric individuals in terms of parental allocation. Offspring fitness can be split into several components: costs of settlement (for either dispersing or philopatric individuals), cost of dispersal (for dispersing individuals), and reproductive success once settled (for either dispersing or philopatric individuals). Estimating these, in order to make a comparison between dispersing and philopatric individual fitness, may be almost impossible in the field for many organisms (Clobert 1995). Costs of dispersal and differential costs of settlement between dispersing and philopatric offspring have been examined (Bélichon et al. 1996). But even the reproductive success of dispersing and philopatric individuals, once settled, may differ (e.g., in *Parus* sp.; Julliard et al. 1996). In addition, the estimates of some fitness components may be misleading because the relationships between fitness traits may not be linear (Bélichon et al. 1996; Lemel et al. 1997). An experimental approach might be more appropriate for controlling several environmental factors in order to compare the fitness of many individuals and the morphological, physiological components that contribute to these fitness differences (see the work on the common lizard, *Lacerta vivipara*; Clobert *et al.* 1994b). Examining the internal resource allocation patterns of maternal parents and the corresponding external environmental cues is important to understand the trade-offs involved in producing dispersing and philopatric offspring.

Costs of kin interaction may be a universal ultimate cause of dispersal (Gandon and Michalakis, chapter 11; Perrin and Goudet, chapter 9). Therefore, according to our model, we should expect the fitness of dispersing individuals to be lower than that of philopatric individuals. Paradoxically, in some cases, it might be advantageous for the parents to invest relatively more in the less profitable dispersing offspring than in the philopatric ones. The evolution of plasticity would reduce the fitness differential between philopatric and dispersing offspring and, at the same time, reduce the potential conflict between mother and offspring. In addition, dispersing and philopatric individuals might become more and more differentiated. Such differentiation could lead to asymmetrical interactions between dispersing and philopatric individuals, conditions favouring frequency dependence at the population scale. In the course of the evolution of dispersal, the main selective pressure may change from kin selection to frequency dependence at the population scale. As a consequence, the ESS dispersal may reach a high enough rate for the cost of kin interaction to become negligible. Finally, one may observe an evolutionarily stable equilibrium between two strategies with equal fitness (see also Holt and Barfield, chapter 6).

If kin interaction is the ultimate cause of dispersal, dispersing individuals should have a lower fitness than non-dispersing ones; otherwise dispersing and non-dispersing individuals should have equal fitness. Reduction of costs associated with dispersal and settlement is beneficial to both offspring and parents. Both trade-offs in the number of each type of offspring and the allocation of resources to each impinge on being able to maximize fitness in both types of offspring (Ims and Hjermann, chapter 14). However, selection does not stop once the decision to disperse or not to disperse is made. The costs associated with movement and settlement may, in many cases, be overcome by the plasticity of various components of the phenotype. Therefore, how particular species overcome these costs, and patterns of resource allocation, will depend on the species-specific life-history characters. Within a species, phenotypic plasticity may be particularly favoured as a means of increasing fitness when dispersal is not an option.

Examples from a wide range of organisms in natural populations are suggestive of the model's results. The plant *Heterosperma pinnatum* produces achenes (single-seeded fruits) from a single inflorescence that may or may not have awns that provide greater dispersibility. In a study of variation among populations, Venable *et al.* (1998) found that populations of *Heterosperma* that produced a higher fraction of achenes with awns came from sites with closed vegetation and either, lower spring precipitation or higher summer precipitation. Aphids, termites, and various Hymenoptera commonly produce wingless and winged offspring. Some populations of the salamander *Ambystoma mexicanum* have both paedomorphic and metamorphic individuals (Semlitsch *et al.* 1990), and even in naked mole-rats there is a morph that appears to be specialized for dispersal (O'Riain and Braude, chapter 10).

Donohue (1997, 1998) examined the annual plant *Cakile edentula*, which has heteromorphic seeds that vary in their dispersability. In this species, however, the pattern of offspring dispersion is controlled to a large extent by the characteristics of the maternal plant. For example, plant height and branching as well as the number of segments per fruit have strong effects on the characteristics of seed dispersal of individual plants. Donohue (1998) concluded that the extent to which each maternal trait influences dispersal, the extent to which it varies as a function of dispersal will all influence phenotypic correlations between maternal parent and offspring. Such differentiation between dispersing and philopatric offspring may lead to asymmetrical interactions betweeen dispersing and philopatric individuals. Additionally, the maternal parent may have a strong influence on the type of environment that the offspring faces (as in *Cakile*), or even on the ability of the offspring to colonize successfully (as with aphids that produce winged offspring with increased density or other stress; e.g. Nunes and Hardie 1996; Weisser, chapter 12b).

As can be seen from the above empirical and theoretical examples, there are numerous instances where selection will favour the evolution of phenotypic plasticity. The key factors favouring plastic alternatives to dispersal, or plasticity following dispersal, are the likelihood of encountering particular environmental conditions and the organism's ability to collect reliable information regarding the environment. Indeed, the ability of an organism to respond appropriately to the prevailing conditions at a philopatric site or a new site, may be a prerequisite for the evolution of dispersal as a plastic strategy in the first place.

As is made clear by the model, there may be a significant conflict between strategies that maximize the fitness of parents and their offspring. This reveals a little-appreciated complexity of the evolution of dispersal: there may be multiple layers of plastic decisions that must be made, and perhaps adjustments made by succeeding generations in response to the choices made by their parents.

Acknowledgements

We thank the organizers of the conference in Roscoff for the opportunity to discuss dispersal from a wide range of perspectives.

19

Dispersal and the genetic properties of metapopulations

Michael C. Whitlock

Abstract

Real populations vary in time and space in demographic parameters such as size, emigration and immigration rate, extinction probability, distance to other populations, etc. Similarly, individuals vary substantially in their properties. In particular, dispersing individuals may differ on average from the others in the population from which they come and from the individuals in the deme to which they go. This individual and populational variation can significantly affect important evolutionary properties of these species, such as effective migration rate, effective population size, rate of positive evolution, probability of accumulating deleterious mutations, population structure, and others.

Keywords: selection, metapopulations, effective migration rate, island model, effective population size, adaptive landscapes, genetic variance

Introduction

Species are typically subdivided into local populations. These populations may be connected by dispersal, causing strong correlations between the evolutionary trajectories of each of the populations. This subdivision and dispersal can also affect the rate and direction of evolution of the species as a whole. Spatial population structure is known to affect the rate of adaptation (both locally and globally), the rate of genetic drift, the maintenance and expression of deleterious alleles, and many other processes and patterns important to evolution.

The genetic effects of dispersal have been studied with models of both discrete subpopulations and a continuous distribution of individuals across space. In the interests of limited space, let us focus on the species in which there are discrete boundaries to spatially separated populations. As such we will not be dealing with hybrid zones, strict isolation by distance models, or other important spatially explicit processes. However, many of the issues that affect discrete populations also affect these more continuous cases.

Discrete models of population structure have traditionally fallen into a small variety of types: the mainland–island model, in which migrants to small populations come from a single source, itself unaffected by the dynamics of the peripheral populations; the island model, in which all subpopulations are equal in size, equal in contribution to the migrant pool, and equal in immigration rate; and stepping-stone models, in which the assumptions of the island model about the random origin of migrants have been relaxed so that migrants can differentially come from nearby populations. Of these, the island model has attracted the most theoretical attention and has formed the basis of many analyses of the effects of population structure on evolutionary biological processes (e.g. Maruyama 1970; Slatkin 1985a; Lande 1992; Barton and Rouhani 1993). This is probably because the island model is one of the simplest to work with; unfortunately the island model is extreme in other ways. This chapter explores the effects of these extreme assumptions and points to some more general results about evolution in metapopulations.

Just as these models make assumptions about the equality and uniformity of populations, many spatially explicit models also assume that individuals are invariable. As is reflected in most of the contributions in this book, all individuals are not equal. Some types of individuals are more likely to disperse than others. Migrant individuals are likely to differ from the average of the deme from which they come, and for a variety of reasons they can differ from the individuals in the deme to which they move. These differences can greatly change the level of gene flow and its evolutionary consequences. This chapter explores the evolutionary effects of this individual variation among dispersers and residents and of the variation among demes in their properties.

I shall use the term *metapopulation* in the sense defined by Hanski and Gilpin (1997): a population of populations. In this sense the term embraces the possibility of extinction and colonization of local populations without requiring these processes. Furthermore, I will use the term *migration* in the usual population genetic sense – the movement of individuals between local populations.

Individual variation and population structure: dispersers differ from residents

The *effective migration rate* is the rate of migration in an ideal population of identical individuals that would accomplish the same rate of gene flow as is present in the population under question (Barton and Bengtsson 1986). There are several reasons why dispersers may be different from resident individuals (see other chapters in this book). They may be of different age, often younger, and with greater reproductive potential. They have travelled and therefore have incurred the fitness costs of that movement. Hungry individuals are more likely to move further, and differential hunger can result from environmental variation or from genetic differences between individuals. Individuals with lower social status are more likely to move further. As a result of these types of factors, dispersing individuals may have a different expected fitness than residents. Depending on which of these factors is most important in a particular species, the effective migration rate may increase or decrease.

Furthermore, and more importantly, migrants can be genetically and phenotypically different from resident individuals in the demes to which they go. Migrant individuals are likely to be genetically differentiated from the local populations, and so may have more fit, less inbred, offspring. On the other hand, if local adaptation is common, then migrant individuals and their offspring are likely to be less fit, because they are unlikely to be locally well adapted. Both of these processes have been observed in crosses at varying distances (Waser 1994). A further boost to the fitness of migrants and their offspring arises from the likelihood that migrants are less sensitive to locally abundant diseases, which are more likely to be well adapted to the local genotypes of hosts (Kaltz and Shykoff 1998).

The overall effect of these factors is not simple, because many of them are genetically determined. If a difference in fitness between migrants and residents is environmental and not inherited, then the effective migration rate is boosted by a fraction equal to the gain in fitness. If the differences in fitness are correlated between parents and offspring, however, then the change to gene flow is compounded over generations. A model of the specific effects of heterosis on the effective migration rate has given rise to an explicit formula for the boost. If the heterosis in crosses between individuals from different populations is H and the harmonic mean recombination rate between loci causing heterosis and a marker locus is $\tilde{r}$, then the effective migration rate is given by:

$$m_e \cong m \exp\left[\frac{H}{\tilde{r}}\right] \qquad (1)$$

where m is the actual rate of movement of individuals between demes (Ingvarsson and Whitlock 2000; see also Barton and Bengtsson 1986, for results on hybrid breakdown). This accounts for the immediate fitness consequences of differences in the fitness of migrants, as well as the diminishing difference in mean fitness of each of the subsequent descendants of the migrants. The mathematics leading to this equation do not make any assumptions that are particular to heterosis, however, so this equation can be used to account for the effective migration rate of any genetic difference in fitness between migrants and residents, either positive or negative. If the difference is large, the effective migration rate may be very different to the actual migration rate, by as much as an order of magnitude. The effective migration rate may be reduced by selection, as has been shown for hybrid zones (Barton and Bengtsson 1986).

Convenient assumptions and inconvenient reality: population-level deviations from the island model

The island model, proposed by Sewall Wright in 1931, imagines a collection of populations, each with N individuals that contribute a fraction m of their individuals in each generation to a migrant pool. Individuals from the migrant pool are then drawn at random, without respect to any of their properties or site of origin, and Nm individuals are sent to each deme to replace those that have been lost to emigration. As such, each deme contributes equally to the next generation, both through its resident population and through its contribution to the migrant pool. Furthermore, this model is spatially implicit; each population is equally connected to the others by migration. There are no edges, no distance biases to dispersal, and no variations in the quality of habitat. The island model also assumes that all of these properties are constant in time, that no population changes its size or migration rates, none becomes extinct, and sufficient time has passed that an equilibrium has been reached among the evolutionary processes in question. These assumptions are convenient mathematically, but are biologically unreasonable.

The real difficulty comes from the fact that the island model is not just a special case, but that it is an extreme case. The island model proposes the minimum amount of variance in reproductive success among populations, a maximum amount of mixing in the movement of migrants, and a minimal variance among populations in their properties. Each of these assumptions, particularly that each deme always contributes equally to the next generation, strongly affects the predicted consequences of population structure on evolution.

Natural populations differ from the island model in a variety of ways (see the other chapters in this book and in Hanski and Gilpin 1997). Local populations go extinct and others form by colonization (McCauley 1993; Giles and Goudet 1997). Populations sometimes split into two or fuse into one deme (Smouse *et al.* 1981). Some demes are larger than others, and some contribute more than others by migration (Dias 1996; Thomas and Hanski 1997). Demes that are closer to one another and not separated by a geographical barrier are more likely to exchange migrants than are those further apart or with some obstacle in between (Neigel 1997). Each of these has effects on the level of genetic differentiation among populations and on the variance among demes in their reproductive success. We will see that these two properties may, in turn, affect many evolutionarily interesting properties of metapopulations.

Evolutionary effects of non-equilibrium population structure and variable dispersal

Population structure creates patterns in gene frequencies across space. This produces correlations within generations in the genetic properties of neighbouring individuals and correlations across generations of both the genetic and environmental properties of populations. These two factors are responsible for many changes in the nature of evolution in subdivided populations.

In much of what follows, it will be seen that two very important descriptions of a metapopulation are its effective population size (N_e) and the correlation of alleles within demes (F_{ST}). Both of these properties are strongly affected by the details of dispersal.

One way in which dispersal affects the evolutionary dynamics of metapopulations is by changing the potential for local adaptation. This chapter does not deal with this important topic; for a review, see chapter 23.

Effective population size and genetic drift

Finite populations are subject to changes in allele frequencies by genetic drift. The rate at which allele frequencies are expected to change depends on the effective population size, which is strongly affected by population structure. Wright (1939) first described the level of genetic drift expected in a subdivided population for the island model. He found that N_e for the whole metapopulation of an island model was given by $Nd/(1 - F_{ST})$, where there are d demes and F_{ST} is Wright's own standardized measure of the genetic differentiation among populations. Since $F_{ST} < 1$, this implies that population subdivision via the island model *always* increases N_e of the species. Intuitively, the reason for this result is that if different alleles drift to high frequency in different populations, then each is preserved longer in the metapopulation as a whole. As a result, the rate of drift becomes lower; therefore N_e is higher. This result has been shown to be general for all forms of population structure in which each deme is the same size and contributes an equal number of individuals to the next generation (Whitlock and Barton 1997). Thus this result holds for the one- and two-dimensional stepping-stone models, although F_{ST} is different in each of these cases.

This increase in N_e with population structure is not general, and most likely it is not even common. In most species, the contribution to subsequent generations is highly variable among demes, a result of either local extinction or variable dispersal from different demes. The result of this variance among demes in reproductive success is that N_e is very likely to be reduced. A general formula for the effective size of a metapopulation was determined by Whitlock and Barton (1997). Variance in reproductive success among demes acts much as variance in reproductive success among individuals: high variance results in low N_e. Again this can be understood qualitatively by considering that high variance in reproductive success implies that only a few demes or individuals are contributing most of the genetic material to subsequent generations. At an extreme, such as with a local extinction, a deme contributes nothing to later generations and therefore does not add to the effective size of the system. Extensions of these metapopulation N_e results to account for local variance in reproductive success have been made by Wang and Caballero (1999) and Nunney (1999).

The effective size of a species affects several very important aspects of its evolution, including the maintenance of genetic variation and fixation of beneficial and deleterious alleles. Barton has shown quite generally that the probability of fixation of deleterious additive alleles in a subdivided population is given by:

$$\frac{2s(1 - F_{ST})(N_e/\bar{N}d)}{\exp[4N_e s(1 - F_{ST})] - 1} \tag{2}$$

where s is the selection coefficient and $\bar{N}$ is the mean deme size (Barton and Whitlock 1997). The probability of fixation can be substantially increased for cases in which the substructure causes N_e to be much less than $\bar{N}d$.

Probability of fixation of beneficial alleles

Ultimately evolution depends on the fixation of beneficial alleles. One of the more striking results from theoretical population genetics is that the probability of fixation of even a beneficial allele is not extremely high, as any allele can be lost stochastically during the phase when it is present in only a few copies. Haldane (1927) found that the probability of fixation of a beneficial allele was approximately $2s$, where s is the heterozygous benefit conveyed by the new allele. A new allele with even a substantial benefit is more likely to be lost than fixed in a population. Kimura (1957) revised this estimate slightly to account for the particulars of non-ideal single populations and found that the probability of fixation of a new beneficial allele was given by $2s(N_e/N)$. Essentially this results from the fact that a new allele is present at a frequency given by $1/2N$, but effectively only $2N_e$ of the alleles have a chance of contributing to the next generation. Neither of these results directly informs us of the probability of fixation of alleles in a subdivided population.

Maruyama (1970) found that the probability of fixation of a beneficial allele in an island model population is given simply by $2s$. He described this as an invariant property of subdivided populations, since the result suggested that population subdivision did not change the probability of fixation. Subsequently, Barton (1993) has shown that for a simple local extinction–colonization model, the probability of fixation of alleles is much changed from that in a panmictic population. Barton suggests that the probability of fixation is not predictable from knowledge of N_e.

However, one general pattern emerges from examination of these few known cases of the effects of subdivision. In all three cases in which the probability of fixation is known – the island model and two cases of extinction–colonization in Barton (1993) – the probability of fixation is given by:

$$u = 2s(N_e/Nd)(1 - F_{ST}) \tag{3}$$

In the case of the island model, N_e is $Nd/(1 - F_{ST})$, so that the last two terms in this equation cancel to leave $2s$. Again, this is merely a function of the idiosyncrasies of the island model and not a general result. With Barton's models, using F_{ST} results from Whitlock and McCauley (1990), equation (3) holds as well. While this equation has not been yet proven to be general, it is likely to be: the expected response to selection, and therefore the effective strength of selection, is discounted by a factor $(1 - F_{ST})$. Thus s in the standard equation may be replaced by $s(1 - F_{ST})$ for subdivided populations. Given that the N_e for subdivided populations is likely to be much reduced relative to the census size and that F_{ST} contributes to a reduction in the effective strength of selection, the rate of adaptation in subdivided populations can be much reduced relative to a panmictic, undivided population.

Adaptive landscapes

The study of population structure was originally motivated in no small part by Wright's interest in the effects of population subdivision on the rate of evolution on complex adaptive landscapes. Wright envisioned that multiple genotypes may be fit, but that other, intermediate, genetic combinations may be less so. Thus Wright suggested that evolution may be limited by the need for the process of evolution to take a population through an unfit intermediate condition, which would be impossible with natural selection alone. Wright's favourite answer to this dilemma is well known: the shifting balance theory. In this view, local populations are able to drift because of the local small size, and some demes, as a result of drift, come to be in the domain of attraction of a new adaptive peak. Wright imagined that selection could then move this population to a new fit state, followed by differential emigration or extinction, which would cause this new genetic combination to spread to other demes in the metapopulation.

This model has been criticized on many fronts (see Coyne et al. 1997). In particular, three criticisms are most telling. First, the probability of a deme making the transition from one adaptive peak to another seems to be very low for landscapes in which there is any real depth or distance between adaptive peaks (Coyne *et al.* 1997; Whitlock 1997). Second, it is not at all clear that evolution becomes limited in the way that Wright envisioned; instead, it is possible that there is always a direction in which evolution may proceed (see letters from Fisher to Wright in Provine 1986; Coyne *et al.* 1997). As a result, species are not likely to be evolutionarily static and therefore can explore a broader range of the adaptive landscape than perhaps Wright envisioned. Finally, there has been significant criticism of the model regarding the final stage of the shifting balance process, that concerned with the export of a new adaptive peak to other demes by differential group fitness (Haldane 1959; Coyne *et al.* 1997).

The attention paid to the shifting balance process in its narrow sense has distracted many from consideration of more important related phenomena which do not share all of these problems. Adaptive evolution may proceed faster in subdivided populations for many reasons, some of which are related to Wright's theory. Local populations experience different selective regimes, and it is known that variable selection is very likely to speed evolution on an adaptive landscape (Whitlock 1997). Slight changes in the strength and nature of selection may result in shifts between alternative adaptive peaks, with a strong bias towards populations reaching the higher peaks. As a result, spatial subdivision creates the situation where each local population has the possibility of reaching a new adaptive peak, in a way in which a large, spread out, panmictic population would not. Wright initially also considered this as a mechanism for local peak shifts (Wright 1932); this is a much more likely mechanism for the initial peak shifts than is genetic drift (Whitlock 1997).

The problems with the spread of new adaptive peaks are more difficult, and the answers depend upon the details of dispersal in the metapopulation. In models with two populations and greater migration from the population with the more fit peak to the population with the less fit peak, a low level of differential migration is capable of spreading the more fit genotype (Crow *et al.* 1990). In contrast, when one considers more realistic population structures with many demes, the probability of the spreading of new peaks is greatly reduced (Gavrilets 1996). This is because the effects of the population at the new peak are diluted across many other demes, but many other demes are consistent in their effects in pulling the population back to the old peak. As a result, the spread of these peaks is enhanced when dispersal is distance limited (Peck *et al.* 1998); in these cases, the population with the new peak is interacting with only a relatively few other demes. These concerns about the ease of spread of new adaptive peaks are problems not only with regard to the shifting balance, but also to any model of evolution on a complex landscape that invokes a spatial element. The feasibility of the export of adaptive peaks is extremely sensitive to the details of the patterns of dispersal.

Maintenance of genetic variance

Genetic variance is the stuff of evolution, without which it could not proceed, and the reasons for the high level of genetic variance that we observe for most characters in most species is one of the most inadequately answered questions of evolutionary biology (Barton and Turelli 1989). Population structure and the nature of dispersal among populations may strongly affect the level of genetic variation maintained in a species. This section reviews some of the variety of ways in which this is true.

The simplest model of genetic variance examines the expected variance maintained in a balance between mutation bringing in new variation and genetic drift losing it from a population. Lynch and Hill (1986) showed that, in an undivided population, the expected variance at equilibrium between drift and mutation is $2N_e\sigma_m^2$, where σ_m^2 is the variance added by mutation per genome per generation and N_e refers to the effective size of the metapopulation as a whole. In a subdivided population, the total amount of variance maintained is given by $(1 + F_{ST}) 2N_e\sigma_m^2$, where N_e is defined as above (Whitlock 2000). The amount of genetic variance within the average deme is given by $(1 - F_{ST}) 2N_e\sigma_m^2$. For an island model, this implies that the amount of variance within any given deme is $2Nd\sigma_m^2$, or an amount equal to that expected in a panmictic population of the same size as the total metapopulation (Lande 1992). Note that this result does not hold for more general and natural population structures, when N_e is not generally $Nd/(1 - F_{ST})$, but is expected to be much less. If, as we expect will be usually the case, $N_e < Nd$, then population structure will reduce the amount of genetic variance maintained at a balance between drift and mutation, both within populations and in the species as a whole.

With limited dispersal and population structure, some local adaptation is possible (Hedrick *et al.* 1976; Hedrick 1986; Linhart and Grant 1996; Lynch *et al.* 1999). One of the strongest mechanisms for the maintenance of selectively important variation is likely to be spatially variable selection. High levels of dispersal prevent selective differentiation among populations and thus are likely to be associated with lower levels of total genetic variance. Extremely low levels of dispersal prevent new alleles from reaching local populations and result in local genetically depauperate populations, even with much diversity at the species level.

Even with spatially uniform selection, populations can diverge by selection to different phenotypic or genotypic combinations (Cohan and Hoffman 1989). With sufficient gene flow, this divergence is expected to be unlikely, but with low levels of migration different populations can occupy different genetic adaptive peaks (Goldstein and Holsinger 1992; Barton and Whitlock 1997). As a result, the metapopulation as a whole maintains more genetic variance, and with migration between populations each local deme has variance introduced by migration at a rate higher than from mutation alone. Substantial increases in variance relative to a panmictic population seem to occur only at relatively low migration rates (say $Nm < 1$) (Barton and Whitlock 1997).

Examples and discussion

Incomplete and variable dispersal causes individuals to be more likely than random to mate with related individuals and allows higher variance in reproductive success. As a result, many important evolutionary patterns can be changed. Under what may be the most common scenario, the effective population size of a species would be reduced by population subdivision, producing an increase in the rate of genetic drift, a decrease in the probability of fixation of beneficial alleles, an increase in the rate of accumulation of deleterious alleles, and lower levels of genetic variance at equilibrium.

These principles may best be described by example. Two models of population structure that emphasize important realistic deviations from the island model will be considered: a model of local extinction and colonization, and an asymmetrical migration model with local sources and sinks.

Many species have reasonably high rates of local population extinction (Hanski and Gilpin 1997). In order for a species with local extinction to not go extinct globally, the rate of local extinction must be balanced by colonization of new populations. This colonization is typically by a smaller number of individuals than the carrying capacity of the new site; therefore this colonization event is a time of higher genetic drift, and successful colonists have very high reproductive success. As a result of these two factors, there tends to be higher F_{ST} (Wade and McCauley 1988; Whitlock and McCauley 1990) and a much lower effective population size (Slatkin 1977; Maruyama and Kimura 1980; Whitlock and Barton 1997) in systems with local extinctions compared with the island model.

As a result of these changes in N_e, metapopulations with frequent local extinction have a higher risk of fixation of deleterious alleles, a lower probability of fixing beneficial alleles, and lower level of equilibrium genetic variance. Local adaptation is impeded by extinction, because local extinction erases previous gains by selection. The effects of extinction and colonization on complex landscape evolution are more complicated. On the one hand the drift associated with colonization (so-called founder effects) can accelerate phase I of the shifting balance process; perhaps more importantly, though, local extinction prevents strong adaptive differentiation to new peaks by limiting the time available for response to selection. Particularly if extinction rates are greater than about one in 100 per generation, local differentiation would on average have only about 100 generations or fewer to reach new genetic combinations and to export them to other populations. This timescale is likely to be too brief (Whitlock 1997).

Another commonly observed phenomenon of real metapopulations is asymmetrical migration (Dias 1996). Some populations have many resources and are well adapted to using them; these populations may act as net sources of migration to other populations. Other populations, in contrast, that are less well locally adapted or with fewer resources, tend to contribute little to the migrant pool and act as population sinks. In fact, some sink populations would not persist without migration from other populations (Holt 1997b); therefore they are genetically very similar to their source populations. As a result, there is not high genetic differentiation among populations, but there is very high variance in the ultimate reproductive success of different populations. (In the extreme case that sink populations do not contribute to the migrant pool at all (so-called 'black-hole sinks'; Holt and Gaines 1992), the sinks do not contribute to the evolutionary genetic future of the species.) This causes a very strong reduction in the effective size of the metapopulation. N_e of source–sink systems is much lower than the census size, and much closer to the combined size of the source populations. Again the probability of fixation of beneficial alleles is reduced, fixation of deleterious alleles is increased, and the overall level of genetic variance should be reduced relative to the symmetrical island model. Adaptation on complex landscapes is compromised, because the small peripheral populations, which by either model of peak shifts are expected to be the main sources of peak shifts, are evolutionarily unstable. As in the single-locus case (Holt 1997b; Holt and Gomulkiewicz 1997), any evolutionary modification in a sink population must by itself at least cause the local population growth rate to become positive in order for that modification to persist.

The details of dispersal between new and extant populations can dramatically change their evolutionary properties. Applying metapopulation biology to practical concerns such as conservation biology or the estimation of dispersal rates must be done in the light of this pleasing complexity.

Acknowledgements

The author is supported by an operating grant from the Natural Science and Engineering Research Council (Canada).

20

Population dynamic consequences of dispersal in local populations and in metapopulations

Ilkka Hanski

Abstract

The two most fundamental population dynamic consequences of dispersal at the level of local populations and metapopulations are, respectively, population regulation via density-dependent emigration and large-scale persistence of classical metapopulations due to the establishment of new populations by immigrants. Dispersal has many additional significant population dynamic ramifications. In the case of small populations, emigration may substantially increase the risk of local extinction, while immigration may reduce it. In a well studied butterfly metapopulation, both demographic and genetic rescue effects operate to reduce the extinction rate of small populations. In metapopulations, dispersal may allow species to exist where conditions are generally unfavourable (sink populations); dispersal may synchronize and stabilize local dynamics, but it may also lead to spatial pattern formation; the rescue effect and the Allee effect may lead to alternative stable equilibria in metapopulation size; and dispersal may facilitate the persistence of prey species and inferior competitors in multispecies communities. Spatiotemporal variability in the availability of suitable habitat is a key determinant of dispersal rate. Evolution of dispersal rate in changing environments remains an area of metapopulation biology where important theoretical and empirical discoveries may be made.

Keywords: population dynamics, population regulation, Allee effect, rescue effect, metapopulation dynamics, multiple equilibria, inbreeding depression, population extinction, spatial pattern formation

Introduction

Population size is affected, as every ecologist knows, by four basic processes: natality and immigration, which increase population size, and mortality and emigration, which decrease it. Because of booming interest in spatial ecology during the past decade (Hanski and Gilpin 1997; McCullough 1997; Tilman and Kareiva 1997; Bascompte and Solé 1998; Hanski 1999a; Dieckmann et al. 2000), the population dynamic and other biological consequences of dispersal, at the level of local populations and entire metapopulations, have received much attention. The days are past when population ecologists ignored dispersal on the assumption that emigration from and immigration to a study plot would balance each other out.

A severe practical problem unfortunately remains. Without great effort in fieldwork, it is generally hard to distinguish between individuals that have become added to a population due to birth versus via immigration, and it is even harder to distinguish between individuals that disappeared because of death versus those that emigrated. If we cannot make these distinctions, it is hard to conduct empirical research to address the role of dispersal in population and metapopulation dynamics. New methods are needed to obtain relevant empirical information, and I will describe one such method on p. 285.

Traditionally, the most widely considered population dynamic consequence of dispersal at the scale of local populations has been population regulation via density-dependent emigration. But, as will be discussed on p. 287, dispersal may also affect the sizes and dynamics of local populations in many other ways (Table 20.1). Spatial variation in the intrinsic growth rate may generate an asymmetrical flow of dispersers that breaks any close correspondence between habitat quality and population density (source–sink dynamics). The reduction that emigration causes in population size becomes an especially important issue in the case of small populations with a substantial risk of extinction. Conversely, immigration may rescue small populations from extinction. The role of dispersal in population extinction is an important issue for conservation, but also has obvious ramifications for the dynamics of metapopulations, which often consist of large numbers of small populations.

At the metapopulation level, dispersal and consequent establishment of new populations are often necessary for the long-term persistence of species, to compensate for local extinctions (p. 291). Dispersal can be viewed as a form of risk-spreading, which enhances the growth rate, and hence the persistence, of metapopulations consisting of local populations with independent dynamics. Dispersal may be the means by which prey species and inferior competitors are able to persist in spite of locally unstable interaction with predators and superior competitors.

While it is obvious that emigration tends to reduce the realized population growth rate and density, and that immigration has the opposite effect, recent theoretical studies have shown that dispersal can do more – it may change the type of population dynamics (Tilman and Kareiva 1997; Bascompte and Solé 1998; Dieckmann *et al.* 2000). Dispersal may synchronize and stabilize the dynamics of coupled local populations that would have complex dynamics in isolation; but it may also generate complex dynamics,

Table 20.1. Population dynamic consequences of dispersal at the level of local populations and in metapopulations. The two key effects of dispersal are printed in bold type

Spatial scale	Effect on population size	Effect on type of dynamics
Local population	**Population regulation** Increased extinction risk Ecological rescue effect Genetic rescue effect	Complex dynamics simplified
Metapopulation	**Persistence of metapopulations** Coexistence of species Source–sink dynamics Multiple equilibria	Complex dynamics simplified Synchronization Spatial pattern formation Supertransients

for instance in spatial pattern formation in uniform environments. Such complexities, for which an already voluminous theoretical literature exists, will be briefly discussed on p. 293. Unfortunately, much of the theory in this area is essentially uncoupled from empirical research, partly because of the problems of measuring the rate and range of dispersal in natural populations, and partly because of the difficulty of ascertaining which type of dynamic complexity is occurring in natural populations, even in the absence of dispersal.

In the concluding section, I return to the question about the influence of dispersal on metapopulation survival. With accelerating loss and fragmentation of many natural habitats, an increasing number of species is forced to conform to a metapopulation structure, which raises, among other things, questions about the evolution of dispersal rate in changing environments. Can we expect the evolution of dispersal rate to rescue metapopulations from imminent extinction due to habitat loss and fragmentation?

Measurement of dispersal rate

For population ecological purposes, we need quantitative estimates of dispersal rate at the timescale of generations. This means that indirect methods of estimating dispersal based on allele frequencies in multiple populations (Slatkin 1994; Rousset, chapter 2) are of limited use for population ecologists. The more recent genetic methods based on highly polymorphic markers and the assignment of individuals to particular populations depending on the combinations of alleles that they carry have more promise, as they allow, under favourable conditions, direct discrimination between immigrants and residents (Waser and Strobeck 1998; see Peacock and Ray, chapter 4, for the use of multilocus minisatellite markers). However, even these methods may not be sufficient to answer many ecological questions about dispersal, because the circumstances are unlikely often to be favourable for the use of the assignment methods at small spatial scales, and because the sample sizes are likely to remain small.

Bennetts *et al.* (chapter 1) and other recent reviews (Ims and Yoccoz 1997; Ruckelshaus *et al.* 1997; Turchin 1998) have discussed the direct measurement of dispersal rate using mark–release–recapture (MRR) and other methods. As this topic is not the primary focus of this chapter, here I will merely outline one recent approach that has been developed for the specific purpose of analysing MRR data collected simultaneously from many local populations in a metapopulation (Hanski *et al.* 2000).

A tempting approach for the ecological study of dispersal in metapopulations is to extend the current methods of modelling individual survival probabilities with MRR data in single populations (Lebreton *et al.* 1993; Ims and Yoccoz 1997, and references therein) to several populations (Arnason 1973; Hestbeck *et al.* 1991; Hilborn 1991; Nichols *et al.* 1993; Bennetts *et al.*, chapter 1). However, the general models for multiple populations include parameters for all pairwise interactions (transfer probabilities) among the local populations, and some simplifying assumptions are both necessary and desirable (Brownie *et al.* 1993; Nichols and Kendall 1995; Spendelow *et al.* 1995) when the number of local populations is large, as is typical in metapopulation studies (Hanski and Gilpin 1997; Hanski 1998, 1999a). Rather than simplifying a general model (Bennetts *et al.*, chapter 1), one might start with a simple model based on the processes that are of interest to ecologists. My own bias is to consider highly fragmented land-scapes consisting of discrete patches of suitable habitat (Hanski 1998, 1999a). Conse-quently, I include in the list of interesting questions about dispersal the scaling of emigration and immigration rates with habitat patch area, the effect of patch isolation on mortality during dispersal, as well as the familiar parameters of survival rate within populations and the rate and range of dispersal.

The model that is briefly sketched below is intended to extract the parameters of survival and dispersal of individuals in a metapopulation based on their capture histories as recorded in a multisite MRR study (Hanski *et al.* 2000). Specifically, the patch-specific but time-independent probability ε_j of emigrating from habitat patch j in unit time is assumed to scale to patch area A_j by the power function:

$$\varepsilon_j = \eta A_j^{-\zeta_{em}} \tag{1}$$

where $\eta > 0$ and $\zeta_{em} > 0$ are two parameters. The connectivity of patch j is measured by:

$$S_j = \sum_{k \neq j} \exp(-\alpha d_{jk}) A_k^{\zeta_{im}} \tag{2}$$

where d_{jk} is some appropriate distance measure between patches j and k, most simply the euclidian distance. The scaling of immigration by patch area is given by the exponent $\zeta_{im} > 0$, and the average range of dispersal is given by $1/\alpha$. As it is reasonable to assume that the probability φ_{mj} of surviving dispersal from patch j increases with the con-nectivity of patch j, we (Hanski *et al.* 2000) have assumed that:

$$\varphi_{mj} = \frac{S_j^2}{\lambda + S_j^2} \tag{3}$$

where $\lambda > 0$ is a parameter. The individuals that survive dispersal are assumed to be distributed among all the target patches in proportion to their contributions to the connectivity of patch j (eqn 2). Thus, the probability of an individual that has left patch j reaching patch k is given by:

$$\psi_{jk} = \frac{\exp(-\alpha d_{jk}) A_k^{\zeta_{im}}}{(\lambda / S_j) + S_j} \tag{4}$$

Thus the probability of surviving migration from patch j to patch k depends on the distance between the two patches (d_{ij}) and the area of the receiving patch k, and also on the density of other patches surrounding patch j, which influences S_j (eqn 2). The latter dependence models 'competition' among the patches for dispersing individuals.

The model has six parameters: φ_p (daily survival rate within habitat patches), η, ζ_{im}, ζ_{em}, α, and λ (it would be conceptually straightforward to include additional factors if necessary). Their values are estimated by numerical optimization using appropriate maximum likelihood estimators (Hanski *et al.* 2000). As an example, Hanski *et al.* (2000) applied the model to a metapopulation of the false heath fritillary butterfly

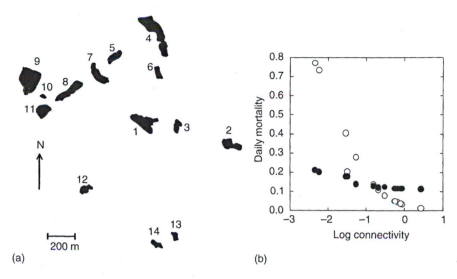

Figure 20.1. (a) The patch network occupied by the false heath fritillary butterfly (*Melitaea diamina*; Wahlberg *et al.* 1996; Wahlberg 1997). (b) Relationship between patch-specific mortality and population connectivity as measured by log S_j (eqn 2). The open symbols give daily mortality during dispersal among emigrants from patch *j*, while the black dots give the total daily mortality of individuals present in patch *j* at the beginning of each day, including mortality in the habitat patch and mortality of the emigrants.

(*Melitaea diamina*) shown in Fig. 20.1, with 14 local populations, 557 marked butterflies, and 1301 recaptures. The estimated parameter values were $\varphi_p = 0.89$, $\eta = 0.13$, $\zeta_{im} = 0.26$, $\zeta_{em} = 0.17$, $\alpha = 4.91$, and $\lambda = 0.18$ (the original paper reports confidence limits for the parameter estimates and goodness-of-fit tests for the model). These parameter values suggest that immigration and emigration scale as patch area to power 0.2, roughly half of the daily losses of individuals from a patch of 1 ha are due to emigration, $<1\%$ of daily dispersal distances are longer than 1 km, and 16% of all deaths are estimated to have occurred during dispersal (Hanski *et al.* 2000). We shall return to this example in the following section (source–sink dynamics), but meanwhile I conclude that this line of statistical modelling of MRR data collected from multiple populations has the potential to produce some of the empirical data on dispersal that have been difficult to obtain using other means.

Consequences of dispersal on local population size and dynamics

Density-dependent emigration may contribute to population regulation. The common wisdom says that this is especially so in territorial species, in which a relatively constant number of individuals has the opportunity to breed within a given area, while the rest are forced to emigrate, which has a strongly stabilizing effect on population dynamics (contest competition). For a review see Lambin *et al.* (chapter 8). Emigration, however, may be inversely density dependent at high density in small mammals, and possibly so in other territorial species, because of a social fence effect (Hestbeck 1982; Lomnicki 1988).

It is thought to be hard for a juvenile disperser to immigrate to a high-density population; hence they stay in the territory of their mother.

Density-dependent emigration is widespread also in non-territorial taxa. For instance, Denno and Peterson (1995) found density-dependent emigration to be common and to contribute to population regulation in sap-feeding phytophagous insects (for reviews on other insect groups see Kowalski and Benson 1978; Lance and Barbosa 1979; Joern and Gaines 1990). On the other hand, in many other species emigration does not appear to be density dependent at the usual densities (Hansson 1991). At the lowest densities, emigration as well as immigration may actually become inversely density dependent (Kuussaari *et al.* 1996; Stamps, chapter 16), leading to an Allee effect that may create an unstable equilibrium below which the population goes deterministically extinct.

Density-independent emigration and immigration will not directly influence population regulation, but such dispersal may lead to lowered and raised population densities, respectively, with potential consequences for the persistence of populations. In extreme cases, species may occur in low-quality sink habitats only because of dispersal from high-quality source habitats. I will next discuss such consequences of dispersal in some detail.

Emigration and increased extinction risk

Empirical metapopulation studies often include small populations living in small habitat patches (Harrison *et al.* 1988; Thomas 1994; Hanski *et al.* 1995a). In the case of vertebrates, even small patches of suitable habitat may support relatively persistent populations, as exemplified by two well studied examples: the American pika (Smith and Gilpin 1997; Moilanen *et al.* 1998; Peacock and Ray, chapter 4) and the European nuthatch (Verboom *et al.* 1991). In contrast, populations of insects and other invertebrates in small habitat patches are likely to lose individuals for no better reason than accidental drift. The smaller the patch, the more frequently individuals encounter the patch boundary and, other things being equal, the higher the loss rate. (See Fig. 20.1 and p. 285, parameters η and ζ_{em}. Thomas and coworkers have presented a quantitative analysis of the risk that emigration poses to small populations of the butterfly *Hesperia comma* in small patches of dry meadow in Britain; see Thomas and Hanski 1997.) Though it is difficult to determine the exact role of emigration because many other processes increase the risk of extinction of small populations of butterflies (Hanski *et al.* 1995a; Lei and Hanski 1997; Saccheri *et al.* 1998) and indeed of other species (Caughley 1994), the role of emigration should not be underestimated either.

Fagan *et al.* (1999), modelling the dynamics of a local population with reaction-diffusion equations, have derived a pleasingly simple deterministic threshold condition for population persistence in a habitat patch with a specific geometry:

$$\lambda_1 < \frac{r}{D} \tag{5}$$

where r is the intrinsic growth rate, D is the diffusion (dispersal) rate, and λ_1 is the leading eigenvalue of a set of equations, capturing the effects of the permeability of patch boundary and patch geometry. The value of λ_1 increases with the size of the core area of the habitat patch (Fagan *et al.* 1999).

The rescue effect: reduced extinction risk

Browan and Kodric-Brown (1977) coined the term 'rescue effect' to describe the reduced risk of extinction of a population whose size has become increased by immigration. Since then, the term has been used to describe practically any kind of (presumed) positive effect of immigration on local population size (Hanski 1999a). Given the apparent potential that immigration has to boost the sizes of, in particular, small populations (Fig. 20.2), there is little doubt that the rescue effect in the original sense would not play an important role especially in the case of small populations located close to large ones. However, as with many other phenomena in spatial ecology, producing conclusive empirical evidence for the rescue effect is difficult.

Table 20.2 gives an example of the Glanville fritillary butterfly (*Melitaea cinxia*), summarizing observations on the annual extinction rate of small populations with 1, 2, 3–5 and > 5 larval groups (the larvae live in large groups of full sibs; Hanski 1999a). The results demonstrate a significant rescue effect in the case of populations that had one or two larval groups, but not in the case of larger populations (Table 20.2; Hanski 1999a). This makes sense, because a given number of immigrants has a larger impact on the dynamics of a small than a large population.

The example in Table 20.2 is probably as conclusive as any example of the existence of the rescue effect, but even this example is open to other complementary interpretations. Saccheri *et al.* (1998) have shown that small populations of the Glanville fritillary suffer from severe inbreeding depression, strong enough to increase the risk of population extinction. Even a small number of immigrants from other populations may significantly reduce inbreeding in small populations (Whitlock, chapter 19), and hence the observed reduction in extinction rate in Table 20.2 may be partly due to such a genetic rescue

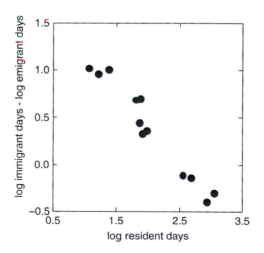

Figure 20.2. The relationship between the number of 'butterfly-days' spent by immigrants in patch *j*, minus the number of butterfly-days spent by emigrants from patch *j* in the other patches, against the number of butterfly-days spent by individuals born to patch *j* as residents in their natal patch. The two smallest patches in Fig. 20.1 have been omitted.

Table 20.2. The rescue effect decreases the annual extinction risk of small populations in the large metapopulation of the Glanville fritillary butterfly (*Melitaea cinxia*) in SW Finland

Population size	Extinct	n	Rescue effect, S	t	P
1	Yes	150	2.55		
	No	76	2.84	−2.97	0.003
2	Yes	46	2.78		
	No	58	3.12	−2.24	0.025
3–5	Yes	46	2.88		
	No	202	2.75	−0.63	0.527
>5	Yes	14	3.31		
	No	204	2.83	1.42	0.155

This table gives the sizes of local populations in terms of the number of larval groups in the autumn (pooled data for several years); the numbers of populations that went extinct versus survived over the next year; a measure of connectivity of the focal population to other nearby populations (Hanski 1994), S, which reflects the expected numbers of immigrants arriving at the population; and t-test for the rescue effect: the effect of S on local extinction (based on a logistic regression model that also includes the effects of patch area and regional trend in population sizes on extinction; Hanski 1999a).

effect, rather than to the more familiar demographic rescue effect. In the Glanville fritillary, both effects are likely to operate simultaneously.

Source–sink dynamics

Dispersal may allow a species to occur at sites where local recruitment alone would not be sufficient for population persistence (Pulliam 1996). Following Pulliam (1988), populations with positive r are called source populations, whereas those with negative r are sink populations. Pseudo-sinks are populations in which deaths exceed births at equilibrium, but which would decline to a positive new equilibrium, rather than to extinction, if they were cut off from other populations (Watkinson and Sutherland 1995). The distinction between sinks and pseudo-sinks is a useful one, even if the pseudo-sink concept itself becomes rather barren in metapopulations with population turnover, in which case the status of any local population is a function of which other habitat patches in the neighbourhood happen to be occupied, and hence how much immigration there is to the focal population.

The butterfly example in Fig. 20.1 illustrates how the flow of individuals to and from populations depends on their sizes and the locations of the respective habitat patches. Examining the numbers of 'butterfly-days' spent by immigrants to a particular patch versus the number of butterfly-days spent as emigrants from that patch shows a clear pattern (Fig. 20.2): in the small patches, in which the pooled number of days spent by butterflies born to that patch is small, immigrant-days greatly exceed emigrant-days, whereas the opposite is true for large patches. In this sense, and regardless of the habitat quality-dependent population growth rate, large patches function as sources and small patches function as (pseudo-)sinks. The ultimate consequences of the pattern of dispersal also depend on at what stage density dependence kicks in; there is no density dependence in emigration or in immigration in the model described on p. 285 and on which the results in Fig. 20.2 are based.

Dispersal and persistence of metapopulations

The population dynamic significance of dispersal is greatest, and most apparent, in classical metapopulations in which all local populations have a substantial risk of extinction. Understandably, the long-term and large-scale persistence of species living under such circumstances is absolutely dependent on a sufficiently high rate of dispersal to generate a sufficiently high rate of establishment of new local populations (Levins 1969; Hanski and Gilpin 1997; Hanski 1999a). O. Ovaskainen and I. Hanski (unpublished manuscript; see also Hanski 1999b) have added a spatially realistic description of the landscape structure to the familiar Levins model, leading to the following deterministic threshold condition for metapopulation persistence:

$$\lambda_1 > \frac{e}{c} \tag{6}$$

where e and c are the extinction and colonization rate parameters, and λ_1 is the leading eigenvalue of a landscape matrix, condensing the effects of habitat patch area and connectivity on extinction and colonization. This simple condition is useful in illustrating how metapopulation persistence depends on both the structure of the landscape (λ_1) and the properties of the species (e/c). Additionally, λ_1, which we have termed the 'metapopulation capacity' of a fragmented landscape (Hanski and Ovaskainen 2000), allows one to rank different fragmented landscapes in terms of their capacity to support a viable metapopulation (Hanski 1999b; Hanski and Ovaskainen 2000).

The influence of dispersal on metapopulation size and persistence is complicated by the adverse effect of emigration on local population size, ignored in the model on which equation (6) is based. The joint effects of emigration, immigration, and colonization have been studied with structured metapopulation models (Gyllenberg and Hanski 1992; Hanski and Zhang 1993; Gyllenberg *et al.* 1997). Assuming that dispersal is costly because of increased mortality of dispersers, a high rate of dispersal may actually lead to metapopulation extinction (Fig. 20.3). An additional twist that may occur when dispersal rate is high is alternative stable equilibria, one of which is metapopulation extinction (Fig. 20.3). It is possible that some highly dispersive species, for instance some butterflies with sparse and scattered host plants, suffer from habitat loss and fragmentation because increased emigration (from habitat patches that have become smaller) greatly increases mortality.

Dispersal as risk-spreading

From the viewpoint of reproducing individuals, dispersal in metapopulations can be seen as a risk-spreading strategy: it is advantageous to 'place' the offspring among several habitat patches in which the conditions influencing fitness vary more or less independently (den Boer 1968; Ronce *et al.*, chapter 24). The significance of such

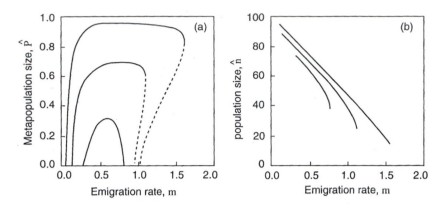

Figure 20.3. Effect of emigration rate on metapopulation size (P; fraction of occupied patches, panel a) and on the size of local populations at equilibrium (panel b) in an island model with identical habitat patches. The three lines are for three different values of the extinction rate: 0.1, 0.05, and 0.01 (from down to top). The dashed line in panel (a) represents unstable equilibria (for details see Hanski and Zhang 1993).

risk-spreading is dramatically exemplified by the possibility of metapopulation persistence of species in which the expected growth rate is negative in all local populations, and which therefore consist of true sink populations only (Kuno 1981; Metz *et al.* 1983; Hanski *et al.* 1996a). The explanation of this apparent paradox is that dispersal among independently fluctuating local populations enhances the overall growth rate in the metapopulation, essentially because the risk of poor reproduction, and the gains of successful reproduction, in a particular population in a particular year are spread via dispersal among the many independently fluctuating populations. More formally, if λ_M and λ_L are the expected *per capita* reproductive rates in the metapopulation as a whole and in a local population, respectively, then in a metapopulation of k independently fluctuating local populations (Metz *et al.* 1983):

$$\lambda_M \approx \lambda_L + 0.5\left(1 - \sum_{i-1}^{k} w_i^2\right)c^2 \tag{7}$$

where w_i is the fixed fraction of immigrants reproducing in the ith local population, and c is the coefficient of variation of r_{ij}, the *per capita* growth rate in population i in year j (in this model all individuals disperse). Clearly, it is possible that $\lambda_M > 0$ while $\lambda_L < 0$, even if there is mortality during dispersal. Although in real metapopulations the positive effect of dispersal in enhancing metapopulation growth rate is diminished by partial dispersal and by some degree of spatial synchrony in local dynamics (Hanski and Woiwod 1993; Pollard and Yates 1993; Sutcliffe *et al.* 1996), dispersal among stochastically fluctuating local populations adds another important element to the study of source and sink populations.

Coexistence of species

Dispersal provides the means for individuals to avoid adverse environmental conditions. A habitat patch may accrue low expected fitness for an individual (or its offspring) not only because of unfavourable abiotic conditions but also because of the presence of natural enemies or superior competitors. One of the successes of spatial ecology is the demonstration, via both theoretical models and empirical studies, of how dispersal of prey and inferior competitors may allow them to coexist with predators (Huffaker 1958; Taylor 1988; Harrison and Taylor 1997) and superior competitors (Hutchinson 1951; Hanski and Ranta 1983; Nee and May 1992; for a review see Hanski 1999a). The key to such coexistence lies in the premise of classical metapopulation ecology (that not all suitable habitat is occupied all the time and hence, for instance, deficient competitive ability can be compensated) for by excellence in locating, via dispersal, the patches of habitat that are currently unused by the superior competitor.

Dispersal and complex spatial dynamics

Two perennial questions about dispersal in metapopulations are to what extent dispersal might stabilize, and to what extent it might synchronize, the dynamics of local populations. The simplest model that allows study of these questions consists of just two local populations connected by dispersal. Let us assume that local dynamics are given by the Ricker model and that a constant fraction of individuals disperse between the two populations.

Potential synchronizing and stabilizing effects of dispersal in metapopulations are dramatically illustrated by the example in Fig. 20.4. Here, the two populations remained

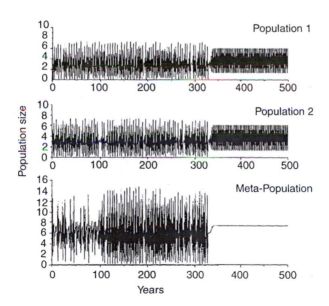

Figure 20.4. An example of the dynamics of a metapopulation consisting of two connected local populations obeying the Ricker model. In both populations $r = 3$. The populations were initially unconnected; hence they oscillate independently and chaotically. At time 100, the populations were connected by assuming 30% emigration rate. See text for discussion.

unconnected until time 100. The value of *r* in the Ricker model was set at 3, which gives rise to chaotic local dynamics in isolated populations (Fig. 20.4). At time 100, the two populations were connected by allowing 30% of individuals to emigrate and by dividing the dispersers equally among the two populations. For a long time, the only difference that dispersal appears to make is an increase in the amplitude of oscillations in the metapopulation as a whole: dispersal has increased the synchrony of local dynamics. However, after approximately 250 time steps, the dynamics change radically: local populations now enter a two-point limit cycle instead of being chaotic, and the two populations cycle out of phase, completely stabilizing the size of the metapopulation (Fig. 20.4). The behaviour of the model between times 100 and 346 in this example represents a transient, a period when the dynamics have not yet settled down to the attractor, represented by the out-of-phase limit cycles.

Variants of this model have been studied in great detail in recent years (McCallum 1992; Gonzales-Andujar and Perry 1993; Gyllenberg *et al.* 1993; Hastings 1993; Bascompte and Solé 1994; Lloyd and May 1996; Ruxton *et al.* 1997, and others). Gyllenberg *et al.* (1993) have conducted an exhaustive mathematical analysis of a model similar to the one used in the example in Fig. 20.4. When the two populations have high growth rates and there is little dispersal, the dynamics remain complex, although with a tendency towards stability (decreased amplitude). With a somewhat higher dispersal rate, the sort of metapopulation stability depicted in Fig. 20.4 emerges, whereas when dispersal rate is even higher, the dominant effect is synchrony in local dynamics, without stability (the two populations oscillate chaotically but in synchrony). Additional complexity is created by alternative stable attractors, which occur for many parameter values (Gyllenberg *et al.* 1993).

It is perhaps not all that surprising that very complex dynamics with long transients ('supertransients'; Hastings and Higgins 1994) and alternative attractors are predicted by models of coupled local populations which by themselves exhibit complex dynamics. The challenge for ecologists is to find ways of relating the model predictions to empirical results. Some model predictions may disappear when stochasticity is added to the model, and such predictions are of only limited or no interest to ecologists. The challenge for theoreticians is to focus on models that are sufficiently well justified ecologically to warrant serious consideration.

The scale of population synchrony

Lande *et al.* (1999; see also Kendall *et al.* 2000) have analysed the spatial scale of population synchrony in a stochastic model assuming homogeneous space and isotropic diffusive dispersal. Using the standard deviation of the relevant function in a given direction as a measure of scale, they derived the following simple and general relationship that is valid for small fluctuations:

$$l_p^2 = l_e^2 + ml^2/\gamma \tag{8}$$

where l_p is the scale of population synchrony, l_e is the scale of regional stochasticity, l is the scale of dispersal, m is dispersal rate, and γ is the strength of population regulation. Thus the spatial scale of population synchrony reflects the scale of regional stochasticity and the scale of dispersal, with the contribution of dispersal decreasing with increasing strength of population regulation. The spatial scale of population synchrony does not depend on the level of environmental stochasticity, whereas the coefficient of variation in population size depends on both the magnitude and the type of stochasticity (Lande *et al.* 1999). The above result offers some opportunity for empirical testing.

Dispersal and spatial pattern formation

Regarding assemblages of many local populations connected by dispersal, one of the principal messages emerging from recent theoretical work is that complex spatial patterns and dynamics may emerge from simple local dynamics, when combined with spatially restricted interactions and restricted dispersal range (Tilman and Kareiva 1997; Bascompte and Solé 1998; Dieckmann *et al.* 2000). In particular, in homogeneous environments without any stochasticity, species may show spatial variation in density solely due to localized population dynamic processes. Thus spatial models of competition predict the emergence of intraspecific aggregation and interspecific segregation of similar species, with the net result of increased likelihood of coexistence (for a review see Pacala and Levin 1997). Similarly, predator–prey dynamics coupled with localized dispersal can generate complex spatial patterns, such as travelling waves, in completely homogeneous environments (for reviews see Kareiva 1990; Nee *et al.* 1997; Turchin *et al.* 1997; Dieckmann *et al.* 2000).

Convincing empirical evidence for spatial pattern formation in populations linked by dispersal is almost completely lacking, partly perhaps because there are not many instances in which the key model assumptions are met, but also because, in most natural environments, spatiotemporal variability greatly influences population dynamics. Spatial patterns in some plant communities provide putative examples of pattern formation in competitive communities (Pacala 1997). One of the more plausible examples of spatial pattern formation due to a predator–prey interaction involves the dynamics of the tussock moth (*Orgyia vetusta*) on the Californian coast, where it has persistent small-scale outbreaks, not explicable by fixed environmental variation (Harrison 1994, 1997). The tussock moth has flightless females and hence a much lower dispersal rate than that of its specialist parasitoids (Harrison 1994). Both observational and experimental studies have shown that the rate of parasitism is especially high just outside the outbreak area, apparently due to dispersal of parasitoids from the outbreak area, which tends to prevent the outbreak from expanding, while experimental populations further away from the outbreak have suffered a low rate of parasitism and have increased in size (Harrison and Wilcox 1995; Brodmann *et al.* 1997; Harrison 1997).

Multiple equilibria

The rescue effect – reduced extinction risk of small populations due to immigration from large populations (Table 20.2) – may provide a sufficiently strong feedback between local and metapopulation dynamics to create two alternative stable states in metapopulation dynamics. Of the two stable equilibria, the lower one corresponds to metapopulation extinction, whereas in the upper equilibrium much of the habitat is occupied (Hanski 1985; Hanski and Gyllenberg 1993). The possibility of alternative stable equilibria has noteworthy practical implications, as a small environmental perturbation may push the metapopulation from the domain of the upper equilibrium to the domain of the lower equilibrium, which would appear as a dramatic collapse of the metapopulation to extinction without any permanent environmental change.

Complex spatial phenomena are difficult to verify in nature, but there is one reasonably good example of alternative stable equilibria in metapopulation dynamics. The Glanville fritillary butterfly (*Melitaea cinxia*) has a very large 'mega-population' of multiple relatively independent metapopulations in the land Islands in SW Finland (Hanski *et al.* 1996b; Hanski 1999a). Though we cannot trace the dynamics of these metapopulations back in time, we can assume that the different metapopulations, as observed at one point in time, represent relatively independent 'snapshots' of habitat occupancy in their respective patch networks.

As there is evidence for the rescue effect in these metapopulations (Table 20.2), we might expect alternative equilibria to occur in their dynamics. Figure 20.5b indeed shows a pattern of habitat occupancy in 65 patch networks that resembles the pattern predicted by a structured metapopulation model (Fig. 20.5a). The seemingly good quantitative correspondence in Fig. 20.5 must be largely coincidental, however, as the model of Hanski and Gyllenberg (1993) cannot be used to make quantitative predictions for spatially realistic metapopulations with a finite number of habitat patches. Nonetheless, the pattern in Fig. 20.5b and other evidence (Hanski *et al.* 1995b; Hanski 1998) give strong qualitative support for the notion of alternative stable equilibria in these butterfly metapopulations.

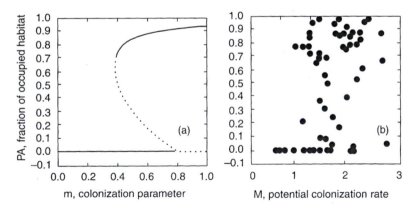

Figure 20.5. Bifurcation diagrams for the fraction of occupied habitat in patch networks (p_A) as a function of the colonization rate (m). Panel (a) gives a theoretical example of alternative stable equilibria from Hanski and Gyllenberg (1993). Continuous lines represent stable equilibria; broken lines are unstable equilibria. Panel (b) gives an empirical example from 65 semi-independent patch networks of the Glanville fritillary butterfly. The potential rate of colonization, M, on the horizontal axis, corresponding to the colonization parameter m in panel (a), gives the expected rate of exchange of butterflies among the patches, with the assumption that all patches are occupied. Each dot gives the level of habitat occupancy in one network. For details see Hanski *et al.* 1995b; Hanski 1999a.

Perspective on the evolutionary consequences of habitat destruction

As the costs and benefits of dispersal are many (Gandon and Michalakis, chapter 11; Ronce *et al.*, chapter 24), it is not surprising that ecologists have put forward two opposing intuitive arguments about the likely course of evolution of dispersal rate in response to increasing habitat loss and fragmentation. According to one view, a high rate of dispersal leads to high mortality during dispersal, and therefore, as the dispersal mortality rate is likely to increase with increasing habitat fragmentation, a lower dispersal rate is expected to evolve. According to the second view, a high dispersal rate is a prerequisite for a high rate of population establishment and, thereby, for long-term metapopulation survival. Therefore, dispersal rate is likely to increase when the environment becomes increasingly fragmented and the rate of extinction becomes increased.

The two viewpoints relate to two extreme situations that are well documented for natural populations. If there is just a single isolated population, selection is expected to operate against emigration. Accordingly, many populations on isolated oceanic islands and mountain tops show a low dispersal rate (Hesse *et al.* 1951; Roff 1990c). At the other extreme, only species with a high dispersal rate may survive in ephemeral habitat patches with fast turnover. Species that occur in naturally patchy habitats do indeed have high dispersal rates (Brown 1951; Southwood 1962; Roff 1994b).

With human-caused habitat fragmentation, the situation is less clear-cut, but, as a broad generalization, the following scenario may often apply (Leimar and Norberg 1997; Hanski 1999a; M. Heino and I. Hanski, unpublished manuscript): with increasing loss and fragmentation of natural habitats, the extinction risk of local populations is expected to increase (because the habitat patches have become smaller) and the colonization rate is expected to decrease (because isolation has become greater). With decreasing habitat patch size, within-population selection against dispersal may be expected to intensify, as the overall rate of emigration is likely to increase (p. 285). On the other hand, reduced population sizes increase the extinction rate regardless of reduced emigration rate, and increased isolation decreases the rate of colonization, ultimately leading to a higher frequency of empty patches, and thereby to selection for increased dispersal at the metapopulation level. As local dynamics are necessarily faster than metapopulation dynamics, we might expect that a sudden episode of habitat loss and fragmentation first enhances local selection against emigration, while the metapopulation-level selection for increased emigration occurs with a delay, in the course of new populations being established, especially by individuals with the highest dispersal propensity (Leimar and Norberg 1997). Furthermore, the metapopulation-level selection will gather in strength only when the fraction of empty patches has become large owing to the shift in the balance between extinctions and colonizations (M. Heino and I. Hanski unpublished manuscript). In brief, in the short term, evolution may deepen the demise of a population in a fragmented landscape, whereas the long-term effect, assuming that the metapopulation survives to experience any long-term effects, may be an increase in dispersal rate, possibly leading to an evolutionary rescue: survival of the metapopulation in the fragmented landscape via evolution of increased dispersal rate.

Empirical results on the evolution of dispersal rate in response to increased habitat loss and fragmentation are practically non-existent. What results there are relate mostly to the questions of whether emigration rate has become reduced in populations that have become isolated and whether individuals in newly established populations exhibit especially high dispersal capacity or propensity (for reviews see Olivieri and Gouyon 1997; Hanski 1999a). Evolution of dispersal rate in changing environments is an area of metapopulation biology where important theoretical and empirical work remains to be done.

21

Dispersal in antagonistic interactions

Minus van Baalen and Michael E. Hochberg

Abstract

The outcome of antagonistic interactions (such as predator–prey and host–parasite interactions) depends sensitively on spatial structure. As the spatial distribution of a population is, at least partially, the result of spatial decisions (which can take the form of habitat selection, foraging tactics, or dispersal) of individuals, there is a close interplay between individual behaviour and population dynamical consequences. We will discuss a number of models to study how natural selection can mould spatial strategies of predator and prey, and how these strategies, in turn, affect their population dynamics. First we will discuss a model to study the consequences of 'ideal and free' patch selection of predators and prey. In general, however, individuals are neither 'ideal' nor 'free'; they do not have prefect information about their environment, and they cannot go wherever they like. Often, spatial strategies amount to adopting 'stay-or-leave' rules. In particular in metapopulation-type interactions (in which new prey populations are founded but subsequently discovered and exterminated by predators), the evolution of dispersal strategies and overall dynamics are tightly linked. We will discuss briefly what may happen if there is no clear distinction between subpopulations. In such systems, patchiness is not imposed by the modeller but arises through spatial self-structuring. Finally, we discuss how dispersal is likely to affect co-evolution over much wider geographical ranges. Previous work showed that when prey and predators were allowed to co-evolve, the highest levels of investment (and least specificity in terms of prey defence and predator offence) occur in habitats most suitable for the prey species; often this is expected to occur towards the centre of the prey's geographical range. We evaluate the robustness of this logic for scenarios in which site-to-site movement, too, can evolve in time and in space. Movement decisions may be determined by a number of factors, including habitat quality, population density, and the distribution of conspecific genotypes in the environment.

Keywords: dispersal, population dynamics, predator–prey systems, metapopulations, habitat selection, ESS, co-evolution

Introduction

Most models for antagonistic interactions between consumers and their resources are variations either of the Nicholson–Bailey (1935) model in discrete time or of the Lotka–Volterra model in continuous time. In their most basic form, these models do not incorporate spatial structure. That is, populations are assumed to be distributed homogeneously in a spatially unstructured environment, and the number of encounters is proportional to the product of the densities of the interacting populations. As a consequence, aspects related to the individuals' locations in space, such as dispersal, habitat selection, and so forth, play no role. However, since the work of Hassell and May (1973) and May (1978), it is now generally accepted that space constitutes an essential dimension of population dynamics and population interactions. Here, we will outline how space and the way individuals use it may affect population dynamics and evolution in interacting populations. In particular we will explore the aspects involved in the interaction between population dynamics and evolution that moulds dispersal strategies of predators and prey. We will not present a complete review of all the possible effects, but will rather confine ourselves to discussing examples that we believe provide exciting new insights into how the evolution of dispersal affects interactions between predators and prey, between hosts and parasitoids, and between hosts and parasites.

Behavioural flexibility

Much research has been devoted to understanding how dispersal may help individuals locate new resources in a spatiotemporally heterogeneous world (e.g., Stephens and Krebs 1986; McPeek and Holt 1992; Holt and McPeek 1996; and many chapters in this book) and how, in addition, it may enable organisms to avoid competition with closely related conspecifics (Hamilton and May 1977; Gandon and Michalakis, chapter 11). In many of these studies, and in particular in those making the link with population dynamics, dispersal is a blind, random process: with a certain constant probability ('dispersal rate') an individual may leave and move to another place, chosen at random. Passive dispersal may be particularly unrealistic when one focuses on antagonistic interactions. Predators will actively search for their prey, hosts may decide to move if local levels of infection rise (Boulinier *et al.*, chapter 12a), and so on.

Active non-random dispersal may require two separate decisions (Stamps, chapter 16). The first is to decide whether to stay or to leave (Stephens and Krebs 1986). The study of 'stay-or-leave' rules has given rise to a rich interplay of theoretical and experimental studies (in particular in the field of host–parasitoid interactions; Bernstein *et al.* 1988, 1992). The second aspect of non-random dispersal is, once on the move, where to go and settle (Stephens and Krebs 1986). For example, predators and parasitoids will try to locate the most profitable habitats. All these decisions may be influenced by available information (internal state, external cues, etc.)

The pattern produced by the integration of these dispersal decisions (when and/or where to go) is a spatial population distribution. The spatial distribution will, in turn, affect other populations, either directly (the risk of predation will be highest where there are most prey) or indirectly through behavioural responses (prey individuals may disperse away from regions with many predators, or avoid them). How these dispersal decisions interact with population dynamics is, in most cases, still an open problem.

Scale

One of the difficult aspects of the relationship between the evolution of dispersal and its population dynamical consequences is the scale of spatial heterogeneities, with respect to individual dispersal capabilities and the information to which they have access (de Roos *et al.* 1991, 1998; Rand and Wilson 1995).

Which dispersal strategies are favoured depends on these scales. For example, a small insect may be able to assess local conditions, but lack the means to explore conditions elsewhere. The 'decision' faced by such an insect is whether or not to venture a jump into the unknown. The decision takes a very different form, however, when the individual's locomotory and information-gathering capabilities are sufficiently sophisticated that they may actually estimate expected fitness across all space and move to those patches offering the highest fitness, such as foraging parasitoids (Bernstein *et al.* 1988; Driessen and Visser 1993; Driessen *et al.* 1995).

However, if the distances between patches are much larger than maximum (lifetime) dispersal distances, then solving the problem of 'where to settle' is not really relevant, because even if an individual knows the best destination, it may simply not be able to go there. Whether dispersal is random or non-random, if only neighbouring patches can be reached, population dynamics become 'diffusion limited' (de Roos *et al.* 1991, 1998). A similar situation may arise when individuals have reliable information only about nearby sites (assuming that unreliable information makes long-range dispersal disadvantageous).

In a metapopulation, the relevant decision may be 'when to leave' rather than 'where to go'. Individuals should stay until their expected fitness elsewhere (including travel costs) is higher than what they can achieve locally. Such dispersal decisions can be complicated by issues related to kin selection (Hamilton and May 1977; Perrin and Goudet, chapter 9). It may be favourable for individuals to leave and face a bleak future if this behaviour benefits their kin (that stay behind). Predators, for example, could, by increasing dispersal rate, reduce predation pressure, thus exploiting patches more efficiently (van Baalen and Sabelis 1995).

Which dispersal strategy is favoured may, in turn, feed back to the dynamics of the system. For example, diffusion-limited dynamics may give rise to such pronounced patchiness that the system will behave like a metapopulation, consisting of a number of weakly coupled populations (Janssen and Sabelis 1992; Jansen and de Roos 2000). This, in turn, may shift the relevant aspect of the dispersal strategy. In a relatively homogeneous system the important issue could be 'where to go', whereas in an inhomogeneous system it could shift to 'when to go'.

Temporal scales are also important to consider, and these may or may not be related to spatial scales. In a diffusion-limited system, spatial scale will be tightly linked with the characteristic spatial scale (Rand and Wilson 1995). But there may be external stochastic or systematic influences (e.g., seasons, underlying environmental heterogeneity) that impose their own spatial scales. Which temporal scale is important may also depend on the behaviour under consideration. In interacting systems, the most important temporal scales may be those associated with colonisation and extinction of populations. Recognition of such spatial scales is important, because it allows one to focus on the cardinal features of the system, sometimes allowing considerable simplification.

In a spatiotemporally heterogeneous setting, the local quality of the environment (in general) will be maximal at one given spatial location at any time. Therefore, according to the most elementary optimisation model for habitat selection, any individual should always move to this place. On the basis of this reasoning one would predict that populations should thus track *en masse* this point of maximal fitness. This is not what is observed in the majority of cases, where populations are characterised by some distribution across space. There are two fundamentally different ways to explain such distributions: (i) individuals are constrained and cannot track the point of maximal fitness, and (ii) individuals are in fact equally adapted and have equal expected fitness. In either case, there is something missing in the simple optimisation framework. Indeed, most biological populations are subject to one or various forms of (local) density dependence, which means that one individual's spatial decisions will influence the pay-off for others. This implies that individuals are involved in a habitat selection game, where the pay-off to a given individual depends not only on its own strategy but also on those of others. In particular in the latter case, the ability to detect the presence of others (prey and/or predators) is likely to convey large fitness benefits, because it enables individuals to select the best habitats with some flexibility.

Outline

First, we outline how behavioural flexibility may shape habitat selection strategies for both trophic levels and explore the consequences for population dynamics. Many parasitoids interact heterogeneously with their host (Pacala *et al.* 1990). For example, many insect herbivores have relatively narrow ranges of host plants. Individual herbivorous hosts and individual parasitoids must decide in which habitat patches to forage. If their locomotory and/or information-gathering capacities are sufficient to explore the whole spatial domain, they will attempt to select the best places to settle in (from their respective points of view). Since all individuals will do so simultaneously, and the profitability of a patch is likely to depend on how many individuals select it, habitat selection takes the form of a game. If all individuals are fitness maximisers then the resulting distribution is known as an 'ideal free distribution' (IFD), characterised by the fact that expected fitness is the same everywhere (Fretwell and Lucas 1970). The reasoning underlying this result is simple but compelling: if (i) individuals have sufficient knowledge about their environment and (ii) there are no constraints to travel, there cannot be differences in fitness because, if there were, at least some individuals would move from bad to good patches, in which case density dependence would ensure that local fitness decreases (Bernstein *et al.* 1988, 1992). It has been shown that such IFDs may stabilise otherwise unstable host–parasitoid interactions, albeit only under very restricted conditions (van Baalen and Sabelis 1993) in the Nicholson–Bailey framework. For a discussion of the consequences of IFDs on predator–prey dynamics in continuous time, see Bernstein *et al.* (1999). We will discuss recent work (van Baalen and Sabelis 1999) that suggests that this conclusion is actually quite sensitive to the details of habitat selection, in particular with respect to the available information.

In the second section, we will investigate the case of limited dispersive ability. We will show that, when there are constraints on dispersal distance, spatial heterogeneity in species interactions may arise even in the absence of an underlying template in spatial heterogeneity. Such 'viscous' predator–prey systems may exhibit a rich variety of behaviours, and dispersal may affect this behaviour in various ways. Although the selective pressures affecting dispersal in such systems have been little investigated, some tantalising results are obtained.

In a third section, we discuss how dispersal and evolution may interact on an even larger scale, that of a network of habitat patches, spanning a range in environmental conditions. The major question here is to know how dispersal may influence large-scale patterns in adaptation of predators and their prey.

Simultaneous ideal free distributions

Many insect species are attacked by parasitoids. Generally, these parasitoids, usually other insects, lay their eggs on or inside their host which is eventually killed, and a new parasitoid emerges. Simple models for this interaction are not persistent; that is, populations fluctuate until one or both of the populations are extinct (Hassell 1978). This has provoked much research to identify those factors that contribute to the persistence of this type of interaction (Bernstein *et al.* 1999). As already suggested by Nicholson and Bailey (1935) and later corroborated by Hassell and May (1973), spatial distributions (and thus the dispersal strategies of hosts and parasitoids) may be a factor that promotes persistence of tightly coupled host–parasitoid associations.

van Baalen and Sabelis (1993) noted that if hosts and parasitoids had sufficiently sophisticated foraging strategies, then they should settle to simultaneous IFDs. One can substitute these distributions into the population dynamical model and investigate the consequences. A first conclusion is that, if there is no underlying spatial heterogeneity, then the simultaneous IFDs will be spatially homogeneous, and population dynamical stability will be lost. This suggests that Hassell and May's (1973) original theory is at odds with the theory of evolution by natural selection, and could mean that a key factor in population persistence is that individuals are unable to select their habitat optimally. Alternatively, essential components may be missing from the model. One such factor could be underlying spatial heterogeneity. van Baalen and Sabelis (1993) showed that IFDs may be sufficiently aggregated to render the population dynamical equilibrium of hosts and parasitoids stable if the environment is subdivided into high- and low-quality patches (for the prey).

This result suggests that natural selection will favour behaviour leading to population stability only under very limited conditions. However, this does not preclude host–parasitoid persistence. The conditions for non-equilibrium persistence, in the form of cyclical or quasi-periodic cycles, are often much wider (May 1978; Hofbauer and Sigmund 1988). Indeed, non-equilibrium persistence is the outcome when another subtle assumption in Hassell and May's (1973) model is relaxed. In Hassell and May's (1973) as well as in van Baalen and Sabelis's (1993) models, spatial distributions are represented by a series of constant parameters. This amounts to the assumption that population distributions are independent of density. But are they? In fact, later research has indicated that the population dynamic outcome sensitively depends on the assumption that individuals can distinguish between patch *types* (different host plants, say), but cannot assess population densities within these patches. When this assumption is relaxed (for the predators and/or the prey), very different population dynamics result (van Baalen and Sabelis 1999). When only the predators are behaviourally flexible (allowing them to assess and respond to within-patch densities), stability is lost and limit cycles result. Thus, the analysis of more realistic behavioural models indicates that the so-called 'aggregative response' (Hassell and May 1973) of predators is not a stabilising mechanism, but rather a destabilising one. Nonetheless, it contributes to persistence, because although they fluctuate, neither the prey nor the predator goes extinct. When both trophic levels are behaviourally flexible, population dynamics become chaotic, but persistent over a much wider range of conditions than under 'rigid' patch selection. The reasons for persistence are not difficult to understand: whenever predator density is high, the high-quality patches become too dangerous, and some prey will seek refuge in lower-quality patches; it is well known that refuge use can be a powerful stabilising mechanism (Hassell 1978; Holt and Hassell 1993; Hochberg and Holt 1995). Murdoch *et al.* (1992), analysing a different type of model, also found that density-dependent parasitoid aggregation had different effects on dynamics than did density-independent aggregation.

Note that, if prey select flexibly, what constitutes a 'patch' will vary with time: distributions expand and shrink when populations fluctuate. This suggests that just considering populations at equilibrium situations may not be sufficient. The presence of very infrequently used refuge patches may be *the* crucial aspect that allows persistence.

It is highly unlikely that individuals are as 'ideal' and 'free' as was assumed in the afore-mentioned simulations, and therefore the range of population dynamics (and associated use of space) produced may never be observed in nature. Nevertheless, the analysis suggests that the population dynamical end-result may be strongly sensitive to assumptions about individual flexibility.

Limitations

Whatever the specific framework, analysis of optimal habitat selection will always predict the IFD to be the end-result of evolution (if indeed the optimum is reached). Although this concept is very useful as a first guess about the consequences of natural selection (see Holt and Barfield, chapter 6, and Stamps, chapter 16), it is a moot point that the IFD is an idealisation. Indeed, the very observation that in some areas populations grow whereas in others they decline refutes the assumption of equal fitness. This is not surprising given that the assumptions of 'idealness' and 'freedom' will be rarely met.

If individuals are not 'ideal', then they lack complete information. They may know local conditions, but have no way to assess whether there exist locations offering higher fitness. And, even if they know of a rich patch, they may not be 'free' to go there. As such, the only decision that individuals can make is whether to stay or to leave. This gives rise to different kinds of population dynamics, where local populations are less tightly coupled, and the metapopulation concept becomes useful.

Exploitation of local populations

In a spatially extended population it is likely that individuals have limited capabilities to assess their environment, either because of limited information or because of limited movement. Then, individuals can, in principle, assess the fitness value of their current spatial location, but their potential fitness elsewhere remains guesswork. Charnov (1976) predicted that under these conditions individuals should leave their patch once its profitability dropped below a certain (environmentally determined) threshold.

This 'marginal value theorem' (MVT) has been repeatedly tested experimentally and theoretically, but mostly within the framework of a single population. Much research has been devoted to understanding how individuals can integrate information about their environment to set the marginal value threshold. Some authors have put the MVT into a population dynamical setting, where the environment (i.e., patch profitabilities) depends on population dynamics (Visser 1991; Driessen and Visser 1993; Driessen *et al.* 1995). All of these approaches are based on the idea that individuals can visit more than one patch in their lifetime. This means that, at the population dynamical level, the population of foragers (predators or parasitoids) is still relatively well mixed. Local population interactions matter, but they are strongly influenced by changes in the global number of dispersing prey and predators. Local dynamics become less strongly coupled to the number of dispersers (and hence to dynamics in other patches) if travel costs (in terms of survival) become so high that searching individuals are lucky to find a single patch at all. Then, dispersal strategies approach the classical 'stay-or-leave' problem. The difference with interacting populations is that there will be a close interaction between ecological (i.e., metapopulation dynamical) and evolutionary processes.

Whether an antagonistic interaction will end up at the 'habitat selection' end of the spectrum or at the 'stay-or-leave' end thus depends on the costs and limitations of dispersal. If information is easy to come by and travel is cheap, the result is a habitat selection game. If information is limited to local conditions and travel is expensive, then a stay-or-leave problem results. In intermediate cases, of course, both aspects may be important. Here we will discuss the latter extreme.

Antagonistic interactions have an intrinsic propensity to cycle (Gomulkiewicz *et al.* 2000; Nuismer *et al.* 1999). If a non-spatial predator–prey system is unstable, interesting phenomena may arise when both populations are distributed across space. Provided individuals move through space relatively slowly, different regions are likely to start to oscillate out of phase (de Roos *et al.* 1991). This means that in some regions prey density will be high, while in other regions prey will be virtually absent. The distribution of the prey will then comprise local 'patches' of high density surrounded by regions devoid of prey; if the predators are limited in their movement, their distribution will reflect the prey distribution only imperfectly. Such distributions will not remain static, of course. Patches of prey will grow and expand. Eventually they will be found and invaded by predators. These predators rapidly multiply, and finally reduce local prey density again. Having depleted the local prey patch, the predators must then search for other patches of prey that may have been founded in the meantime.

This kind of situation, populations of discrete individuals distributed over continuous space, is difficult to analyse, because it violates a crucial assumption allowing the derivation of differential equations to describe the dynamics (de Roos *et al.* 1991; Matsuda *et al.* 1992; Durrett and Levin 1994). On the other hand, ignoring space means that we cannot treat the underlying question in any depth. An alternative way to assess the ecological and evolutionary consequences of limited dispersal is to assume population subdivision over a pre-existing set of patches. This is the classic predator–prey metapopulation setting.

If predators extinguish their prey locally, then prey and predator can persist only in a larger spatial framework. Such a dynamic ensemble of local populations that go extinct and become recolonised is called a 'metapopulation'. In a predator–prey metapopulation, natural selection may favour other traits than in a single 'well mixed' system. In a well mixed system, natural selection will favour predators that maximise their prey capture rate even if this spoils things for the population (the 'tragedy of the commons' principle; Hardin 1968). In a metapopulation setting, however, this may be different (Gilpin 1975; van Baalen and Sabelis 1995). Consider a predator that has just located a new, expanding, prey population. Were this predator and its descendants to maximise prey capture rate, they would all stay within the prey patch until all prey were consumed. However, there are other ways to exploit a prey patch. Consider a predator whose descendants have an innate tendency to leave the patch before the prey go extinct. From the viewpoint of the individual, dispersing is disadvantageous; after all, why abandon your food if it is uncertain whether you will find food elsewhere? The possible resolution of this problem is that selection on dispersal would appear to be associated with some relatives staying behind (Hamilton 1964; Hamilton and May 1977; Perrin and Goudet, and Gandon and Michalakis, chapters 9 and 11, respectively), meaning that the local prey population is exploited less heavily and will persist for a longer time. By the time such 'milkers' have finished with the local prey population, the milker foundress will have many more descendants roaming the area than will a 'killer' foundress whose descendants all stayed within the patch as long as possible.

The potential advantages of milking are thus considerable. But there are disadvantages too: if a milker patch is invaded by a killer, then the killer and its descendants quickly consume all prey, spoiling things for the milkers in the patch. It therefore pays to adopt the milker strategy only if the risk of invasion is rather low. However, this risk is low only when the density of killers is low but their frequency is high, meaning that the conditions for milkers to grow increases. This is followed by the growth of killers and the subsequent shrinking of milker and killer populations, and the cycle starts over again. Thus there is a balance determining optimal strategies for the exploitation of a prey patch. (Actually, a balanced polymorphism might arise as well, but this possibility has not been explored.)

A similar analysis can be carried out for the prey. As it turns out, the underlying dilemma for the prey is rather different than that for the predator (Egas *et al.* 2000). In absence of predators, the dilemma for the prey resembles that of the predators, and involves the optimal exploitation of the prey's own resources. In the presence of predators, all local populations will eventually be found and exterminated by predators. This means that any descendent prey population will be founded by individuals that dispersed before the population had gone extinct. To the prey, the dilemma is therefore to adopt a dispersal rate that is not too high (to exploit the local potential for population growth), nor too low (*some* individuals should disperse to establish new populations).

Of course, this example applies only to 'stay-or-leave' rules for the prey in a metapopulation. For an extensive review of some aspects of patch selection under predation risk, see Lima (1998).

Environmental gradients

On an even larger scale, local environments will vary across space. The ecological and evolutionary implications of how dispersal may create or destroy biological patterns are only starting to be appreciated. For example, Hochberg and Ives (1999) developed a metapopulation model of a tightly coupled antagonistic interaction (e.g., between a predator and its prey), with the goal of understanding how patch-to-patch movement could influence restrictions in the geographical range of both species. They assumed that central habitat patches tended to be the most productive for the prey (i.e., with the highest prey intrinsic growth rates), whereas edge patches tend to be demographic sinks for the predator. They assumed that dispersal was passive (following a geometrical distribution), and that hosts and/or parasitoids could go locally extinct if their densities dropped below a threshold. They found that predators that (1) disperse readily and (2) have high potential rates of increase can indeed restrict the geographical boundaries of their prey (see also Hastings 2000). When these qualitative features are relaxed, predators can still have important impacts on their prey, the former readily fragmenting the geographical distribution of the latter. It is predicted that such fragmentation may be a precursor of allopatric speciation in predator–prey associations, and it is speculated that the incredible species diversity of insect parasitoids, the most tightly coupled of predator–prey type interactions (Godfray 1994), may be due in part to this phenomenon.

In a related evolutionary vein, we analysed a model to predict where along a gradient in habitat productivity one should expect long-term evolutionary interactions between antagonistics to be most intense (Hochberg and van Baalen 1998). We assumed that individual prey can invest (at a cost) in defence, diminishing the attack rate of a monophagous predator. Similarly, the predators can invest (also at a cost) in 'offence', increasing their attack rates on the prey.

The analysis of cases without dispersal yields the general result that, at an evolutionary equilibrium, the greatest (most costly) investments in the interaction occur in the most productive habitats; although not necessarily the case, such patches will tend to be found towards the centre of the prey's geographical range (Lawton 1996).

When passive dispersal occurs, adaptation in the most productive patches tends to dominate that in the least productive patches, and the final distribution of prey and predator investments will depend on (1) the relative frequencies and productivities of patches, (2) the relative dispersal rates of prey and predators, and (3) the explicit spatial configuration of habitat types. Because dispersal in the Hochberg and van Baalen models is passive and the different investment types of prey and predator always eventually attain equilibrium densities, dispersal generates what would appear to be maladaptation of prey and predator with respect to scenarios in which the prey and predator remain sedentary, and cases where prey and/or predators make active decisions as whether or not to disperse (P. Commenville, unpublished results). This means that one should tend differentially to find individuals that are not adapted to local conditions as one proceeds from productive to marginal habitats (source–sink dynamics; Holt 1985; see also Holt 1997b; Jansen and Yoshimura 1998). That is, under prey dispersal we observe that in marginal habitats there are too few predators to warrant high levels of prey defence. Similarly, since 'locally adapted' prey invest less in defence, immigrant predators from more productive sites tend to invest too much in offence. These high-investment strains persist due to migration; they would otherwise be rapidly out-competed by less aggressive, locally adapted strains.

One of the key features of the underlying predator–prey model we considered was the fact that, locally, populations will always attain a stable equilibrium. Increases in patch-to-patch movement tended to create maladaptation with respect to cases where there was no movement at all. Gandon *et al.* (1996) investigated a different meaning of local adaptation: the propensity of local strains of a host (or a parasite) to resist (or infect) local strains of the antagonist more readily than strains at other localities. They simulated a model in which local population dynamics were unstable and fluctuations were so intense that strains would eventually go extinct locally if it were not for migration. They found that, as long as one species migrated at sufficiently low levels, then the faster migrating species tended to be locally adapted when averaging over all sites in the system. When both species move at sufficiently high rates, neither host nor parasite tends to be better adapted (on average) at local as compared to distant sites.

Discussion

To understand the evolution of dispersal, it may be necessary to integrate trophic interactions into future models, especially in cases where there is little competition for space. In such a setting individuals may decide to move because their local resources are insufficient or individuals may avoid certain places because the risk of attack (or infection; see Boulinier *et al.*, chapter 12a) is too high. Often, one includes all of these effects in the catch-all term of 'density dependence', but this may oversimplify reality. Often, models for trophic interactions *do* predict density dependence, but often in highly complex, non-linear forms.

There are many ways in which dispersal strategies affect antagonistic interactions, and equally many eco-evolutionary feedback loops that affect these dispersal strategies. How the basic strategy of random, constant and non-directional dispersal may affect interacting populations is reasonably well known (although surprises lurk even in simple, well studied models; see for example Jansen 1995). However, the set of possible feedbacks that govern the ecology and evolution of more sophisticated dispersal strategies has yet even to be sketched.

Patchiness

Any spatial decision will be significantly affected by the degree of temporal and spatial heterogeneity. Such heterogeneity may be externally caused, but in systems composed of interacting populations it may also be internally generated. Particularly interesting feedback loops arise if the patchiness of the system depends on individual dispersal strategies.

For example, if dispersal is only a local process, then the population structure will be heterogeneous in space, and determined by the dispersal characteristics of the individuals and the incipient ecological interactions (see for example Hochberg *et al.* 1996). Determining these characteristic spatial scales is not trivial, but they can yield powerful insights into the role of space in a given ecological interaction (Rand and Wilson 1995). If, for example, a system under consideration is much smaller than its characteristic spatial scale, then the system is essentially 'well mixed' and simple non-spatial models may suffice. In contrast, if the characteristic scale of movement is much smaller than the spatial extent of the system, then space *must* explicitly be taken into account. This is because under these conditions the system will self-structure, creating dynamic behaviours unproducible by non-spatial ecological (de Roos *et al.* 1991, 1998) and evolutionary (Boerlijst *et al.* 1993) models.

Demographic stochasticity

If local prey and predators are sufficiently abundant so that individual movement can be represented by a steady flow of individuals (rather than departures and arrivals occurring in discrete jumps), then diffusion approximations become an accurate method for linking local populations. This classical setting has been studied extensively (Murray 1980). In the purely ecological domain, it is known that, depending upon diffusion parameters (i.e., dispersal of predators and prey), a variety of new dynamic behaviours can emerge. For example, if local prey densities are regulated mainly by the local predators, and if the predators diffuse faster than the prey, 'diffusive instability' can arise, creating complex spatial patterns (Murray 1980). As yet, it is not well known whether such special dynamics follow from evolutionarily stable dispersal. To address such evolutionary questions, we need to know the dynamics of rare mutants, and it is here that the diffusion approach often breaks down. Even if population densities are high, a mutant initially will be represented by a small number of individuals. Assuming that these few individuals are 'spread out' much like the resident population can produce erroneous results. For example, when there is a region of high fitness, a diffusing mutant will certainly reach this region (because a continuous low density means, effectively, that fractions of individuals can move across space; see Mollison 1991), whereas in reality the probability that a single individual will find it may be vanishingly small. Taking into account the discreteness of individuals leads to more complicated analysis, as dynamics (both of resident and mutant) are strongly affected by demographic stochasticity. Such stochasticity may render host–parasitoid systems less persistent, but it may facilitate parasitoid coexistence (Wilson and Hassell 1997).

Other trophic levels

We have seen that there are a multitude of ways in which dispersal strategies will affect antagonistic interactions, and that there are probably equally many feedbacks of population dynamics on the selective pressures moulding these dispersal strategies.

But to understand why populations do indeed have the dispersal strategies they have may require considering antagonistic interactions in an even larger setting. As an example, consider a three-trophic-level system of a predator–prey interaction where the prey are actually herbivores, so that the entire interaction is supported by a population of plants. Individual plants may have a considerable interest in interfering with the dispersal strategies of either prey (herbivores) or predators; that is, a plant should try to induce its herbivores to leave, and it should try to attract predators from elsewhere (Egas *et al.* 2000). There is increasing evidence to suggest that plants often do play such active roles (Vet *et al.* 1991; Sabelis *et al.* 1999; Vet and Dicke 1999). Thus the next stage in understanding the evolution of dispersal strategies in interacting populations is to include the dynamic role of information.

Acknowledgements

Two anonymous referees are thanked for their valuable comments on an earlier version of this chapter. M.v.B is a Research Fellow at the University of Amsterdam, supported by the Royal Dutch Academy of Sciences (KNAW). M.E.H is supported by the French National Centre for Scientific Research (CNRS) and the French Ministry of Education and Research (programmes SRETIE and ACCSV7).

22

The properties of competitive communities with coupled local and regional dynamics

Nicolas Mouquet, Gerard S.E.E. Mulder,
Vincent A.A. Jansen, and Michel Loreau

Abstract

In this chapter we review the results from models that link local and regional scales, and illustrate some of the mechanisms that can maintain biodiversity. First, we present a model of local competition for space in plant communities. The model is a classical metapopulation model, but we apply it to a population of individuals rather than to a population of populations, and we add an external source of immigration by a propagule rain. In such a system, we show that local coexistence is possible, and that the number of coexisting species is a growing function of immigration intensity. To explain the origin of the propagule rain, we next present a model of a meta-community, defined as a regional set of communities linked by dispersal. Assuming environmental heterogeneity at the regional scale, we show that the number of coexisting species cannot be greater than the number of different communities or habitats. A new concept arises from these conditions of coexistence, which we call regional similarity. To persist in a metacommunity, all species must have the same mean competitive ability at the regional (metacommunity) scale. In both models, communities change from low-diversity communities when the local dynamic is important to high-diversity communities when the regional dynamic is the most important process. Moreover, switching to local from regional dynamics generates very different relationships between diversity and ecosystem processes such as plant productivity. Our second model describes the evolution of the properties of communities. Generally, the properties of communities, like their composition, diversity, and productivity that develop over very long periods of time, will be determined by two constraints. As illustrated above, a first constraint is that the component species have to be able to coexist in a community. Second, a community will adapt under evolutionary processes and therefore a second constraint is that a community is evolutionarily stable. Only a subset of all communities that can coexist will be evolutionarily stable. To study the properties of evolutionarily stable assemblages, we present an evolutionary model for competitive communities. Our model is based on a framework to describe evolution in metapopulations. The local dynamics are specified by competition for a single resource, the regional dynamics by seasonal harvesting and redistribution of dispersing propagules. We let the characteristics of the competitive communities adapt through selection and small mutations. The properties of the evolutionarily stable ecosystem, in particular its biodiversity, the composition of the community, and biomass composition depend on the maximum productivity and the length of the productive season. Our different models illustrate how the composition and properties of competitive communities are structured through the population and evolutionary dynamics, and how these, in turn, depend on the local and regional dynamics.

Keywords: altruism, assemblage, biodiversity, biogeography, coexistence, competition for space, competitive exclusion, community properties, diversity, ecosystem processes, ESS, fitness, haystack model, immigration, inclusive fitness, kin selection, metacommunity, plant community, propagule rain, single-resource competition, source–sink

Introduction

In some natural plant communities, more than 80 species can be found in a square metre (Zobel 1992). How such a large number of species can coexist on a limited number of resources is still not completely understood (for reviews see Tilman and Pacala 1993; Bengtsson *et al.* 1994b). A possible solution to this paradox lies in the different spatial scales on which ecological interactions work (Holt 1993; Loreau and Mouquet 1999).

Individuals predominantly interact with other individuals in their local environment. This results in a small characteristic spatial scale. However, the dynamics at the local scale also depends on the regional population through immigration (Holt 1993; Ricklefs and Schluter 1993; Zobel 1997; Cornell and Karlson 1998; Loreau and Mouquet 1999). This introduces two distinct spatial scales into the interaction: a local scale and a regional one (for procedures to estimate the characteristic local spatial scale see de Roos *et al.* 1991; Rand and Wilson 1995; Wilson and Keeling 1999). To understand the effects on population dynamics, both experimental and theoretical works have concentrated on cases with a clear separation between these two scales. Typically, local interactions take place in sites or patches; the influence of regional interactions on local interaction is through immigration of individuals, which will be a function of some average of the density over all patches in the region.

Populations that cannot persist in a single location can persist in a collection of different patches. This observation, which lies at the heart of the theories of island biogeography (MacArthur and Wilson 1967) and metapopulations (Levins 1969; for reviews see Gilpin and Hanski 1991; Hastings and Harrison 1994; Hanski 1999a), may also explain the coexistence of many species on a limited number of resources. If all competitors inhabited a single location, the number of resources would limit the number of species. In a collection of patches, local populations may disappear from sites, but as long as each local population on average colonizes at least one new site, the metapopulation will persist. Many species can coexist in systems with local and regional dynamics, even on a single resource. In their simplest form, metapopulation models describe only empty and occupied patches, and thus completely ignore local dynamics, the colonization of new patches depending only on the number of occupied patches (Levins 1969).

The competitive dominance of one species is a robust result of models of competition for a single resource. With resources so limited and competition so fierce, how can species-rich communities nevertheless be explained? One possible explanation is that there exist underlying regional differences so that different species dominate in different sites. Another possible explanation is that there are no regional differences, but different species have different strategies to use space: some species go for local competitive advantage, whereas others go for spatial expansion. This idea has received much attention in the framework of the competition–colonization trade-off hypothesis (Hastings 1980; Tilman 1994). The basic assumptions of models including this trade-off are that each patch can be occupied by only one individual; being a superior competitor implies being a weaker colonizer (see Tilman 1990); and if a superior competitor invades an already occupied patch, it immediately replaces any inferior competitor. Therefore, an inferior competitor (better colonizer) survives only in those patches that are left open by its superior competitors.

We shall study these two hypotheses with two different types of models. The first type of model relaxes the assumption of a trade-off between competitive and colonizing abilities, and focuses on the influence, through immigration, of regional process on local dynamics. The second type of model relaxes the assumption of immediate replacement of the inferior by the superior competitor and studies the properties of competitive communities from an evolutionary perspective. In this chapter, we ask how species richness is maintained and how it is generated. Both types of models yield results on coexistence far more general than those described previously. As an extension, we will use them to explore the relationships between ecosystem functioning and species diversity, which emerge from these two approaches to species coexistence.

Source–sink dynamics in competitive plant communities

The propagule rain model

We first present a model that incorporates the influence of immigration from a regional pool on a community of plants competing for space (for more details, see Loreau and Mouquet 1999). As in metapopulation theory (Levins 1969, 1970), the habitat is assumed to be discrete and we model the dynamics of extinction and colonization of patches. Our model differs from metapopulation models (Levins and Culver 1971; Horn and MacArthur 1972; Slatkin 1974; Hastings 1980) because each site can support only one plant (Goldwasser *et al.* 1994; Tilman 1994; Pacala and Rees 1998), and because we consider only indirect competition for space (i.e. a plant will release a site only at its death). The establishment of plants in vacant sites obeys a competitive lottery (Chesson and Warner 1981; Sale 1982).

We define P_i as the proportion of sites occupied by species i in the community. There are S such species that compete for a limited proportion of vacant sites, V. For each species i, we include seed production, short-distance dispersal, germination, and seedling establishment in its local reproduction rate, c_i. All forms of natural death are encapsulated in its mortality rate, m_i. As in island–continent models (Gotelli 1991), we add an external source of immigrants in the form of a propagule rain. Parameter I_i describes the species-specific potential immigration rate of species i, which is determined by its long-distance dispersal capacity and its relative abundance in the regional source. Immigration intensity depends on the size of, and distance from, the regional source, and is defined by a parameter α. Because only vacant sites can be occupied, potential reproduction (local reproduction plus immigration) is not fully realized.

This model, which we call the propagule rain model, reads:

$$\frac{dP_i}{dt} = f_i(P_1, P_2, \ldots P_s) = (\alpha I_i + c_i P_i)V - m_i P_i \qquad V = 1 - \sum_{j=1}^{s} P_j \qquad (1)$$

First, consider a closed community in the absence of immigration ($\alpha = 0$), and let us define $r_i = c_i/m_i$ as species i's local basic reproductive rate (Fagerström and Westoby 1997; Loreau 1998b) which is equivalent to its lifetime reproductive success in a vacant environment. The community reaches an equilibrium ($f_i = 0$) when $V = 1/r_i$. However, because there are as many different values of r_i as there are species in the system, only the species with the highest basic reproductive rate can persist at equilibrium.

Next, consider the general case of a community open to immigration of all species ($\alpha > 0$ and $I_i > 0$ $\forall i$). All species coexist because individuals arrive continuously from outside the community. However, in this deterministic model many species are maintained at unrealistically low densities. In reality, rare species are subject to the risk of extinction by demographic stochasticity. We have studied the consequences of such extinction on coexistence in communities, with different immigration intensities. To mimic demographic stochasticity, we used a numerical approximation of equation (1) with a threshold proportion of sites, below which a species is considered to be extinct. Our results show that species richness at equilibrium increases continuously with immigration intensity α (Fig. 22.1a). This increase is steepest at an intermediate value of α, when the contribution of immigration allows the potential recruitment of a number of individuals greater than the extinction threshold.

To describe community composition we measured equitability based on the classical Shannon diversity index (see Fig. 22.1b). When equitability is equal to 1 all the species have the same proportion, and when equitability is low there is one dominant species and many rare species. The increase in the number of species, and the increase in the proportion of sites occupied by new species, occur at different immigration intensities, therefore equitability first decreases. At higher immigration intensities, the amount of space occupied by the various species becomes more and more similar, and equitability increases until it reaches a ceiling. Therefore, there is a switch from a species-poor, non-equitable community to a species-rich, equitable community as the intensity of immigration increases.

The metacommunity model

The propagule rain model shows that immigration from an external source is able to maintain a high local diversity in a system that would otherwise experience competitive exclusion of all but one species. It does, however, raise the questions: Where does the propagule rain come from? And what about emigration? To answer these questions we formulated a model for a metacommunity, in which immigration to a community is a function of emigration from other communities. Our approach fits in with 'mesoscale ecology', which seeks to link local and regional processes (Roughgarden *et al.* 1988; Holt 1993). This type of approach has already been developed, both in population

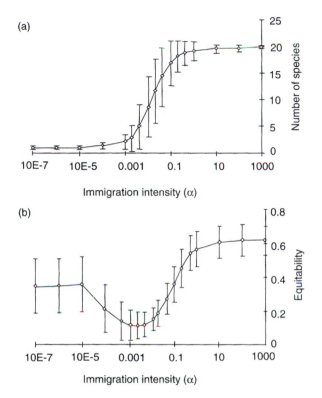

Figure 22.1. Variation of species richness and equitability (measured by Shannon's index: $E^i = -\sum P_i \ln P_i / \ln S$) at equilibrium, as a function of the immigration intensity α. Means and standard deviations were obtained from 1000 simulations. The potential species pool comprises 20 species. The extinction threshold is fixed at 0.001. For each simulation parameters were sampled randomly from a potential species pool, and these species were allowed to compete for space until equilibrium was reached. We selected parameters randomly between 0 and 1 in a uniform distribution.

ecology (Levin 1974; Iwasa and Roughgarden 1986; Kishimoto 1990; Holt 1997a) to explain the coexistence of competing species, and in population genetics (e.g. Levene 1953) to explain the maintenance of polymorphism.

We define a metacommunity as a regional set of communities linked by dispersal. As mentioned in chapter 1, defining dispersal is not a trivial challenge. A metacommunity model encompasses two levels of dispersal: long-distance (i.e. inter-community) and short-distance (i.e. intra-community) dispersal. In the rest of this section we use the word dispersal when referring to inter-community dispersal. Let P_{ij} be the proportion of sites occupied by species i in community j. There are S species that compete for a limited proportion of vacant sites, V_j, in N communities. Immigration of species i to community j, I_{ij}, is a function of emigration from other communities. For each community, there is a proportion of reproductive progeny that disperses, a, and a proportion of non-dispersers, $1 - a$. Because a is taken to be the same for all species, it can be interpreted as the relative importance of local, compared with regional, dynamics (depending on the size of and distance between communities). Following Levin (1974), we assume heterogeneity of environmental conditions at the regional scale by changing species-specific parameters in each community.

This model, which we call the metacommunity model, reads:

$$\frac{dP_{ij}}{dt} = f_{ij}(P_{1j}, P_{2j}, \ldots P_{sj}) = (I_{ij} + (1 - a)c_{ij}P_{ij})V_j - m_{ij}P_{ij}$$

$$V_j = 1 - \sum_{k=1}^{s} P_{kj} \quad \text{and} \quad I_{ij} = \frac{a}{N-1}\sum_{k \neq j}^{N} c_{ik}P_{ik} \tag{2}$$

This model is complex, and we present only general results here. A full analysis will be presented elsewhere (N. Mouquet and M. Loreau, unpublished manuscript). This model has allowed us to extend and generalize Iwasa and Roughgarden's (1986) results: at equilibrium, the number of species coexisting in each community cannot be greater than the number of different communities (habitats) at the regional scale. Each community acts as a source of immigrants for other communities in the region, provided that communities are different enough for different species to be competitively dominant there. This is a strict application of niche theory, because it is not possible to have more coexisting species than the number of limiting factors. However, we have shifted the scale of heterogeneity from the local to the regional scale, which may be more relevant to natural systems.

Further, to persist in a metacommunity, all species must have the same mean competitive ability at the regional (metacommunity) scale. We have called this condition the regional similarity rule. Coexistence is then possible even if species are *locally different*, because they are *regionally similar*. Let the basic reproductive rate of species i in community j be $r_{ij} = c_{ij}/m_{ij}$. As an extreme case, we can define regional competitive ability for a species i as the product of its local reproductive rates in all communities ($r_i = \prod_{j=1}^{N} r_{ij}$). Therefore, a sufficient condition for regional similarity is obtained when all values of r_i are equal. We have called this condition strict regional similarity.

We will now illustrate this result with the case of two species in two communities. Coexistence is then possible if $r_{11}r_{12} = r_{21}r_{22}$. What are the consequences of increasing the dispersal rate between the communities (i.e. increasing the relative importance of regional versus local dynamics)? For the case of strict similarity, we analysed the stability of the equilibrium: stable coexistence is possible only for a value of dispersal smaller than 0.5. We also studied the concept of regional similarity, for divergence from this case. Because constraints on parameters are very strong (strict parameters combination), strict regional similarity between coexisting species is biologically unrealistic. So, let us define δ as a function of divergence from the strict similarity case:

$$\delta = \frac{r_{11}r_{12}}{r_{21}r_{22}} \tag{3}$$

When $\delta = 1$, we find the condition of strict similarity as defined above. To diverge from that case we fixed all parameters except c_{11} (the local reproductive ability of species 1 in community 1), which varied from 0 to 2. As in the propagule rain model, we solved the equations numerically and included an extinction threshold. Figure 22.2 shows that, for low values of dispersal, coexistence is achieved even when species are not strictly similar. Therefore, regional similarity is defined as a product of species parameters,

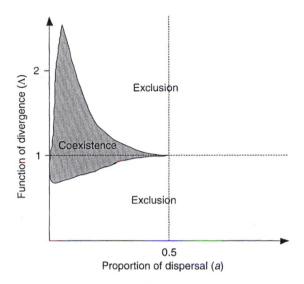

Figure 22.2. Zone of local coexistence (grey) as a function of divergence from the strict similarity case, δ (y-axis), and dispersal, a (x-axis). This result is obtained from a metacommunity of two species competing in two communities. Simulations were performed until equilibrium was reached. Parameters are $c_{12} = 0.5$, $c_{21} = 0.5$, $c_{22} = 0.8$, and $m_{11} = m_{12} = m_{21} = m_{22} = 0.3$. Divergence from the strict similarity case is obtained by varying parameter c_{11}. The extinction threshold is set at 0.01.

equilibrium values of coexisting species, and the level of dispersal between communities. Furthermore, we have shown that the range of parameters that allows coexistence is wider at intermediate levels of dispersal, but as yet we have not found an analytical expression for this range.

Species richness is only one aspect of species diversity, which we complemented by measuring equitability. We performed simulations in the simple case of strict similarity, as discussed above, for two species in two communities. When dispersal is zero, communities are closed and there is local exclusion by the most competitive species (i.e. the species with the highest local competitive ability, r_{ij}). Species richness is then equal to 1 and equitability is zero. When dispersal increases, local species richness becomes maximal (here, 2) and communities become more and more similar, until equitability reaches its maximum when dispersal is 0.5. It is possible to generalize this result to a meta-community of S species and N communities. When dispersal is low, the most competitive species dominates in each community, and others, maintained by dispersal from other communities, are rare and therefore under constant extinction risk; thence species diversity is low. As dispersal increases, more and more species are maintained above the extinction threshold by immigration, and thus species richness and equitability increase.

Community properties

We now explore some consequences of increasing the relative importance of regional versus local dynamics (increasing dispersal) on community properties such as total space occupation and primary productivity at equilibrium.

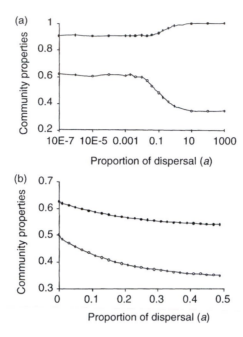

Figure 22.3. Community properties as a function of the proportion of dispersal, which measures the relative importance of local versus regional dynamics: space occupation (filled circles) and primary productivity (open circles). (a) Results from the propagule rain model (means from 1000 simulations), in which the immigration rate I is a decreasing function of the potential reproduction rate c ($I_i = 1 - c_i$). (b) Results from the meta-community model with two species competing in two communities, in the case of strict similarity. Simulations are performed until equilibrium is reached. Parameters are $c_{11} = 0.8$, $c_{12} = 0.5$, $c_{21} = 0.5$, $c_{22} = 0.8$, and $m_{11} = m_{12} = m_{21} = m_{22} = 0.3$.

In the propagule rain model the proportion of vacant sites in the community, on average, decreases as immigration intensity increases (Fig. 22.3a). Assuming that a species' productivity is correlated with both the number of sites it occupies and its potential reproduction rate in a site, we approximate total plant (primary) productivity by:

$$\Phi = \sum_i c_i P_i \tag{4}$$

The effect of immigration intensity on productivity depends on the relationship between the immigration (I) and potential reproduction (c) rates. For example, consider the classical case of a trade-off between competitive and dispersal abilities. In our model, competitive ability is determined by r_i, which is proportional to c_i. Thus, this trade-off can be represented by a negative relationship between I and c. In this case, productivity decreases as diversity increases (Fig. 22.3a). For a low immigration intensity (α), the species that persist at equilibrium have a high potential reproduction rate, c, hence productivity is high. As α increases, the immigration rate I becomes more important in determining dominance; because dominant species have a low potential reproductive rate (c), productivity declines. Productivity was found to be roughly constant when immigration and local reproduction are positively correlated, and slightly decreasing with increasing immigration intensity when there is no correlation between immigration and local reproduction.

In the metacommunity model, we have used an equivalent to equation (4) summed over all j to approximate local plant productivity in a metacommunity. Figure 22.3b shows that both productivity and space occupation decrease as dispersal increases. Let us calculate the equilibrium values of P_{ij} from equation (2):

$$\hat{P}_{ij} = \frac{\hat{I}_{ij}\hat{V}_j}{m_{ij}(1 - (1 - a)r_{ij}\hat{V}_j)} \tag{5}$$

When dispersal (a) is low, the dominant species in each community is the one with the highest local competitive ability r_{ij} (i.e. a high local reproduction parameter, c_{ij}). Thus, it can be understood from equation (5) that space occupation ($\sum \hat{P}_{ij}$) and productivity ($\sum c_{ij}\hat{P}_{ij}$) are high. As dispersal increases, species with low c_{ij} are maintained by immigration, occupying more sites, whereas dominant species occupy fewer sites, which results in decreasing productivity and space occupation. As discussed above, productivity and space occupation are then negatively correlated with diversity, because diversity is positively correlated with dispersal.

Evolution in competitive communities

In the real world, the properties of species evolve. In most ecological models, species characteristics are chosen arbitrarily, as if they were formed by some creation event. This would also lead to equally arbitrary communities, as other characteristics could result in entirely different communities. For instance, in classical metapopulation models for a competitive community (Hastings 1980; Tilman 1994), a major assumption is that a superior competitor immediately and completely replaces an inferior competitor. In order for two species to coexist, the inferior competitor needs to be a much better colonizer to make up for the patches lost to the superior competitor. Therefore, a good competitor casts a competitive shadow in trait space (Nowak and May 1994). If the best competitor happens to have a small population size, this shadow will be short and many species can coexist; if it happens to have a high population size, only a few species can coexist. The composition of the community thus depends strongly on the arbitrarily chosen properties of the dominant competitor.

Although the idea that the natural world has arbitrarily chosen characteristics has a respectable pedigree reaching back several millennia, here we let species characteristics be determined by evolution. We consider a system in which mutants occasionally appear that may replace the wild type if they have a selective advantage, so that a species' traits gradually evolve. Such a process of gradual evolution leads to an assemblage of species with a typical community structure in which evolution of a species depends on the characteristics of the other species in the community.

Under gradual evolution the assumption of immediate replacement in metapopulation models has important consequences for biodiversity and community structure, of competitive communities. Because a superior competitor will immediately replace marginally less-competitive types, a superior competitor grows as if it has the environment for itself. A marginally less competitive type will be in the competitive shadow cast by a superior competitor, and will disappear altogether. Through this mechanism the successive appearance (through mutations) of marginally better competitors will lead to an increase in the competitiveness of the superior competitor. A better competitor is a weaker colonizer (Tilman 1990, 1994); therefore the equilibrium densities decrease with increased competitiveness. The evolutionary process will stop only if the superior competitor has reached a very low density. All other competitors will be subject to a similar evolutionary process and will evolve towards maximal competitiveness and minimal population sizes. In these models biodiversity is therefore not maintained unless new, less competitive, species are continuously created through mutation.

Model formulation

Here, we describe a model in which we relax the assumption of immediate replacement. Our model is of the haystack type (Maynard Smith 1964; Cohen and Eshel 1976): a large number patches are simultaneously seeded with various numbers of individuals. We assume a random distribution of seeds according to a Poisson distribution. These local populations 'incubate' for some time, after which the local populations produce a new generation of individuals that are then redistributed. Although our model can easily be generalized to suit many biological scenarios, we will interpret the dispersing units as seeds and the incubation period as a growing season in which annual plants compete within a site. Our model has an explicit description of the within-patch dynamics that we solve by separation of timescales. This approximate solution enables us to analyse the evolutionarily stable assemblage (i.e. a community of species with evolutionarily stable growth rates). To keep the technical details to a minimum, we will present only the typical structure of our model and some results. The details will be presented elsewhere (Jansen and Mulder 1999).

We assume that the different types interact within a patch through competition for a single resource. This is described by a set of non-linear differential equations of the form (Hofbauer and Sigmund 1988):

$$\frac{dX}{dt} = X\left(z - \frac{X + X^*}{k}\right) \qquad \frac{dX^*}{dt} = X^*\left(z^* - \frac{X + X^*}{k}\right) \qquad (6)$$

The amount of biomass of each type is given by X and X^*. The traits z and z^* represent the growth rates of the respective types, and k the carrying capacity or quality of the local environment. We assume that the environment affects only the carrying capacity and not the growth rates. Because competition is for a single resource, no two different types can coexist within a patch indefinitely. The fraction of a type ($F = X^*/(X + X^*)$) changes over time as:

$$\frac{dF}{dt} = (z^* - z)F(1 - F)$$

in which the logistic equation can be recognized. The initial value of this fraction depends on the precise number of seeds of each type that this patch has received. In this way, the amount of biomass at the end of the season, which is the fraction times the total amount, can be approximated. Locally, species with a higher growth rate, z, will tend to replace species with a low z, so that the growth rate z is therefore a measure of competitive ability.

Note that if the types are very different ($z^* \gg z$ or $z^* \ll z$), one type is replaced quickly by the other and the immediate replacement scenario is recovered. However, if the two types are very similar, the fraction changes slowly, slower than the growth of the total amount of biomass, which we have assumed to be a fast process. The total amount of biomass therefore goes to a quasi-steady state, which is a linear combination of the biomass equilibria that the respective types would have had if they had had a patch completely to themselves. This procedure can be made mathematically precise for a large class of local dynamics, and allows us to approximate the amount of biomass of each type at the end of the season (Jansen and Mulder 1999).

At the end of the season all biomass is harvested so that all patches are empty again. Biomass is converted into seeds according to a constraint that relates the growth rate to fecundity. The trade-off is chosen such that a good competitor has a low fecundity. The particular function for the fecundity we use here is $\phi(z) = (z_{max} - z)/z$, where z_{max} is the maximum trait value at which no seeds are produced. Such a trade-off between growth rate and fecundity arises if plants can store part of their assimilates for later seed production. To avoid this function reaching infinite values, we also introduce a minimum value for z: $z > z_{min} > 0$. After all patches are emptied, the seeds are redistributed over the patches, according to a Poisson distribution, before the next season starts.

A description of the global dynamics of a rare mutant

To describe a globally rare mutant we need to know the dynamics across the seasons. Again, we first consider a single type, with a trait with value z. In all patches that have received at least one seed, the biomass at the end of the season will be $\tilde{X} = zk$ ($\tilde{X}$ is the equilibrium amount of biomass of type z in a patch where no z^* is present). The number of seeds produced in these patches is $\phi(z)\tilde{X}$. If the average seed number per patch is given by N, the fraction of patches with at least one seed is $1 - e^{-N}$; hence, in the next season the average number of seeds per patch is:

$$N' = \phi(z)\tilde{X}(1 - e^{-N}) \tag{7}$$

This defines the seasonal dynamics of the seed number that goes to a unique and stable equilibrium $\tilde{N}$ if $\phi(z)\tilde{X} = k(z_{max} - z) > 1$.

To find out which mutants are selected in an environment dominated by this type, we consider a rare mutant type with growth rate z^* and a background population of residents with trait z at equilibrium density $\tilde{N}$, and ask whether or not the average number of mutant seeds in a patch, N^*, increases. If so, the mutant can invade the resident population. The within-patch dynamics can be approximated as outlined above: the fraction of mutants in a patch, F, changes logistically. If the mutant is rare, the fraction of patches with two or more mutant seeds is negligible, while the probability of receiving a single seed is approximately linear in N^*. The dynamics of a rare mutant is therefore approximately:

$$N^* = W(z,z^*)N^* \tag{8}$$

The invasion rate of the mutant, $W(z,z^*)$, is a measure of the mutant's fitness (Metz *et al.* 1992). A mutant's fitness depends both on its own trait and on the trait of the resident. It can be derived from the population dynamics; because the mutant is rare and the resident is at equilibrium, this can be done through a linearization around $N = \tilde{N}$ and $N^* = 0$:

$$W(z,z^*) = \sum_{i=0}^{\infty} \phi(z^*)F_t[\tilde{X} + F_t(\tilde{X}^* - \tilde{X})]e^{-\tilde{N}}\tilde{N}i/i! \tag{9}$$

where i is the number of resident seeds in a patch, and $i!$ is the factorial of i. The number of seeds produced in a patch is $\phi(z)[\tilde{X} + F_t(\tilde{X}^* - \tilde{X})]$ where $\tilde{X}^* = z^*k$ is the amount of biomass a mutant produces in a patch without residents, and F_t is the fraction of mutants at the end of the season. A fraction, F_t, of the total seeds are mutant seeds. This fitness is a canonical form of a rare and similar mutant for a large class of haystack models with different local dynamics. The total amount of biomass (the term in square brackets) is determined both by the mutant and by the resident, and reflects how the types interact.

With this expression for fitness, the evolutionarily stable traits $\tilde{z}$ can be derived in the usual way by finding the fitness maximum for z^* (Maynard Smith 1982; Metz *et al.* 1992; Geritz *et al.* 1998). For every value of k, z_{min} is evolutionarily stable (this is due to the particular form of the fecundity trade-off). Figure 22.4a shows that there is a second evolutionarily stable trait (thick line) whose value depends on the quality of the environment, k. For very poor environments z_{min} is the only evolutionarily stable growth rate; in richer environments the second evolutionarily stable growth rate exists, which corresponds to a stronger competitor. Generally, an increase in environmental quality causes the evolutionarily stable type to be even more competitive. This is because in richer environments the population densities tend to be larger, patches receive more seeds, and competition within a patch becomes more important so that more competitive types evolve.

Similar types cannot invade populations with an evolutionarily stable trait. However, for types that are very different we found immediate replacement; therefore, types that are much weaker competitors and that fall outside the competitive shadow cast by the evolutionarily stable traits can invade and coexist. Whenever these two types have seeds in the same patch, the better competitor excludes the weaker competitor in this patch almost completely within a season. Weaker competitors, therefore, are dependent on those patches that the good competitor leaves unoccupied. The good competitor's dynamics remain virtually unchanged by the presence of the weaker competitors; therefore also its evolutionary stability properties remain unchanged, and much weaker competitors do not change the value of the best competitor's evolutionarily stable trait.

Weaker competitors perceive the presence of the good competitor only as a reduction in the number of available patches. In this smaller number of patches, the weaker competitors can evolve independently of the strong competitor. By applying a similar

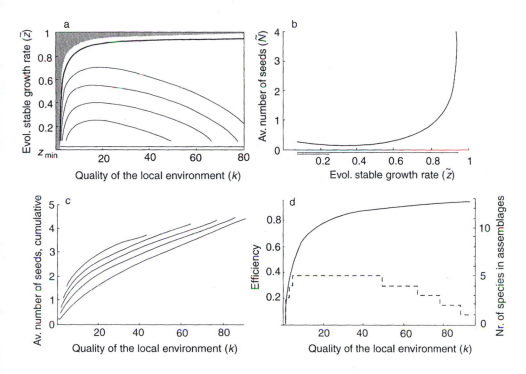

Figure 22.4. (a) The values of the evolutionarily stable growth rate as a function of the quality of the environment, k, in the evolutionary model. (b) The number of dispersers as a function of the evolutionarily stable growth rate. Note that this relation is increasing for large values of the growth rates but decreasing for very small values. The domains of the evolutionarily stable growth rates of the most (top bar) and the least (lower bar) competitive types are indicated with bars under the axis. (c) The cumulative number of dispersers as a function of the quality of the environment. The spacing between the different lines represents the number of dispersers that the next species adds to the total. (d) The efficiency, that is, the amount of biomass at the end of the season before seed conversion, divided by the quality of the environment k, as a function of k. The dashed line represents the number of different types in the evolutionarily stable assemblage.

argument as in the previous section, a third evolutionarily stable growth rate can be found. Because the third type is independent of the second type, apart from a reduction in the number of patches available, the evolutionarily stable growth rate of the third type is the same as that of the second type in a smaller number of patches. Because the fitness (equation 9) is linear in the quality of the environment, this reduced number of patches can also be expressed as a reduction in the quality of the environment. The evolutionarily stable growth rate of the third type is therefore the same as that of a superior competitor in a poorer environment. In this way the evolutionarily stable growth rates of successively weaker competitors, which live in the patches unoccupied by superior competitors, can be found; these types are represented in Fig. 22.4a by thinner lines under the thick curve. We can repeat this procedure until the number of available patches is so far reduced that no further types can be added. In this way the types that constitute the evolutionarily stable assemblage can be determined.

Properties of the evolutionarily stable assemblage

Figure 22.4a gives the trait values of all types in the stable assemblage. The assemblage has well defined types, which differ from other types by a limiting similarity and consist of an integer number of species. This is a feature that is caused by evolution: in our model any number of species can coexist, but not every community is evolutionarily stable. The number of species initially increases with increased environmental quality, but eventually decreases, causing a humped curve in the number of species (Fig. 22.4d, dashed line). An intuitive explanation for this fact is that the local production of seeds is higher in richer environments. For the best competitor, this will result in an increase in equilibrium population density. For the remaining species in the assemblage, there is a second effect of increased richness: because of the increase in the superior competitor's density, they have to survive in fewer patches. The combination of these opposing effects will result in an optimum number of species for an intermediate quality of the environment. Such relationships are well documented: the average number of species tends to have a humped curve when plotted against productivity (Rosenzweig 1995). Although our results resemble those from metapopulation models (May and Nowak 1994; Nowak and May 1994; Tilman 1994; Tilman *et al.* 1994) there is a fundamental difference, in the sense that in these models gradual evolution would lead to selection for even more competitive types (Nowak and May 1994; Lehman and Tilman 1998), which have even-smaller population densities.

All types in a stable assemblage have a locally evolutionarily stable growth rate. Although the types are different, their evolutionarily stable growth rate is formed through the same processes for all types. Therefore, a link between the number of dispersers and the growth rate emerges, through the evolutionary interaction (Fig. 22.4b). This link is the same for all types, but the qualitative relationship can appear to be very different for different types. For the most competitive type, the number of produced seeds mostly increases with the growth rate; for the least competitive type, it decreases.

Evolutionarily stable assemblages can be used to study the properties of competitive communities. The total number of dispersers in the community increases sharply with the quality of the environment; for better environments this increase is less steep (Fig. 22.4c). In Fig. 22.4d we relate the diversity to ecosystem functioning: ecosystem efficiency is determined as the total amount of biomass at the end of the season relative to the quality of the environment, k. This measure can be also linked to ecosystem productivity. We found an increase in ecosystem efficiency with increasing environmental quality.

Discussion

Our models show two important ways in which dispersal can mitigate competition for space as a single common limiting resource, and hence maintain local species coexistence.

In our metacommunity model, this is obtained through habitat heterogeneity at the regional scale: different sites support different dominant species, which creates source–sink dynamics at the metacommunity scale and maintains local diversity in all sites. Numerous field studies have shown an influence of regional processes on local species richness. Keddy (1981, 1982), studying plants on a sand dune, showed that the density of a species on the poor habitat quality side was a function of its density on the rich side. Field studies by Shmida and collaborators (Shmida and Ellner 1984; Shmida and Wilson 1985; Auerbach and Shmida 1987; Kadmon and Shmida 1990) provided similar results at the community level. Seed addition experiments, too, have stressed the influence of immigration on local diversity (Houle and Phillips 1989; Tilman 1997). This concerns especially the maintenance of rare species that correspond, in the source–sink models, to intermediate intensity of immigration or low dispersal between communities. These results are related to the 'mass effect' theory (Shmida and Wilson 1985) and to the source–sink hypothesis (Pulliam 1988), since all species but one would go extinct if the local community was closed (although the latter result depends on the assumption of a constant environment; Jansen and Yoshimura 1998).

In our evolutionary model, coexistence is obtained through a difference in the strategies by which space is exploited: some species go for local competitive advantage whereas others go for spatial expansion. Several earlier studies have reported the same result (Hastings 1980; Tilman 1994), but the results of those models are dominated by the properties that the dominant competitor happens to have. Our result is novel in that an assemblage of species is the outcome of evolution. In the assemblage, competition works very much as is assumed in the models of Hastings (1980) and Tilman (1994), in that superior competitors quickly replace inferior competitors. However, this same assumption will lead to spurious results if it is applied to evolutionary processes in which competition between similar strains plays a predominant role.

Our evolutionary model shows another consequence of dispersal: the more dispersers there are, the larger the average number of seeds per patch and the larger the chance of meeting competitors in a patch. Therefore, under conditions that give rise to more dispersers, more competitive types evolve. Dispersal from a patch is often considered a form of altruism, because it reduces competition between kin (Hamilton and May 1977). Here, more dispersers will increase immigration, which reduces relatedness, so that dispersal will cause less altruistic types to evolve.

In the two types of model, the community properties that emerge lead to some counter-intuitive patterns in the relation between species diversity and ecosystem processes. In the propagule rain and metacommunity models, the increased species diversity is accompanied by a decreased average local competitive ability, which generates a negative relationship between diversity and primary productivity or total space occupation (in the metacommunity model), as predicted by Loreau (1998a). In the evolutionary model, a positive relation between efficiency and species diversity emerges for poor environments. But, in richer environments, diversity decreases through the dominance of the most competitive species, and a negative correlation with efficiency appears. In this case, variations in diversity and variations in productivity are both determined by variations in the quality of the environment among different sites. This provides another example of across-site comparisons that lead to counter-intuitive patterns, as predicted by Loreau (1998a). Both types of model stress the role of dominant species on community properties. Species diversity increases when the competitive ability of the dominant species decreases, either because of increased dispersal in sink areas or because of lower quality of the environment. Thus, counter-intuitively, the magnitude of ecosystem processes (such as primary productivity, efficiency, or space occupation) tends to decrease as diversity increases.

This emphasizes the critical importance of dispersal between habitats in understanding the structure and functioning of communities and ecosystems. Local communities and ecosystems are shaped not only by local ecological factors, but also by exchanges with neighbouring or more distant systems, and by the evolutionary history of these spatial interactions. Dispersal links local and regional processes, and thus contributes to regulating the composition, diversity, and functional properties of local competitive communities. While the effect of regional processes on local communities has been increasingly recognized recently (see for example chapters in Ricklefs and Schluter 1993), the reciprocal effect of local processes on regional diversity has received much less attention. A comprehensive multiscale theory of species diversity and ecosystem properties incorporating this feedback still waits to be developed. We hope the models presented in this chapter will contribute to this goal.

Acknowledgements

The authors thank A. Gonzalez and three anonymous reviewers for comments on the manuscript. V.A.A.J. gratefully acknowledges the support of the Wellcome Trust.

Part V

Perspectives

23

The evolutionary consequences of gene flow and local adaptation: future approaches

Nick H. Barton

Abstract

Almost all natural populations are subdivided and experience spatially heterogeneous selection. This article reviews the ways in which such structuring of populations may affect evolution, and then discusses the various means by which we can find the actual role of population structure and gene flow in evolution. Divergence may occur through random sampling drift. The drift of strictly neutral alleles is of no evolutionary importance in itself, although it does provide a valuable tool for measuring gene flow. Random drift and population subdivision interfere with directional selection; however, this delays the spread of adaptations only slightly. Drift plays a positive role in Wright's 'shifting balance' model of evolution, by allowing local populations to escape from suboptimal 'adaptive peaks'. Spatially heterogeneous selection is a much more powerful force for divergence, and can be effective within a spatially extended population. However, gene flow may interact with population dynamics to impede local adaptation, and may limit the species' range. How can we set about assessing the effects of subdivision and environmental heterogeneity? Most genetic studies of population structure have followed a research programme established by Wright and Dobzhansky: to infer the number of migrants between subpopulations (Nm) from the balance between gene flow and random drift. However, this approach suffers from several weaknesses; most important, it does not measure the rate of gene flow, m, which determines the ability of local populations to respond to selection. Genealogies inferred from DNA sequence variation are in principle more informative; however, many of the same limitations apply. Direct measurements of traits that influence fitness, and of fitness itself, can give information more relevant to divergence and adaptation, and also relate more closely to understanding the optimal patterns of dispersal and habitat choice. Do genotypes do better in their own habitat or in their native site? Are crosses between different populations more or less fit? Such observations, while laborious and as yet sparse, tell us about geographical divergence in genes under selection, as opposed to neutral markers. Such work is difficult, but essential for an understanding of the adaptive role of dispersal.

Keywords: dispersal, local adaptation, gene flow, genealogies, isolation by distance, clines, genetic drift, F_{ST}, linkage disequilibria

Introduction

Almost all natural populations are structured, such that individuals are most likely to mate with their neighbours; furthermore, they experience selection pressures that vary from place to place. To the extent that individuals, and the genes they carry, are distributed over a spatially extended habitat, and experience selection pressures that vary in space, gene flow is an important evolutionary process.

It is a measure of our current ignorance that we do not know to what extent a drastic increase in dispersal and gene flow would alter evolution. An extreme but defensible view is that species can for the most part be treated as if they were single panmictic populations. Within the 'classical' view (Provine 1971; Lewontin 1974), most genetic variation is either neutral or due to deleterious mutations. The fate of alleles that are destined to be lost is then little affected by population subdivision, provided that they nowhere increase fitness. Those few neutral alleles that become common do so slowly; therefore, they may diffuse across the entire species' range long before they reach fixation, and so appear spatially homogeneous. Adaptation and divergence occur through occasional favourable mutations. The chance that these will fix is largely independent of population structure, provided that they nowhere decrease fitness (Maruyama 1970). Even if genetic variation is maintained by selection, and even if selection varies from place to place, it is still not obvious that gene flow need alter the long-term pattern of evolution. This approximation of species as large panmictic populations is close to Fisher's (1930) view, although Fisher (1930, 1950) did model clinal speciation in a spatially extended habitat.

In this article, I will first set out the various ways in which population structure, and hence gene flow, might affect evolution. I will then discuss how we can best gather evidence as to their actual role in evolution. While the theoretical possibilities are for the most part well established, we still have only a rudimentary understanding of the actual evolutionary consequences of population structure. The recent availability of multiple genetic markers, and a convergence between the fields of evolution and ecology, opens up a variety of novel approaches to resolving this key issue.

Theoretical possibilities

Divergence through random drift

The consequences of population structure depend on the causes of divergence. First, consider random sampling drift. The classical works of Wright (1931, 1943) and Malécot (1948) established the basic results for the 'island model' (a set of populations, each exchanging migrants with a common pool), and for populations spread over one and two dimensions. Provided that mutation, weak balancing selection, or long-range migration maintains polymorphism, the variance in allele frequency across populations approaches an equilibrium that is proportional to the product of allele frequencies: $\text{var}(p) = \overline{pq}F_{ST}$ (Malécot 1948). In the island model, the standardized coefficient $F_{ST} = 1/(1 + 4Nm)$, where N is the local deme size and m the proportion of immigrants in each generation. This is independent of the strength of mutation, μ (or analogous stabilizing force), provided that $m \gg \mu$. In a continuous one-dimensional habitat, the relation is qualitatively different, and does depend on the mutation rate: $F_{ST} = 1/(1 + 4\rho\sigma\sqrt{\mu})$, where ρ is the population density and σ the standard deviation of distance between parent and offspring. In two dimensions, however, results are similar to the island model, and depend only weakly on mutation rate: $F_{ST} = 1/(1 + 4\pi\rho\sigma^2/\log[1/(2\mu)])$ (Malécot 1948). (For a stepping-stone model, where a proportion $m/2$ of migrants is exchanged with each neighbouring deme, density ρ corresponds to N, and dispersal rate σ^2 to m; Malécot 1948; Kimura and

Weiss 1964.) The key result is that, in a balance between gene flow and random drift, the variance in allele frequency between populations depends in the island model on the number of migrants exchanged per generation (Nm) or, in two dimensions, on the number of individuals within a generation's dispersal distance – (Wright's (1943) 'neighbourhood size', $Nb = 4\pi\rho\sigma^2$). This equilibrium is approached rapidly, over approximately $1/m$ generations; the theory is robust, provided one observes over local scales, and provided geographical differentiation is in fact due to random drift. (See Whitlock, chapter 19, for metapopulation models that include fluctuating population structure.)

Patterns of neutral divergence are a useful tool for investigating population structure (see below). However, what is important for evolution is the effect of random drift on selection in subdivided populations. With simple directional or balancing selection, random drift interferes with natural selection. In a large population with a constant and symmetrical pattern of gene flow, such that dispersal does not create or destroy alleles, the probability of fixation of an allele with advantage s is $2s$, regardless of population subdivision (Maruyama 1970). This is because, under these assumptions, the number of offspring of a gene does not depend on location. With fluctuating population density, fixation probabilities are reduced. For example, in the island model a rate of extinction $\lambda \gg m$ reduces the fixation probability considerably, to $2sm/(\lambda[(1 + 2N\lambda)])$ (Barton 1993). Once it is established, the rate of spread of an allele through an island population is also slowed down, by a factor $(1 - F_{ST})$; this is because drift reduces the genetic variance within demes by that factor. Similarly, the load due to additive deleterious mutations is increased by a factor $1/(1 - F_{ST})$. In an abundant species, drift can overwhelm only extremely weak fitness differences ($N_e s[1 - F_{ST}] < 10$, according to Barton and Whitlock 1996). (Here, F_{ST} refers to the locus in question. However, it is likely to be close to that observed for neutral markers; Barton and Whitlock 1996.) Thus, random drift in subdivided populations is unlikely to have a significant effect on evolution under these straightforward forms of selection; consequently, gene flow cannot influence evolution through the spatial differentiation generated by drift.

Drift can play a positive role if populations can evolve towards alternative stable states; then, selection alone may keep them at suboptimal 'adaptive peaks', and random fluctuations allow them to shift to new and fitter peaks. This is the basis of Wright's (1931) 'shifting balance' theory of evolution, which has motivated much of the work on structured populations over the last half-century. However, it is hard to find evidence that such random peak shifts have contributed to either speciation or adaptation (Coyne *et al.* 1997). The most important theoretical obstacles to adaptation via the 'shifting balance' are, first, that chance factors such as local extinctions and range expansions are likely to fix arbitrary adaptive peaks, and, second, that separate adaptations established in different populations cannot easily be brought together. It is more plausible that random shifts between adaptive peaks have contributed to reproductive isolation. However, it is hard to test this possibility against alternatives, for example that reproductive isolation is a side-effect of divergence driven entirely by selection.

Spatial variation in selection

Selection is likely to be a much stronger force than drift within subdivided populations. Indeed, it must be so if the species is to remain adapted to existing conditions, and to respond to changing conditions. Even if a species is subdivided into demes so small that selection is weaker than drift within each ($Ns \ll 1$), the theory outlined in the previous section shows that the aggregate effect of selection across all the demes prevails provided that $N_{tot}s \gg 1$, and Nm or Nb is not too small (Barton and Whitlock 1996). We therefore turn to consider deterministic models, in which selection acts in different directions in different places.

There has long been controversy over whether geographical isolation is required for the divergence of lineages in a heterogeneous environment. Darwin and Wallace held that populations could adapt to gradually varying conditions across their range and that this could lead to speciation. In contrast, Wagner emphasized the importance of barriers to migration in allowing speciation (see Mayr 1982, p. 562). More recently, Ehrlich and Raven (1969) argued that, because genes diffuse slowly through most species, gene flow cannot be a significant factor impeding divergence. Population genetic theory strongly supports this view, at least if one considers adaptations that can be built up by individual alleles or by continuous changes in additive quantitative traits.

Whether populations can adapt to local conditions depends essentially on the relative rates of gene flow and selection. An allele with selective advantage s in the heterozygote can be established in a single population despite gene flow, provided that the rate of immigration of individuals carrying the alternative allele is lower than the rate of selection ($m < s$; Haldane 1931). This generalizes to an island model in which selection varies from deme to deme (Barton and Whitlock 1996, Fig. 4). If genes diffuse across a continuous habitat, then an allele can become established, provided that it is favoured in a region larger than a critical distance set by the characteristic scale $l = \sigma/\sqrt{2s}$ (Slatkin 1973; Nagylaki 1975). New mutations have a probability of fixation approaching $2s$ if they arise within the appropriate region, but the probability falls to zero outside this region (Barton 1987). A similar condition applies to adaptations based on quantitative traits; here, the characteristic scale is $\sigma\sqrt{V_s/2V_g}$, where V_s is a measure of the strength of stabilizing selection and V_g is the additive genetic variance (Slatkin 1978b). Since the dispersal range σ is typically much smaller than the species' range, weak selection can allow a species to adapt to diverse local conditions. This can be seen directly. For example, the grass *Agrostis tenuis* has evolved tolerance to heavy metals on mines only a few metres wide (MacNair 1987). Similarly, narrow clines approximately 10 km wide separate races of *Heliconius* butterflies adapted to different mimicry rings (Mallet 1993). Such examples leave little doubt that gene flow need not prevent divergence over very short scales.

Almost all of the population genetic theory on gene flow assumes random dispersal. However, if the dispersal rate is regarded as a trait that has itself evolved under both direct and indirect selection (Ronce *et al.*, chapter 24), it is apparent that individual dispersal should come to depend on an individual's state (age, condition, etc.): that is, it should evolve plasticity (Murren *et al.*, chapter 18; Ims and Hjermann, chapter 14). Moreover, since the state of an individual depends on its genotype, one expects the rate of dispersal also to evolve some dependence on that genotype (Roff and Fairbairn, chapter 13). Even if dispersal evolves as a side-effect of direct selection to increase immediate fitness (e.g. through choice of breeding site or optimal foraging), it may still come to depend on phenotype and genotype. Most theoretical treatments have concentrated on the consequences of within-deme habitat preference for genetic variation (e.g. Maynard Smith and Hoekstra 1980), and on sympatric speciation through habitat specialization (e.g. Kawecki 1997; Dieckmann and Doebeli 1999; Kondrashov and Kondrashov 1999). The theoretical framework for modelling genotype-dependent dispersal through a spatially extended population has been developed by Nagylaki and Moody (1980), but there has been relatively little examination of the biological consequences. However, theory alone has little to say about just how finely adapted an individual's dispersal behaviour should be: that depends on constraints on dispersal, and on the costs of an elaborate state-dependent behaviour.

Reproductive isolation may accumulate as a result of spatially varying selection in several ways. It may arise as a side-effect of habitat preference, a possibility emphasized by Bush (1994) as the most plausible mechanism of sympatric speciation. If populations diverge across a cline, then further differences may accumulate if alleles are commonly favourable in one genetic background but not in the other (Clarke 1966). Once a set of clines is subject to strong overall selection, the consequent barrier to gene flow may promote divergence. Selection may favour a reinforcement of assortative mating and habitat specialization within the cline, strengthening reproductive isolation and ecological specialization (Wallace 1889; Fisher 1930; Dobzhansky 1940). Similarly, selection favours a reduced dispersal rate under circumstances where gene flow is acting against natural selection. These various mechanisms of speciation in sympatry or parapatry depend on gene flow in different ways. Gene flow is a strong force countering habitat specialization within single populations. It also tends to swamp changes that are favoured within narrow clines, but are disadvantageous elsewhere (Sanderson 1989). However, the rate of divergence due to accumulation of alleles from either side depends to what extent selection is conditional on genetic background or environment, and is independent of the rate of gene flow.

Population dynamics

Since fitness determines the rate of population growth as well as the rate of gene frequency change, population dynamics must interact with natural selection (Hanski, chapter 20). There has been much recent interest in this interface between evolution and ecology, stimulated in particular by conservation biology. On the one hand, dispersal is crucial to the dynamics of spatially extended populations. On the other hand, a dense extended population subject to uniform conditions could remain at uniform density, in which case dispersal would have no net effect (although a large dispersal rate could still be selected as a result of competition between kin; Hamilton and May 1977). However, demographic or environmental stochasticity, or chaotic instability could all generate density variation, which would then be ameliorated by dispersal. In principle, dispersal itself could cause unstable population dynamics, by analogy with the Turing model (Rohani *et al.* 1997). Dispersal has a particularly strong effect on population dynamics when it is required for colonization of empty patches in a metapopulation (Hanski, chapter 20) or for advance into empty habitat (Mollison 1977; Shaw 1995). These kinds of dynamic population structure have a strong effect in reducing neutral genetic variability: temporal variation in population density implies correlated variation in fitness of all the constituent individuals, which amplifies random genetic drift (Whitlock and Barton 1997; Whitlock, chapter 19).

If relative as well as absolute fitnesses vary substantially from place to place, then there may be a strong interaction between population dynamics and dispersal. Within a metapopulation, new colonies formed by random dispersal will tend to be maladapted, and will rely on new mutation or subsequent immigration to recover local adaptations. Transient maladaptation reduces the viability of young populations, and the mean fitness of the whole population (Endler 1979; Holt and Gomulkiewicz 1997). Similar effects apply in a spatial continuum. Suppose that some quantitative trait is under stabilizing selection, towards an optimum that changes steadily across the species' range. There is always an equilibrium in which the mean tracks the optimum perfectly. However, if the gradient in the environment is too steep, this becomes unstable. The population is then adapted to conditions in a limited region, and collapses to low density outside this region. Because gene flow is predominantly from dense to less dense regions, this collapse is irreversible: adaptation at the edge of the species' range is swamped by gene flow from the centre (Kirkpatrick and Barton 1997).

Future approaches

How can we find the importance of dispersal and gene flow in evolution? Much of empirical evolutionary biology has concentrated on questions relating to population structure. This is partly because of their potential importance (especially in relation to Wright's (1931) 'shifting balance' and the geography of speciation), but also because there are striking spatial patterns to be explained, and because variation in space can be used as a surrogate for unobservably slow changes through time. In this section, I will argue that much of this research has been misdirected, and outline ways of addressing more directly the key evolutionary questions.

Measuring the balance between drift and gene flow: the classical research programme

Wright *et al.* (1942) applied Wright's theory of F coefficients to describe variations in the frequency of lethal alleles in natural populations of *Drosophila pseudoobscura*, and to infer the numbers of migrants (Nm), on the assumption that these variations are due to random drift. This paper spawned a major research programme, stimulating both theory and many empirical surveys of F_{ST} (see Provine, in Lewontin *et al.* 1981). Geographical patterns in morphology, allozymes, and DNA-based markers have been interpreted in terms of F_{ST}, and there has been an extensive development of alternative statistical estimators (see Rousset, chapter 2). There has been no comprehensive review of F_{ST} in nature, but a rough impression would be that such values are typically around 0.05–0.1, but can be as high as 0.7 in species such as salamanders that have limited dispersal (see Slatkin 1981; McCauley and Eanes 1987). Within the island model, this implies a typical Nm of 2–5, but occasionally much lower.

The theoretical analysis given above, however, shows that many of these attempts may be misguided (see Rousset, chapter 2; Whitlock and McCauley 1998). First, inference from the variance in allele frequency is statistically unreliable: as well as the sampling error in estimating the actual allele frequencies in the population, there is considerable random variation in the evolutionary history of different loci, and there may be great variation in the value of Nm from population to population. A large number of neutral loci is therefore needed to give reliable estimates of Nm, and a large number of samples to give its distribution across populations. Second, inferences assume an equilibrium between random drift and gene flow. Such an equilibrium is reached rapidly over small regions, and is robust to the action of uniform selection and mutation. However, over longer scales in space and time, equilibrium is unlikely: the pattern of allele frequencies then reflects a unique population history, rather than a steady-state balance between two processes. Third, inferences are sensitive to the model of migration. Usually, the island model is assumed. If, in fact, the population is two-dimensional, then inferences on Nm (or $\rho\sigma^2$) hold approximately. However, the theory breaks down entirely in one dimension, or with more complex structures (for example, a metapopulation or a branching river system; see Whitlock, chapter 19). The method also cannot be applied if spatial variation is due to historical population movements or to selection, rather than drift. It is disturbing that there are hardly any cases where indirect estimates of Nm have been checked against direct measures (see Rousset, chapter 2; Ferrière and Le Galliard, chapter 5).

The most serious criticism of inference from F_{ST}, however, is that it does not measure gene flow, but rather measures the balance between gene flow and random drift. Differences in F_{ST} might be due to differences in the rate of drift ($1/2N$ or $1/\rho$) rather than in the rate of gene flow (m or σ^2). In principle, one can estimate the effective population density directly, and divide through to find gene flow. However, it is hard to estimate the effective numbers from the actual numbers, especially if genetic diversity is influenced by selection on linked loci (Santiago and Caballero 1998) or correlations in fitness across generations (Whitlock and Barton 1997). If one is interested in the relative strengths of selection and drift, then it is m/s or σ^2/s that matters: a population might have high levels of gene flow relative to drift (Nm or $Nb \gg 1$), and yet selection might nevertheless dominate over gene flow. There are questions for which Nm is the key parameter. Wright's motivation was to find whether Nm is typically small enough to allow frequent shifts between adaptive peaks through random drift, which does require that $Nm < 1$. Even here, however, there are problems: Nm need be small only within part of a species' range to drive the 'shifting balance', and randomly changing selection is more likely than drift to cause peak shifts (Coyne *et al.* 1997; Whitlock 1997).

Almost all analyses of population structure infer the number of migrants (Nm) or neighbourhood size (Nb) from F_{ST}, or from some equivalent measure of variation in allele frequency (Slatkin 1985b; Barton and Slatkin 1986). Is it possible to go beyond this classical method, to make inferences about other aspects of past and present gene flow? In two dimensions, the spatial pattern of allele frequencies contains additional information. If (as is usually the case) F_{ST} is small, allele frequencies will fluctuate rather little, and the whole distribution can be approximated by a multivariate Gaussian, which is defined by its mean and covariance. This covariance depends on two scales: the scale K which describes local population structure, and the scale $l = \sigma/\sqrt{2s}$, which describes the balance between mutation and gene flow. The local scale K depends on deme spacing in a discrete model (Malécot 1948; Kimura and Weiss 1964) and on the dispersal range σ in a continuum (Malécot 1948). If the mutation rate is known, or if the analysis is concentrated over scales much smaller than l, then this local scale can be estimated (see Barton and Wilson 1995). This approach was first used by Sokal and Wartenberg (1983), and has been explored more recently by Epperson and Allard (1989) and Epperson (1993b).

If all demes in a population are sampled, then one can, in principle, estimate the full matrix of migration rates between demes (Smith 1969). Slatkin (1981) has suggested that this might be done using Nm estimated from the variance in allele frequency between pairs of adjacent populations. However, an extremely large number of loci is needed to give enough statistical power to estimate each element of the migration matrix, and, moreover, correlations between pairs of loci depend strongly on the surrounding network in which the pair is embedded. If one has a model that predicts the migration matrix from the pattern of dispersal, then it is better to estimate parameters of that model, rather than the migration matrix itself (Hanski, chapter 20; Tufto *et al.* 1996). Similarly, if the appropriate model is one of admixture of divergent populations, then historical patterns of dispersal can be estimated from allele frequency. This approach has been particularly successful for European humans, for whom many genetic markers are available (e.g. Sokal *et al.* 1990; Cavalli-Sforza *et al.* 1994).

Inference from linkage disequilibria

When genetically distinct populations mix, gene flow generates linkage disequilibria (Li and Nei 1974). These rapidly equilibrate to a balance between gene flow and recombination, and so the level of linkage disequilibrium gives a robust method for estimating rates of gene flow (Barton and Gale 1993). It is important to note that such estimates are of the actual rate of gene flow (m or σ), rather than of the balance between drift and gene flow (Nm or Nb); the former is often of more interest, for example, in finding the strength of selection needed to maintain genetic divergence. This method can be applied in essentially the same way to nuclear genetic markers (e.g. Barton 1982; Szymura and Barton 1991; Lenormand *et al.* 1999), cytonuclear associations (Asmussen *et al.* 1987), and morphological traits (Nürnberger *et al.* 1995; Kruuk 1997).

The availability of highly polymorphic microsatellite markers makes it possible to identify from which population an individual originates, even when those populations differ only moderately in allele frequency (e.g. Bowcock *et al.* 1994). This seems to offer a straightforward and direct method for estimating the current rate of immigration, and its source (Waser and Strobeck 1998). It is inherently more efficient than the allele frequency-based methods discussed above, because it incorporates additional information from linkage disequilibria. This approach is most straightforward when there is introgression due to a low rate of hybridization with a genetically distinct population: an nth generation backcross hybrid then carries a proportion $2^{-(n+1)}$ of its genes from the donor population. While it is not feasible to ascertain to which backcross generation an individual belongs (Boecklen and Howard 1997), both the rate of introgression and any selection against introgression can be estimated (e.g. Goodman *et al.* 1999). Although this method has been applied to genetically similar populations (Peacock and Ray, chapter 4; Dias *et al.* 1996), it is most effective for recent and rare hybridizations. This is because, first, the number of backcrossed individuals increases exponentially, weakening statistical power, and, second, as introgression continues, the allele frequency differences on which the method depends dissipate (see Goodman *et al.* 1999). The approach might be made much more powerful, however, if linked markers were included, since associations with these would dissipate over a longer timescale (see Baird 1995).

Inference from genealogies

DNA sequencing now presents us with data that are more naturally represented by genealogies than by allele frequencies. Every sequence may be unique, in which case allele frequencies tell us nothing; all the information is contained in the genealogical relationship between sequences. If recombination is rare enough relative to mutation, the genealogy can be seen more or less directly, as, for example, in bacteria where recombination occurs via occasional transformation (Maynard Smith 1990) or in viruses, where mutations are frequent (Sharp 1995). Even if we cannot be sure of the relationships, the theory may still best be described in terms of the underlying genealogies. For example, if we can calculate the likelihood of parameters such as the population size or the rate of gene flow for a particular genealogy, the overall likelihood can be found as a weighted sum over the set of plausible genealogies (Felsenstein 1992; Kuhner *et al.* 1995). There has been much recent attention, both theoretical and empirical, to the use of genealogies to investigate gene flow and population history. Nevertheless, it remains unclear how best to use such data. How much more valuable is it to use information from the full set of genealogies, rather than from allele frequencies or linkage disequilibria? What is the best sampling scheme? Is it better to have small genealogies across many loci, or genealogies for a few loci, but each containing many individuals?

A simple approximation of the distribution of genealogies in two dimensions was introduced by Wright (1943). As one looks backwards, the pool of potential ancestors diffuses out over an ever-widening region, with area proportional to time. Wright suggested that one could take the probability that two lineages coalesce in a single common ancestor to be $1/(2Nbt)$, and so derived identity coefficients under 'isolation by distance' (Fig. 23.1). This approximation is equivalent to Malécot's (1948) model in which genes reproduce and diffuse independently of one another. Both models suffer from a fundamental inconsistency, in that if the population density is to remain even, some kind of local interaction must reduce the fitness of individuals in dense regions (Felsenstein 1975). There is a further difficulty, in that the diffusion approximation of gene flow that is so successful in deterministic models breaks down over small scales in two or more dimensions (Nagylaki 1989). These theoretical problems have inhibited the development of 'isolation by distance' models, and their application to genealogies. However, analysis and simulations show that the Wright/Malécot approximation is accurate over all but small scales (Barton and Wilson 1995), and so offers a framework for estimating gene flow and population density from spatially structured samples of genealogies (Epperson 1993a; Barton and Wilson 1995).

Neigel and Avise (1993) propose a simple method for estimating gene flow: to plot the squared distance between genes against the time since they diverged, for all pairs in the sample, and to use the slope of this relationship to estimate $4\sigma^2$; divergence time is estimated by assuming some kind of molecular clock. There are several problems. First, Neigel and Avise's simulations are based on a neighbourhood size much smaller than is typical of natural populations. Simulations of the coalescent process for $Nb = 6.28$ show that, although $|x - x'|^2$ does initially increase approximately linearly with time, the slope of the relation is much smaller than $4\sigma^2$, and even nearby genes are separated by a long coalescence time (Barton and Wilson 1995). Second, the distribution of $|x - x'|^2$ is determined by the sampling geometry, and so $|x - x'|^2$ should be treated as the

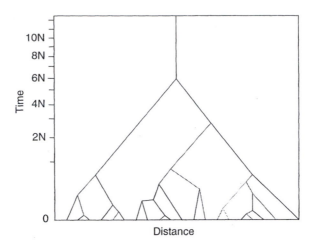

Figure 23.1. A typical genealogy for one locus in one panmictic population. Initially, there are many pairs of lineages, and so coalescence events are very frequent; as lineages coalesce, further coalescence becomes rarer. For this example, it takes 2.77N generations for 20 lineages to coalesce into two, and then a further 3.14N generations for these two to meet in the common ancestor. Note that time is plotted on a square-root scale: on a linear scale, early coalescences are indistinguishable.

independent variable; the coefficient of regression of $|\underline{x} - \underline{x}'|^2$ against time will depend on the distribution of distances sampled, whereas the converse regression of time against $|\underline{x} - \underline{x}'|^2$ would not. Third, because the distribution of coalescence times decreases with $1/t$ in two-dimensional populations, the expected coalescence time at equilibrium is infinite, for any $|\underline{x} - \underline{x}'|$. In any actual sample, the mean must be finite; however, if the distribution has infinite moments, the results will be very variable.

These statistical problems can all be dealt with, at the cost of additional complexity. However, a fundamental problem is that in two dimensions, and with a large neighbourhood size, only a small fraction of nearby lineages coalesce within a few generations, whilst most trace back to common ancestors that are so ancient that they carry no geographical information (Fig. 23.2). It may therefore be more efficient to use allele frequency data from many genetic loci than to attempt to reconstruct genealogies. Of course, if one's concern is with the ancient history rather than with current gene flow, then genealogies are appropriate. Nevertheless, the wide sampling variability across loci, due to both chance and the potential for selection of particular genes, makes reliance on one locus unwise. The fact that there are unresolved arguments over what can be inferred from human mitochondrial DNA illustrates this point (Templeton 1992; Brookfield 1997; Hey 1997).

Genealogical methods may be particularly powerful when applied to multiple loci and when gene flow is between highly divergent groups. One of the clearest examples is of the transfer of resistance genes between bacterial 'species' (Maynard Smith 1990). Within predominantly sexual organisms, gene flow is usually between genetically similar species or populations, and so its consequences are less easy to detect. However, differential introgression is frequently observed for mitochondrial DNA across animal hybrid zones (Harrison 1989); it is not clear whether this is due to generally weaker linkage of

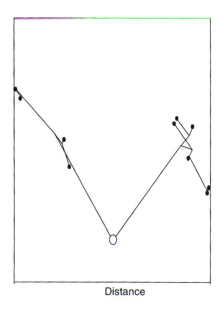

Distance

Figure 23.2. A genealogy connecting 10 genes in two dimensions, plotting locations in space, and drawing lines to indicate the relationships. Closed circles indicate genes, and the open circle the common ancestor. For simplicity, these trees are drawn assuming random reproduction, with no local density regulation.

mitochondrial DNA to the rest of the genome, or to selection of favourable mitochondrial variants. To take another example, Wang *et al.* (1997) have demonstrated significantly different levels of gene flow between three loci in the *Drosophila pseudoobscura* group: the genealogy estimated for *Hsp82* shows distinct clusters, corresponding to the morphological species, whereas *Adh* shows substantial gene flow. This suggests that there is a trickle of gene flow between the taxa; this is consistent with Dobzhansky's (1973) direct observation of a low rate of hybridization. However, it is not clear whether the variation across loci is due to chance, to selection maintaining divergence at or near *Hsp82*, or to a selective sweep across species at *Adh*. In general, there is a pressing need for incorporation of spatial structure into analyses of the effects of selection on sequence variability (e.g. Berry and Kreitman 1993; Aquadro *et al.* 1995; Charlesworth *et al.* 1997).

Evidence from traits that affect fitness

Most of the theoretical and empirical literature on population structure centres on variation that is neutral (or assumed to be so). That is appropriate if the aim is to make inferences about the rates of gene flow and the kind of population subdivision. However, if one wishes to make a direct judgement as to the effects of population structure on speciation and adaptation, then what matters are genes and traits that affect fitness. Geographical variation provides a powerful way of investigating selection. In particular, clines that are maintained by a balance between selection and gene flow have a characteristic width $l = \sigma/\sqrt{2s}$, and so measurements of dispersal and cline width can give indirect estimates of selection (Haldane 1948; Slatkin 1973; Endler 1977).

Comparison of cline widths for different traits can show the relative selection acting on each; for example, in the hybrid zone between *Chorthippus parallelus parallelus* and *C. p. erythropus* in the Pyrenees, enzyme clines are several kilometres wide, suggesting neutral mixing following secondary contact, whereas the clines in nucleolar organizer and stridulatory peg number are much narrower, suggesting strong selection (Butlin *et al.* 1991). A problem with this approach, however, is that when overall selection is strong relative to recombination linkage disequilibria pull all clines together, regardless of the selection acting on each (Nürnberger *et al.* 1995).

The most direct approach is to measure fitness and its components directly. In principle, the extent to which gene flow interferes with adaptation can then be measured directly. One can ask whether crosses between populations show a reduced fitness compared with crosses within local populations ('outbreeding depression'). Transplant experiments can then be used to determine whether this is due to the establishment of incompatible gene combinations in different demes (as in Wright's (1931) 'shifting balance'), or to adaptation to local conditions (see Waser and Price 1994 for a review; Dias *et al.* 1996). The spatial pattern of fitness interaction can give further information. Burt (1995) has pointed out that the reduction in fitness between individuals separated by one dispersal range gives a direct estimate of the additive genetic variance in fitness that drives adaptation to the local environment or genetic background; in each generation, natural selection increases mean fitness by an amount equal to the additive variance, and gene flow then returns it to the original value. To take another example, the distance over which fitness is reduced in a hybrid zone reflects the selection on each gene responsible for that fitness deficit, and so allows a rough estimate of the number of genes involved (Barton and Hewitt 1981; Szymura and Barton 1991).

Conclusions

Despite great efforts, clear theory, and many examples of local adaptation and divergence, we still have little understanding of the actual importance of population structure for evolution as a whole. Novel methods, based on highly variable genetic markers and on the possibility of sampling genealogies, hold much promise. However, the theoretical basis of these is as yet unclear, and there may be fundamental limitations in our ability to disentangle the complex history of gene movements from present-day spatial patterns. In particular, most indirect methods of inference from genetic markers estimate the relative strengths of genetic drift and gene flow, rather than gene flow itself. Although more laborious, it may be most profitable to examine fitness, and traits that affect fitness, directly, so as to discover just how far much selection varies from place to place. It is this that determines the importance of gene flow in evolution.

Acknowledgements

This work was supported by the Darwin Trust of Edinburgh, the Biotechnology and Biological Sciences Research Council, and the Natural Environment Research Council. I would like to thank J. Clobert and the referees for helpful suggestions.

24

Perspectives on the study of dispersal evolution

Ophélie Ronce, Isabelle Olivieri, Jean Clobert, and Etienne Danchin

Abstract

In this chapter, we stress the gaps in our knowledge about dispersal evolution and suggest several promising research directions. (i) Both empirical and theoretical studies illustrate how different causes may result in the evolution of dispersal behaviour. Integrative approaches that study the interactions between these causes are, however, rare. Theoretical expectations must be clarified by a careful comparison of existing models and their specific assumptions. The importance and evolutionary significance of plasticity for dispersal has also been largely neglected. (ii) The evolution of dispersal cannot be understood simply as a change in a single trait. Changes in dispersal rate alter the demographic functioning of populations and their genetic structure, which, in turn, modify the selective pressures on dispersal. Theoretical studies incorporating both types of feedback are rare, and empirical investigations of such interactions almost absent. Similarly, selective interactions between the evolution of dispersal and life histories are yet largely unexplored. (iii) Knowledge about the determinism and ontogeny of dispersal behaviour is likely to alter predictions concerning its evolution. This includes information ranging from the genetic determinism of dispersal, the effect of hormones on physiology and behaviour, and maternal effects, to the particular environmental cues used by individuals to assess habitat quality. (iv) Experimental evolution should provide information about the relative importance of different causes of the evolution of dispersal, the interaction between demography and the evolution of dispersal, and the joint evolution of other traits. These experiments may also indicate, from comparison of lines before and after selection, which are the genes and/or physiological mechanisms involved in dispersal behaviour.

Keywords: conditional dispersal, stochastic models, deterministic models, bet-hedging, causes of dispersal, genetics of dispersal, ontogeny of dispersal, experimental evolution, genetic structure, demography, community evolution, evolutionary feedback, habitat selection, cues

Introduction

Dispersal is considered to be a central life-history trait in evolutionary biology, ecology, and conservation biology. In addition to the multiple demographic consequences of dispersal, its evolutionary consequences range from the evolution of altruism, and therefore of any form of sociality, to speciation processes and evolution of community structure. Dispersal has been a difficult trait to study, and most often has been approached by indirect methods. In this chapter, we try, based on previous chapters, to summarize what is emerging from our present knowledge about the causes of the evolution of dispersal, the various types of feedback that affect its evolution, and the mechanisms by which it develops. We also stress the gaps in our knowledge about dispersal evolution. We suggest several research directions concerning these various points. Finally, we define the research areas that seem to us the most promising.

Are there multiple causes of the evolution of the same dispersal behaviour or multiple dispersal behaviours associated with each cause?

Previous chapters of this book have revealed that there are still many ongoing controversies about which major cause has promoted the evolution of dispersal in a given group of organisms. Both empirical and theoretical studies support a variety of alternatives (Gandon and Michalakis, chapter 11). We could classify the different causes of dispersal evolution in three broad classes. First, the evolution of dispersal may be understood in the light of habitat selection theory (Holt and Barfield, chapter 6). By 'habitat selection', as in Danchin *et al.* (chapter 17), we mean all behaviours at departure or arrival that help individuals to select and establish themselves in certain types of habitats (where their fitness is supposedly improved). In particular, by allowing escape from crowding or inbreeding, dispersal may help individuals to reach sites of better quality. This definition of habitat selection is broader than the classical one (Stamps, chapter 16), where habitat selection involves only settlement behaviours. Second, dispersal can be viewed as an altruistic trait, and its evolution understood in the light of kin selection theory (Lambin *et al.*, chapter 8; Perrin and Goudet, chapter 9; Gandon and Michalakis, chapter 11). Third, in fluctuating environments, dispersal can be seen as a bet-hedging strategy (see below, p. 343).

In wild populations, several factors probably act on the evolution of dispersal. Unfortunately, theoretical treatments have generally not included the simultaneous effect of different forces and their interactions (but see Gandon and Michalakis, chapter 11; Perrin and Goudet, chapter 9; Gandon and Michalakis 1999). A clear discrimination between different causes has also rarely been achieved in empirical studies, for example see the discussions about the role of inbreeding versus sex-specific competition for the evolution of dispersal in mammals (Moore and Ali 1984; Dobson and Jones 1985). Thus, we need to: (1) clarify theoretical expectations about cases when dispersal evolves as a unique answer to different problems, and (2) develop criteria to distinguish the relative importance of different forces selecting for dispersal in a given ecological setting.

We can identify several reasons why this task remains unfulfilled. First, the lack of clarity about the forces included in previous models has blurred the interpretation of theoretical results (p. 343). Second, ignoring the fact that different causes may act at different timescales is a source of confusion (p. 344). Third, evolutionary causes of dispersal may vary greatly with the spatial scale (p. 345). Accounting more carefully for spatial scales should increase our understanding of the interaction between the various forces that lead to dispersal evolution. It should also help us to discriminate between different dispersal behaviours within the same species. Similarly, a more careful study of conditional dispersal should allow us to assess the relative importance of different causes in dispersal evolution (p. 346).

Deterministic versus stochastic models

Confusion may arise from the same environmental factor having several consequences, although a given modelling framework might allow the observation of only one of these consequences. In particular, deterministic models sometimes obscure those aspects of dispersal evolution that are linked to stochastic processes. For instance, local extinction creates open sites that promote the evolution of dispersal as a mechanism to escape crowding since only dispersing individuals can reach open sites. But local extinction is also associated with environmental instability, thus promoting the evolution of dispersal as a bet-hedging strategy that reduces temporal variance in fitness (Philippi and Seger 1989). Indeed, in temporally variable environments, dispersal dampens the fluctuations in fitness at the scale of the whole metapopulation, by distributing individuals more evenly in space. Then, even though dispersal does not necessarily move individuals to currently better sites, its effect on the global temporal variance in fitness has been shown to be beneficial (Kuno 1981; Venable and Brown 1988). In several models, the evolutionarily stable (ES) dispersal rate increases with extinction rate (van Valen 1971; Comins *et al.* 1980; Levin *et al.* 1984; Olivieri *et al.* 1995). This could result from increased colonization opportunities or be explained by a bet-hedging argument. However, the latter interpretation, which states that dispersal evolves to 'escape local extinction', is inadequate in those models because it requires that dispersal decreases the global temporal variance in fitness, at the scale of the metapopulation. When local variations occur independently, such yearly variations in fitness occur only if genotypes occupy a finite number of sites. Since the four above-mentioned models (van Valen 1971; Comins *et al.* 1980; Levin *et al.* 1984; Olivieri *et al.* 1995) assumed that the various genotypes occupy an infinity of sites, the evolution of dispersal in these models results from escape from competition (with both relatives and non-relatives) and not bet-hedging. In absence of competition for space (no density dependence), regardless of the instability of the habitat, the predicted value for the ES dispersal strategy is strict philopatry in those models, as soon as the cost of dispersal is not null (Venable and Brown 1988). This paradoxical result rests entirely on the assumption of an infinite number of occupied sites in the metapopulation.

Stochastic models (e.g. Karlson and Taylor 1992, 1995) have investigated the consequences of stochasticity on dispersal evolution and sometimes led to predictions contrasting with those of deterministic models. Stochastic models predict larger dispersal rates and stronger parent–offspring conflicts if there is any cost of dispersal. If local disturbance probability is homogeneous, dispersal does not allow an individual to escape local extinction, but dispersal of several offspring increases the probability that some of them will settle in an undisturbed site. Qualitative predictions may also differ. For instance, Karlson and Taylor (1992) predicted that optimal dispersal rates should decrease when local extinction rates increase, contrary to deterministic model predictions (but see Ronce *et al.* 2000).

A second source of stochasticity is linked to the necessarily finite local population sizes. This stochasticity is ignored in deterministic models that do not manipulate discrete numbers of individuals but fractions of individuals (Levin *et al.* 1984; Olivieri *et al.* 1995). Such models behave as though local population sizes were infinite. Finite population size generates some variation in allelic frequencies (genetic drift) and variation in population sizes or age structure (demographic stochasticity). Because of the former effect, those deterministic models underestimate the importance of kin selection in the evolution of dispersal (see Hamilton and May 1977; Frank 1986; Gandon and Michalakis 1999; Ozaki 1995). Higher dispersal rates are thus expected in stochastic models. Selective pressures on dispersal are also changed by the latter effect (Cadet 1998; O. Ronce, S. Gandon, and F. Rousset, unpublished manuscript). Finally, a third source of stochasticity, caused by the finite number of migrants, remains poorly studied. A simulation model by Travis and Dytham (1998) suggests that varying the number of immigrants per site selects against dispersal.

Deterministic and stochastic models imply very different assumptions about why dispersal evolves. The comparison of these two approaches may be difficult, but most fruitful. Contrary to deterministic models, studies on the risk-spreading function of dispersal have generally assumed no density dependence (Kuno 1981; Venable and Brown 1988) and no frequency dependence. The optimized fitness criteria also differ between classes of models. For instance, Karlson and Taylor (1992, 1995) used the dispersal rate that minimizes the long-term probability of extinction of a lineage. As noted by these authors, this dispersal rate is likely to differ from the dispersal rate that would maximize the short-term rate of increase of the lineage. More generally, we need more thoughts about the appropriate fitness criteria in the presence of stochasticity (and in particular demographic stochasticity, which makes the invasion of rare mutants a more complex issue; see Ferrière and Le Galliard, chapter 5). Recent work also suggests that the standard evolutionarily stable strategy definition may have limited interest in a context where genetic drift is important, and criteria other than the invasion by a rare mutant should be used to predict the direction of phenotypic evolution (Rousset and Billiard 2000). Similarly, much knowledge would be gained by a more careful comparison of simulation and analytical results. Relaxing assumptions in a hierarchical way would allow us to identify the relative impact of various evolutionary forces.

Long-term versus short-term selective forces

Distinguishing between long-term and short-term selective forces is another challenge that relates to the problem of the fitness function. Just as sexual reproduction allows maximization of a species' evolutionary potential through the continuous production of new genotypes, long-distance dispersal allows a species to escape unfavourable conditions and colonize new environments. It is, however, unclear and, in fact, unlikely that sex has evolved and is maintained because it increases species' evolutionary potential. Similarly, how much of the evolution of dispersal is due to long-term benefits such as colonizing ability? How much of long-distance dispersal is a byproduct of selection for short-distance dispersal? For instance, in plants, dispersal curves are often leptokurtic, with most seeds dispersed around the mother plant, and a few seeds dispersed further. This was observed even for species in which fruits are equipped with a pappus, facilitating wind dispersal (Olivieri and Gouyon 1985; Colas *et al.* 1997). Similar dispersal distributions were observed in bird species (Clobert and Lebreton 1991) and mammals (Waser 1987). Although, from a species point of view, the important offspring are those dispersed away from the local (ephemeral) population, it is unclear whether long-distance dispersal in these species has actually evolved to colonize new sites.

Different dispersal behaviour at different spatial scales: evolution of dispersal distance

Depending on the distance moved, dispersal has very different consequences (see Fig. 24.1) and may be selected-for for very different reasons. Moving a short distance (e.g. moving to the next adjacent territory in territorial mammals) may allow escape from direct competition with parents without losing the benefit of local adaptation and without much cost. Dispersal to neighbouring groups only, however, does not prevent kin competition (Gandon and Rousset 1999). Greater dispersal distances may be selected-for in viscous populations when sib competition (or, more generally, kin competition) is involved. For similar reasons, escaping inbreeding may require greater dispersal distances (e.g. changing social group). The distances required to attain a vacant territory or a less-crowded patch of habitat are highly variable depending on the disturbance and demographic dynamics that determine patterns of occupancy. Nonetheless, distances travelled to colonize newly opened habitat are expected to be larger than those travelled to avoid interactions with relatives. Escaping deteriorating conditions, predation, or parasitism implies moving distances determined by the spatial autocorrelation in habitat quality. For predator avoidance, the movements of the predator determine these spatial autocorrelations. Finally, the various costs linked with dispersal behaviour (loss of the help received from parents, loss of familiarity, loss of local adaptation, increased energetic costs) will change with the distance.

Few theoretical studies have focused on dispersal distances (Ezoe 1998). Most empirical studies on plants have dealt with dispersal curves as a function of distance (see for instance Colas *et al.* 1997). Understanding how dispersal distances evolve might allow us to understand how organisms compromise between the different forces acting on this trait (Waser 1987). This might also elucidate some aspects of parent–offspring conflict. We clearly need to take into account several spatial scales of dispersal, from localized to metapopulation level. Similarly, we need to consider the spatial patterns of heterogeneity in the environment (Lavorel *et al.* 1995; Wiens, chapter 7) and how different selective forces change with scale.

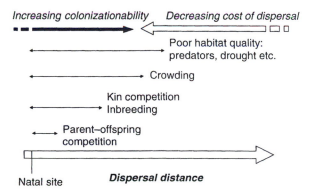

Figure 24.1. The different dispersal distances needed to escape negative characteristics of the natal habitat.

Conditional dispersal

Causes of dispersal may change in relative intensity over time and space. When the same individual is likely to experience a variety of conditions, a fixed dispersal phenotype might be maladaptive. According to various theoretical models, condition-dependent dispersal strategies appear dominant over fixed dispersal strategies in a large set of situations (see references in Olivieri and Gouyon 1997; Boulinier and Danchin 1997; Danchin *et al.*, chapter 17). Conditional dispersal may be thought of as a way to decrease conflicts between the different selective forces acting on dispersal. Conditional dispersal has been reported in very different groups of animals (Ims and Hjermann, chapter 14; Danchin *et al.*, chapter 17) but the importance of this kind of phenotypic plasticity is largely unexplored as yet. This is true especially for plant species. Theoretical studies considering the evolution of plastic dispersal strategies are still rare.

Much knowledge could be gained by studying conditional dispersal more carefully. Patterns of phenotypic variation in dispersal may reveal the main causes of dispersal evolution, how their relative importance varies in time and space, and what their interactions are. A differential expression of dispersal for the same genotype in different environments may provide more immediate access to the factors favouring its expression than would the comparisons of different genotypes or species in different habitats. For instance, when dispersal rates increase at high local densities, it is tempting to conclude that escape of crowding is an important cause of dispersal. This rests on the assumption that observed plastic changes in dispersal are truly adaptive and not the simple expression of constraints.

Ideally, the pattern of phenotypic variation in dispersal should help us to define diagnostic criteria about the relative importance of different selective forces. However, this is not a simple task. For instance, declining dispersal with maternal age in a species that undergoes senescence could reveal the influence of strong parent–offspring competition on the evolution of natal dispersal (Ronce *et al.* 1998). Further theoretical studies have, however, shown that age–dependent dispersal could be adaptive in other situations even in the absence of direct parent–offspring interactions (S. Brachet, O. Ronce, I. Olivieri, J. Clobert and P.-H. Gouyon, unpublished manuscript). Patterns of variation associated with only one environmental cue are unlikely to be conclusive, as this cue can be indirectly associated with different selective forces. Nonetheless, the accumulation of investigations concerning different environmental effects on dispersal behaviour can help us to build a fairly consistent picture of the causes of dispersal in one species. Empirical studies of the common lizard illustrate the potentialities of such an approach. In this species, several maternal effects (maternal age, food condition, hormonal level) consistently suggest that parent–offspring competition is an important cause of natal dispersal (Clobert *et al.* 1994b; Sorci *et al.* 1994; Massot and Clobert 1995; Léna and de Fraipont 1998; Ronce *et al.* 1998; de Fraipont *et al.* 2000). However, the examination of conditional dispersal strategies has also revealed that several causes were probably responsible for dispersal evolution (Léna *et al.* 1998; Massot and Clobert 2000).

Finally, we claim that, in many situations, dispersal probably evolved for several different reasons. It is thus pointless to look for a unique force causing dispersal evolution. Rather, more effort should be devoted in the future to understanding, both theoretically and empirically, how different forces combine and select for dispersal.

Furthermore, the different chapters in this book reveal that there is not a unique dispersal behaviour, but multiple dispersal strategies depending on temporal spatial scales and conditions experienced by individuals. Ignoring the multiple aspects of dispersal is not only misleading, but it detracts from key information about evolutionary forces contained in the variability of dispersal.

In the following section, we further suggest that dispersal evolution can hardly be understood if dispersal is considered as an independent trait. Its evolution cannot be disentangled from the evolution of other traits (at individual, population, and community levels).

Evolutionary feedback

Interactions with demography

Usually, models either focus on the evolution of dispersal without examining the demographic consequences, or focus on the genetic or demographic consequences of colonization and fragmentation, keeping dispersal fixed, ignoring its evolution (e.g. Gyllenberg and Hanski 1992; Gyllenberg *et al.* 1997). For instance, evolutionary models often assume that local population sizes do not depend on dispersal rate (Comins *et al.* 1980; Gandon and Michalakis 1999). Only a few studies have considered the interactions between the evolution of dispersal and demographic dynamics (Olivieri *et al.* 1995; Holt and McPeek 1996; Doebeli and Ruxton 1997; Ronce *et al.* 2000). These studies suggest that such interactions can modify model predictions. However, methodological difficulties arise when including more sophisticated demography in evolutionary models. It is often unclear which density-dependent function should be used, and what the consequences of this choice are. Empirical data concerning interactions between dispersal evolution and demography are severely lacking. Models predict that lower dispersal rates should evolve in metapopulations in which local populations grow slowly after foundation (Olivieri *et al.* 1995; Ronce *et al.* 2000). Comparison among habitats that differ in productivity could allow us to test such predictions. More convincing arguments could be gained by manipulating local carrying capacity or productivity within metapopulations. Several field experiments have investigated the ecological consequences of productivity and disturbances in plant communities (e.g. Campbell and Grime 1992). However, the spatial and temporal scales of these experiments are probably inadequate to witness changes in dispersal (only plastic changes in dispersal may be detected). Recent experiments on *Tribolium* (Constantino *et al.* 1995, 1997) show that this genus constitutes a promising model. By manipulating adult survival rates, the demographic dynamics were changed, and populations reached an equilibrium size or fluctuated chaotically. By maintaining such populations under various demographic regimes for sufficiently long to witness selective changes in life histories, we might be able to test some theoretical predictions. Similar experiments could be performed to witness plastic changes in dispersal.

Interactions with genetic structure

The same kind of feedback occurs between the evolution of dispersal and genetic structure in metapopulations. The evolution of dispersal in finite populations depends on genetic identities, that is, the probability that two interacting individuals bear the same alleles at loci determining dispersal behaviour. Conversely, the genetic identities depend on the level of gene flow among populations (and thus on dispersal). Thus, studies considering the genetic structure as if it were fixed are misleading. A few theoretical studies have analysed the joint evolution of dispersal and genetic probabilities of identity in a metapopulation (Taylor 1988; Gandon and Michalakis 1999).

The evolution of dispersal may also depend on the genetic structure at loci other than those determining dispersal. These loci may contribute directly to fitness through their effect on either inbreeding depression (e.g. when they carry deleterious recessive alleles) or outbreeding depression (e.g. when they carry locally adapted alleles). A few genetic models (Wiener and Feldman 1993) have considered the evolution of allelic frequencies at the two types of loci (those determining fitness in different environments and those determining dispersal among environments). These models concluded that, because linkage evolves between loci, dispersal should be counter-selected, but their extremely restrictive assumptions (no temporal variation, overdominance) cast doubts on the generality of this result. More theoretical work is needed to understand how inbreeding depression and local adaptation evolve jointly with dispersal. As for the interactions between demography and dispersal, experiments should be set up where the degree of genetic identity of competing individuals is manipulated (as in Taylor and Crespi 1994) and the consequences for dispersal evolution examined.

A general problem with analytical models of joint evolution of dispersal and genetic structure is their peculiar assumptions about demography: they ignore demographic dynamics and assume constant population sizes. The consequences of more realistic demographic dynamics on the organization of genetic variability are poorly known. Currently, there is a kind of theoretical gap, with, on the one hand, models that incorporate more or less sophisticated demographic dynamics but ignore genetic aspects such as kin selection and, on the other hand, models that deal with the genetic structure of metapopulations but ignore demography. Each theoretical approach has separately generated predictions that may be contradictory. Without a common framework, we are unable to make predictions about the resultant of such conflicting forces. Each approach separately has proven the prominent role of kin selection and demographic effects on the evolution of dispersal. Only the incorporation of these two aspects into the same model will lead to reliable conclusions and predictions about the relative impact of these different forces.

Interactions with other life-history traits and the evolution of syndromes

It may be incorrect to consider the evolution of dispersal independently from that of other attributes. Associations among traits or syndromes can be examined at different scales, by comparing species, populations, or individuals within populations. The meaning of these associations is likely to depend on the scale. Dispersers and residents are likely to differ in many characteristics other than movement ability (Bélichon *et al.* 1996; Roff and Fairbairn, chapter 13). Therefore, dispersal behaviour should be considered to be a complex association of traits, and its evolution understood in light of these associations (migratory syndromes; Dingle 1996; Roff and Fairbairn, chapter 13). A rarely investigated question is how genetic correlation between dispersal and other traits evolves. Some of these correlations are clearly adaptive (such as syndromes associating functional wings with behavioural propensity to disperse). Others are interpreted as physiological constraints (such as syndromes associating decreased fecundity with functional wings). Are costs of dispersal (in terms of differences in survival or fecundity between migrants and residents) completely constrained by energetic trade-offs, or are they likely to evolve? The evolution of costs of drug and insecticide resistance in bacteria and insects shows that costs associated with physiological characteristics or behaviour may also evolve (Guillemaud *et al.* 1998).

Even in the absence of differences between migrants and residents, various dispersal strategies may evolve in species differing in life histories or reproductive systems (Olivieri and Gouyon 1997). In turn, dispersal alters the selective pressures acting on life-history traits (Ronce and Olivieri 1997; Pen 2000). A better understanding of observed patterns would be gained by studying the joint evolution of dispersal and other traits, incorporating the selective interactions between them. Examples are the evolution of dispersal and dormancy (Venable and Brown 1988), dispersal and reproductive effort (O. Ronce, F. Perret, and I. Olivieri, unpublished manuscript), dispersal and mating system. In plants with a seed heteromorphism, dispersed seeds usually are less inbred than non-dispersed ones (reviewed in Olivieri and Berger 1985). Various hypotheses have been proposed, but none has been tested formally. Ronfort and Couvet (1995) showed that localized dispersal might lead to biparental inbreeding and affect inbreeding depression so that intermediate selfing rates might evolve. However, in their model, dispersal distances were not allowed to evolve. It would be very interesting to let inbreeding avoidance, inbreeding depression, and dispersal evolve jointly (for a first approach see Gandon 1999; Perrin and Goudet, chapter 9; Gandon and Michalakis, chapter 11). Models should try further to integrate coevolution between the various causes and effects of dispersal.

Interactions between species: co-evolution of dispersal rates in communities

Dispersal evolves not only in relation to life history or demography, but also depends on the dispersal of other species (Boulinier *et al.*, chapter 12a). The relative mobility of predator and prey was shown early on to have profound impacts on the dynamics of predator–prey systems (Huffaker 1958). The co-evolution of predator and prey dispersal rates has now been described theoretically (van Baalen and Hochberg, chapter 21): the evolution of predator dispersal may promote prey dispersal in habitats of decreasing quality. The evolution of resistance and local adaptation of hosts to their parasites depends on their relative dispersal rates (Gandon *et al.* 1996). However, it is unknown whether hosts or parasites are more likely to be locally adapted when both can evolve to their optimal dispersal rate. In community ecology, theoretical studies of the coexistence of competing species often consider the attributes of coexisting species as if they were fixed (Lavorel *et al.* 1994). Thus, little is known about factors maintaining multiple dispersal strategies in the same community, when competing species are allowed to evolve their own dispersal rates. Recent theoretical work has shown the possibility of evolutionary branching, a process by which several dispersal strategies evolve from a monomorphic initial state and coexist more or less transiently in the same habitat (Doebeli and Ruxton 1997; A. Mathias, E. Kishidi, and I. Olivieri, unpublished manuscript;

Mouquet *et al.*, chapter 22). If assortative mating can evolve at the same time (Doebeli 1996), how often will this lead to speciation? Fitting more evolutionary thinking into community ecology is a promising field of research.

Determinism and ontogeny of dispersal behaviour – does it help us to understand its evolution better?

When is the dispersal phenotype determined?

There are several possible levels of determinism for an individual's dispersal behaviour (see Fig. 24.2). Very little is known about each of them except in a few model organisms.

Genetic effects

The dispersal phenotype of an individual can be determined by its genotype. A complex trait such as dispersal is probably influenced by many loci. Consequently, quantitative genetics has often been used to study genetic variation in dispersal-related traits (references in Dingle 1996; Roff and Fairbairn, chapter 13). There is also evidence that the dispersal phenotype may be determined even before the zygote stage, for example

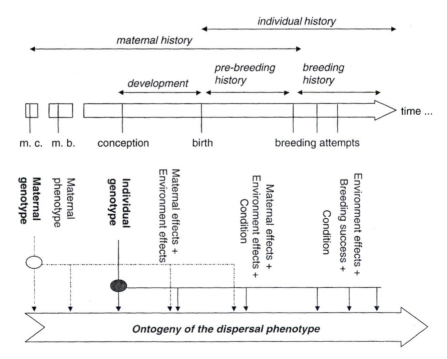

Figure 24.2. When is dispersal phenotype determined? The ontogeny of dispersal phenotype may be under the influence of various genetic and environmental effects throughout the individual and maternal history. Maternal and/or individual genotypes ultimately determine the dispersal phenotype and its sensitivity to environmental effects. m.c., maternal conception; m.b. maternal birth.

when the maternal genotype determines the dispersal phenotype of its offspring. In plants, the presence or size of seed dispersal structures is controlled by the expression of maternal genes. For instance, in some *Medicago* species it has been suggested that a single mutation can change spiny pods into spineless pods (Lesins *et al.* 1970, cited in Bena *et al.* 1998). Maternal genes may control the proportion of dispersers among the offspring (e.g. seed heteromorphic plants; Venable and Burquez 1989). In animals, there are cases where the mother may manipulate the dispersal behaviour of its offspring. This might occur through the expression of maternal genes affecting embryo development, or by more direct aggressive interactions between parent and offspring, enforcing their dispersal.

The mother may determine the propensity of each offspring to disperse, but this propensity is usually modulated by the offspring genotype and environment. Very little is known about the relative variation in dispersal behaviour explained by offspring and maternal genotypes respectively. Only experimental reciprocal crosses would allow us to have access to this information. Theoretical studies have suggested repeatedly that dispersal should differ depending on whether it is controlled by maternal or offspring genes (Hamilton and May 1977; Motro 1983; Roitberg and Mangel 1993), resulting in a parent–offspring conflict. What theory does not tell us is who is most likely to win the conflict if both types of genes are involved in the determinism of dispersal.

Environmental effects

Few theoretical studies have considered other sources of variation in dispersal than genetic ones. Environmental effects have a large impact on the evolutionary dynamics of traits: they can either accelerate or slow down selection on these traits. Maternal effects occur in aphids, where crowded mothers produce more winged offspring than isolated ones (Mousseau and Dingle 1991). In the common lizard, *Lacerta vivipara*, juvenile dispersal depends on maternal food condition (Massot and Clobert 1995), maternal age (Ronce *et al.* 1998), and maternal parasitism (Sorci *et al.* 1994). The annual plant *Crepis sancta* produces a larger fraction of seeds equipped with dispersal structures when the mother plant is stressed by nutrient depletion or simulated herbivory (E. Imbert and O. Ronce, unpublished manuscript). Maternal effects on dispersal are documented in quite a few taxonomic groups (Ims and Hjermann, chapter 14). Empirical studies should, however, be subject to caution when studies of maternal effects are correlative rather than experimental.

The father and grandparents (Reznick 1981) may also influence the dispersal phenotype of offspring (Dingle 1994). MacKay and Wellington (1977) reported a grand-maternal effect in pea aphids: females born early in their mother's reproductive life produced more winged offspring than those born later (effect of grand-maternal age). Moreover, this effect depends on the grand-maternal wing morph. Environmental and genetic influences are thus not confined to single generations.

Natal dispersal has been shown to depend on several natal environmental parameters other than maternal condition (Ims and Hjermann, chapter 14). Breeding dispersal has been reported to vary with environmental features such as climate, food, predators, parasites, or conspecifics. The fact that breeding dispersal is determined by the individual's breeding history in a patch has been discussed in several instances (Danchin *et al.* 1998a). It has also been suggested that the breeding history of other individuals in the population could influence dispersal decisions in a focal individual (Boulinier and Danchin 1997; Danchin *et al.*, chapter 17; Stamps, chapter 16).

What are the consequences and evolutionary significance of earlier or later determination of dispersal behaviour?

Little is known about the relative importance of genetic and environmental effects on proximal dispersal determinism. Knowing more would imply understanding the physiological and cognitive pathways that lead to changes in dispersal. Dufty and Belthoff (chapter 15) review some evidence about the role of hormones in the ontogeny of dispersal behaviour (see also Roff and Fairbairn, chapter 13). Identifying genes involved in the determination of dispersal would allow us better to understand dispersal ontogeny. A first step in this direction could be Quantitative Trait Loci (QTL) mapping. We do not know which genes are turned off or on by environmental influences (including maternal influences). How easy is it for a different dispersal behaviour to evolve? Bena *et al.* (1998) and Andersen (1993) have addressed this question through the use of phylogeny.

Current information suggests a large amount of variation among species. For instance, significant maternal effects have never been found in *Gryllus firmus* (Roff 1986), while they stand as a rule in aphids and plants, persisting through several generations. Do these differences reveal developmental constraints or adaptation? Indeed, the mechanisms determining dispersal are themselves likely to evolve. For instance, maternal effects may be heritable, thus raising the question of why would maternal effects be selected for in aphids and not in *G. firmus*? *G. firmus* lives in extremely unpredictable marshes and has one generation per year, so the probability that philopatric offspring will experience the same conditions as the mother is low. In the absence of temporal autocorrelation of the environment, as well as in the absence of direct interaction between mother and offspring, maternal effects have no particular adaptive value. Conversely, in aphids, generations succeed rapidly so that the mother and its offspring experience similar environments (in *Rhopalosiphum padi*, age at maturity is reached about 8 days after birth while mothers survive about 40 days; M. Hullé and J.-C. Simon, personal communication). Moreover, for aphids exploiting a host plant, there is a predictable pattern of deterioration in habitat quality through time, due to increasing crowding and host plant senescence (Sutherland 1969a,b). More generally, we predict that early environmental determination of dispersal is adaptive only if early conditions are predictive of local conditions in the future. Conversely, fixed genetic determinism for dispersal (i.e. a flat reaction norm for dispersal) would evolve in extremely unpredictable habitats where environmental cues convey very little information about future prospects (Danchin *et al.*, chapter 17).

Empirical data suggest that the dispersal phenotype or dispersal decisions are modulated by the information acquired at various times, even late in life ('stage-dependent dispersal' according to Ims and Hjermann, chapter 14). Very little is known about how conflicting information acquired at different times combines to produce the dispersal phenotype. Does the information accrue, as in some aphids where both pre-natal and postnatal crowding are necessary to produce the winged morph, or does only the current information matter, cancelling past information? In the first case, we might consider whether cumulative effects make the information more reliable. This depends entirely on the patterns of temporal autocorrelation of the environment at different timescales (Danchin *et al.*, chapter 17). Environmental information acquired at different times also provides information about the temporal patterns of variability in habitat quality. Dispersal decisions could indeed depend on how organisms perceive their habitat variability pattern. This possibility has not been investigated yet, either empirically or theoretically.

Finally, the early versus late determination of dispersal phenotype may be linked to a number of constraints. As plants cannot disperse once settled, information they have acquired about their habitat quality is of little utility for their own habitat choice, contrary to species that might disperse between breeding attempts. However, maternal history transmitted to offspring through maternal effects might affect seed dispersal. We would thus expect maternal effects to be important in any organism with a sessile adult phase.

Are there good and bad cues for habitat quality?

Identifying the environmental factors (including individual condition) that trigger changes in dispersal, and assuming that receptivity to these factors is adaptive, raises the question of why organisms evolve receptivity to one environmental cue and not to another (see the discussion in Stamps, chapter 16; Danchin *et al.*, chapter 17). More generally, the evolution of conditional dispersal strategies introduces questions about the information value of cues for habitat quality (Danchin *et al.*, chapter 17). As for the timing of dispersal determination, the reliability of cues depends on whether the spatio-temporal variation in these cues matches that of habitat quality. Very few empirical studies have documented such patterns. Theoretical results also suggest that the quality of a cue for dispersal may be subject to frequency dependence, so that the cues used by a large fraction of the population are devalued (B. Doligez, E. Danchin, and T. Boulinier, unpublished manuscript). For all these reasons, quantifying information costs may be extremely difficult.

The ecological and evolutionary consequences of mismatches between habitat quality and cues for habitat selection need to be investigated. We can identify two lines of research on this subject. First, we may expect deviations from the ideal free distribution (Danchin and Wagner 1997), the extent and consequences of which should be investigated (Holt and Barfield, chapter 6; Stamps, chapter 16). Second, occasional mismatches between habitat quality and cues could lead to the coexistence of several habitat selection strategies (S. Brachet, O. Ronce, I. Olivieri, J. Clobert, and P.-H. Gouyon, unpublished manuscript; B. Doligez, E. Danchin, and T. Boulinier, unpublished manuscript). The question of the diversity of habitat selection strategies has rarely been studied.

Using complementary cues may reduce the mismatch between environmental cues and habitat quality. Different environmental cues may be associated with different reasons to disperse. Using several cues may then solve conflicts between the different causes of dispersal evolution. Experimental studies varying two factors at a time would help in understanding how organisms perceive and weigh the information corresponding to different selective forces. For instance, in the common lizard, *Lacerta vivipara*, the dispersal rate of offspring depends on the interaction between the age of the mother and its level of stress (Meylan *et al.* 1999; de Fraipont *et al.* 2000): stress hormones in the mother promote offspring dispersal in young mothers and offspring philopatry in old mothers. A similar pattern is observed in manna ash trees, *Fraxinus ornus*, where the relationship between the size of the tree and the dispersal ability of its seeds depends on the successional stage of the population (S. Brachet, O. Ronce, I. Olivieri, J. Clobert, and P.-H. Gouyon, unpublished manuscript). Empirical studies that have examined the response of dispersal to several environmental cues simultaneously are extremely rare. Theoretical studies considering the combination of several sources of information about habitat quality are lacking, making the interpretation of empirical evidence difficult.

Concepts and methods: where shall we go now?

The significant increase in theoretical investigations, new data sets, and pioneering experiments has considerably improved the definition of the framework within which the evolution of dispersal should be investigated. From all the chapters in this book, we feel that five main complementary areas need to be developed simultaneously. They can be summarized as follows: theoretical unification, artificial evolution, constraints or adaptations, and demographic consequences.

Theoretical unification

To consider simultaneously the multiple forces affecting dispersal evolution, we need models incorporating both genetics (accounting for both kin competition and inbreeding) and demographic dynamics. Analytical advances are severely limited by the complexity of the task. An alternative approach could be to build spatially explicit and individually based models to extract relevant synthetic concepts and measures (pair correlation dynamics; Harada and Iwasa 1996). In some instances, ESS models, by ignoring the underlying genetic structure (epistatic effects, linkage disequilibrium) or by assuming arbitrary trade-offs among traits, may perform very poorly. Short-term quantitative predictions may be achieved more successfully by using quantitative genetical models. We might also wonder how detailed and realistic the models should be. This is also linked to the question of how we should test models' predictions. There is probably no single answer to these questions. By 'theoretical unification', we do not suggest that we should adopt a single theoretical approach, but rather that more effort should be devoted to the comparison of different models.

Selection experiments and evolution in experimental systems

Experiments on organisms with short generation times, in controlled environments, may provide complementary information about the relative importance of different causes in the evolution of dispersal (by selecting under different environments). It may also provide information about the interaction between demography and the evolution of dispersal (by recording change in dynamics and in dispersal but also by manipulating densities), and about co-evolution between dispersal and other traits (by recording correlated changes). Comparisons of lines before and after selection may also indicate which genes and/or physiological mechanisms are involved in dispersal behaviour. More complex to perform, but probably of wider interest, analysis of selection on dispersal reaction norms would improve our understanding of the adaptive value of plastic dispersal. The interplay between dispersal and phenotypic plasticity for other traits could also be investigated through the same type of settings.

Conditional dispersal: mechanisms, identification, and hierarchy of causes of dispersal evolution

Under which circumstances will a plastic dispersal strategy develop, and how is conflicting information integrated (if it is) into the phenotype? A general theory on the ontogeny of the dispersal phenotype is currently lacking (see Fig. 24.3 for an attempt to formalize this question). Experiments might help us to build such a theory. How and at which stage the dispersal phenotype is produced are still open questions. Manipulating

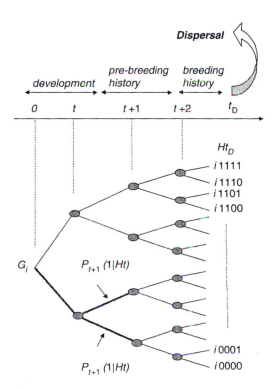

Figure 24.3. How to formalize the ontogeny of dispersal. At time 0 (conception) the genotype i is determined. Ascending (or 1) and descending (or 0) lines indicate whether the dispersal phenotype is sensitive or not to environmental effects at time t. Ht describes the history of a given individual in terms of its sensitivity to environmental factors at different times until time t. This 'history' is described by a sequence of 0 and 1. The probability of being sensitive to environmental factors at time $t+1$, $P^i_{t+1}(1|Ht)$ may depend on the genotype i, and on the past history at time t, Ht. For instance, sensitivity to environmental effects at time t may change the sensitivity to events at time $t+1$. From an evolutionary point of view, that probability and the type of sensitivity to the environment at time t should be a function of the autocorrelation of the environmental between time t and t_D, the time of departure. The complete history until dispersal is Ht_D. $Ht_D = i0000$ describes a strategy with strict genetic determinism for dispersal and no phenotypic plasticity.

information gathered at different times during phenotype development may prove useful. We could create situations where some forces would tend to favour dispersal behaviour (e.g. increased crowding) and other forces would discourage it (e.g. the presence of cooperating relatives) and observe how organisms plastically solve the conflict. Such experiments would also help us to build a hierarchy of the causes that promote dispersal behaviour. In the same way, identifying proximate cues and hormonal mechanisms is of primary importance.

The study of conditional dispersal strategies not only provides insights into dispersal mechanisms, but also helps us better to understand the causes of dispersal. Models involving different ultimate causes of dispersal share many predictions. The presence of adaptive plasticity for dispersal can allow us to go beyond purely descriptive approaches based on correlation. If dispersal varies plastically, we can manipulate the factors thought to drive dispersal evolution, disentangle the patterns of covariation, and observe phenotypic responses.

Constraints and adaptation

Many of the above considerations assume that changes in dispersal phenotype are adaptive. Patterns of variation in dispersal that depend on habitat characteristics or individual condition have been interpreted repeatedly as evidence for adaptive plasticity in animal species and as the expression of developmental constraints in plants (E. Imbert and O. Ronce, unpublished manuscript). In both cases empirical demonstrations are lacking. Two ideas are linked to the concept of adaptation: that the currently observed behaviour has been moulded by natural selection, and that it is optimal. Demonstrating that behaviour is adaptive requires us to identify variants for this trait and fitness differences between them. In the case of dispersal, this is an extremely difficult task. Measuring genotype-by-environment interactions or some heritability for dispersal reaction norms would be a first step in this direction. The optimality of a dispersal reaction norm could be studied experimentally by measuring fitness differences between various dispersal phenotypes within or among environments. When phenotypic variants are naturally rare in a given environment, one could produce them experimentally by manipulating the proximate cues triggering changes in dispersal phenotype. Carefully designed experiments, supported by theoretical treatments, would be essential here.

Demographic consequences

How dispersal evolves and is determined also has consequences for conservation biology (see chapter 25). The very few models that examine the interplay between dispersal evolution and demography reveal important interactions between them. Predicting evolutionary changes in dispersal and their potential consequences on demography is of interest, especially when evolution of dispersal is rapid, as documented by Cody and Overton (1996). However, in many situations these changes may occur on very long timescales. On short timescales, dispersal strategies and habitat selection strategies can be considered to be fixed. Taking into account a detailed description of dispersal behaviour is important for predicting population demography and, especially, extinction probabilities. For this purpose, we need both models and experiments that compare the demographic consequences of various dispersal strategies that have evolved for different reasons. In particular, the rescue effect of dispersal depends largely on the spatial scale at which each dispersal type is operating, and the likelihood that the movement will end up in colonization or population reinforcement. Experiments on distance moved and rate of colonization with respect to the cause of dispersal are needed, as well as studies on the correlation between cues used for departure and cues used for settlement (Stamps, chapter 16; Danchin *et al.*, chapter 17).

Almost nothing is known on the above topics, which are central to understanding the evolution and consequences of dispersal. Although some theoretical developments are still needed, we claim that data collection and, particularly, experiments are urgently needed. We now have at our disposal mathematical tools that enable us to encapsulate biological realism (multiple factors, interaction at the local scale) and therefore to make more precise predictions. Pioneering experiments have indicated that dispersal behaviour is a complex trait under multiple influences. In turn, such advances will help to construct more relevant experiments and more finely tuned models. This is expected to lead to important discoveries and to enhance the generality of models in the years to come.

Acknowledgements

We thank P.-H. Gouyon and A. Mathias for discussion, S. Gandon, N. Perrin, and two anonymous referees for useful comments on the chapter, F. Saunier for her help and patience. I.O. acknowledges support from E.E.C. TMR ('Fragland' project) allocated to I. Hanski, as well as support from the French Ministère de l'Aménagement du Territoire et de l'Environnement, through both the Bureau des Ressources Génétiques (contract no. 97/117 to I.O.) and the National Program Diversitas, Fragmented Populations network (contract no. 98/153 to I.O.). O.R. acknowledges a PhD grant from the French Ministry of Research and Education. This is publication ISEM ISEM2000-046 of the Institut des Sciences de l'Evolution de Montpellier.

25

Dispersal in theory and practice: consequences for conservation biology

David W. Macdonald and Dominic D.P. Johnson

Abstract

Dispersal is a fundamental parameter of population processes. In small and isolated populations, linked only when individuals are able to disperse between them, it becomes a particularly critical one which ultimately determines whether a species will become extinct. Predictions for these populations are limited because estimates of dispersal variables from the field are difficult to obtain and therefore scarce. Models are therefore also limited because they cannot be parameterized accurately and have, until recently, ignored behavioural and spatial variation in dispersal. This problem is compounded for endangered species which most desperately need accurate viability analyses and management plans, but about which little is usually known and opportunities for research may be few. Threatened populations are not the only conservation application for dispersal. Predators and pests are a significant conservation concern, and advances in the management of these populations also require detailed understanding of dispersal behaviour to forecast their expansions. More detailed behavioural research into dispersal is clearly needed. Aside from their academic merit, the chapters within this book provide an excellent basis for furthering the application of research on dispersal to conservation biology.

Keywords: conservation, viability analyses, models, parameters, extinction, management, fragmentation, habitat

Introduction

Dispersal is one of the crucial parameters in even the most simplistic population models, providing (along with the birth/death ratio) one of two principal modes of population change (Hanski 1999a). It is automatically, therefore, an influential variable in population dynamics, and especially so in subpopulations (Levins 1970; Hanski 1999a) prone to extinction within a metapopulation. This is because, among subpopulations increasingly isolated by habitat change and fragmentation, a trickle of dispersal, the 'rescue effect', may stave off extinction. Indeed, if local extinction does occur, the only prospect for reversing it lies in the arrival of dispersers (Brown and Kodric-Brown 1977; Hanski 1999a). This will be successful only if the proximate cause of the extinction can be identified and eradicated, which may require significant conservation efforts or large-scale habitat restoration. The ability to model, to predict, and ultimately to manage these eventualities thus depends on knowledge of the processes and parameters of dispersal. Not only does dispersal generate considerable theoretical interest, but its direct consequences and repercussions for population dynamics are of vital importance to the conservation of threatened species world-wide (Simberloff 1988). This is widely relevant across the globe, from the small and isolated habitats and populations in logged tropical rainforests areas in the third world (Whitmore 1997), to those of the fragmented agricultural landscapes of western Europe (Macdonald and Smith 1990, 1991). Global climate change also threatens species with specific habitat requirements, where their dispersal is not sufficient to adjust their distributions accordingly (Davis 1989).

The many examples of species for which dispersal is a critical matter would fill a book by themselves. However, examples that illustrate the central importance of dispersal to a diversity of issues include the viability of fragmented populations of water voles, *Arvicola terrestris*, and the expansion of their introduced predator, the American mink, *Mustela vison*, in European waterways (Barreto *et al.* 1998a; Macdonald and Strachan, 1999); the mechanisms of inbreeding avoidance amongst Ethiopian wolves, *Canis simensis*, (Sillero-Zubiri *et al.* 1996) and African hunting dogs, *Lycaon pictus* (McNutt 1996); limited dispersal ability of plants in the face of climate change (Primack and Miao 1992), and the epizootiology and control of rabies in red foxes, *Vulpes vulpes* (Macdonald 1995). Dispersal is, in the words of John Wiens (chapter 7): 'the glue that binds populations together'. This glue is an indispensable item in the toolbox for maintaining dwindling populations and managing pests over the many damaged landscapes across the globe.

We first discuss the range of problems that dispersal poses for conservation. Then, we discuss the importance of, and some caveats with, population viability models, which are particularly important as tools for conservation since empirical field tests are rare. We then explore the problems that the process of environmental change itself is likely to cause for dispersal, followed by how knowledge about dispersal may be applied in conservation biology. We conclude with some remarks emphasizing the urgency of the conservation of species, and how much remains to be discovered about dispersal.

The problems of dispersal for conservation

Lack of data and reliance on models

A decade ago, Johnson and Gaines (1990) noted a 'prodigious' number of models of dispersal evolution. Unfortunately, as three recent reviews identify, there remains an almost complete absence of empirical data on dispersal and other relevant behavioural parameters in recent population extinction studies (Reed and Dobson 1993; Doak and Mills 1994; Caro 1999), and there are difficulties in modelling and understanding evolutionary mechanisms of dispersal (Dobson 1985). Much of the recent work on extinction processes has focused on the metapopulation paradigm (Hanski and Gilpin 1991, 1997), upon which, according to Doak and Mills (1994, p. 619): 'essentially all of the most influential papers on the subject, have been modelling efforts'. In fact, evidence has not always been forthcoming to support the population turnovers and extinctions predicted by metapopulation dynamics in patchy habitats (Harrison *et al.* 1988; Taylor 1990, 1991; Harrison 1991). Yet, empirical studies are constrained by the difficulties of obtaining information about dispersing individuals (North 1988; Verhulst *et al.* 1997) and, as a consequence, many models for population viability analysis (PVA) remain largely untested with field data. Previous models (including 'packaged' PVA models) have treated dispersal as a fixed trait, whereas it is becoming increasingly clear that there is considerable phenotypic plasticity in dispersal behaviour (van Vuren and Armitage 1994; Reed 1999; Murren *et al.*, chapter 18; O'Riain and Braude, chapter 10) and that it may be condition or situation dependent (Dufty and Belthoff, chapter 15; Ims and Hjermann, chapter 14; Stamps, chapter 16).

Lack of behavioural detail in dispersal studies

Reed (1999) specifically targets the lack of inclusion of behavioural mechanisms in population modelling, a damaging omission because behaviour is an important influence on population persistence for many taxonomic groups spanning birds, mammals, insects, and marine animals. This is especially pertinent given the renaissance in recognizing the importance of 'Allee' effects, which result in decreased reproductive rate or survival at low densities (Courchamp *et al.* 1999). This negative feedback obviously has significant implications for the conservation of small populations (Stephens and Sutherland 1999). The mechanisms by which the Allee effect occurs are many, but social species are likely to be particularly disadvantaged as many of them rely heavily on conspecifics for cooperative behaviours, and this adds to other compounding problems (Woodroffe and Ginsberg 1998). Worse still, Reed argues, by omitting behavioural aspects of dispersal behaviour, predictions made for surviving population numbers may be over-optimistic. This delusion could arise where models assume equal dispersal tendency, whereas the reality might encompass behaviour patterns that result in resistance to dispersing.

Variation in dispersal behaviour

Dispersal is generally incorporated into population models using a random diffusion/ random walk paradigm, where animals disperse in random directions and stay in the first suitable habitat (Wiens, chapter 7). In reality, however, behaviour has been shown to operate as a selective process in dispersal (references in Reed 1999; Stamps, chapter 16; Danchin *et al.*, chapter 17). For example, some bird species with strong habitat specificity do not venture from their patch into the intervening landscape and therefore are constrained in dispersing whereas others readily fly from one fragment of habitat to another (Wiens 1994, chapter 7; Johnson and Mighell 1999). Thus, dispersal estimates from natural habitats may be markedly different from those when the animal is faced with a matrix of unfamiliar habitat types. Landscape structure influences not only the sites between which animals may disperse, but also the permeability of a given route. Understanding the constraints upon dispersal demands knowledge of the species' natural history: expansive water may be a barrier to colonization of islands by American mink (Herteinsson 1992), whereas dormice use thin hedgerows in agricultural landscapes to disperse through expanses of open field that would otherwise be impenetrable (Bright and Morris 1996). Frank and Woodroffe (2000) draw attention to how species that disperse readily, such as lions (*Panthera leo*) and coyotes (*Canis latrans*), are relatively resilient in the face of population control, while some species (e.g. badgers (*Meles meles*) and spotted hyaenas (*Crocuta crocuta*)) are remarkably reluctant to disperse and therefore recolonize cleared areas extremely slowly. Tuyttens and Macdonald (2000) emphasize the likelihood that populations perturbed by attempted management are likely to behave differently than their undisturbed counterparts; one of the behaviour patterns likely to be perturbed by management is dispersal. Clearly, versatile species with generalist habitat requirements are likely to be able to disperse through a wide array of habitats, and those that are most mobile may do so over great distances; for example, red foxes may disperse over tens, or even hundreds, of kilometres (Lloyd 1980; Macdonald 1984). Nonetheless, behaviour holds many surprises. In the UK, the home ranges of water voles are generally confined to the water's edge, but recent evidence reveals that individuals may disperse overland across watersheds (X. Lambin, personal communication).

The complications outlined above, which reveal the shortcomings of progress in dispersal studies for practical conservation planning, are widely and profitably discussed in the chapters of this book. Several authors note significant failings: inconclusive tests of hypotheses (Lambin *et al.*, chapter 8), lack of data and appropriate tests (Ronce *et al.*, chapter 24; Wiens, chapter 7), problems with analyses (Peacock and Ray, chapter 4), and lack of knowledge of proximate factors (Dufty and Belthoff, chapter 15). Happily, however, solutions are emerging, with: proposals for how better to measure dispersal (Bennetts *et al.*, chapter 1; Ross, chapter 3); methods of separating interactive effects (Gandon and Michalakis, chapter 11; Perrin and Goudet, chapter 9; Ronce *et al.*, chapter 24); and reviews of possible complexities that need to be controlled for in future studies, such as variation in dispersal due to developmental conditions, environmental cues (Stamps, chapter 16), behaviour, or physiology (O'Riain and Braude, chapter 10; Murren *et al.*, chapter 18). These latter topics have further consequences for metapopulations and the concept of effective population size, because of differences in genotype and fitness in those individuals that tend to disperse (Whitlock, chapter 19).

The urgency of conservation and dependency on models of dispersal

Clearly, we are in an extinction crisis, with species disappearing at 100-fold, or even 1000-fold, background rates of extinction. The practical goal of the science of conservation biology is to illuminate ways of slowing this trend. Dispersal is a crucial element of population dynamics and, consequently, a central issue in conservation biology. Our ability to forestall this crisis will depend, in some measure, on better understanding dispersal. There are, however, significant limitations to this end. The principal problem appears to be lack of empirical data (Reed 1999), but also a lack of experiments that have been able to disentangle the multiple causes of dispersal and the relevant spatial scale (Ims and Hjermann, chapter 14; Ronce *et al.*, chapter 24) as well as problems associated with assumptions of hypotheses and interpretations of data (Peacock and Ray, chapter 4; Lambin *et al.*, chapter 8). Ims and Hjermann (chapter 14) suggest that failure to account for the spatial scale of environmental and related variables, relative to a species' dispersal capacity, may explain inconsistent results concerning condition-dependent dispersal in the literature.

Biological conservationists do not enjoy the luxury of waiting for academic developments to arise that will eventually resolve these difficulties. They must be urgently stimulated to produce, at the least, approximate expectations and remedies on which to base conservation actions. Models have therefore become very important in targeting the correct variables and identifying the principal processes. Of course, in some circumstances an extreme paucity of information means that realistic models cannot be constructed. Nevertheless, models may very usefully arise in two general types. The first are theoretical models that formalize or examine, qualitatively or quantitatively, particular processes and variables in order to understand dispersal behaviour and its population consequences. The second type are those that specifically aim to forecast a species' population dynamics given certain specific parameters and starting conditions, and, in some cases, the details of the landscape – population viability models. It is the latter class that has come to be particularly associated with conservation projects, and on which we shall concentrate here.

As a consequence of the above, many conservation management schemes have only model projections on which to base decisions. Such an example comes from the much-debated population viability models developed for the northern spotted owl, *Strix occidentalis caurina*, in diminishing old-growth coniferous forests (e.g. Lande 1988; Lamberson *et al.* 1992). Empirical studies for each species are desperately needed (ideally case by case) as many of these models go largely unverified, or, at best, they rely on substitution of data collected from similar species, or 'off-the-shelf' PVA models that make unrealistic assumptions about dispersal behaviour. Bennetts *et al.* (chapter 1) set out some practical guidelines to achieve dispersal estimates, and were surprised by the lack of empirical tests of models despite the advent of new methods and new technologies for tracking and modelling animal movements using Global Positioning System (GPS) or satellites and Geographic Information Systems (GIS) (see also Wiens, chapter 7). That it is substantially easier to develop a model than to test it is a fact that should be noted by those prioritizing research funding. It is a poignant inevitability that the lack of data is particularly apparent for endangered species. They, by definition, are few and limited in distribution, and disturbing them to obtain such life-history data may jeopardize the remaining individuals (Doak and Mills 1994; Reed 1999).

Consequences of dispersal on genetic structure of populations

The genetic structure of the population also depends, among other things, on dispersal rates and dispersal distances (Endler 1977; Hanski and Gilpin 1997); these parameters, therefore, ultimately impose constraints within which local populations can evolve (see also Roff and Fairbairn, chapter 13). Most theoretical studies of dispersal have focused on the consequences for population genetical structure and the mechanisms of its evolution (Johnson and Gaines 1990). Authors such as Peacock and Ray (chapter 4) and Ross (chapter 3) discuss advances in studying dispersal through genetic data, although Rousset (chapter 2) concludes that molecular approaches have provided only few success stories, against an undeveloped theoretical background and using methods largely untested in the field (see also Peacock and Ray, chapter 4). Barton (chapter 23) also draws attention to problems associated with using genetic markers as estimates of dispersal and associated gene movements. He suggests greater emphasis on measuring fitness directly in the field; clearly, this is difficult, but critical in order to clarify the consequences of dispersal. Ross (chapter 3) warns that different genetic methods can give different results, but encourages the development of detailed understanding of dispersal and gene flow from genetic approaches within a 'predictive framework' developed from the natural history of the species, which can be done only with long-term and large-sample studies. Peacock and Ray (chapter 4) and Bennetts *et al.* (chapter 1) also suggest integration of genetic and demographic approaches in dispersal studies, not only during data collection but also throughout the process of model development, right through to their testing. Peacock and Ray (chapter 4) reassert the point that hypotheses for dispersal (or anything else) cannot be tested properly in the absence of adequate data, and in this case that includes data on mating patterns and parentage.

Some further complications of dispersal for conservation

In addition to noting the problem of lack of data on dispersal, Lambin *et al.* (chapter 8) draw our attention to the fact that convincing evidence is scarce for certain beliefs about the functional significance of dispersal, for example kin competition. As another example, evidence is mounting for negative density-dependent dispersal rates (Lambin *et al.*, chapter 8; Ims and Hjermann, chapter 14). Amongst carnivores, for instance, there is no evidence that dispersal is less common in controlled populations because there are more vacancies for youngsters at, or close to, home. If anything, control seems only to increase dispersal (see Woodroffe 2000) and risk social perturbation, with potentially counter-productive effects amongst survivors (Swinton *et al.* 1997; Tuyttens and Macdonald 1998).

Another current, but poorly understood, issue vital for conservation planning concerns feedback mechanisms in dispersal behaviour (Ronce *et al.*, chapter 24; van Baalen and Hochberg, chapter 21). While the multiple causes of dispersal are difficult enough to disentangle (Gandon and Michalakis, chapter 11; Ronce *et al.*, chapter 24), feedback from dispersal consequences on the population level may further cloud interpretation of empirical results of mechanisms and measurements of dispersal. This is especially pertinent at present, as many animal populations are undergoing declines (or increases) in densities following anthropogenic change and therefore exhibit dynamic dispersal parameters as the perturbed populations experience shifting selective pressures.

A final word of caution before proceeding further is that we have thus far heralded dispersal as a sort of panacea for species conservation, especially in fragmented habitats and metapopulations. There are, however, several facets of dispersal that, conversely, may help to drive a species to its own extinction, especially in fast-changing and altered landscapes. For example, dispersing individuals are vectors for pathogens to move between otherwise isolated populations of susceptible conspecifics, the effects of which may become more risky to small populations (Caro 1999). Alternatively, a high dispersal rate can, in theory, cause the establishment of synchronous local population fluctuations, exposing the whole metapopulation to similar dynamics where stochastic effects are correlated (Hanski 1989; see also Lande *et al.* 1999 for a recent treatment). Thus, the probability of extinction of the whole metapopulation at a high dispersal rate may exceed the product of the independent probabilities of all local extinctions when the dispersal rate is low. Because of such adverse consequences, it is vital to keep to the fore an attention to the multitude of ramifications that dispersal may have for populations. Management of damaging side-effects may be as important as administering the cure.

Progress in models of dispersal in conservation

Empirical knowledge can often usefully be complemented by modelling. As has already been emphasized, this opportunity becomes a somewhat frustrating necessity in many investigations of dispersal, due to the scarcity of empirical data. Indeed, in assessing the relative importance of potential variables to population viability, models may themselves be useful in identifying which parameters most merit expending further effort in the field to obtain accurately. A synergy between reality and simulation is nevertheless important in conservation biology for at least two other reasons. First, conservation planning generally involves evaluation of alternative management options; it may be feasible to make a preliminary evaluation by harmlessly manipulating the virtual world of a model before risking expensive or irreversible action on the ground (if, and only if, the model is realistic). Second, fieldwork is almost always sufficiently labour-intensive as to be confined to a small scale, whereas policies are often implemented on a wide scale, sometimes regionally or nationally. Therefore, models can be useful if they facilitate extrapolation to wider geographical scales. However, models that make 'predictions' of population viability can be dangerous because they may entertain an illusion of broad application across a range of conservation management programmes before they have undergone rigorous testing or validation. As an example, a predictive model (developed with empirical data) for mountain sheep populations (*Ovis canadensis*) (Berger 1990), which had specific management implications, sparked a vehement debate about whether its predictions were valid (Berger 1999; Wehausen 1999), possibly partly because of assumptions made about dispersal. As another indication of the perils of generalizing with models, an experimental study of two closely related species of *Drosophila* showed how the division of single populations into small subpopulations had very different effects on persistence depending on the species, even though they were otherwise very similar (Ims and Stenseth 1989). What is observed to happen in one species is not necessarily, therefore, a good model for another, even if they are closely related.

Modelling population persistence for species of particular conservation concern has now become a routine process, and there are now several widely available population viability analysis (PVA) programs (Boyce 1992). These provide a useful framework within which to identify priority causes of decline for management. However, although they generally rank the influence of input variables consistently, different versions may give different predictions with identical problems and data (Macdonald *et al.* 1998). Since they contain prescribed algorithms (Lima and Zollner 1996) they are essentially a 'black box', and application to a novel species may be biased from the outset. Of course, they cannot reasonably be designed to incorporate all situation-specific details across the great diversity of novel problems. Given that they are often the only solution, an ideal compromise would be for programmers and conservationists to work together towards a customized adaptation of the original.

Apart from these problems, PVAs have hitherto also been limited by the general absence of spatial information. Along with the development of GIS technologies and sophisticated computer-generated prediction models based on two- or three-dimensional outputs on digitized landscapes, the scope for creating ever more 'realistic' (and complicated) models is increasing. Several chapters in this book also provide new extensions to older models by specifically incorporating dispersal parameters describing variation between individuals or habitats or interactions (e.g. Murren *et al.*, chapter 18; Hanski, chapter 20; Mouquet *et al.*, chapter 22; Perrin and Goudet, chapter 9), while van Baalen and Hochberg (chapter 21) introduce the complexity of intertrophic interactions, which may have been a source of bias in former models. However, as with any statistical routine, and process-based spatially explicit population modelling in particular, the increase in complexity goes hand in hand with more parameters and a decrease in interpretability and confidence in results.

However, to the extent that the models are reliable, they allow us to explore various options for each species. For example, work might be undertaken to improve the habitat for water voles by restoring riparian banks, or for dormice by planting and coppicing hazel. These activities might be approached in various ways, each with different cost implications. How many hazel copses should there be, and distributed in which pattern? We have illustrated elsewhere how our models, or their refined successors, could inform just such decisions (Macdonald *et al.* 1998; D.W. Macdonald and S.P. Rushton, unpublished manuscript; Rushton *et al.* 2000). Of course, it would be folly to base action on such models without some reassurance that there are grounds for confidence in them. For example, in simulations of spatial spread, Macdonald and Strachan (1999) present tentative evidence of similarities between simulated and observed water vole populations (see also Rushton *et al.* 2000). However, these are only correlations, and so again we urge caution. At the least, however, such explorations highlight the topics on which field research might usefully be focused.

Despite the absence of parameter estimates for even simple models, spatially explicit models of populations used in conservation biology are becoming something of a cottage industry, so caution in their applicability is crucial, particularly regarding dispersal. However, a recent paper reports that some of the errors claimed to be associated with such models were grossly exaggerated (Mooij and DeAngelis 1998, and references therein). Indeed, at least as a didactic, exploratory tool, the power added to individual, process-based models by advances in GIS appears to offer a useful practical guide for conservation planning and a good means of identifying gaps in field data (e.g. Macdonald *et al.* 1998, 2000b; Rushton *et al.* 2000; South *et al.* 2000). However, while the apparent realism of this GIS-based approach strikes us as useful, we realize it can also be beguiling, and so we echo Doak and Mills' (1994) observation of a magnetism between conservationists, who are often keen (and sometimes desperate) to apply new theoretical techniques to urgent problems, and modellers, who are equally keen to find applications for their models.

Influence of environmental change on dispersal

There are many direct effects of human influence on dispersal, for example the extinction of natural seed dispersers or destruction of habitat through which individuals are willing to disperse. The many indirect effects, though, can be separated into two broad categories. First, global climate change means that many sedentary species, and especially plants, will need to disperse latitudinally (or vertically, as in altitude) in accordance with shifting suitable habitat, perhaps 10 times faster than during the last ice-age retreat (Davis 1989; Primack and Miao 1992; Wilson 1992). Primack and Miao (1992) report that this may be too fast for certain plant species to persist and warn that a similar situation may exist for aquatic species of isolated lakes as well as, particularly, for vertebrate species of montane peaks. This is of particular concern, given that such areas contain endemic species especially prone to extinction (Balmford and Long 1994; Stattersfield *et al.* 1998) and whose distributions appear to be closely linked to environmental variables (Johnson *et al.* 1998). Such isolated populations of restricted-range species may not have a suitable habitat to move into even if they are able to disperse. Large-scale environmental stochasticity is likely to be a particularly problematic influence because, of course, correlated extinctions of populations increase the risk of metapopulation extinction (Hanski 1999a). Indeed, it may be that today's rare species have become rare as a result of low dispersal and high correlation of between-population extinctions (Hanski 1989). The synchrony in dynamics between populations of a species may become especially critical. In addition to these indirect global climate change effects, the deterministic influences of humans are increasingly on a sufficiently large scale to be labelled as 'global' with regard to those species with limited geographical distributions. This is enough to warrant concern that entire metapopulations will be annihilated in one fell swoop, denying an opportunity for even some fragments to persist.

The second broad category is that of the local knock-on effects of habitat destruction and fragmentation, and the implications for the connectivity of the populations. Wiens (chapter 7) warns that there is not necessarily any reason to suppose that movement pathways bear a 'close relationship to theoretical optima'. This introduces a considerable problem that may have to be faced by the study of dispersal in the future. While this volume has documented many likely evolutionary and mechanistic influences on dispersal tendency and characteristics, these may not operate as they formerly did, because of human-induced change resulting in species under considerable stress from habitat destruction and contamination. For example, migratory orientation has been shown to be influenced by organophosphorus pesticides (Vyas *et al.* 1995). There is a danger, therefore, that species' dispersal decisions may now be maladaptive if based upon cues of the environment that have been changed (Stamps, chapter 16), or that they may select habitats according to conspecific attraction (Stamps 1988; Reed 1999), which may be inappropriate if population densities reflect habitat disturbance rather than an otherwise approximately ideal free distribution (IFD) (Reed and Dobson 1993). Furthermore, new studies indicate that animals use conspecific reproductive success, rather than just presence, as a cue (Boulinier and Danchin 1997; Danchin and Wagner 1997), in which case depressed reproduction due to environmental degradation could directly alter the normal patterns of dispersal. Ims and Hjermann (chapter 14) recognize a need for studies to test whether it remains optimal for animals to employ condition-dependent dispersal when environmental change may directly or indirectly alter the cues and methods that animals use to disperse. Such effects could have broad influences, as there is evidence that individual variation in dispersal may arise due to a multitude of condition- or situation-dependent factors which human activities may have caused to change, including prior experience (Stamps, chapter 16), degree of parasitism (Boulinier *et al.*, chapter 12a), predation pressure (Weisser, chapter 12b), hormones, body condition (Dufty and Belthoff, chapter 15), conspecific attraction, and patch choice criteria (Danchin *et al.*, chapter 17).

Another complication of environmental changes may be their interaction with other pressures to make dispersal more critical than usual. For example, Macdonald *et al.* (1999) explored the effects of predation (either naturally or in the form of human activities) on small populations of prey. A simulation model illustrated that, in small populations, stochastic factors always caused significant impacts while, in general, predation could only limit populations rather than extinguishing them. However, these simulations suggested that, in theory, predation pressure on already small, isolated populations by generalist predators can cause local extinctions, particularly if these populations are also threatened by other factors. Thus, an interactive effect of depressed populations (for environmental reasons) can expose isolated populations of species, unable to disperse between habitat fragments, to extinction from factors that are not usually lethal to a large connected population (Barreto *et al.* 1998b; Macdonald and Strachan 1999). Similarly, movement and dispersal patterns relative to the size and configuration of protected areas emerge as especially important in the planning of nature reserves for the protection of wide-ranging carnivores (Woodroffe and Ginsberg 1998; Woodroffe 2000), because anthropogenic causes of mortality are intensified at reserve edges. Indeed, Woodroffe (2000) found a positive correlation between dispersal distance and proneness to extinction among large carnivores.

Among four species of conservation concern, recent spatially explicit models (D.W. Macdonald and S.P. Rushton, unpublished manuscript) have shown that interactions between life histories and the environment resulted in dispersal having very different influences on persistence depending on the species. For two rodent species, water voles and hazel dormice, *Muscardinus avellanarius*, maximum dispersal distance (among other factors) was a significant partial correlation coefficient of the number of populations that persisted, as well as of dispersal mortality in the dormouse. However, for two mustelid species, American mink and pine marten, *Martes martes*, neither dispersal distance nor dispersal mortality emerged as a significant correlate of the total population size. Adult mortality was a significant correlate for both predators, as well as home range size for the mink and litter size and juvenile mortality for the pine marten. These species-specific differences do not necessarily hint at a generalization for their taxanomic group, but they do emphasize that the role of dispersal in population dynamics may vary substantially between species. Of course, these are dependent on the characteristics of the environment and the parameters used, but it serves here as a brief example of how dispersal appears differentially to affect populations depending on the details of the environment, its relative spatial scale, and the particular life-history parameters of the species.

As a concluding caveat, for species facing extinction, which will often stem from human impacts on their environment (Balmford and Long 1994), dispersal may have unexpected interactive influences on persistence. The behaviour of populations is an emergent property of the reactions of individuals to their circumstances, which are now changing rapidly. Dispersal is likely to be a crucial element of this behavioural response. It is such threatened species that most urgently require models (albeit less urgently than they require data) to determine management plans, as they are also the ones for which mistakes would bear the greatest cost.

Conservation applications

Because dispersal favours metapopulation survival, it is sometimes a specific goal of conservation management to boost a species' dispersal ability through habitat recreation, stepping stones, or corridors. The observation that dispersal is more complicated than previously thought is almost a truism, but one that nevertheless may account for difficulties in reaching consensus in conservation biology. For example, underestimating the complexity of dispersal may explain why, despite evidence that linear links improve population viability in some studies (e.g. chipmunk populations in hedgerows; Henderson *et al.* 1985), the value of corridors linking protected areas remains controversial (Simberloff *et al.* 1992; Gonzalez *et al.* 1998). It may also explain failures when attempting to model population dynamics with field data from another population of the same species (Hanski and Thomas 1994).

Conservation programmes increasingly involve the translocation of animals to reinforce failing populations (once, of course, the cause of initial decline has been identified and stopped). Indeed, reintroduction could be necessary (and hopefully would still be possible) if all of these other measures have failed. This requires reliable information about dispersal. Previous reintroductions have often been poorly managed, and there is a lack of scientific follow-up studies to learn from mistakes, improve methods, and detail dispersal behaviour (Sarrazin and Barbault 1996). Settlement behaviour in particular may be critical to success and may be directly assisted if understood (Sarrazin *et al.* 1996). Introduction experiments offer a controlled environment in which to identify key features of settlement behaviour (Massot *et al.* 1994); however, the conditions of departure from the original territory and of the transient stage during dispersal are far harder to investigate. Nevertheless, conservation programmes could be doing more to establish specific hypothesis-testing research as part of their reintroduction schemes in order to answer questions about, among other things, dispersal (Sarrazin and Barbault 1996). There are, of course, many ways in which knowledge of dispersal might potentially be applied to practical action in conservation. Two examples strike us as particularly relevant to the discussion within this book, and are discussed below.

Models of metapopulation extinctions regard populations as essentially approximating to some variant of the IFD (see Holt and Barfield, chapter 6). However, it has been suggested, also, that individuals may aggregate in relatively large numbers within a metapopulation, using conspecifics as a cue of habitat quality (Smith and Peacock 1990). This asymmetry in distribution, with a component independent of the IFD, may increase the likelihood of extinction of some populations insofar as 'eggs' are distributed between fewer 'baskets', representing higher losses and population extinctions as well as fewer sources for dispersal (e.g. Ray *et al.* 1991). However, Danchin *et al.* (chapter 17) suggest that, where settlement is problematic, conservationists may be able to employ this phenomenon, perhaps with strategically placed dummies or captives, in order to encourage aggregation in suitable habitats, which can then be targeted for protection (Reed and Dobson 1993; Sarrazin *et al.* 1996).

Dispersal has a critical influence on social species, and, reciprocally, sociality may influence dispersal and resulting gene flow. A mechanism of group formation common amongst social species is offspring philopatry (Ross, chapter 3). There is a possibility that social species with particularly strong philopatric tendencies might be extinction-prone under modern circumstances. For example, E. I. Krutchenkova, T. M. Goltsman and D.W. Macdonald (unpublished manuscript) have described an island population of Arctic foxes amongst which dispersal is a rarity, with the dangerous consequence that young adults remain in their natal range, exposed to social suppression of reproduction, while neighbouring territories lie vacant. The African hunting dog has a declining distribution across Africa, and efforts by conservation biologists to reintroduce them have been hindered by, in addition to many logistical factors, complex social factors apparently stemming from rather specific social prerequisites (Woodroffe, Ginsberg and Macdonald 1997). An understanding of dispersal in this context may be crucial for their conservation. Indeed, in southern Africa 'metapopulation management' in the form of translocating wild dogs between reserves may be a fruitful conservation tool (Mills *et al.* 1998), and such interventive management mimicking dispersal may increasingly, if tragically, be necessary as a last resort in human-dominated landscapes (Macdonald *et al.* 2000a; Woodroffe and Ginsberg 2000).

Some caveats for future interpretation of dispersal

The significance of various parameters to a population's viability is contingent upon the interaction between its life history, its competitors, and the landscape. In particular, dispersal plays a different role in the population processes of small restricted-range species from that in large wide-ranging ones. In the modern landscape, long journeys are highly likely to involve encounters with human development, and such encounters generally involve risk. Thus, it is notable, perhaps, that larger species at the top of the food chain are typically in greatest danger of extinction and often disappear first (Woodroffe 2000). This may provide some clues to the importance of spatial scale as a factor in dispersal success and survival probability (Wiens, chapter 7). While habitat loss is continuous, fragmentation is a 'threshold phenomenon', occurring specifically at certain critical percentages of loss (Wiens, chapter 7); this will differentially affect species according to the fractal dimension relevant to their movements. Thus, we must be cautious in interpreting at what stage differently mobile species are in danger of isolation and at what point dispersal really becomes the limiting factor for species persistence. For example, it has been difficult to predict with accuracy the values of certain threshold population sizes, below which populations have been shown in models to decline rapidly (Lamberson *et al.* 1992), but also within which a small number of dispersers can drastically change the genetic structure of populations (Whitlock, chapter 19).

As well as identifying conservation priorities, it may sadly become necessary to avoid squandering effort in attempting the conservation of lost causes. Hanski (chapter 20) reminds us that some populations exist in unsuitable habitats as sinks, simply because immigration rates exceed those of death and emigration (Harrison *et al.* 1988; Paradis 1995). Are such populations worth conserving if they could not survive once in isolation? They could be sources for re-colonization of temporarily vacated source habitats

(Thomas and Jones 1993), and could be deliberately used as such if maintained as zoo populations for captive breeding. Even if reintroduction were not possible, however, when all else had failed there would seem to be great virtue in a living museum, who would not wish now for a bank of dodos or thylacines in captivity? Nevertheless, it remains important to discover which wild populations should be prioritized for investment.

More generally, three further considerations may be of note in the context of dispersal and conservation. Hamilton and May (1977) found that the evolutionarily stable dispersal rate was 0.5 (due to kin competition effects) even when the dispersal mortality rate was 100%. A wide range of further and extended models also reduced to the same underlying rate (Johnson and Gaines 1990). Thus, if these models are correct, high dispersal mortality observed in nature may not necessarily be construed as a cause for concern. Second, although the literature on inbreeding and gene flow and the consequences for conservation is vast, there is little evidence from wild populations of demographic decline due to inbreeding (Caro and Laurenson 1994). This may, however, be partly due to lack of detailed data from populations with which to test for this; such evidence that has emerged for increasing extinction risk with inbreeding are from detailed, long-term study populations (Saccheri *et al.* 1998; Westemeier *et al.* 1998). Third, while there is a clear emphasis on habitat destruction and fragmentation, some species may have distributions at least partly defined by competitive exclusion (Hassell *et al.* 1994; see also Mouquet *et al.*, chapter 22). A thorough knowledge of the whole community, therefore, may be necessary to predict species dynamics where other species, and especially predators, have already become extinct.

What is the likely course of dispersal evolution in altered environments?

Very generally speaking, according to models, spatial variation selects against dispersal, while temporal variation predicts selection for dispersal (Johnson and Gaines 1990, and references therein). Thus, habitats undergoing change as a result of human should encourage dispersal, whereas the resulting static mosaic should decrease it. However, while this may be true for flexible dispersers, habitat fragmentation presumably halts the dispersal of 'cautious' species, alluded to before (see Wiens, chapter 7). Woodroffe (2000) similarly points out that, amongst carnivores, the Nee–May model predicts that good dispersers should do well in fragmented habitats.

A high rate of dispersal is not always a panacea for fragmented metapopulation persistence. Too high a rate of dispersal might result in the source itself being deprived of a viable population. For this reason, and due to the increased mortality of dispersers in more fragmented habitats and the consequently increased metapopulation extinction probability (Hanski, chapter 20), one may predict the evolution of short dispersal in endangered populations. Indeed, Thomas and Jones (1993) showed that more butterflies dispersed out of small patches (with long edges), with the result that populations living on small patches tended to go extinct purely through dispersal behaviour. Conversely, one may also expect the opposite possibility of selection for greater dispersal to areas where fitness is higher and thus for selection of longer dispersal distances as suitable habitat patches become ever more distant (Hanski, chapter 20). In the latter case, those who disperse furthest will have a disproportionate influence on the survival of meta-populations and gene flow (Wiens, chapter 7). Roff and Fairbairn (chapter 13) review evidence that many physiological and behavioural traits defining migratory propensity are at a level of heritability enabling rapid response to selection pressure. They also report that different migratory and fitness traits are genetically correlated; thus, selection on one may have considerable pleiotropic effects on the species' life history and behaviour (see the experimental references in Roff and Fairbairn, chapter 13). Thus, the evolution of dispersal in altered habitats seems both to be feasible and to have important potential consequences (is it also possible that selection might simply favour behavioural flexibility?). Nevertheless, Hanski (chapter 20) tells us that empirical studies on the evolution of dispersal rates in areas experiencing increasing fragmentation are 'practically non-existent', and theoretical work also remains to be done in this area. One can only, therefore, make a plea to fill this most important of gaps in our knowledge. What if we were to succeed in developing models and empirical results for 'model' populations (Stenseth and Lidicker 1992), only to find that, when applied to endangered animals, their long exposure to new selection pressures in an anthropogenic landscape has altered their dispersal behaviour?

Conclusions: crisis management?

Several authors have observed that a comprehensive review on dispersal is no longer a practical possibility because of the enormity of relevant literature. Yet, as has been repeatedly pointed out in this book, there are still large gaps in our knowledge, and disappointingly few available estimates of dispersal variables from natural populations. The contributions to this book have both highlighted these gaps and made significant efforts to fill them, or at least pointed the way for future studies to do so. Academic and conservation-based interest in the subject, because of the increasingly critical state of the latter, should be combined. 'The sixth great extinction spasm of geological time is upon us' (Wilson 1992, p. 327); furthering our understanding of dispersal is by no means an idle concern. Future work for conservation can focus on three principal angles of attack: detailed field studies of metapopulations (Hanski *et al.* 1995a), experimental manip-ulations that can control for the biases of field studies and allow detailed hypothesis testing (Massot *et al.* 1994; Sarrazin and Barbault 1996), and, finally, detailed individ-ual process-based modelling (D.W. Macdonald and S.P. Rushton, unpublished manuscript).

Conservation has two principal problems: first, reducing deterministic habitat destruction by humans and contamination leading to reduced species populations and, second, minimizing the effects of stochastic processes on these resulting small and iso-lated populations. Dispersal is particularly crucial to determining a species' ability to cope with both of these processes. Dispersal may be the last saving grace sustaining dwindling populations above a threshold under which demographic or genetic stochastic events become lethal. Conservation often involves active management, and in particular habitat restoration and damage minimization. These types of management activities often tend to be on a large scale, partly because of national environmental policy, and therefore are undertaken without time or opportunities for experimentation, yet with considerable costs attached to making mistakes (Doak and Mills 1994). So, despite the current paucity of parameter estimates, once equipped with reliable models in the future it might be possible to confine some of these mistakes to the virtual reality of the computer, thereby better informing decisions that are required in the field. Because conservation decisions ultimately have to be enacted in real landscapes, process-based models conducted in GIS provide an arena for potentially powerful predictive analyses. Obviously, models cannot precisely simulate the behaviour of the model species. Indeed, models are only models, and provide projections rather than predictions, so we do not expect even the best of them to capture every nuance of behaviour that could be observed in the field. Rather, we would judge them useful if they approximated reality sufficiently to help identify the main parameters that decision-makers must bear in mind, and on which conservationists might focus attention (Verboom *et al.* 1993). One can cautiously (Mills *et al.* 1999) examine, through sensitivity analysis, specifically how dispersal ranks in importance as a predictor of population size or persistence, and thus implement management of the key variable responsible for decline.

Accurate parameterization remains a vital prerequisite for all these models, so reliable results are still limited by poor field data. The plight of endangered species poses a thrilling impetus for dispersal modelling, offering the chance for researchers to make a significant contribution. With that opportunity comes responsibility: in the practical arena, models face a daunting test, insofar as the cost of misjudgements may be measured in extinctions.

Acknowledgements

We are grateful to Drs W. Cresswell, J. McNutt, and R. Woodroffe for helpful comments on an earlier draft, and to Jean Clobert for his invitation to attend the Roscoff workshop which led to this review.

References

Aars, J. and Ims, R.A. (1999). The effect of habitat corridors on rates of transfer and interbreeding between vole demes. *Ecology*, **80**, 1648–55.

Aars, J. and Ims, R.A. (2000). Spatial density dependent transfer and gene frequency changes in patchy Vole populations. *American Naturalist*, **155**, 252–65.

Aars, J., Andreassen, H.P., and Ims, R.A. (1995). Root voles: litter sex ratio variation in fragmented habitat. *Journal of Animal Ecology*, **64**, 459–72.

Aars, J., Johannesen, E., and Ims, R.A. (1999). Demographic consequences of movements in subdivided root vole populations. *Oikos*, **85**, 204–17.

Abrams, P.A. (1999). The adaptive dynamics of consumer choice. *American Naturalist*, **153**, 83–97.

Addicott, J.F., Aho, J.M., Antolin, M.F., Padilla, D.K., Richardson, J.S., and Soluk, D.A. (1987). Ecological neighborhoods: scaling environmental patterns. *Oikos*, **49**, 340–6.

Adkins-Regan, E. (1987). Sexual differentiation in birds. *Trends in Neurosciences*, **10**, 517–22.

Adkins-Regan, E., Ottinger, M.A., and Park, J. (1995). Maternal transfer of estradiol to egg yolk alters sexual differentiation of avian offspring. *Journal of Experimental Zoology*, **271**, 466–70.

Adriaensen, F., Verwimp, N., and Dhondt, A.A. (1998). Between cohort variation in dispersal distance in the European kestrel *Falco tinnunculus* as shown by ringing recoveries. *Ardea*, **86**, 147–52.

Agren, J. and Schemske, D.W. (1993). Outcrossing rates and inbreeding depression in two annual monoecious herbs, *Begonia hirsuta* and *B. semiovata*. *Evolution*, **47**, 125–35.

Ahima, R.S., Prabakaran, D., Mantzoros, C., Qu, D., Lowell, B., Maratos-Flier, E., and Flier, J.S. (1996). Role of leptin in the neuroendocrine response to fasting. *Nature*, **382**, 250–2.

Ahima, R.S., Prabakaran, D., and Flier, J.S. (1998). Postnatal leptin surge and regulation of circadian rhythm of leptin by feeding: implications for energy homeostasis and neuroendocrine function. *Journal of Clinical Investigation*, **101**, 1020–7.

Akaike, H. (1973). Information theory and an extension of the maximum likelihood principle. In *Proceedings of the 2nd International Symposium on Information Theory* (eds. B. Petrov and F. Czakil), pp. 267–81. Akademiai Kiado Budapest.

Alatalo, R.V., Lundberg, A., and Björklund, M. (1982). Can the song of male birds attract other males? An experiment with the pied flycatcher *Ficedula hypoleuca*. *Bird Behaviour*, **4**, 42–5.

Alatalo, R.V., Lundberg, A., and Glynn, C. (1986). Female pied flycatchers choose territory quality and not male characteristics. *Nature*, **323**, 152–3.

Allee, W.C. (1951). *The social life of animals*. Beacon Press, Boston, MA.

Allison, P.D. (1995). *Survival analysis using the SAS system: a practical guide*. SAS Institute, Cary, NC.

Andersen, N.M. (1993). The evolution of wing polymorphism in water striders (gerridae), a phylogenetic approach. *Oikos*, **67**, 433–43.

Anderson, P.K. (1989). Dispersal in rodents: a resident fitness hypothesis. *American Society of Mammalogists Special Publication*, **9**, 141.

Andersson, M. (1994). *Sexual selection. Monographs in Behavior and Ecology*. Princeton University Press, Princeton, NJ.

Andow, D.A., Kareiva, P.M., Levin, S.A., and Okubo, A. (1990). Spread of invading organisms. *Landscape Ecology*, **4**, 177–88.

Andreassen, H.P. and Ims, R.A. (1990). Responses of female grey-backed voles *Clethrionomys rufocanus* to malnutrition: a combined laboratory and field experiment. *Oikos*, **59**, 107–14.

Andreassen, H.P., Hertzberg, K., and Ims, R.A. (1998). Space-use responses to habitat fragmentation and connectivity in the root vole *Microtus oeconomus*. *Ecology*, **79**, 1223–35.

Andrén, H. (1992). Corvid density and nest predation in relation to forest fragmentation: a landscape perspective. *Ecology*, **73**, 794–804.

Andrén, H. (1994). Effects of habitat fragmentation in birds and mammals in landscapes with different proportions of suitable habitat: a review. *Oikos*, **71**, 355–66.

Andrén, H., Delin, A., and Seiler, A. (1997). Population response to landscape changes depends on specialization to different landscape elements. *Oikos*, **80**, 193–6.

Andrew, R.J. and Rogers, L. (1972). Testosterone, search behaviour, and persistence. *Nature*, **237**, 343–6.

Angelstam, P. (1992). Conservation of communities. The importance of edges, surroundings and landscape mosaic structure. In *Ecological principles of nature conservation* (ed. L. Hansson), pp. 9–70. Elsevier Applied Science, London.

Anholt, B.R. (1990). Size-biased dispersal prior to breeding in a damselfy. *Oecologia*, **83**, 385–7.

Applebaum, S.W. and Heifetz, Y. (1999). Density-dependent physiological phase in insects. *Annual Review of Entomology*, **44**, 317–41.

Aquadro, C.F., Begun, D.J., and Kindahl, E.C. (1995). Selection, recombination and DNA polymorphism in *Drosophila*. In *Non-neutral evolution* (ed. B. Golding), pp. 46–56. Chapman and Hall, London.

Arcese, P. (1987). Age, experience and defense of territories against floaters in the song sparrow. *Animal Behaviour*, **35**, 773–84.

Arcese, P. (1989a). Intrasexual competition, mating system and natal dispersal in song sparrows. *Animal Behaviour*, **38**, 958–79.

Arcese, P. (1989b). Territory acquisition and loss in male song sparrows. *Animal Behaviour*, **37**, 45–55.

Armes, N.J. and Cooter, R.J. (1991). Effects of age and mated status on flight potential of *Helicoverpa armigera* (Lepidoptera: Noctuidae). *Physiological Entomology*, **16**, 131–44.

Armitage, K.B. (1986). Individual differences in the behavior of juvenile yellow-bellied marmots. *Behavioral Ecology and Sociobiology*, **18**, 419–24.

Arnason, A.N. (1973). The estimating of population size, migration rates and survival in a stratified population. *Researches on Population Ecology*, **15**, 1–8.

Arvedlund, M. and Nielsen, L.E. (1996). Do the anemonefish *Amphiprion ocellaris* (Pisces: Pomacentridae) imprint themselves to their host sea anemone *Heteractis magnifica* (Anthozoa: Actinidae)? *Ethology*, **102**, 197–211.

Ashton, P.M.S. and Berlyn, G.P. (1992). Leaf adaptations in some *Shorea* species to sun and shade. *New Phytologist*, **121**, 587–96.

Ashwell, C.M., Czerwinski, S.M., Brocht, D.M., and McMurtry, J.P. (1999). Hormonal regulation of leptin expression in broiler chickens. *American Journal of Physiology*, **276**, R226–R232.

Asmussen, M.A., Arnold, J., and Avise, J.C. (1987). Definition and properties of disequilibrium statistics for associations between nuclear and cytoplasmic genotypes. *Genetics*, **115**, 755–68.

Astheimer, L.B., Buttemer, W.A., and Wingfield, J.C. (1992). Interactions of corticosterone with feeding, activity and metabolism in passerine birds. *Ornis Scandinavica*, **23**, 355–65.

Audy, M.C. (1976). Le cycle sexuel saisonnier du mâle des Mustélidés européens. *General and Comparative Endocrinology*, **30**, 117–27.

Auerbach, M. and Shmida, A. (1987). Spatial scale and the determinants of plant species richness. *Trends in Ecology and Evolution*, **2**, 238–42.

Augspurger, C.K. and Katijama, K. (1992). Experimental studies of seedling recruitment from contrasting seed distributions. *Ecology*, **73**, 1270–84.

Avise, J.C. (1994). *Molecular markers, natural history and evolution*. Chapman and Hall, New York.

Badyaev, A.V., Martin, T.E., and Etges, W.J. (1996). Habitat sampling and habitat selection by female wild turkeys: ecological correlates and reproductive consequences. *Auk*, **113**, 636–46.

Bahlo, M. and Griffiths, R.C. (2000). Inference from gene trees in a subdivided population. *Theoretical Population Biology*, **57**, 79–95.

Baird, S.J.E. (1995). The mixing of genotypes in hybrid zones: a simulation study of multi-locus clines. *Evolution*, **49**, 1038–45.

Baker, A.J., Dietz, J.M., and Kleiman, D.G. (1993). Behavioral evidence for monopolization of paternity in multi-male groups of golden lion tamarins. *Animal behaviour*, **46**, 1091–103.

Balkau, B.J. and Feldman, M.W. (1973). Selection for migration modification. *Genetics*, **74**, 171–4.

Ball, D.E. and Vinson, S.B. (1983). Mating in the fire ant, *Solenopsis invicta* Buren: evidence that alates mate only once. *Journal of the Georgia Entomological Society*, **18**, 287–91.

Balloux, F., Brünner, H., Lugon-Moulin, N., Hausser, J., and Goudet, J. (2000). Microsatellites can be misleading: an empirical and simulation study. *Evolution*, **54**, 1414–22.

Balmford, A. and Long, A. (1994). Avian endemism and forest loss. *Nature*, **372**, 623–4.

Barker, R.J. (1995). *Open population mark–recapture models including ancillary sightings*. PhD thesis, Massey University, Palmerston North, New Zealand.

Barker, R.J. (1997). Joint modeling of live recapture, tag resight, and tag recovery data. *Biometrics*, **53**, 666–77.

Barnard, C.J. and Behnke, J.M. (ed.) (1990). *Parasitism and host behaviour*. Taylor and Francis, London.

Barreto, G.R., Rushton, S.P., Strachan, R., and Macdonald, D.W. (1998a). The role of habitat and mink predation in determining the status and distribution of declining populations of water voles in England. *Animal Conservation*, **1**, 129–37.

Barreto, G.R., Macdonald, D.W., and Strachan, R. (1998b). The tightrope hypothesis: an explanation for plummeting water vole numbers in the Thames catch-

ment. In *United Kingdom Floodplains* (ed. R. Bailey, P.V. Vose and B.R. Sherwood), pp. 311–27. Westbury academic and scientific publishing, London.

Barrett, G.W. and Peles, J.D. (ed.) (1999). *Landscape ecology of small mammals*. Springer, New York.

Barton, N.H. (1979). The dynamics of hybrid zones. *Heredity*, **43**, 341–59.

Barton, N.H. (1982). The structure of the hybrid zone in *Uroderma bilobatum* (Chiroptera: Phyllostomatidae). *Evolution*, **36**, 863–6.

Barton, N.H. (1987). The probability of establishment of an advantageous mutation in a subdivided population. *Genetical Research*, **50**, 35–40.

Barton, N.H. (1993). The probability of fixation of a favoured allele in a subdivided population. *Genetical Research*, **62**, 149–58.

Barton, N.H. and Bengtsson, B.O. (1986). The barrier to genetic exchange between hybridising populations. *Heredity*, **56**, 357–76.

Barton, N.H. and Clark, A. (1990). Population structure and process in evolution. In *Population biology: ecological and evolutionary viewpoints* (ed. K. Wöhrmann and S.K. Jain), pp. 115–73. Springer, Berlin.

Barton, N.H. and Gale, K.S. (1993). Genetic analysis of hybrid zones. In *Hybrid zones and the evolutionary process* (ed. R.G. Harrison), pp. 13–45. Oxford University Press, Oxford.

Barton, N.H. and Hewitt, G.M. (1981). The genetic basis of hybrid inviability between two chromosomal races of the grasshopper *Podisma pedestris*. *Heredity*, **47**, 367–83.

Barton, N.H. and Rouhani, S. (1993). Adaptation and the 'shifting balance'. *Genetical Research*, **61**, 57–74.

Barton, N.H. and Slatkin, M. (1986). A quasi-equilibrium theory of the distribution of rare alleles in a subdivided population. *Heredity*, **56**, 409–16.

Barton, N.H. and Turelli, M. (1989). Evolutionary quantitative genetics: how little do we know? *Annual Review of Genetics*, **23**, 337–70.

Barton, N.H. and Whitlock, M. (1996). The evolution of metapopulations. In *Metapopulation biology* (ed. I. Hanski and

M. Gilpin), pp. 183–210. Academic Press, San Diego, CA.

Barton, N.H. and Whitlock, M.C. (1997). The evolution of metapopulations. In *Metapopulation biology: ecology, genetics and evolution* (ed. I.A. Hanski and M.E. Gilpin), pp. 183–210. Academic Press, San Diego, CA.

Barton, N.H. and Wilson, I. (1995). Genealogies and geography. *Philosophical Transactions of the Royal Society of London B*, **349**, 49–59.

Bartz, S.H. (1979). Evolution of eusociality in termites. *Proceedings of the National Academy of Sciences*, **76**, 5764–8.

Bascompte, J. and Solé, R.V. (1994). Spatially induced bifurcations in single-species population dynamics. *Journal of Animal Ecology*, **63**, 256–64.

Bascompte, J. and Solé, R.V. (1998). *Modeling spatiotemporal dynamics in ecology*. Springer, New York.

Bateson, P.P.G. (1983). Optimal outbreeding. In *Mate choice* (ed. P. Bateson), pp. 257–77. Cambridge University Press, Cambridge.

Baum, M.J., Brand, T., Ooms, M., Vreeburg, J.T., and Slob, A.K. (1988). Immediate postnatal rise in whole body androgen content in male rats: correlation with increased testicular content and reduced body clearance of testosterone. *Biology of Reproduction*, **38**, 980–6.

Bayliss, D.E. (1993). Spatial distribution of *Balanus amphitrite* and *Elminius adelaidae* on mangrove pneumatophores. *Marine Biology*, **116**, 251–6.

Bazykin, A.D. (1969). Hypothetical mechanism of speciation. *Evolution*, **23**, 685–7.

Bazzaz, F.A. (1991). Habitat selection in plants. *American Naturalist*, **137**: 116–30.

Bazzaz, F.A. (1996). *Plants in changing environments*. Cambridge University Press, Cambridge.

Bazzaz, F.A. and Sultan, S.E. (1987). Ecological variation and the maintenance of plant diversity. In *Differentiation patterns in higher plants* (ed. K.M. Urbanska), pp. 69–93. Academic Press, London.

Beaumont, M.A. and Nichols, R.A. (1996). Evaluating loci for use in the genetic analysis of population structure. *Proceedings of the Royal Society of London B*, **263**, 1619–26.

Beck, B.B., Rapaport, L.G., Price, S., Wilson, M.R., and Wilson, A.C. (1994). Reintroduction of captive-born animals. In *Creative conservation: interactive management of wild and captive animals* (ed. P.J. Olney, G.M. Mace, and A.T. Feistner), pp. 265–86. Chapman and Hall, London.

Beerli, P. and Felsenstein, J. (1999). Maximum likelihood estimation of migration rates and effective population numbers in two populations using a coalescent approach. *Genetics*, **152**, 763–73.

Begon, M., Harper, J.L., and Townsend, C.R. (1996). *Ecology*, 3rd edn. Blackwell Science, Cambridge, MA.

Bekoff, M. (1977). Mammalian dispersal and the ontogeny of individual behavioural phenotypes. *American Naturalist*, **111**, 715–32.

Beletsky, L.D. and Orians, G.H. (1991). Effects of breeding experience and familiarity on site fidelity in female red-winged blackbirds. *Ecology*, **72**, 787–96.

Bélichon, S., Clobert, J., and Massot, M. (1996). Are there differences in fitness components between philopatric and dispersing individuals? *Acta Oecologica*, **17**, 503–17.

Bell, D.A., Trail, P.W., and Baptista, L.F. (1998). Song learning and vocal tradition in nuttall's white-crowned sparrows. *Animal Behaviour*, **55**, 939–56.

Bellamy, P.E., Brown, N.J., Enoksson, B., Firbank, L.G., Fuller, R.J., Hinsley, S.A., and Schotman, A.G.M. (1998). The influences of habitat, landscape structure and climate on local distribution patterns of the nuthatch (*Sitta europaea L.*). *Oecologia*, **115**, 127–36.

Belthoff, J.R. and Dufty, A.M., Jr. (1995). Activity levels and the dispersal of western screech-owls, *Otus kennicottii*. *Animal Behaviour*, **50**, 558–61.

Belthoff, J.R. and Dufty, A.M., Jr. (1998). Corticosterone, body condition, and locomotor activity: a model for natal dispersal. *Animal Behaviour*, **54**, 405–15.

Belthoff, J.R. and Ritchison, G. (1989). Natal dispersal of eastern screech owls. *Condor*, **91**, 254–65.

Bena, G., Prosperi, J.-M., Lejeune, B., and Olivieri, I. (1998). Evolution of annual species of the genus Medicago: a molecular phylogenetic approach. *Molecular Phylogenetics and Evolution*, **9**, 552–9.

Bengtsson, B.O. (1978). Avoiding inbreeding: at what cost? *Journal of Theoretical Biology*, **73**, 439–44.

Bengtsson, G., Hedlund, K., and Rundgren, S. (1994a). Food- and density-dependent dispersal: evidence from a soil collembolan. *Journal of Animal Ecology*, **63**, 513–20.

Bengtsson, J., Fargerstrom, T., and Rydin, H. (1994b). Competition and coexistence in plant communities. *Trends in Ecology and Evolution*, **9**, 246–50.

Bennett, A.F. (1999). *Linkages in the landscapes: the role of corridors and connectivity in wildlife conservation*. IUCN Publications, Cambridge.

Bennetts, R.E. and Kitchens, W.M. (1997). *The demography and movements of snail kites*. Florida U.S. Geological Survey/Biological Resources Division, Cooperative Fish and Wildlife Research Unit, Technical Report, **56**. Gainesville, FL.

Bennetts, R.E., Sparks, S.A., and Jansen, D. (2000). Factors influencing movement probabilities of Florida Tree Snails *Liguus fasciatus* (Muller) in Big Cypress National Preserve following hurricane Andrew. *Malacologia*, **42**, 31–7.

Bennun, L. (1994). The contribution of helpers to feeding nestlings in grey capped social weavers, *Pseudonigrita arnaudi*. *Animal Behaviour*, **47**, 1047–56.

Bensch, S., Hasselquist, D., and von Schantz, T. (1994). Genetic similarity between parents predicts hatching failure: nonincestuous inbreeding in the great reed warbler. *Evolution*, **48**, 317–26.

Berger, J. (1990). Persistence of different-sized populations: an empirical assessment of rapid extinctions in bighorn sheep. *Conservation Biology*, **4**, 91–8.

Berger, J. (1999). Intervention and persistence in small populations of bighorn sheep. *Conservation Biology*, **13**, 432–5.

Bernstein, C. (1984). Prey and predator emigration responses in the acarine system *Tetranychus urticae–Phytoseiulus persimilis*. *Oecologia*, **61**, 134–42.

Bernstein, C., Kacelnik A., and Krebs, J.R. (1988). Individual decisions and the distribution of predators in a patchy environment. *Journal of Animal Ecology*, **57**, 1007–26.

Bernstein, C., Krebs, J.R., and Kacelnik, A. (1991). Distribution of birds among habitats: theory and relevance to conservation. In *Bird population studies. Relevance to conservation and management* (ed. C.M. Perrins, J.D. Lebreton, and G.J.M. Hirons), pp. 317–45. Oxford University Press, Oxford.

Bernstein, C., Kacelnik, A., and Krebs, J.R. (1992). Individual decisions and the distribution of predators in a patchy environment. II. The influence of travel costs and the structure of the environment. *Journal of Animal Ecology*, **60**, 205–25.

Bernstein, C., Auger, P., and Poggiale, J.C. (1999). Predator migration decisions, the ideal free distribution, and predator–prey dynamics. *American Naturalist*, **153**, 267–81.

Berry, A. and Kreitman, M. (1993). Molecular analysis of an allozyme cline: alcohol dehydrogenase in *Drosophila melanogaster* on the east coast of N America. *Genetics*, **134**, 869–93.

Berthold, P. (1988). Evolutionary aspects of migratory behavior in European warblers. *Journal of Evolutionary Biology*, **1**, 195–209.

Berthold, P. and Pulido, F. (1994). Heritability of migratory activity in a natural bird population. *Proceedings of the Royal Society of London B*, **257**, 311–15.

Bertness, M.D. and Leonard, G.H. (1997). The role of positive interactions in communities: lessons from intertidal habitats. *Ecology*, **78**, 1976–89.

Bertram, B.C.R. (1975). Social factors influencing reproduction in wild lions. *Journal of Zoology*, **177**, 463–82.

Bertram, B.C.R. (1978). *Pride of lions*. Scribners, New York.

Biebach, H. (1983). Genetic determination of partial migration in the European robin (*Erithacus rubecula*). *Auk*, **100**, 601–6.

Biggins, D.E., Godbey, J.L., Hanebury, L.R., Luce, B., Marinari, P.E., Matchett, M.R., and Vargas, A. (1998). The effect of rearing methods on survival of reintroduced black-footed ferrets. *Journal of Wildlife Management*, **62**, 643–53.

Bjørnstad, O.N., Ims, R.A., and Lambin, X. (1999). Spatial population dynamics: analyzing patterns and processes of population synchrony. *Trends in Ecology and Evolution*, **14**, 427–32.

Blackburn, G.S., Wilson, D.J., and Krebs, C.J. (1998). Dispersal of juvenile collared lemmings (*Dicrostonys groenlandicus*) in a high-density population. *Canadian Journal of Zoology*, **76**, 2255–61.

Blaustein, L. (1998). Influence of the predatory backswimmer, *Notonecta maculata*, on invertebrate community structure. *Ecological Entomology*, **23**, 246–52.

Blaustein, L. (1999). Oviposition site selection in response to risk of predation: evidence from aquatic habitats and consequences for population dynamics and community structure. In *Evolutionary theory and processes: modern perspectives* (ed. S.P. Wasser), pp. 441–56. Kluwer Academic, The Netherlands.

Blem, C.R. (1976). Patterns of lipid storage and utilization in birds. *American Zoology*, **16**, 671–84.

Blem, C.R. (1990). Avian energy storage. In *Current Ornithology* (ed. D.M. Power), **7**, pp. 59–113. Plenum Press, New York.

Blouin, M.S., Yowell, C.A., Courtney, C.H., and Dame, J.B. (1995). Host movement and the genetic structure of populations of parasitic nematodes. *Genetics*, **141**, 1007–14.

Blouin, S.F. and Blouin, M. (1988). Inbreeding avoidance behaviors. *Trends in Ecology and Evolution*, **3**, 230–3.

Blower, S.M. and Roughgarden, J. (1989). Parasites detect host spatial pattern and density: a field experimental analysis. *Oecologia*, **78**, 138–41.

Blumstein, D.T. and Arnold, W. (1998). Ecology and social behavior of golden marmots (*Marmota caudata aurea*). *Journal of Mammalogy*, **79**, 873–86.

Boecklen, W.J. and Howard, D.J. (1997). Genetic analysis of hybrid zones: number of markers and power of resolution. *Ecology*, **78**, 2611–16.

Boerlijst, M.C., Lamers, M., and Hogeweg, P. (1993). Evolutionary consequences of spiral patterns in a host–parasitoid system. *Proceedings of the Royal Society of London B*, **253**, 15–18.

Boersma, M., Spaak, P., and De Meester, L. (1998). Predator-mediated plasticity in morphology, life history, and behavior of *Daphnia:* the uncoupling of responses. *American Naturalist*, **152**, 237–48.

Bogliani, G., Sergio, F., and Tavecchia, G. (1999). Woodpigeons nesting in association with hobby falcons: advantages and choice rules. *Animal Behaviour*, **57**, 125–31.

Bohonak, A.J. (1999). Dispersal, gene flow, and population structure. *Quarterly Review of Biology*, **74**, 21–45.

Boissin, J., Nouguier-Soulé, J., and Assenmacher, I. (1969). Circannual and circadian rhythms of adrenal cortical functions in birds. *Indian Journal of Zootomy*, **10**, 187–96.

Bolker, B.M. and Pacala S.W. (1999). Spatial moment for plant competition: understanding spatial strategies and the advantages of short dispersal. *American Naturalist*, **153**, 575–602.

Bolker, B.M., Altmann, M., Aubert, M., Ball, F., Barlow, N.D., Bowers, R.G., Dobson, A.P., Elkington, J.S., Garnett, G.P., Gilligan, C.A., Hassell, M.P., Isham, V., Jacquez, J.A., Kleczkowski, A., Levin, S.A., May, R.M., Metz, J.A.J., Mollinson, D., Morris, M., Real, L., Sattenspiel, L., Swinton, J., White, P., and Williams, B.G. (1995). Group report: spatial dynamics of infectious diseases in natural populations. In *Ecology of infectious diseases in natural populations* (ed. B.T. Grenfell and A.P. Dobson), pp. 399–420. Cambridge University Press, Cambridge.

Bollinger, E.K. and Gavin, T.A. (1989). The effect of site quality on breeding-site fidelity in Bobolinks. *Auk*, **106**, 584–94.

Bondrup-Nielsen, S. (1985). An evaluation of the effects of space use and habitat patterns on dispersal in small mammals. *Annales Zoologici Fennici*, **22**, 373–83.

Bondrup-Nielsen, S. (1992). Emigration of meadow voles, *Microtus pennsylvanicus*: the effect of sex ratio. *Oikos*, **65**, 358–60.

Bondrup-Nielsen, S. and Ims, R.A. (1988). Predicting stable and cyclic populations of *Clethrionomys*. *Oikos*, **52**, 178–85.

Boonstra, R. (1989). Life history variation in maturation in fluctuating meadow vole populations (*Microtus pennsylvannicus*). *Oikos*, **54**, 265–74.

Booth, D.J. (1992). Larval settlement patterns and preferences by domino damselfish *Dascyllus albisella* Gill. *Journal of Experimental Marine Biology and Ecology*, **155**, 85–104.

Boots, M. and Sasaki, A. (1999). 'Small worlds' and the evolution of virulence: infection occurs locally and at a distance. *Proceedings of the Royal Society of London B*, **266**, 1933–8.

Bornstein, S.R. (1997). Is leptin a stress related peptide? *Nature Medicine*, **3**, 937.

Bornstein, S.R., Licinio, J., Tauchnitz, R., Engelmann, L., Negrao, A.B., Gold, P., and Chrousos, G.P. (1998). Plasma leptin levels are increased in survivors of acute sepsis: associated loss of diurnal rhythm in cortisol and leptin secretion. *Journal of Clinical Endocrinology*, **83**, 280–3.

Boulinier, T. (1996). On breeding performance, colony growth and habitat selection in buff-necked ibis. *Condor*, **98**, 440–1.

Boulinier, T. and Danchin, E. (1996). Population trends in kittiwake *Rissa tridactyla* colonies in relation to tick infestation. *Ibis*, **138**, 326–34.

Boulinier, T. and Danchin, E. (1997). The use of conspecific reproductive success for breeding patch selection in territorial migratory species. *Evolutionary Ecology*, **11**, 505–17.

Boulinier, T. and Lemel, J.-Y. (1996). Spatial and temporal variations of factors affecting breeding habitat quality in colonial birds: some consequences for dispersal and habitat selection. *Acta Oecologica*, **17**, 531–52.

Boulinier, T., Danchin, E., Monnat, J.-Y., Doutrelant, C., and Cadiou, B. (1996a). Timing of prospecting and the value of information in a colonial breeding bird. *Journal of Avian Biology*, **27**, 252–6.

Boulinier, T., Ives, A.R., and Danchin, E. (1996b). Measuring aggregation of parasites at different host population levels. *Parasitology*, **112**, 581–7.

Boulinier, T., Sorci, G., Monnat, J.-Y., and Danchin, E. (1997). Parent–offspring regression suggests heritable susceptibility to ectoparasites in a natural population of kittiwake

Rissa tridactyla. *Journal of Evolutionary Biology*, **10**, 77–85.

Boulinier, T., Danchin, E., and Durand, S. (1999). Conspecific attraction and breeding site selection in kittiwakes: an experiment. In *Proceedings of the 22nd International Ornithological Congress* (ed. N.J. Adams and R.H. Slotow), pp. 1315–27. University of Natal Press, Durban.

Bourke, A.F.G. and Franks, N.R. (1995). *Social evolution in ants*. Princeton University Press, Princeton, NJ.

Bowcock, A.M., Ruiz-Linares, A., Tomfohrde, J., Minch, E., Kidd, J.R., and Cavalli-Sforza, L.L. (1994). High resolution of human evolutionary trees with polymorphic microsatellites. *Nature*, **368**, 455–7.

Bowne, D.R., Peles, J.D., and Barrett, G.W. (1999). Effects of landscape spatial structure on movement patterns of the hispid cotton rat (*Sigmodon hispidus*). *Landscape Ecology*, **14**, 53–65.

Boyce, M.S. (1992). Population viability analysis. *Annual Review of Ecology and Systematics*, **23**, 481–506.

Boyce, M.S. and Perrins, C.M. (1987). Optimizing great tit clutch size in a fluctuating environment. *Ecology*, **68**, 142–53.

Boyd, S.K. (1991). Effect of vasotocin on locomotor activity in bullfrogs varies with developmental stage and sex. *Hormones and Behavior*, **25**, 57–69.

Brandt, C.A. (1985). The evolution of sexual differences in natal dispersal: tests of Greenwood's hypothesis. *Contributions in Marine Science (Suppl.)*, **27**, 386–96.

Brandt, C.A. (1992). Social factors in immigration and emigration. In *Animal dispersal: small mammals as a model* (ed. N.C. Stenseth and W.Z.J. Lidicker), pp. 96–141. Chapman and Hall, London.

Braude, S. (1991). Which naked mole-rats volcano? In *The biology of the naked mole-rat* (ed. P.W. Sherman, J.U.M. Jarvis, and R.D. Alexander), pp. 185–94. Princeton University Press, Princeton, NJ.

Braude, S. (2000). Dispersal and new colony formation in wild naked-mole-rats: evidence against inbreeding as the system of mating. *Behavioral Ecology*, **11**, 7–12.

Bray, M.M. (1993). Effect of ACTH and glucocorticoids on lipid metabolism in the Japanese quail, *Coturnix coturnix japonica*. *Comparative Biochemistry and Physiology*, **105A**, 689–96.

Brett, R.A. (1986). *The ecology and behaviour of the naked mole-rat* (Heterocephalus glaber *Ruppell*) (*Rodentia: Bathyergidae*). PhD thesis, University of London, London.

Brett, R.A. (1991). The population structure of naked mole-rat colonies. In *The biology of the naked mole-rat* (ed. P.W. Sherman, J.U.M. Jarvis, and R.D. Alexander), pp. 97–136. Princeton University Press, Princeton, NJ.

Breuner, C.W., Greenberg, A.L., and Wingfield, J.C. (1998). Noninvasive corticosterone treatment rapidly increases activity in Gambel's white-crowned sparrows (*Zonotrichia leucophrys gambelii*). *General and Comparative Endocrinology*, **111**, 386–94.

Brewer, B.A., Lacy, R.C., Foster, M.L., and Alaks, G. (1990). Inbreeding depression in insular and central populations of *Peromyscus* mice. *Journal of Heredity*, **81**, 257–66.

Briggs, S.E., Godin, J.-G.J., and Dugatkin, L.A. (1996). Mate choice copying under predation risk in the Trinidadian guppy (*Poecilia reticulata*). *Behavioral Ecology*, **7**, 151–7.

Bright, P.W. and Morris, P.A. (1996). Why are dormice rare? A case study in conservation biology. *Mammal Review*, **26**, 157–87.

Brockman, H.J. and Penn, D. (1992). Mating tactics in the horseshoe crab, *Limulus polyphemus*. *Animal Behaviour*, **44**, 653–65.

Brodmann, P.A., Wilcox, C.V., and Harrison, S. (1997). Mobile parasitoids may restrict the spatial spread of an insect outbreak. *Journal of Animal Ecology*, **66**, 65–72.

Bronson, F.H. (1989). *Mammalian reproductive biology*. Chicago University Press, Chicago, IL.

Brooker, M.G., Rowley, M.A., Adams, M., and Baverstock, P.R. (1990). Promiscuity: an inbreeding avoidance mechanism in a socially monogamous species? *Behavioral Ecology and Sociobiology*, **26**, 191–9.

Brookfield, J.F.Y. (1997). Importance of ancestral DNA ages. *Nature*, **388**, 134.

Brooks, R. (1996). Copying and the repeatability of mate choice. *Behavioral Ecology and Sociobiology*, **39**, 323–9.

Brooks, R. (1998). The importance of mate copying and cultural inheritance of mating preferences. *Trends in Ecology and Evolution*, **13**, 45–6.

Brown, C.R. and Brown, M.B. (1986). Ectoparasitism as a cost of coloniality in cliff swallows (*Hirundo pyrrhonota*). *Ecology*, **67**, 1206–18.

Brown, C.R. and Brown, M.B. (1992). Ectoparasitism as a cause of natal dispersal in cliff swallows. *Ecology*, **73**, 1718–23.

Brown, C.R. and Brown, M.B. (1996). *Coloniality in the cliff swallow*. University of Chicago Press, Chicago, IL.

Brown, C.R., Bomberger Brown, M., and Danchin, E. (2000). The effect of conspecific reproductive success on colony choice in cliff swallows. *Journal of Animal Ecology*, **69**, 133–42.

Brown, E.S. (1951). The relation between migration rate and type of habitat in aquatic insects, with special reference to certain species of Corixidae. *Proceedings of the Zoological Society of London*, **121**, 539–45.

Brown, J. (1987). *Helping and communal breeding in birds. Ecology and evolution*. Princeton University Press, Princeton, NJ.

Brown, J.H. and Kodric-Brown, A. (1977). Turnover rates in insular biogeography: effect of immigration on extinction. *Ecology*, **58**, 445–9.

Brown, J.L., and Vleck, C.M. (1998). Prolactin and helping in birds: has natural selection strengthened helping behavior? *Behavioral Ecology*, **9**, 541–5.

Brown, J.S. (1998). Game theory and habitat selection. In *Game theory and animal behavior* (ed. L.A. Dugatkin and H.K. Reeve), pp. 188–220. Oxford University Press, Oxford.

Brownie, C., Anderson, D.R., Burnham, K.P., and Robson, D.S. (1985). *Statistical influence from band recovery data: a handbook*, 2nd edn. Resource Publication 156, U.S.D.I. Fish and Wildlife Service, Washington, D.C.

Brownie, C., Hines, J.E., Nichols, J.D., Pollock, K.H., and Hestbeck, J.B. (1993).

Capture–recapture studies for multiple strata including non-Markovian transition probabilities. *Biometrics*, **49**, 1173–87.

Bujalska, G. (1973). The role of spacing behaviour among females in the regulation of reproduction in the bank vole. *Journal of Reproduction and Fertility (Suppl.)*, **19**, 465–74.

Bulmer, M.G. (1986). Sex ratios in geographically structured populations. *Trends in Ecology and Evolution*, **1**, 35–8.

Bungo, T., Shimojo, M., Masuda, Y., Tachibanab, T., Tanaka, S., Sugahara, K., and Furuse, M. (1999). Intracerebroventricular administration of mouse leptin does not reduce food intake in the chicken. *Brain Research*, **817**, 196–8.

Burger, J. (1982). The role of reproductive success in colony-site selection and abandonment in black skimmers (*Rynchops niger*). *Auk*, **99**, 109–15.

Burger, J. (1988). Social attraction in nesting least terns: effects of numbers, spacing, and pair bonds. *Condor*, **90**, 575–82.

Burke, T., Davies, N.B. Bruford, M.W., and Hatchwell, B.J. (1989). Parental care and mating behaviour of polyandrous dunnocks *Prunella modularis* related to paternity by DNA fingerprinting. *Nature*, **338**, 249–251.

Burnham, K.P. (1993). A theory for combined analysis of ring recovery and recapture data. In *The use of marked individuals in the study of bird population dynamics*. (ed. J.D. Lebreton and P.M. North), pp. 199–213. Birkhauser, Basel.

Burnham, K.P. and Anderson, D.R. (1998). *Model selection and inference: a practical information – theoretic approach*. Springer, New York.

Burt, A. (1995). The evolution of fitness. *Evolution*, **49**, 1–8.

Bush, A.O., Lafferty, K.D., Lotz, J.M., and Shostak, A.W. (1997). Parasitology meets ecology on its own terms: Margolis *et al.* revisited. *Journal of Parasitology*, **83**, 575–83.

Bush, G.L. (1994). Sympatric speciation in animals: new wine in old bottles. *Trends in Ecology and Evolution*, **9**, 285–9.

Butlin, R.K., Ritchie, M.G., and Hewitt, G.M. (1991). Comparisons among morphological characters and between localities in the *Chorthippus paralellus* hybrid zone (Orthoptera: Acrididae). *Philosophical Transactions of the Royal Society of London B*, **334**, 297–308.

Bygott, J.D., Bertram, B.C., and Hanby, J.P. (1979). Male lions in large coalitions gain reproductive advantages. *Nature*, **282**, 838–40.

Cadet, C. (1998). Dynamique adaptative de la dispersion dans une métapopulation: modèles stochastiques densité-dépendants. Diplôme d'Etudes Approfondies, Université Paris VI, Paris.

Cadiou, B. (1999). Attendance of breeders and prospectors reflects the quality of colonies in the kittiwake *Rissa tridactyla*. *Ibis*, **141**, 321–6.

Cadiou, B., Monnat, J.-Y., and Danchin, E. (1994). Prospecting in the kittiwake, *Rissa tridactyla*: different behavioural patterns and the role of squatting in recruitment. *Animal Behaviour*, **47**, 847–56.

Caldwell, R.L. and Hegmann, J.P. (1969). Heritability of flight duration in the milkweed bug *Lygaeus kalmii*. *Nature*, **223**, 91–2.

Campbell, B.D. and Grime, J.P. (1992). An experimental test of plant strategy theory. *Ecology*, **73**, 15–29.

Caro, J.F., Sinha, M.K., Kolaczynski, J.W., Zhang, P.L., and Considine, R.V. (1996). Leptin: the tale of an obesity gene. *Diabetes*, **45**, 1455–62.

Caro, T.M. (1999). The behaviour–conservation interface. *Trends in Ecology and Evolution*, **14**, 366–9.

Caro, T.M. and Laurenson, M.K. (1994). Ecological and genetic factors in conservation: a cautionary tale. *Science*, **263**, 485–6.

Carriere, Y. and Roitberg, B.D. (1995). Evolution of host-selection behaviour in insect herbivores: genetic variation and covariation in host acceptance within and between populations of the obliquebanded leafroller, *Choristoneura rosaceana* (family: Tortricidae). *Heredity*, **74**, 357–68.

Caswell, H. (1989). *Matrix population models: construction, analysis, and interpretation*. Sinauer, Sunderland, MA.

Caughley, G. (1994). Directions in conservation biology. *Journal of Animal Ecology*, **63**, 215–44.

Cavalli-Sforza, L.L., Menozzi, P., and Piazza, A. (1994). *The history and geography of human genes*. Princeton University Press, Princeton, NJ.

Chambers, J.C. and MacMahon, J.A. (1994). A day in the life of a seed: movements and fates of seeds and their implications for natural and managed systems. *Annual Review of Ecology and Systematics*, **25**, 263–92.

Chapman, B.R. and George, J.E. (1991). The effects of ectoparasites on cliff swallow growth and survival. In *Bird–parasite interactions. Ecology, evolution and behaviour* (ed. J.E. Loye and M. Zuk), pp. 69–92. Oxford University Press, Oxford.

Chapman, C.A., Fedigan, L.M., Fedigan, L., and Chapman, L.J. (1989). Post-weaning resource competition and sex ratios in spider monkeys. *Oikos*, **54**, 315–19.

Charlesworth, B. (1994). *Evolution in age-structured populations*, 2nd edn. Cambridge University Press, Cambridge.

Charlesworth, B. (1998). Measures of divergence between populations and the effect of forces that reduce variability. *Molecular Biology and Evolution*, **15**, 538–43.

Charlesworth, B., Nordborg, M., and Charlesworth, D. (1997). The effects of local selection, balanced polymorphism and background selection on equilibrium patterns of genetic diversity in subdivided populations. *Genetical Research*, **70**, 155–74.

Charlesworth, D., Morgan, M.T., and Charlesworth, B. (1990). Inbreeding depression, genetic load, and the evolution of outcrossing rates in a multilocus system with no lineage. *Evolution*, **44**, 1469–89.

Charnov, E.L. (1976). Optimal foraging: the marginal value theorem. *Theoretical Population Biology*, **9**, 129–36.

Chase, M.E. and Bailey, R.C. (1996). Recruitment of *Dreissena polymorpha*: does the presence and density of conspecifics determine the recruitment density and pattern in a population? *Malacologia*, **38**, 19–31.

Chastel, C. (1988). Tick-borne virus infections of marine birds. In *Advances in disease vector research* (ed. K.F. Harris), pp. 25–60. Springer, New York.

Chazdon, R.L. (1986). Light variation and carbon gain in rainforest understory palms. *Journal of Ecology*, **74**, 995–1012.

Chazdon, R.L. and Kaufmann, S. (1993). Plasticity of leaf anatomy of two rain forest shrubs in relation to photosynthetic light acclimation. *Functional Ecology*, **7**, 385–94.

Chazdon, R.L., Pearcy, R., Lee, D., and Fetcher, N. (1996). Photosynthetic responses of tropical forest plants to contrasting light environments. In *Tropical Forest Plant Ecophysiology* (ed. S.S. Mulkey, R. Chazdon, and A.P. Smith). Chapman and Hall, New York.

Chen, X. (1993). Comparison of inbreeding and outbreeding in hermaphroditic *Arianta arbustorum* (L.) (land snail). *Heredity*, **71**, 456–61.

Chepko-Sade, B.D. and Halpin, Z.T. (1987). *Mammalian dispersal patterns*. University of Chicago Press, Chicago, IL.

Cherel, Y., Robin, J.-P., and Le Maho, Y. (1988). Physiology and biochemistry of long-term fasting in birds. *Canadian Jounal of Zoology*, **66**, 159–66.

Chesser, R.K. (1991a). Gene diversity and female philopatry. *Genetics*, **127**, 437–47.

Chesser, R.K. (1991b). Influence of gene flow and breeding tactics on gene diversity within populations. *Genetics*, **129**, 573–83.

Chesser, R.K. and Ryman, N. (1986). Inbreeding as a strategy in subdivided populations. *Evolution*, **40**, 616–24.

Chesser, R.K., Sugg, D.W., Rhodes, O.E., Nowak, J.M., and Smith, M.H. (1993). Evolution of mammalian social structure. *Acta Theriologica*, **38 (Suppl. 2)**, 163–74.

Chesson, P.L. and Warner, R.W. (1981). Environmental variability promotes coexistence in lottery competitive systems. *American Naturalist*, **117**, 923–43.

Christian, J.J. (1970). Social subordination, population density and mammalian evolution. *Science*, **168**, 84–90.

Ciszek, D. (2000). New colony formation in the 'highly inbred' eusocial naked mole-rat: outbreeding is preferred. *Behavioral Ecology*, **11**, 1–6.

Clark, A.G. (1990). Genetic components of variation in energy storage in *Drosophila melanogaster*. *Evolution*, **44**, 637–50.

Clark, C.W. and Mangel, M. (1984). Foraging and flocking strategies: information in an uncertain environment. *American Naturalist*, **123**, 626–41.

Clarke, A.B. (1978). Sex ratio and local resource competition in a prosimian primate. *Science*, **201**, 163–5.

Clarke, A.L., Saether, B.E., and Roskaft, E. (1997). Sex biases in avian dispersal: a reappraisal. *Oikos*, **79**, 429–38.

Clarke, B.C. (1966). The evolution of morph ratio clines. *American Naturalist*, **100**, 389–400.

Clayton, D.H. and Moore, J. (ed.) (1997). *Host–parasite evolution: general principles and avian models*. Oxford University Press, Oxford.

Clobert, J. (1995). Capture–recapture and evolutionary ecology: a difficult wedding? *Journal of Applied Statistics*, **22**, 989–1008.

Clobert, J. and Lebreton, J.-D. (1991). Estimation of demographic parameters in bird populations. In *Bird population studies: their relevance to conservation and management* (ed. C.M. Perrins, J.-D. Lebreton, and G.J.M. Hirons), pp. 75–104. Oxford University Press, Oxford.

Clobert, J., Perrins, C.M., McCleery, R.H., and Gosler, A.G. (1988). Survival rate in the great tit *Parus major* in relation to sex, age, and status. *Journal of Animal Ecology*, **57**, 287–306.

Clobert, J., Lebreton, J.-D., Allainé, D., and Gaillard, J.M. (1994a). The estimation of age-specific breeding probabilities from recaptures or resightings in vertebrate populations: II. Longitudinal models. *Biometrics*, **50**, 375–87.

Clobert, J., Massot, M., Lecomte, J., Sorci, G., de Fraipont, M., and Barbault, R. (1994b). Determinant of dispersal behavior: the common lizard as a case study. In *Lizard Ecology, historical and experimental perspectives* (ed. L.J. Vitt and E.R. Pianka), pp. 182–206. Princeton University Press, Princeton, NJ.

Clutton-Brock, T.H. (1989). Female transfer and inbreeding avoidance in social mammals. *Nature, 337*, 70–2.

Clutton-Brock, T.H., and McComb, K. (1993). Experimental tests of copying and mate choice in fallow deer (*Dama dama*). *Behavioral Ecology*, **4**, 191–3.

Clutton-Brock, T.H., Guiness, F., and Albon, S. (1982). *Red deer: behaviour and ecology of two sexes*. Chicago University Press, Chicago, IL.

Clutton-Brock, T.H., MacColl, A.D.C., Chadwick, P., Gaynor, D., Kansky, R., and Skinner, J.D. (1999). Reproduction and survival of suricates (*Suricata suricatta*) in the Southern Kalahari. *African Journal of Ecology* **37**, 69–75.

Cockburn, A. (1988). *Social behaviour in fluctuating populations*. Croom Helm, London.

Cockburn, A. (1992). Habitat heterogeneity and dispersal: environmental and genetic consequences. In *Animal dispersal: small mammals as a model* (ed. N.C. Stenseth and W.Z. Lidicker, Jr.), pp. 61–95. Chapman and Hall, London.

Cockburn, A., Scott, M.P., and Scotts, D.J. (1985). Inbreeding avoidance and male-biased dispersal in *Antechinus* ssp (Marsupialia, Dasyuridae). *Animal Behaviour*, **33**, 908–15.

Cockerham, C.C. (1973). Analyses of gene frequencies. *Genetics*, **74**, 679–700.

Cody, M.L. and Overton, J.M. (1996). Short-term evolution of reduced dispersal in island plant populations. *Journal of Ecology*, **84**, 53–61.

Cohan, F.M. and Hoffmann, A.A. (1989). Uniform selection as a diversifying force in evolution: evidence from *Drosophila*. *American Naturalist* **134**, 613–37.

Cohen, D. (1993). Fitness in random environments. In *Adaptation in stochastic environments* (ed. J. Yoshimura and C.W. Clark), pp. 8–25. Springer, Berlin.

Cohen, D. and Eshel, I. (1976). On the founder effect and the evolution of altruistic traits. *Theoretical Population Biology*, **10**, 276–302.

Colas, B., Olivieri, I., and Riba, M. (1997). *Centaurea corymbosa*, a cliff dwelling species tottering on the brink of extinction: a demographic and genetic study. *Proceedings of the National Academy of Sciences of the USA*, **94**, 3471–6.

Colvin, J. and Gatehouse, A.G. (1993). The reproduction–flight syndrome and the inheritance of tethered-flight activity in the cotton-bollworm moth *Heliothis armigera*. *Physiological Entomology*, **18**, 16–22.

Combes, C. (1991). Ethological aspects of parasite transmission. *American Naturalist*, **138**, 866–80.

Combes, C. (1995). *Interactions durables. Ecologie et évolution du parasitisme*. Masson, Paris.

Comeau, M., Conan, G.Y., Maynou, F., Robichaud, G., Therriault, J.C., and Starr, M. (1998). Growth, spatial distribution and abundance of benthic stages of the snow crab (*Chionoectes opilio*) in Bonne Bay, Newfoundland, Canada. *Canadian Journal of Fisheries and Aquatic Sciences*, **55**, 262–79.

Comins, H.N., Hamilton, W.D., and May, R.M. (1980). Evolutionarily stable dispersal strategies. *Journal of Theoretical Biology*, **82**, 205–30.

Commenville, P. (1998). *L'évolution des stratégies de dispersion dans les systèmes proies–prédateurs: perspectives nouvelles apportées par la modélisation*. Diplôme d'Etudes approfondies, Université Paris 6, Paris.

Condon, M.A. and Steck, G.J. (1997). Evolution of host use in fruit flies of the genus *Blepharoneura* (Diptera: Tephritidae): cryptic species on sexually dimorphic host plants. *Biological Journal of the Linnean Society*, **60**, 443–66.

Connolly, K. (1966). Locomotor activity in *Drosophila*. II. Selection for active and inactive strains. *Animal Behaviour*, **14**, 444–9.

Conroy, M.J., Anderson, J.E., Rathbun, S.L., and Krementz, D.G. (1996). Statistical inference on patch-specific survival and movement rates from marked animals. *Environmental and Ecological Statistics*, **3**, 99–118.

Constantino, R.F., Cushing, J.M., Dennis, B., and Desharnais, R.A. (1995). Experimentally induced transitions in the dynamic behavior of insects populations. *Nature*, **375**, 227–30.

Constantino, R.F., Desharnais, R.A., Cushing, J.M., and Dennis, B. (1997).

Chaotic dynamics in an insect population. *Science*, **275**, 389–91.

Cook, J.M. and Crozier, R.H. (1995). Sex determination and population biology in the Hymenoptera. *Trends in Ecology and Evolution*, **10**, 281–6.

Cornell, H.V. and Karlson, R.H. (1998). Local and regional processes as controls of species richness. In *Spatial Ecology* (ed. D. Tilman and P. Kareiva), pp. 250–68. Princeton University Press, Princeton, NJ.

Cornuet, J.M., Piry, S., Luikart, G., Estoup, A., and Solignac, M. (1999). New methods employing multilocus genotypes to select or exclude populations as origins of individuals. *Genetics*, **153**, 1989–2000.

Courchamp, F., Clutton-Brock, T., and Grenfell, B. (1999). Inverse density dependence and the Allee effect. *Trends in Ecology and Evolution* **14**, 405–10.

Cox, D.R. (1970). *The analysis of binary data*. Chapman and Hall, London.

Cox, D.R. (1972). Regression models and life tables. *Journal of the Royal Statistical Society B*, **34**, 187–220.

Cox, D.R. (1975). Partial likelihood. *Biometrika*, **62**, 209–76.

Cox, D.R. and Hinkley, D.V. (1974). *Theoretical Statistics*. Chapman and Hall, London.

Cox, D.R. and Oakes, D. (1984). Analysis of survival data. *Monographs on Statistics and Applied Probability*, **21**. Chapman and Hall, New York.

Coyne, J.A., Barton, N.H., and Turelli, M. (1997). Perspective: a critique of Sewall Wright's shifting balance theory of evolution. *Evolution*, **51**, 643–71.

Crawley, M.J. (1983). *Herbivory. The dynamics of animal–plant interactions*. Blackwell Scientific, Oxford.

Crawley, M.J. (1992). Seed predators and plant population dynamics. In *The ecology of regeneration in plant communities*. (ed. M. Fenner), pp. 157–92. CAB International, Wallingford, UK.

Crisp, D.J. (1974). Factors influencing the settlement of marine invertebrate larvae. In *Chemoreception of marine organisms* (ed. P.T. Grant and A.N. Mackie), pp. 177–265. Academic Press, New York.

Crist, T.O., Guertin, D.S., Wiens, J.A., and Milne, B.T. (1992). Animal movements in heterogeneous landscapes: an experiment with *Eleodes* beetles in shortgrass prairie. *Functional Ecology*, **6**, 536–44.

Crnokrak, P. and Roff, D.A. (1998). The genetic basis of the trade-off between calling and wing morph in males of the cricket, *Gryllus firmus. Evolution*, **52**, 1111–18.

Crow, J.F. and Kimura, M. (1970). *An introduction to population genetics theory*. Harper and Row, New York.

Crow, J.F., Engels, W.R., and Denniston, C. (1990). Phase three of Wright's shifting-balance theory. *Evolution*, **44**, 233–47.

Curio, E. (1993). Proximate and developmental aspects of antipredator behavior. *Advances in the Study of Behavior*, **22**, 135–238.

Curtsinger, J.W. and Laurie-Ahlberg, C.C. (1981). Genetic variability of flight metabolism in *Drosophila melanogaster*. I. Characterization of power output during tethered flight. *Genetics*, **98**, 549–64.

Dale, V.H. (1997). The relationship between land-use change and climate change. *Ecological Applications*, **7**, 753–69.

Dallman, M.F., Strack, A.M., Akana, S.F., Bradbury, M.J., Hanson, E.S., Scribner, K.A., and Smith, M. (1993). Feast and famine: critical role of glucocorticoids with insulin in daily energy flow. *Frontiers in Neuroendocrinology*, **14**, 303–47.

Danchin, E. (1992). The incidence of the tick parasite *Ixodes uriae* in kittiwake *Rissa tridactyla* colonies in relation to the age of the colony, and a mechanism of infecting new colonies. *Ibis*, **134**, 134–41.

Danchin, E. and Monnat, J.-Y. (1992). Population dynamics modeling of two neighbouring kittiwake *Rissa tridactyla* colonies. *Ardea*, **80**, 171–80.

Danchin, E. and Wagner, R.H. (1997). The evolution of coloniality: the emergence of new perspectives. *Trends in Ecology and Evolution*, **12**, 342–7.

Danchin, E., Cadiou, B., Monnat, J.-Y., and Rodriguez Estrella, R. (1991). Recruitment in long-lived birds: conceptual framework and behavioural mechanisms. *Acta XX Congressus Internationalis Ornithologici*,

pp. 1641–56. Hutcheson, Bowman and Stewart, Wellington, New Zealand.

Danchin, E., Boulinier, T., and Massot, M. (1998a). Conspecific reproductive success and breeding habitat selection: implications for the study of coloniality. *Ecology*, **79**, 2415–28.

Danchin, E., Wagner, R.H., and Boulinier, T. (1998b). The evolution of coloniality: does commodity selection explain it all? Reply. *Trends in Ecology and Evolution*, **13**, 76.

Danielson, B.J. (1991). Communities in a landscape: the influence of habitat heterogenity on the interactions between species. *American Naturalist*, **138**, 1105–20.

Danielson, B.J. (1992). Habitat selection, interspecific interactions and landscape composition. *Evolutionary Ecology*, **6**, 399–411.

Darwin, C. (1859). *On the origin of species by means of natural selection*. Murray, London.

Davies, N.B., Brooke, M. de L., and Kacelnik, A. (1996). Recognition errors and probability of parasitism determine whether reed warblers should accept or reject mimetic cuckoo eggs. *Proceedings of the Royal Society of London B*, **263**, 925–31.

Davies, S.J.J.F. (1984). Nomadism as a response to desert conditions in Australia. *Journal of Arid Environments*, **7**, 183–95.

Davis, A.R. and Campbell, D.J. (1996). Two levels of spacing and limits to local population density for settled larvae of the ascidian *Clavelina moluccensis*: a nearest neighbour analysis. *Oecologia*, **108**, 701–7.

Davis, M.B. (1989). Lags in vegetation response to greenhouse warming. *Climatic Change*, **15**, 75–82.

Day, T. and Taylor, P.D. (1998). Unifying genetic and game theoretic models of kin selection for continuous traits. *Journal of Theoretical Biology*, **194**, 391–407.

de Fraipont, M., Clobert, J., John-Alder, H., and Meylan, S. (2000). Increased prenatal maternal corticosterone promotes philopatry of offspring in common lizards *Lacerta vivipara. Journal of Animal Ecology*, **69**, 404–13.

de Kloet, E.R. and Reul, J.M.H.M. (1987). Feedback action and tonic influence of corticosteroids on brain function: a concept arising from the heterogeneity of brain receptor systems. *Psychoneuroendocrinology*, **12**, 83–105.

de Kloet, E.R., Ratka, A., Reul, J.M.H.M., Sutanto, W., and Van Eekelen, J.A.M. (1987). Corticosteroid receptor types in brain: regulation and putative function. *Annals of the New York Academy of Sciences*, **512**, 351–61.

de Laet, J.V. (1985). Dominance and aggression in juvenile great tits, *Parus major major* L. in relation to dispersal. In *Behavioral ecology: ecological consequences of adaptive behavior* (ed. R.M. Sibly and R.H. Smith), pp. 375–80. Blackwell Scientific, Oxford.

de Roos, A.M., McCauley, E., and Wilson, W.G. (1991). Mobility versus density-limited predator–prey dynamics on different spatial scales. *Proceedings of the Royal Society of London B*, **246**, 117–22.

de Roos, A.M., McCauley, E., and Wilson, W.G. (1998). Pattern formation and the spatial scale of interaction between predators and their prey. *Theoretical Population Biology*, **53**, 108–30.

DeAngelis, D.L. and Gross, L.J. (ed.) (1992). *Individual-based models and approaches in ecology: populations, communities and ecosystems*. Chapman and Hall, New York.

DeHeer, C.J., Goodisman, M.A.D., and Ross, K.G. (1999). Queen dispersal strategies in the multiple-queen form of the fire ant *Solenopsis invicta. American Naturalist*, **153**, 660–75.

Delestrade, A., McCleery, R.H., and Perrins, C.M. (1996). Natal dispersal in a heterogeneous environment: the case of the great tit in Wytham. *Acta Oecologica*, **17**, 519–29.

Delin, A.E. and Andrén, H. (1999). Effects of habitat fragmentation on Eurasian red squirrel (*Sciurus vulgaris*) in a forest landscape. *Landscape Ecology*, **14**, 67–72.

den Boer, P.J. (1968). Spreading of risk and stabilization of animal numbers. *Acta Biotheoretica*, **18**, 165–94.

Denno, R.F. and Peterson, M.A. (1995). Density-dependent dispersal and its consequences for population dynamics. In *Population dynamics: new approaches and synthesis* (ed. N. Cappuccino and

P.W. Price), pp. 113–130. Academic Press, San Diego, CA.

Denno, R.F., Olmstead, K.L., and McCloud, E.S. (1989). Reproductive cost of flight capability – a comparison of life-history traits in wing dimorphic planthoppers. *Ecological Entomology*, **14**, 31–44.

Denno, R.F., Roderick, G.K., Olmstead, K.L., and Dobel, H.G. (1991). Density-related migration in planthoppers (homoptera, delphacidae) – the role of habitat persistence. *American Naturalist*, **138**, 1513–41.

Denno, R.F., Cheng, J., Roderick, G.K., and Perfect, T.J. (1994). Density-related effects on the components of fitness and population dynamics of planthoppers. In *Planthoppers: their ecology and management* (ed. R.F. Denno and T.J. Perfect), pp. 257–81. Chapman and Hall, New York.

Denno, R.F., Roderick, G.K., Peterson, M.A., Huberty, A.F., Döbel, H.G., Eubanks, M.D., Losey, J.E., and Langellotto, G.A. (1996). Habitat persistence underlies intraspecific variation in the dispersal strategies of planthoppers. *Ecological Monographs*, **66**, 389–408.

Desrochers, A. and Magrath, R. (1993). Environmental predictability and remating in European blackbirds. *Behavioral Ecology*, **4**, 271–5.

Dias, P.C. (1996). Sources and sinks in population biology. *Trends in Ecology and Evolution*, **11**, 326–30.

Dias, P.C., Verheyen, G.R., and Raymond, M. (1996). Source–sink populations in Mediterranean blue tits: evidence using single-locus minisatellite probes. *Journal of Evolutionary Biology*, **9**, 965–78.

Dickinson, J.L. and Akre, J.J. (1998). Extrapair paternity, inclusive fitness, and within-group benefits of helping in western bluebirds. *Molecular Ecology*, **7**, 95–105.

Dieckmann, U. and Doebeli, M. (1999). On the origin of species by sympatric speciation. *Nature*, **400**, 354–7.

Dieckmann, U. and Law, R. (1996). The dynamical theory of coevolution: a derivation from stochastic ecological processes. *Journal of Mathematical Biology*, **34**, 579–612.

Dieckmann, U. and Law, R. (2000). Relaxation projections and the method of moments. In *The geometry of ecological interactions: simplifying spatial complexity* (ed. U. Dieckmann, R. Law., and J.A.J. Metz), pp. 412–55. Cambridge University Press, Cambridge.

Dieckmann, U., O'Hara, B., and Weisser, W.W. (1999). The evolutionary ecology of dispersal. *Trends in Ecology and Evolution*, **14**, 88–90.

Dieckmann, U., Law, R., and Metz, J.A.J. (ed.) (2000). *The geometry of ecological interaction: simplifying spatial complexity.* Cambridge University Press, Cambridge.

Diffendorfer, J.E. (1998). Testing models of source–sink dynamics and balanced dispersal. *Oikos*, **81**, 417–33.

Dingle, H. (1984). Behavior, genes, and life histories, complex adaptations in uncertain environments. In *A new ecology: novel approaches to interactive systems* (ed. C.N. Slobodchikoff, W.S. Gaud, and P.W. Price), pp. 169–84. John Wiley, New York.

Dingle, H. (1994). Genetic analyses of animal migration. In *Quantitative genetic studies of behavioral evolution* (ed. C.R.B. Boake), pp. 145–64. University of Chicago Press, Chicago, IL.

Dingle, H. (1996). *Migration: the biology of life on the move.* Oxford University Press, Oxford.

Dingle, H. and Winchell, R. (1997). Juvenile hormone as a mediator of plasticity in insect life histories. *Archives of Insect Biochemistry and Physiology*, **35**, 359–73.

Diss, A.L., Kunkel, J.G., Montgomery, M.E., and Leonard, D.E. (1996). Effects of maternal nutrition and egg provisioning on parameters of larval hatch, survival and dispersal in the gypsy moth, *Lymantria dispar* L. *Oecologia*, **106**, 470–7.

Dixon, A.F.G. (1985). *Aphid ecology.* Blackie, Glasgow.

Dixon, A.F.G. (1998). *Aphid ecology.* Chapman and Hall, London.

Djieto-Lordon, C. and Dejean, A. (1999). Tropical arboreal ant mosaics: innate attraction and imprinting determine nest site selection in dominant ants. *Behavioral Ecology and Sociobiology*, **45**, 219–25.

Doak, D.F. and Mills, L.S. (1994). A useful role for theory in conservation. *Ecology*, **75**, 615–26.

Dobson, A.P. (1988). The population biology of parasite-induced changes in host behavior. *Quarterly Review of Biology*, **64**, 139–65.

Dobson, F.S. (1979). An experimental study of dispersal in the Californian ground squirrel. *Ecology*, **60**, 1103–9.

Dobson, F.S. (1982). Competition for mates and predominant juvenile male dispersal in mammals. *Animal Behaviour*, **30**, 183–92.

Dobson, F.S. (1985). Multiple causes of dispersal. *American Naturalist*, **126**, 855–8.

Dobson, F.S. (1994). Measures of gene flow in the columbian ground squirrel. *Oecologia*, **100**, 190–5.

Dobson, F.S. and Jones, W.T. (1985). Multiple causes of dispersal. *American Naturalist*, **126**, 855–8.

Dobson, F.S., Chesser, R.K., Hoogland, J.L., Sugg, D.W., and Foltz, D.W. (1997). Do black-tailed prairie dogs minimize inbreeding? *Evolution*, **51**, 970–8.

Dobzhansky, T. (1940). Speciation as a stage in evolutionary divergence. *American Naturalist*, **74**, 312–21.

Dobzhansky, T. (1970). *Genetics of the evolutionary process.* Columbia University Press, New York.

Dobzhansky, T. (1973). Is there gene exchange between *Drosophila pseudoobscura* and *D. persimilis* in their natural habitats? *American Naturalist*, **107**, 312–14.

Dobzhansky, T. and Wright, S. (1941). Genetics of natural populations. V. Relations between mutation rate and accumulation of lethals in populations of *Drosophila pseudoobscura. Genetics*, **26**, 23–51

Dodson, S. (1989). Predator-induced reaction norms. *BioScience*, **39**, 447–52.

Doebeli, M. (1996). A quantitative genetic competition model for sympatric speciation. *Journal of Evolutionary Biology*, **9**, 893–909.

Doebeli, M. and Knowlton N. (1998). The evolution of interspecific mutualisms. *Proceedings of the National Academy of Sciences*, **95**, 8676–80.

Doebeli, M. and Ruxton, G.D. (1997). Evolution of dispersal rates in metapopulation models: branching and cyclic dynamics in phenotype space. *Evolution*, **51**, 1730–41.

Doligez, B., Danchin, E., Clobert, J., and Gustafsson, L. (1999). The use of conspecific reproductive success for breeding habitat selection in a non colonial, hole nesting species, the collared flycatcher *Ficedula albicollis*. *Journal of Animal Ecology*, **68**, 1–15.

Doncaster, C.P., Clobert, J., Doligez, B., Gustafsson, L., and Danchin, E. (1997). Balanced dispersal between spatially varying local populations: an alternative to the source–sink model. *American Naturalist*, **150**, 425–45.

Donohue, K. (1997). Seed dispersal in *Cakile edentula* var. *lacustris*: decoupling the fitness effects of density and distance from the home site. *Oecologia*, **110**, 520–7.

Donohue, K. (1998). Maternal determinants of seed dispersal in *Cakile edentula*: fruit, plant, and site traits. *Ecology*, **79**, 2771–88.

Doolan, S.P. and MacDonald, D.W. (1996). Dispersal and extra-territorial prospecting by slender-tailed meerkats (*Suricata suricata*) in south western Kalahari. *Journal of Zoology*, **240**, 59–73.

Dow, H. and Fredga, S. (1985). Selection of nest sites by a hole-nesting duck, the goldeneye *Bucephala clangula*. *Ibis*, **127**, 16–30.

Doyle, R.W. (1975). Settlement of planktonic larvae: a theory of habitat selection in varying environments. *American Naturalist*, **109**, 113–26.

Drake, V.A. and Gatehouse, A.G. (ed.) (1995). *Insect migration: tracking resources through space and time*. Cambridge University Press, Cambridge.

Drent, P.J. (1984). Mortality and dispersal in summer and its consequences for the density of great tits *Parus major* at the onset of autumn. *Ardea*, **72**, 27–162.

Drickamer, L. (1996). Intra-uterine position and anogenital distance in house mice: consequences under field conditions. *Animal Behaviour*, **51**, 925–34.

Driessen, G. and Visser, M.E. (1993). The influence of adaptive foraging decisions on spatial heterogeneity of parasitism and parasitoid population efficiency. *Oikos*, **67**, 209–17.

Driessen, G., Bernstein, C., van Alphen, J.J.M., and Kacelnik, A. (1995). A count-down mechanism for host search in the parasitoid *Venturia canescens*. *Journal of Animal Ecology*, **64**, 117–25.

Duffy, D.C. (1983). The ecology of tick parasitism on densely nesting peruvian seabirds. *Ecology*, **64**, 110–19.

Dufty, A.M., Jr. and Belthoff, J.R. (1998). Corticosterone and the stress response in young western screech owls: effects of captivity, gender, and activity period. *Physiological Zoology*, **70**, 143–9.

Dufty, A.M., Jr. and Wingfield, J.C. (1986). Temporal patterns of circulating LH and steroid hormones in a brood parasite, the brown-headed cowbird. I. Males. *Journal of Zoology*, **208**, 191–203.

Dugatkin, L.A. (1992). Sexual selection and imitation: females copy the mate choice of others. *American Naturalist*, **139**, 1384–9.

Dugatkin, L.A. (1996a). Interface between culturally based preferences and genetic preferences: female mate choice in *Poecilia reticulata*. *Proceedings of the National Academy of Sciences*, **93**, 2770–3.

Dugatkin, L.A. (1996b). Copying and mate choice. In *Social learning in animals. The roots of culture* (ed. C.M. Heyes and B.G. Galef), pp. 85–105. Academic Press, New York.

Dugatkin, L.A. and Godin, J.-G.J. (1992). Reversal of female mate choice by copying in the guppy (*Poecilia reticulata*). *Proceedings of the Royal Society of London B*, **249**, 179–84.

Dugatkin, L.A. and Godin, J.-G.J. (1993). Female mate choice copying in the guppy (*Poecilia reticulata*): age-dependent effects. *Behavioral Ecology*, **4**, 289–92.

Dugatkin, L.A. and Godin, J.-G.J. (1998). Effects of hunger on mate-choice copying in the guppy. *Ethology*, **104**, 194–202.

Dugatkin, L.A. and Höglund, J. (1995). Delayed breeding and the evolution of mate copying in lekking species. *Journal of Theoretical Biology*, **174**, 261–7.

Dunning, J., Stewart, D.J., Danielson, B.J., Noon, B.R., Root, T.L., Lamberson, R.H., and Stevens, E.E. (1995). Spatially explicit population models: current forms and future uses. *Ecological Applications*, **5**, 3–11.

Durrett, R. and Levin, S. (1994). The importance of being discrete (and spatial). *Theoretical Population Biology*, **46**, 363–94.

Dusseau, J.W. and Meier, A.H. (1971). Diurnal and seasonal variations of plasma adrenal steroid hormone in the white-throated sparrow, *Zonotrichia albicollis*. *General and Comparative Endocrinology*, **16**, 399–408.

Eason, P. and Stamps, J.A. (1993). An early warning system for detecting intruders in a territorial animal. *Animal Behaviour*, **46**, 1105–9.

Eden, S.F. (1987). Natal philopatry of the magpie *Pica pica*. *Ibis*, **129**, 477–90.

Efron, B. (1988). Logistic regression, survival analysis, and the Kaplan–Meier curve. *Journal of the American Statistic Association*, **83**, 414–25.

Egas, M., van Baalen, M., and Sabelis, M.W. (2000). Plant defence and herbivore deterrents: do plants advertise the presence of bodyguards to herbivores. *Oikos*, in press.

Ehrlich, P.R. and Raven, P.H. (1969). Differentiation of populations. *Science*, **165**, 1228–32.

Ekman, J.B. and Askenmo, C.E.H. (1984). Social rank and habitat use in willow tit groups. *Animal Behaviour*, **32**, 508–14.

Ekman, J.B., Bylin, A., and Tegelström, H. (1999). Increased lifetime reproductive success for Siberian jay (*Perisoreus infaustus*) males with delayed dispersal. *Proceeding of the Royal Society of London B*, **266**, 911–15.

Elliot, J.K., Elliot, J.M., and Mariscal, R.N. (1995). Host selection, location, and association behaviors of anemonefishes in field settlement experiments. *Marine Biology*, **122**, 377–89.

Ellner, S. (1986). Germination dimorphism and parent–offspring conflict in seed germination. *Journal of Theoretical Biology*, **123**, 173–85.

Ellner, S. (1987). Competition and dormancy: a reanalysis and review. *American Naturalist*, **130**, 798–803.

Ellner, S.P., Sasaki, A., Haraguchi, Y., Matsuda, H. (1998). Speed of invasion in lattice population models: pair-edge approximation. *Journal of Mathematical Biology*, **36**, 469–84.

Ellsworth, D.S., and Reich, P.B. (1993). Canopy structure and vertical patterns of photosynthesis and related leaf traits in a deciduous forest. *Oecologia*, **96**, 169–78.

Ellsworth, E.A. and Belthoff, J.R. (1999). Effects of social status on the dispersal behaviour of juvenile western screech-owls. *Animal Behaviour*, **57**, 883–92.

Emlen, S.T. (1991). Evolution of cooperative breeding in birds and mammals. In *Behavioural ecology: an evolutionary approach* (ed. J.R. Krebs and N.B. Davies), pp. 301–35. Blackwell, Oxford.

Emlen, S.T. (1994). Benefits, constraints, and the evolution of the family. *Trends in Ecology and Evolution*, **9**, 282–5.

Emlen, S.T. and Wrege, P.H. (1991). Breeding biology of white-fronted bee-eaters at Nakuru, the influence of helpers on breeder fitness. *Journal of Animal Ecology*, **60**, 309–26.

Endler, J.A. (1977). *Geographic variation, speciation, and clines*. Princeton University Press, Princeton, NJ.

Endler, J.A. (1979). Gene flow and life history patterns. *Genetics*, **93**, 263–84.

Englund, G. (1993). Effects of density and food availability on habitat selection in a net-spinning caddis larva, *Hydropsyche siltalai*. *Oikos*, **68**, 473–80.

Engstrom, R.T. and Mikusinski, G. (1998). Ecological neighborhoods in red-cockaded woodpecker populations. *Auk*, **115**, 473–8.

Ens, B.J. (1992). *The social prisoner*. PhD thesis, University of Groningen, Groningen.

Ens, B.J., Weissing, F.J., and Drent, R.H. (1995). The despotic distribution and deferred maturity: two sides of the same coin. *American Naturalist*, **146**, 625–50.

Epperson, B.K. (1993a). Recent advances in correlation studies of spatial patterns of genetic variation. *Evolutionary Biology*, **27**, 95–155.

Epperson, B.K. (1993b). Spatial and space-time correlations in systems of subpopulations with genetic drift and migration. *Genetics*, **133**, 711–27.

Epperson, B.K. and Allard, R.W. (1989). Spatial autocorrelation analysis of the distribution of genotypes within populations of lodgepole pine. *Genetics*, **121**, 369–77.

Epperson, B.K. and Li, T.-Q. (1997). Gene dispersal and spatial genetic structure. *Evolution*, **51**, 672–81.

Epple, A., Gower, B., ten Busch, M., Gill, T., Milakofsky, L., Piechotta, R., Nibbio, B., Hare, T., and Stetson, M.H. (1997). Stress responses in avian embryos. *American Zoologist*, **37**, 536–45.

Erckmann, W.J., Beletsky, L.D., Orians, G.H., Johnsen, T., Sharbaugh, S., and D'Antonnio, C. (1990). Old nests as cues for nest-site selection: an experimental test with red-winged blackbirds. *Condor*, **92**, 113–17.

Evans, A.S. and Cabin, R.J. (1995). Can dormancy affect the evolution of post germination traits? The case of *Lesquerella fendeleri*. *Ecology*, **76**, 344–56.

Evans, R.M. and Arriza, J.L. (1989). A molecular framework for the actions of glucocorticoid hormones in the nervous system. *Neuron*, **2**, 1105–12.

Ewens, W.J. (1972). The sampling theory of selectively neutral alleles. *Theoretical Population Biology*, **3**, 87–112.

Ewens, W.J. (1989). The effective population size in the presence of catastrophes. In *Mathematical evolutionary theory* (ed. M.W. Feldman), pp. 9–25. Princeton University Press, Princeton, NJ.

Excoffier, L., Smouse, P.E., and Quattro, J.M. (1992). Analysis of molecular variance inferred from metric distances among DNA haplotypes: applications to human mitochondrial DNA restriction data. *Genetics*, **131**, 479–91.

Ezoe, H. (1998). Optimal dispersal range and seed size in a stable environment. *Journal of Theoretical Biology*, **190**, 287–93.

Fadamiro, H.Y., Wyatt, T.D., and Birch, M.C. (1996). Flight activity of *Prostephanus truncatus* (Horn) (Coleoptera: Bostrichidae) in relation to population density, resource quality, age, and sex. *Journal of Insect Behavior*, **9**, 339–51.

Fagan, W.F., Cantrell, R.S., and Cosner, C. (1999). How habitat edges change species interactions? *American Naturalist*, **153**, 165–82.

Fagerström, T. and Westoby, M. (1997). Population dynamics in sessile organisms: some general results from three seemingly different theory lineages. *Oikos*, **80**, 588–94.

Fahrig, L. (1997). Relative effects of habitat loss and fragmentation on population extinction. *Journal of Wildlife Management*, **61**, 603–10.

Fairbairn, D.J. and Roff, D.A. (1990). Genetic correlations among traits determining migratory tendency in the sand cricket, *Gryllus firmus*. *Evolution*, **44**, 1787–95.

Farmer, A.H. and Wiens, J.A. (1998). Optimal migration schedules depend on the landscape and the physical environment: a dynamic modeling view. *Journal of Avian Biology*, **29**, 405–15.

Faulkes, G.C., Abbott, D.H., and Mellor, A.L. (1990). Investigation of genetic diversity in wild colonies of naked mole-rats (*Heterocephalus glaber*) by DNA fingerprinting. *Journal of Zoology*, **221**, 87–97.

Faulkes, G.C., Abbott, D.H., Liddell, C.E., George, L.M., and Jarvis, J.U.M. (1991). Hormonal and behavioural aspects of reproductive suppression in female naked mole-rats. In *The biology of the naked mole-rat* (ed. P.W. Sherman, J.U.M. Jarvis, and R.D. Alexander), pp. 426–45. Princeton University Press, Princeton, NJ.

Faulkes, G.C., Abbott, D.H., O'Brien, H.P., Lau, L., Roy, M.R., Wayne, R.K., and Bruford, M.W. (1997). Micro- and macrogeographical genetic structure of colonies of naked mole-rats *Heterocephalus glaber*. *Molecular Ecology*, **6**, 615–28.

Favre, L., Balloux, F., Goudet, J., and Perrin, N. (1997). Female-biased dispersal in the monogamous mammal *Crocidura russula*: Evidence from field data and microsatellite patterns. *Proceedings of the Royal Society of London B*, **264**, 127–32.

Feare, C.J. (1976). Desertion and abnormal development in a colony of sooty terns *Sterna fuscata* infested by virus-infested ticks. *Ibis*, **118**, 112–15.

Felsenstein, J. (1975). A pain in the torus: some difficulties with the model of isolation by distance. *American Naturalist*, **109**, 359–68.

Felsenstein, J. (1982). How can we infer geography and history from gene frequencies? *Journal of Theoretical Biology*, **96**, 9–20.

Felsenstein, J. (1992). Estimating effective population size from samples of sequences: inefficiency of pairwise and segregating sites as compared to phylogenetic estimates. *Genetical Research*, **59**, 139–47.

Ferrer, M. (1992). Natal dispersal in relation to nutritional condition in Spanish imperial eagles. *Ornis Scandinavica*, **23**, 104–7.

Ferrer, M. (1993). Ontogeny of dispersal distances in young Spanish imperial eagles. *Behavioral Ecology and Sociobiology*, **32**, 259–63.

Ferrière, R. and Clobert, J. (1992). Evolutionarily stable age at first reproduction in a density-dependent model. *Journal of Theoretical Biology*, **157**, 253–67.

Ferrière, R. and Gatto, M. (1995). Lyapunov exponents and the mathematics of invasion in oscillatory or chaotic populations. *Theoretical Population Biology*, **48**, 126–71.

Ferrière, R. and Michod, R.E. (1995). Invading wave of cooperation in a spatial iterated prisoner's dilemma. *Proceedings of the Royal Society of London B*, **259**, 77–83.

Ferrière, R. and Michod, R.E. (1996). The evolution of cooperation in spatially heterogeneous populations. *American Naturalist* **147**, 692–717.

Ferrière, R., Belthoff, J.R., Olivieri, I., and Krackow, S. (2000). Evolving dispersal: where to go next? *Trends in Ecology and Evolution*, **15**, 5–7.

Finnimore, S.A., Nyquist, W.E., Shaner, G.E., Myers, S.P., and Foley, M.E. (1998). Temperature response in wild oat (*Avena fatua* L.) generations segregating for seed dormancy. *Heredity*, **81**, 674–82.

Fisher, E.A., and Peterson, C.W. (1987). The evolution of sexual patterns in the sea basses. *BioScience*, **37**, 482–9.

Fisher, R.A. (1930). *The genetical theory of natural selection*. Oxford University Press, Oxford.

Fisher, R.A. (1950). Gene frequencies in a cline determined by selection and diffusion. *Biometrics*, **6**, 353–61.

Fleming, T.H. (1991). Fruiting plantfrugivore mutualism: the evolutionary theater and the ecological play. In *Plant–animal interactions*. (ed. P.W. Price, T.M. Lewinsohn, G.W. Fernanades, and W.W. Benson), pp. 119–44. John Wiley, New York.

Fletcher, D.J.C. (1986). Triple action of queen pheromones in the regulation of reproduction in fire ant (*Solenopsis invicta*) colonies. *Advances in Invertebrate Reproduction*, **4**, 305–16.

Fletcher, D.J.C. and Michener, C.D. (1987). *Kin recognition in animals*. John Wiley, Chichester.

Folstad, I., Nilssen, A., Halvorsen, C., and Andersen, J. (1991). Parasite-avoidance: the cause of post-calving migrations in *Rangifer*? *Canadian Journal of Zoology*, **69**, 2423–9.

Fonseca, D.M. and Hart, D.D. (1996). Density-dependent dispersal of black fly neonates is mediated by flow. *Oikos*, **75**, 49–58.

Forbes, L.S. (1989). Coloniality in herons: Lack's predation hypothesis reconsidered. *Colonial Waterbirds*, **12**, 24–9.

Forbes, L.S. and Kaiser, G.W. (1994). Habitat choice in breeding seabirds: when to cross the information barrier. *Oikos*, **70**, 377–84.

Forman, R.T.T. (1995). *Land mosaics: the ecology of landscapes and regions*. Cambridge University Press, Cambridge.

Forsgren, E., Karlsson, A., and Kvarnemo, C. (1996). Female sand gobies gain direct benefits by choosing males with eggs in their nest. *Behavioral Ecology and Sociobiology*, **39**, 91–6.

Forsman, J.T., Monkkonen., M, Helle, P., and Inkeroinen, J. (1998). Heterospecific attraction and food resources in migrants' breeding patch selection in northern boreal forest. *Oecologia*, **115**, 278–86.

Frame, L.H. and Frame, G.W. (1976). Female African wild dogs emigrate. *Nature*, **263**, 227–9.

Frank, L.G. (1986). Social organization of the spotted hyaena (*Crocuta crocuta*). I. Demography. *Animal Behaviour*, **34**, 1500–9.

Frank, L. and Woodroffe, R.B. (2000). Behaviour of carnivores in exploited and controlled populations. In *Carnivore conservation* (ed. J.L. Gittleman, R.K. Wayne, D.W. Macdonald, and S. Funk). Cambridge University Press, Cambridge, in press.

Frank, S.A. (1986). Dispersal polymorphism in subdivided populations. *Journal of Theoretical Biology*, **122**, 303–9.

Frank, S.A. (1997). Multivariate analysis of correlated selection and kin selection, with an ESS maximization method. *Journal of Theoretical Biology*, **189**, 307–416.

Frank, S.A. (1998). *Foundations of social evolution*. Princeton University Press, Princeton, NJ.

Freeland, W.J. (1980). Mangabey (*Cercocebus albigena*) movement patterns in relation to food availability and fecal contamination. *Ecology*, **61**, 1297–1303.

Fretwell, S.D. (1972). *Populations in a seasonal environment. Monographs in Population Biology*, Vol. **5**, pp. 1–217. Princeton University Press, Princeton, NJ.

Fretwell, S.D. and Lucas, H.L. (1970). On territorial behaviour and other factors influencing habitat distribution in birds. I. Theoretical development. *Acta Biotheoretica*, **19**, 16–36.

Friedman-Einat, M., Boswell, T., Horev, G., Girishvarma, G., Dunn, I. C., Talbot, R.T., and Sharp, P.J. (1999). The chicken leptin gene: has it been cloned? *General and Comparative Endocrinology*, **115**, 354–63.

Fryxell, J.M. and Lundberg, P. (1998). *Individual behavior and community dynamics*. Chapman and Hall, London.

Futuyma, D.J. and Moreno, G. (1988). The evolution of ecological specialization. *Annual Review of Ecology and Systematics*, **19**, 207–33.

Gadgil, M. (1971). Dispersal: population consequences and evolution. *Ecology*, **52**, 253–61.

Gaines, M.S. and McClenaghan, L.R., Jr. (1980). Dispersal in small mammals. *Annual Review of Ecology and Systematics*, **11**, 163–96.

Galef, B.G., Jr. and Whiskin, E.E. (1998). Limits on social influence on food choices of Norway rats. *Animal Behaviour*, **56**, 1015–20.

Gandon, S. (1999). Kin competition the cost of inbreeding and the evolution of dispersal. *Journal of Theoretical Biology*, **200**, 345–64.

Gandon, S. and Michalakis, Y. (1999). The evolution of dispersal in a metapopulation with extinction and kin competition. *Journal of Theoretical Biology*, **199**, 275–90.

Gandon, S. and Rousset, F. (1999). The evolution of stepping stone dispersal rate.

Proceedings of the Royal Society of London B, **266**, 2507–13.

Gandon, S., Capowiez, Y., Dubois, Y., Michalakis, Y., and Olivieri, I. (1996). Local adaptation and gene-for-gene coevolution in a metapopulation model. *Proceedings of the Royal Society of London B*, **263**, 1003–9.

Garbutt, K., Bazzaz, F.A., and Levin, D.A. (1985). Population and genotype niche width in clonal *Phlox paniculata*. *American Journal of Botany*, **72**, 640–8.

Gardner, R.H., O'Neill, R.V., Turner, M.G., and Dale, V.H. (1989). Quantifying scale-dependent effects of animal movement with wimple percolation models. *Landscape Ecology*, **3**, 217–27.

Gatehouse, A.G. (1986). Migration in the African armyworm *Spodoptera exempta*: genetic determination of migratory capacity and a new synthesis. In *Insect flight* (ed. W. Danthanarayana), pp. 128–44. Springer, Berlin.

Gatehouse, A.G. and Zhang, X.X. (1995). Migratory potential in insects: variation in an uncertain environment. In *Insect migration: tracking resources through space and time* (ed. V.A. Drake and A.G. Gatehouse), pp. 193–242. Cambridge University Press, Cambridge.

Gavrilets, S. (1996). On phase three of the shifting-balance theory. *Evolution*, **50**, 1034–41.

Gavrilets, S. (1997). Evolution and speciation on holey adaptive landscapes. *Trends in Ecology and Evolution*, **12**, 307–12.

Geritz S.A.H., Metz J.A.J., Kisdi E., and Meszéna G. (1997). The dynamics of adaptation and evolutionary branching. *Physical Review Letters*, **78**, 2024–7.

Geritz, S.A.H., Kisdi, E., Meszéna, G., and Metz, J.A.J. (1998). Evolutionary singular strategies and the adaptive growth and branching of the evolutionary tree. *Evolutionary Ecology*, **12**, 35–57.

Gese, E.M., Ruff, R.L. and Crabtree, R.L. (1996). Social and nutritional factors influencing the dispersal of resident coyotes. *Animal Behaviour*, **52**, 1025–43.

Gibbs, H.L. and Grant, P.R. (1989). Inbreeding in Darwin's medium ground finches (*Geospiza fortis*). *Evolution*, **43**, 1273–84.

Gibson, R.M. and Höglund, J. (1992). Copying and sexual selection. *Trends in Ecology and Evolution*, 7, 229–32.

Gibson, R.M., Bradbury, J.W., and Vehrencamp, S.L. (1991). Mate choice in lekking sage grouse: the roles of vocal display, female site fidelity and copying. *Behavioral Ecology*, 2, 165–80.

Gilbert, D.A., Lehman, N., O'Brien, S.J., and Wayne, R.K. (1990). Genetic fingerprinting reflects population differentiation in the California Channel Island fox. *Nature*, 344, 764–6.

Gilchrist, G.W. (1995). Specialists and generalists in changing environments. I. Fitness landscapes of thermal sensitivity. *American Naturalist*, 146, 252–70.

Giles, B.E. and Goudet, J. (1997). A case study of genetic structure in a metapopulation. In *Metapopulation dynamics: ecology, genetics and evolution* (ed. I.A. Hanski and M.E. Gilpin), pp. 429–54. Academic Press, New York.

Gillespie, J.H. (1975). The role of migration in the genetic structure of populations in temporally and spatially varying environments. 1. Conditions for polymorphism. *American Naturalist*, 109, 127–35.

Gillespie, J.H. (1977). Natural selection for variance in offspring number: a new evolutionary principle. *American Naturalist*, 111, 1010–4.

Gilpin, M.E. (1975). *Group selection in predator–prey communities*. Princeton University Press, Princeton, NJ.

Gilpin, M.E. (1991). The genetic effective size of a metapopulation. In *Metapopulation dynamics: empirical and theoretical investigations* (ed. M.E. Gilpin and I. Hanski), pp. 165–75. Academic Press, London.

Gilpin, M.E. and Hanski, I. (1991). *Metapopulation dynamics: empirical and theoretical investigations*. Academic Press, London.

Ginzburg, L.R. and Taneyhill, D.E. (1994). Population-cycles of forest lepidoptera – a maternal effect hypothesis. *Journal of Animal Ecology*, 63, 79–92.

Girman, D.J., Mills, M.G.L., Geffen, E., and Wayne, R.K. (1997). A molecular genetic analysis of social structure, dispersal, and interpack relationships of the African wild dog (*Lycaon pictus*). *Behavioral Ecology and Sociobiology*, 40, 187–98.

Glickman, S.E., Frank, L.G., Pavgi, S., and Licht, P. (1992). Hormonal correlates of 'masculinization' in female spotted hyaenas (*Crocuta crocuta*). 1. Infancy to sexual maturity. *Journal of Reproduction and Fertility*, 95, 451–62.

Godfray, H.C.J. (1994). *Parasitoids: behavioural and evolutionary ecology*. Princeton University Press, Princeton, NJ.

Goldstein, D.B. and Holsinger, K.E. (1992). Maintenance of polygenic variation in spatially structured populations: roles for local mating and genetic redundancy. *Evolution*, 46, 412–29.

Goldtschmidt, T., Bakker, T.C.M., and Feuth-de Bruijn, E. (1993). Selective copying in mate choice of female sticklebacks. *Animal Behaviour*, 45, 541–7.

Goldwasser, L., Cook, J., and Silverman, E.D. (1994). The effects of variability on metapopulation dynamics and rates of invasion. *Ecology*, 75, 40–7.

Gomulkiewicz, R., Thompson, J.N., Holt, R.D., Nuismer, S.L., and Hochberg, M.E. (2000). Hot spots, cold spots, and the geographic mosaic theory of coevolution. *American Naturalist*, 156, 156–74.

Gonzalez, A., Lawton, J.H., Gilbert, F.S., Blackburn, T.M., and Evans-Freke, I. (1998). Metapopulation dynamics, abundance and distribution in a microecosystem. *Science*, 281, 2045–7.

Gonzalez-Andujar, J.L. and Perry, J.N. (1993). Chaos, metapopulations and dispersal. *Ecological Modelling*, 65, 255–63.

Goodisman, M.A.D. and Asmussen, M.A. (1997). Cytonuclear theory for haplodiploid species and X-linked genes. I. Hardy–Weinberg dynamics and continent–island, hybrid zone models. *Genetics*, 147, 321–38.

Goodisman, M.A.D., DeHeer, C.J., and Ross, K.G. (2000). Behavior of polygyne fire ant queens on nuptial flights. *Journal of Insect Behavior*, in press.

Goodman, D. (1987). The demography of chance extinction. In *Viable populations for conservation* (ed. M.E. Soulé), pp. 11–34. Cambridge University Press, Cambridge.

Goodman, S.J. (1998). Patterns of extensive genetic differentiation and variation among European harbor seals (*Phoca vitulina vitulina*) revealed using microsatellite DNA polymorphisms. *Molecular Biology and Evolution*, **15**, 104–18.

Goodman, S.J., Barton, N.H., Swanson, G., Abernethy, K., and Pemberton, J.M. (1999). Introgression through rare hybridisation: a genetic study of a hybrid zone between red and sika deer (genus *Cervus*), in Argyll, Scotland. *Genetics*, **152**, 355–71.

Gorman, M.L. and Stone, R.D. (1990). *The natural history of moles*. Comstock, Ithaca, NY.

Gosselin, L.A. and Qian, P.Y. (1997). Juvenile mortality in benthic marine invertebrates. *Marine Ecology Progress Series*, **146**, 265–82.

Gotelli, N.J. (1991). Metapopulation models: the rescue effect, the propagule rain, and the core-satellite hypothesis. *American Naturalist*, **138**, 769–76.

Grafen, A. (1979). The hawk–dove game played between relatives. *Animal Behaviour*, **27**, 905–7.

Grant, J.W.A. and Green, L.D. (1996). Mate copying versus preference for actively courting males by female japanese medaka (*Oyzias latipes*). *Behavioral Ecology*, **7**, 165–7.

Greenwood, P.J. (1980). Mating systems, philopatry and dispersal in birds and mammals. *Animal Behaviour*, **28**, 1140–62.

Greenwood, P.J. (1983). Mating systems and evolutionary consequences of dispersal. In *The ecology of animal movement* (ed. I.R. Swingland and P.J. Greenwood), pp. 116–31. Clarendon Press, Oxford.

Greenwood, P.J. and Harvey, P.H. (1982). The natal and breeding dispersal of birds. *Annual Review of Ecology and Systematics*, **13**, 1–21.

Grenfell, B.T. and Dobson, A.P. (ed.) (1995). *Ecology of infectious diseases in natural populations*. Cambridge University Press, Cambridge.

Griffiths, R.C. and Tavaré, S. (1994). Simulating probability distributions in the coalescent. *Theoretical Population Biology*, **46**, 131–49.

Grime, J.P. (1979). *Plant strategies and vegetation processes*. John Wiley, New York.

Grinnell, J. (1922). The role of the "accidental". *The Auk*, **39**, 373–380.

Groeters, F.R. and Dingle, H. (1987). Genetic and maternal influences on life history plasticity in response to photoperiod by milkweed bugs (*Oncopeltus fasciatus*). *American Naturalist*, **129**, 332–46.

Gröndahl, F. (1989). Interactions between polyps of *Aurelia aurita* and planktonic larvae of scyphozoans: an experimental study. *Marine Ecology Progress Series*, **45**, 87–93.

Grosberg, R.K. and Quinn, J.F. (1986). The genetic control and consequences of kin recognition by the larvae of a colonial marine invertebrate. *Nature*, **322**, 456–9.

Grosholz, E.D. (1993). The influence of habitat heterogeneity on host–pathogen population dynamics. *Oecologia*, **96**, 347–53.

Grubb, T.C., Jr. and Doherty, P.F., Jr. (1999). On home-range gap-crossing. *Auk*, **116**, 618–28.

Gu, H. and Barker, J.S.F. (1995). Genetic and phenotypic variation for flight ability in the cactophilic *Drosophila* species, *D. aldrichi* and *D. buzzatii*. *Entomologia Experimentalis et Applicata*, **76**, 25–35.

Gu, H. and Danthanarayana, W. (1992). Quantitative genetic analysis of dispersal in *Epiphyas postvittana*. I. Genetic variation in flight capacity. *Heredity*, **68**, 53–60.

Guillemaud, T., Lenormand, T., Bourguet, D., Chevillon, C., Pasteur, N., and Raymond, M. (1998). Evolution of resistance in *Culex pipiens*: allele replacement and changing environment. *Evolution*, **52**, 443–53.

Gyllenberg, M. and Hanski, I. (1992). Single-species metapopulation dynamics: a structured model. *Theoretical Population Biology*, **42**, 35–61.

Gyllenberg, M., Söderbacka, G., and Ericsson, S. (1993). Does migration stabilize local population dynamics? Analysis of a discrete metapopulation model. *Mathematical Biosciences*, **118**, 25–49.

Gyllenberg, M., Hanski, I., and Hastings, A. (1997). Structured metapopulation models. In: *Metapopulation biology* (ed. I. Hanski and

M. Gilpin), pp. 93–122. Academic Press, San Diego, CA.

Haccou, P. and Iwasa, Y. (1995). Optimal mixed strategies in stochastic environments. *Theoretical Population Biology*, **47**, 212–43.

Haddad, N.M. (1999). Corridor use predicted from behaviors at habitat boundaries. *American Naturalist*, **153**, 215–27.

Hafner, D.J. and Sullivan R.M. (1995). Historical and ecological biogeography of near-arctic pikas (Lagomorpha: Ochotonidae). *Journal of Mammalogy*, **76**, 302–21.

Haldane, J.B.S. (1927). A mathematical theory of natural and artificial selection. V. Selection and mutation. *Proceedings of the Cambridge Philosophical Society*, **23**, 838–44.

Haldane, J.B.S. (1931). A mathematical theory of natural selection VI. Isolation. *Transactions of the Cambridge Philosophical Society*, **26**, 220–30.

Haldane, J.B.S. (1948). The theory of a cline. *Journal of Genetics*, **48**, 277–84.

Haldane, J.B.S. (1959). Natural selection. In *Darwin's biological work: some aspects reconsidered* (ed. P.R. Bell), pp. 101–49. Wiley, New York.

Hamilton, W.D. (1964). The genetical evolution of social behaviour I, II. *Journal of Theoretical Biology*, **7**, 1–52.

Hamilton, W.D. (1967). Extraordinary sex ratios. *Science*, **156**, 477–88.

Hamilton, W.D. (1970). Selfish and spiteful behaviour in an evolutionary model. *Nature*, **228**, 1218–20.

Hamilton, W.D. and May, R.M. (1977). Dispersal in stable habitats. *Nature*, **269**, 578–81.

Hanski, I. (1985). Single-species spatial dynamics may contribute to long-term rarity and commonness. *Ecology*, **66**, 335–43.

Hanski, I. (1989). Metapopulation dynamics: does it help to have more of the same. *Trends in Ecology and Evolution*, **4**, 113–14.

Hanski, I. (1994). A practical model of metapopulation dynamics. *Journal of Animal Ecology*, **63**, 151–62.

Hanski, I. (1998). Metapopulation dynamics. *Nature*, **396**, 41–9.

Hanski, I. (1999a). *Metapopulation ecology*. Oxford University Press, Oxford.

Hanski, I. (1999b). Habitat connectivity, habitat continuity, and metapopulations in dynamic landscapes. *Oikos*, **87**, 209–19.

Hanski, I. and Gilpin, M. (1991). Metapopulation dynamics: a brief history and conceptual domain. *Biological Journal of the Linnean Society*, **42**, 3–16.

Hanski, I. and Gilpin, M.E. (ed.) (1997). *Metapopulation biology: ecology, genetics and evolution*. Academic Press, San Diego, CA.

Hanski, I. and Gyllenberg, M. (1993). Two general metapopulation models and the core-satellite species hypothesis. *American Naturalist*, **142**, 17–41.

Hanski, I. and Ovaskainen, O. (2000). The metapopulation capacity of a fragmented landscape. *Nature*, **404**, 755–58.

Hanski, I. and Ranta, E. (1983). Coexistence in a patchy environment: three species of *Daphnia* in rock pools. *Journal of Animal Ecology*, **52**, 263–79.

Hanski, I. and Thomas, C.D. (1994). Metapopulation dynamics and conservation: a spatially explicit model applied to butterflies. *Biological Conservation*, **68**, 167–80.

Hanski, I. and Woiwod, I.P. (1993). Spatial synchrony in the dynamics of moth and aphid populations. *Journal of Animal Ecology*, **62**, 656–68.

Hanski, I. and Zhang, D.-Y. (1993). Migration, metapopulation dynamics and fugitive coexistence. *Journal of Theoretical Biology*, **163**, 491–504.

Hanski, I., Peltonen, A., and Kaski, L. (1991). Natal dispersal and social dominance in the common shrew *Sorex araneus*. *Oikos*, **62**, 48–58.

Hanski, I., Pakkala, T., Kuussaari, M., and Lei, G. (1995a). Metapopulation persistence of an endangered butterfly in a fragmented landscape. *Oikos*, **72**, 21–8.

Hanski, I., Pöyry, J., Pakkala, T., and Kuussaari, M. (1995b). Multiple equilibria in metapopulation dynamics. *Nature*, **377**, 618–21.

Hanski, I., Foley, P., and Hassell, M.P. (1996a). Random walks in a metapopulation: how much density dependence is necessary for long-term persistence? *Journal of Animal Ecology*, **65**, 274–82.

Hanski, I., Moilanen, A., Pakkala, T., and Kuussaari, M. (1996b). The quantitative incidence function model and persistence of an endangered butterfly metapopulation. *Conservation Biology*, **10**, 578–90.

Hanski, I., Alho, J., and Moilanen, A. (2000). Estimating the parameters of survival and migration of individuals in metapopulations. *Ecology*, **81**, 239–51.

Hansson, L. (1991). Dispersal and connectivity in metapopulation. In: *Metapopulation dynamics* (ed. M. Gilpin and I. Hanski), pp. 89–103. Academic Press, San Diego, CA.

Hanzawa, F.M., Beattie, A.J., and Culver, D.C. (1988). Directed dispersal: demographic analysis of an ant–seed mutualism. *American Naturalist*, **131**, 1–13.

Harada, Y. and Iwasa, Y. (1994). Lattice population-dynamics for plants with dispersing seeds and vegetative reproduction. *Researches on Population Ecology*, **36**, 237–49.

Harada, Y. and Iwasa, Y. (1996). Analyses of spatial patterns and population processes of clonal plants. *Researches on Population Ecology*, **38**, 153–64.

Harada, Y., Ezoe, H., Iwasa, Y., Matsuda, H., and Sato, K. (1995). Population persistence and spatially limited social interaction. *Theoretical Population Biology*, **48**, 65–91.

Hardin, G. (1968). The tragedy of the commons. *Science*, **162**, 1243–7.

Hardy, O.J. and Vekemans, X. (1999). Isolation by distance in a continuous population: reconciliation between spatial autocorrelation analysis and population genetics models. *Heredity*, **83**, 145–54.

Harris, S. and Trewhella, W. (1988). An analysis of some of the factors affecting dispersal in an urban fox (*Vulpes vulpes*) population. *Journal of Applied Ecology*, **25**, 409–22.

Harrison, R.G. (1989). Animal mitochondrial DNA as a genetic marker in population and evolutionary biology. *Trends in Ecology and Evolution*, **4**, 6–12.

Harrison, S. (1991). Local extinction in a metapopulation context: an empirical evaluation. *Biological Journal of the Linnean Society*, **42**, 73–88.

Harrison, S. (1994). Resources and dispersal as factors limiting a population of the tussock moth (*Orgyia vetusta*), a flightless defoliator. *Oecologia*, **99**, 27–34.

Harrison, S. (1995). Lack of strong induced or maternal effects in tussock moths (*Orgyia vetusta*) on bush lupine (*Lupinus arboreus*). *Oecologia*, **103**, 343–8.

Harrison, S. (1997). Persistent, local outbreaks in the western tussock moth (*Orgyia vetusta*): roles of resource quality, predation and poor dispersal. *Ecological Entomology*, **22**, 158–66.

Harrison, S. and Taylor, A.D. (1997). Empirical evidence for metapopulation dynamics. In *Metapopulation dynamics* (ed. M. Gilpin and I. Hanski), pp. 27–42. Academic Press, San Diego, CA.

Harrison, S. and Wilcox, C. (1995). Evidence that predator satiation may restrict the spatial spread of a tussock moth (*Orgyia vetusta*), a flightless defoliator. *Oecologia*, **101**, 309–16.

Harrison, S., Murphy, D.D., and Ehrlich, P.R. (1988). Distribution of the bay checkerspot butterfly, *Euphydryas editha bayensis*: evidence for a metapopulation model. *American Naturalist*, **132**, 360–82.

Harvell, C.D. (1990). The ecology and evolution of inducible defenses. *Quarterly Review of Biology*, **65**, 323–40.

Harvell, C.D. (1998). Genetic variation and polymorphism in the inducible spines of a marine bryozoan. *Evolution*, **52**, 80–6.

Harvey, P.H. and Ralls, K. (1986). Do animals avoid incest? *Nature*, **320**, 575–6.

Hassell, M.P. (1978). *The dynamics of arthropod predator–prey systems*. Princeton University Press, Princeton, NJ.

Hassell, M.P. and May, R.M. (1973). Stability in insect host–parasite models. *Journal of Animal Ecology*, **42**, 693–736.

Hassell, M.P., Comins, H.N., and May, R.M. (1994). Species coexistence and self-organizing spatial dynamics. *Nature*, **370**, 290–2.

Hastings, A. (1980). Disturbance, coexistence, history and the competition for space. *Theoretical Population Biology*, **18**, 363–73.

Hastings, A. (1983). Can spatial variation alone lead to selection for dispersal? *Theoretical Population Biology*, **24**, 244–51.

Hastings, A. (1993). Complex interactions between dispersal and dynamics: lessons from coupled logistic equations. *Ecology*, **74**, 1362–72.

Hastings, A. (2000). Parasitoid spread: lessons for and from invasion biology. In *Parasitoid population biology* (ed. M.E. Hochberg and A.R. Ives, pp. 70–82). Princeton University Press, Princeton, NJ.

Hastings, A. and Harrison, S. (1994). Metapopulation dynamics and genetics. *Annual Review of Ecology and Systematics*, **25**, 167–88.

Hastings, A. and Higgins, K. (1994). Persistence of transients in spatially structured models. *Science*, **263**, 1133–6.

Heath, D.D., Devlin, R.H., Heath, J.W., and Iwama, G.K. (1994). Genetic, environmental and interaction effects on the incidence of jacking in *Oncorhynchus tshawytscha* (chinook salmon). *Heredity*, **72**, 146–54.

Hedrick, P.W. (1986). Genetic-polymorphism in heterogeneous environments – a decade later. *Annual Review of Ecology and Systematics*, **17**, 535–66.

Hedrick, P.W. and Gilpin, M.E. (1997). Genetic effective size of a metapopulation. In *Metapopulation biology: ecology, genetics and evolution* (ed. I.A. Hanski and M.E. Gilpin), pp. 165–81. Academic Press, San Diego, CA.

Hedrick, P.W., Ginevan, M.E., and Ewing, E.P. (1976). Genetic polymorphism in heterogeneous environments. *Annual Review of Ecology and Systematics*, **7**, 1–32.

Heeb, P., Werner, I. Kolliker, M., and Richner, H. (1998). Benefits of induced host responses against an ectoparasite. *Proceedings of the Royal Society of London B*, **265**, 51–6.

Heeb, P., Werner, I., Mateman, A.C., Kölliker, M., Brinkhof, M.W.G., Lessels, C.M., and Richner, H. (1999). Ectoparasite infestation and sex-biased local recruitment of hosts. *Nature*, **400**, 63–5.

Heg, D. (1999). *Life history decisions in oystercatchers*. PhD thesis. University of Groningen, Groningen.

Heiman, M.L., Ahima, R.S., Craft, L.S., Schoner, B., Stephens, T.W., and Flier, J.S. (1997). Leptin inhibition of the hypothalamic–pituitary–adrenal axis in response to stress. *Endocrinology*, **138**, 3859–63.

Heisey, D.M. and Fuller, T.K. (1985). Evaluation of survival and cause-specific mortality rates using telemetry data. *Journal of Wildlife Management*, **49**, 668–74.

Helldin, J.O. and Lindström, E.R. (1995). Late winter social activity in pine marten (*Martes martes*): false heat or dispersal? *Annales Zoologici Fennici*, **32**, 145–9.

Henderson, M.T., Merriam, G., and Wegner, J. (1985). Patchy environments and species survival: chipmunks in an agricultural mosaic. *Biological Conservation*, **31**, 95–105.

Henry, C., Arsaut, J., Arnauld, E., and Demotes-Mainard, J. (1996). Transient neonatal elevation in hypothalamic estrogen receptor mRNA in prenatally-stressed male rats. *Neuroscience Letters*, **27**, 141–5.

Hersteinsson, P. (1992). Mammals of the Thingvallavatn area. *Oikos*, **64**, 396–404.

Herzig, A.L. (1995). Effects of population density on long-distance dispersal in the goldenrod beetle *Trirhabda virgata*. *Ecology*, **76**, 2044–54.

Hess, G.R. (1994). Conservation corridors and contagious disease: a cautionary note. *Conservation Biology*, **8**, 256–62.

Hesse, R., Allee, W.C., and Schmidt, K.P. (1951). *Ecological animal geography*, 2nd edn. Chapman and Hall, New York.

Hestbeck, J.B. (1982). Population regulation of cyclic mammals: the social fence hypothesis. *Oikos*, **39**, 157–63.

Hestbeck, J.B., Nichols, J.D., and Malecki, R.A. (1991). Estimates of movement and site fidelity using mark–resight data of wintering Canada geese. *Ecology*, **72**, 523–33.

Hey, J. (1997). Mitochondrial and nuclear genes present conflicting portraits of human origins. *Molecular Biology and Evolution*, **14**, 166–72.

Hilborn, R. (1990). Determination of fish movement patterns from tag recoveries using maximum likelihood estimators. *Canadian Journal of Fisheries and Aquatic Sciences*, **47**, 635–43.

Hilborn, R. (1991). Modeling the stability of fish schools: exchange of individual fish between schools of skipjack tuna (*Katsuwonus pelamis*). *Canadian Journal of Fisheries and Aquatic Sciences*, **48**, 1081–91.

Hildén, O. (1965). Habitat selection in birds: a review. *Annales Zoologici Fennici*, **2**, 53–75.

Hitching, S.P. and Beebee, T.J.C. (1998). Loss of genetic diversity and fitness in common toad (*Bufo bufo*) populations isolated by inimical habitat. *Journal of Evolutionary Biology*, **11**, 269–83.

Hobbs, R.J. (1994). Landscape ecology and conservation: moving from description to application. *Pacific Conservation Biology*, **1**, 170–6.

Hochberg, M.E. and Holt, R.D. (1995). Refuge evolution and the population dynamics of coupled host–parasitoid associations. *Evolutionary Ecology*, **9**, 633–61.

Hochberg, M.E. and Ives, A.R. (1999). Can natural enemies enforce geographical range limits? *Ecography*, **22**, 268–76.

Hochberg, M.E. and van Baalen, M. (1998). Antagonistic coevolution over productivity gradients. *American Naturalist*, **152**, 620–34.

Hochberg, M.E., Elmes, G.W., Thomas, J.A., and Clarke, R.T. (1996). Mechanisms of local persistence in coupled host–parasitoid associations. The case model of *Maculinea rebeli* and *Ichneumon eumerus*. *Philosophical Transactions of the Royal Society of London B*, **351**, 1713–24.

Hodgson, D.J. and Godfray, H.C.J. (1999). The consequences of clustering by *Aphis fabae* foundresses on spring migrant production. *Oecologia*, **118**, 446–52.

Hoeck, H.N. (1982). Population dynamics, dispersal and genetic isolation in two species of hyrax (*Heterohyrax brucei* and *Procavia johnstoni*) on habitat islands in the Serengeti. *Zeitschrift für Tierpsychologie*, **59**, 117–210.

Hoelzel, A.R. (1998). Genetic structure of cetacean populations in sympatry, parapatry, and mixed assemblages: implications for conservation policy. *Journal of Heredity*, **89**, 451–8.

Hoelzel, A.R. and Dover, G.A. (1991). Genetic differentiation between sympatric killer whale populations. *Heredity*, **66**, 191–5.

Hofbauer, J. and Sigmund, S.K. (1988). *The theory of evolution and dynamical systems*. Cambridge University Press, Cambridge.

Höglund, J., Alatalo, R.V., and Lundberg, A. (1990). Copying the mate choice of others?

Observations on female black grouse. *Behaviour*, **114**, 221–36.

Hoi, H. and Hoi-Leitner, M. (1997). An alternative route to coloniality in the bearded tit: females pursue extra-pair fertilizations. *Behavioral Ecology*, **8**, 113–19.

Holberton, R.L., Parrish, J.D., and Wingfield, J.C. (1996). Modulation of the adrenocortical stress response in neotropical migrants during autumn migration. *Auk*, **113**, 558–64.

Holbrook, N.M. and Lund, C.P. (1995). Photosynthesis in forest canopies. In *Forest Canopies* (ed. M.D. Lowman and N.M. Nadkarni) pp. 411–30. Academic Press, San Diego, CA.

Holekamp, K.E. (1984). Natal dispersal in Belding's ground squirrels (*Spermophilus beldingi*). *Behavioral Ecology and Sociobiology*, **16**, 21–39.

Holekamp, K.E. (1986). Proximal causes of natal dispersal in Belding's ground squirrels (*Spermophilus beldingi*). *Ecological Monographs*, **56**, 365–91.

Holekamp, K.E. and Sherman, P.W. (1989). Why male ground squirrels disperse? *American Scientist*, **77**, 232–9.

Holekamp, K.E. and Smale, L. (1995). Rapid change in offspring sex-ratios after clan fission in the spotted hyena. *American Naturalist*, **145**, 261–78.

Holekamp, K.E. and Smale, L. (1998). Dispersal status influences hormone and behavior in the male spotted hyena. *Hormones and Behavior*, **33**, 205–16.

Holekamp, K.E., Smale, L., Simpson, H.B., and Holekamp, N.A. (1984). Hormonal influences on natal dispersal in free-living Belding's ground squirrels (*Spermophilus beldingi*). *Hormones and Behavior*, **18**, 465–83.

Holekamp, K.E., Ogutu, J.O., Dublin, H.T., Frank, L.G., and Smale, L. (1993). Fission of a spotted hyena clan – consequences of prolonged female absenteeism and causes of female emigration. *Ethology*, **93**, 285–99.

Holmes, J.C. and Bethel, W.M. (1972). Modification of intermediate host behaviour by parasites. In *Behavioural aspects of parasite transmission* (ed. E.U. Canning and C.A. Wright), pp. 123–49. Academic Press, London.

Holopainen, I., Aho, J., Vornanen, M., and Huuskonen, H. (1997). Phenotypic plasticity and predator affects on morphology and physiology of crucian carp in nature and in the lab. *Journal of Fish Biology*, **50**, 781–98.

Holsinger, K.E. (1988). Inbreeding depression doesn't matter: the genetic basis of mating-system evolution. *Evolution*, **42**, 1235–44.

Holsinger, K.E. (1991). Inbreeding depression and the evolution of plant mating systems. *Trends in Ecology and Evolution*, **6**, 307–8.

Holt, R.D. (1985). Population dynamics in two-patch environments: some anomalous consequences of an optimal habitat distribution. *Theoretical Population Biology*, **28**, 181–208.

Holt, R.D. (1993). Ecology at the mesoscale: the influence of regional processes on local communities. In *Species diversity in ecological communities: historical and geographical perspectives* (ed. R.E. Ricklefs and D. Schluter), pp. 77–88. University of Chicago Press, Chicago, IL.

Holt, R.D. (1996). Demographic constraints in evolution: towards unifying the evolutionary theories of senescence and niche conservatism. *Evolutionary Ecology*, **10**, 1–11.

Holt, R.D. (1997a). From metapopulation dynamics to community structure: some consequences of spatial heterogeneity. In *Metapopulation biology: ecology, genetics, and evolution* (ed. I. Hanski and M.E. Gilpin), pp. 149–64. Academic Press San Diego, CA.

Holt, R.D. (1997b). On the evolutionary stability of sink populations. *Evolutionary Ecology*, **11**, 723–31.

Holt, R.D. and Gaines, M.S. (1992). Analysis of adaptation in heterogeneous landscapes: implications for the evolution of fundamental niches. *Evolutionary Ecology*, **6**, 433–47.

Holt, R.D. and Gomulkiewicz, R. (1997). How does immigration influence local adaptation? A reexamination of a familiar paradigm. *American Naturalist*, **149**, 563–72.

Holt, R.D. and Hassell, M.P. (1993). Environmental heterogeneity and the stability of host–parasitoid interactions. *Journal of Animal Ecology*, **62**, 89–100.

Holt, R.D. and McPeek, M.A. (1996). Chaotic population dynamics favors the evolution of dispersal. *American Naturalist*, **148**, 709–18.

Honda, J.Y. and Walker, G.P. (1996). Olfactory response of *Anagrus nigriventris* (Hymenoptera: Mymaridae): effects of host plant chemical cues mediated by rearing and oviposition experience. *Entomophaga*, **41**, 3–13.

Honeycutt, R.L., Nelson, K., Schlitter, D.A., and Sherman, P.W. (1991). Genetic variation within and among populations of the naked mole-rat: evidence from nuclear and mitochondrial genomes. In *The biology of the naked mole-rat* (ed. P.W. Sherman, J.U.M. Jarvis, and R.D. Alexander), pp. 195–208. Princeton University Press, Princeton, NJ.

Hoogland, J.L. (1982). Prairie dogs avoid extreme inbreeding. *Science*, **215**, 1639–41.

Hoogland, J.L. (1995). *The black-tailed prairie dog*. University of Chicago Press, Chicago, IL.

Horn, H.S. and MacArthur, R.H. (1972). Competition among fugitive species in a harlequin environment. *Ecology*, **53**, 749–52.

Hosmer, D.W. and Lemeshow, S.L. (1989). *Applied logistic regression*. John Wiley, New York.

Hosmer, D.W. and Lemeshow, S.L. (1999). *Applied survival analysis*. John Wiley, New York.

Houle, G. and Phillips, D.L. (1989). Seed availability and biotic interactions in granite outcrop plant communities. *Ecology*, **70**, 1307–16.

Houston, A. and McNamara, J. (1986). The influence of mortality on the behaviour that maximizes reproductive success in a patchy environment. *Oikos*, **47**, 267–74.

Houston, A.I. and McNamara, J.N. (1988). The IFE when competitive abilities differ: approach based on statistical mechanics. *Animal Behaviour*, **36**, 166–74.

Howard, W.E. (1960). Innate and environmental dispersal of individual vertebrates. *American Midland Naturalist*, **63**, 152–61.

Howe, H.F. (1986). Seed dispersal by fruit-eating birds and mammals. In *Seed dispersal* (ed. D.R. Murray), pp. 123–90. Academic Press, New York.

Howe, H.F. and Smallwood, J. (1982). Ecology of seed dispersal. *Annual Review of Ecology and Systematics*, **13**, 201–28.

Hudson, R.R. (1998). Island models and the coalescent process. *Molecular Ecology* **7**, 413–18.

Hudson, R.R., Slatkin, M., and Maddison, W.P. (1992). Estimation of levels of gene flow from DNA sequence data. *Genetics*, **132**, 583–9.

Huffaker, C.B. (1958). Experimental studies on predation: dispersion factors and predator–prey oscillations. *Hilgardia*, **27**, 343–83.

Hughes, C. (1998). Integrating molecular techniques with field methods in studies of social behavior: a revolution results. *Ecology*, **79**, 383–99.

Hugie, D.M. and Grand, T.C. (1998). Movement between patches, unequal competitors and the ideal free distribution. *Evolutionary Ecology*, **12**, 1–19.

Husband, B.C. and Barrett, S.C.H. (1996). A metapopulation perspective in plant population biology. *Journal of Ecology*, **84**, 461–9.

Hutchinson, G.E. (1951). Copepodology for the ornithologist. *Ecology*, **32**, 571–7.

Hutson, V. and Vickers, G.T. (1992). Travelling waves and dominance ESSs. *Journal of Mathematical Biology*, **30**, 457–71.

Hyatt, L.A. and Evans, A.S. (1998). Is decreased germination fraction associated with risk of sibling competition? *Oikos*, **83**, 29–35.

Ims, R.A. (1989). Kinship and origin effects on dispersal and space sharing in *Clethrionomys rufocanus*. *Ecology*, **70**, 607–16.

Ims, R.A. (1990). Determinants of natal dispersal and space use in grey-sided voles, *Clethrionomys rufocanus*: a combined field and laboratory experiment. *Oikos*, **57**, 106–13.

Ims, R.A. (1995). Movement patterns related to spatial structures. In *Mosaic landscapes and ecological processes* (ed. L. Hansson, L. Fahrig, and G. Merriam), pp. 85–109, Chapman and Hall, London.

Ims, R.A. and Andreassen, H.P. (1999). Demographic synchrony in fragmented populations. In *The ecology of small mammals at the landscape level: experimental approaches* (ed. G. Barrett and J. Peles), pp. 129–45. Springer, New York.

Ims, R.A. and Stenseth, N.C. (1989). Divided the fruitflies fall. *Nature*, **342**, 21–2.

Ims, R.A. and Yoccoz, N.G. (1997). Studying transfer processes in metapopulations: emigration, migration, and colonization. In *Metapopulation biology: ecology, genetics and evolution* (ed. I. Hanski and M. Gilpin), pp. 247–66. Academic Press, San Diego, CA.

Ims, R.A., Rolstad, J., and Wegge, P. (1993). Predicting space use responses to habitat fragmentation: can voles (*Mictorus oeconomus*) serve as an experimental model system (EMS) for capercaillie grouse, *Tetrao urogallus*, in boreal forests? *Biological Conservation*, **63**, 261–8.

Ingvarsson, P.K. and Whitlock, M.C. (2000). Heterosis increases the effective migration rate. *Proceedings of the Royal Society of London*, **267**, 1321–26.

Inui, A. (1999). Feeding and body-weight regulation by hypothalamic neuropeptides – mediation of the actions of leptin. *Trends in Neurosciences*, **22**, 62–7.

Iwasa, Y. and Roughgarden, J. (1986). Interspecific competition among metapopulations with space-limited subpopulation. *Theoretical Population Biology*, **30**, 194–214.

Iwasaki, K. (1995). Dominance order and resting site fidelity in the intertidal pulmonate limpet siphonaria-sirius (pilsbry). *Ecological Research*, **10**, 105–15.

Jackson, C.H.N. (1938). The analysis of an animal population. *Journal of Animal Ecology*, **8**, 238–246.

Jackson, W.M., Rohwer, S., and Nolan, V.J. (1989). Within-season breeding dispersal in prairie warblers and other passerines. *Condor*, **91**, 233–41.

Jacquard, A. (1975). Inbreeding: one word, several meanings. *Theoretical Population Biology*, **7**, 338–63.

Jacquot, J. and Vessey, S. (1995). Influence of the natal environment on dispersal of white-footed mice. *Behavioral Ecology and Sociobiology*, **37**, 407–12.

Jaenike, J., Benway, H., and Stevens, G. (1995). Parasite-induced mortality in mycophagous *Drosophila*. *Ecology*, **76**, 383–91.

Janetos, A.C. (1980). Strategies of female choice: a theoretical analysis. *Behavioral Ecology and Sociobiology*, **7**, 107–12.

Jannett, F. (1978). The density-dependent formation of extended maternal families of the montane vole, *Microtus montanus nanus*. *Behavioral Ecology and Sociobiology*, **3**, 245–63.

Jansen, V.A.A. (1995). Regulation of predator–prey systems through spatial interactions: a possible solution to the paradox of enrichment. *Oikos*, **74**, 383–90.

Jansen, V.A.A. and de Roos, A.M. (2000). The predator–prey role of space in reducing cycles. In *The geometry of ecological interactions: simplifying spatial complexity* (ed. R. Law, U. Dieckmann, and J.A.J. Metz), 183–208. Cambridge University Press, Cambridge.

Jansen, V.A.A. and Mulder, G.S.E.E. (1999). Evolving biodiversity. *Ecology Letters*, **2**, 379–86.

Jansen, V.A.A. and Sabelis, M.W. (1992). Prey dispersal and predator persistence. *Experimental and Applied Acarology*, **14**, 215–31.

Jansen, V.A.A. and Yoshimura, J. (1998). Populations can persist in an environment consisting of sink habitats only. *Proceedings of the National Academy of Sciences of the USA*, **95**, 3696–8.

Janssen, A. and Sabelis, M.W. (1992). Phytoseiid life-histories, local predator–prey dynamics, and strategies for control of tetranychid mites. *Experimental and Applied Acarology*, **14**, 233–50.

Janssen, A., Pallini, A., Venzon, M., and Sabelis, M.W. (1998). Behaviour and indirect interactions in food webs of plant-inhabiting arthropods. *Experimental and Applied Acarology*, **22**, 497–521.

Jarosz, A.M. and Burdon, J.J. (1991). Host–pathogen interactions in natural populations of *Linum marginale* and *Melampsora lini*. II. Local and regional variation in patterns of resistance and racial structure. *Evolution*, **45**, 1618–27.

Jarvis, J.U.M. (1985). Ecological studies on *Heterocephalus glaber*, the naked mole-rat, in Kenya. *National Geographic Society of Research Report*, **20**, 429–37.

Jarvis, J.U.M. (1991). Reproduction of naked mole-rats. In *The biology of the naked mole-rat* (ed. P.W. Sherman, J.U.M. Jarvis, and R.D. Alexander). pp. 384–425. Princeton University Press, Princeton, NJ.

Jarvis, J.U.M., O'Riain, M.J., and McDaid, E. (1991). Growth and factors affecting body size in naked mole-rats. In *The biology of the naked mole-rat* (ed. P.W. Sherman, J.U.M. Jarvis, and R.D. Alexander), pp. 358–83. Princeton University Press, Princeton, NJ.

Jarvis, J.U.M., O'Riain, M.J., Bennett, N.C., and Sherman, P.W. (1994). Eusociality: a family affair. *Trends in Ecology and Evolution*, **9**, 47–51.

Jenkins, P.F. (1978). Cultural transmission of songpatterns and dialect development in a free-living bird population. *Animal Behaviour*, **25**, 50–78.

Jimenez, J.A., Hughes, K.A., Alaks, G., Graham, L., and Lacy, R.C. (1994). An experimental study of inbreeding depression in a natural habitat. *Science*, **266**, 271–3.

Joern, A. and Gaines, S.B. (1990). Population dynamics and regulation in grasshoppers. In *Biology of grasshoppers* (ed. R.F. Chapman and A. Joern), pp. 415–82. Wiley, New York.

Johannesen, E. and Andreassen, H. (1998). Survival and reproduction of resident and immigrant root vole (*Microtus oeconomus*) females. *Canadian Journal of Zoology*, **76**, 763–6.

Johnson, C.G. (1969). *Migration and dispersal of insects by flight*. Methuen, London.

Johnson, D.D.P. and Mighell, J.S. (1999). Dry-season bird diversity in tropical rainforest and surrounding habitats in north-east Australia. *Emu*, **99**, 108–20.

Johnson, D.D.P., Hay, S.I., and Rogers, D.J. (1998). Contemporary environmental correlates of endemic bird areas derived from meteorological satellite sensors. *Proceedings of the Royal Society B*, **265**, 951–9.

Johnson, E.O., Kamilaris, T.C., Chrousos, G.P., and Gold, P.W. (1992). Mechanisms of stress: a dynamic overview of hormonal and behavioral homeostasis. *Neuroscience and Behavioral Reviews*, **16**, 115–30.

Johnson, M.L. and Gaines, M.S. (1990). Evolution of dispersal: theoretical models and empirical tests using birds and mammals. *Annual Review of Ecology and Systematics*, **21**, 449–80.

Johnson, M.S. and Black, R. (1995). Neighbourhood size and the importance of barriers to gene flow in an intertidal snail. *Heredity*, 75, 142–54.

Johnston, D.W. (1970). Caloric density of avian adipose tissue. *Comparative Biochemistry and Physiology*, 34, 827–32.

Johnston, J.S. (1982). Genetic variation for anemotaxis (wind-directed movement) in laboratory and wild-caught populations of *Drosophila*. *Behaviour Genetics*, 12, 281–93.

Johnstone, R.A. (1997). The tactics of mutual mate choice and competitive search. *Behavioral Ecology and Sociobiology*, 40, 51–9.

Jones, W. (1986). Survivorship in philopatric and dispersing kangaroo rats (*Dipodomys spectabilis*). *Ecology*, 67, 202–7.

Jones, W., Waser, P., Elliott, N., Link, N., and Bush, B. (1988). Philopatry, dispersal, and habitat saturation in the banner-tailed kangaroo rat, *Dipodomys spectabilis*. *Ecology*, 69, 1466–73.

Jordano, P. (1992). Fruits and frugivory. In *The ecology of regeneration in plant communities* (ed. M. Fenner), pp. 105–56. CAB International, Wallingford, UK.

Julliard, R., Perret, P., and Blondel, J. (1996). Reproductive strategies of immigrant and philopatric Corsican blue tit. *Acta Oecologica*, 17, 487–501.

Kacelnik, A., Krebs, J.R., and Bernstein, C. (1992). The ideal free distribution and predator–prey populations. *Trends in Ecology and Evolution*, 7, 50–5.

Kadmon, R. and Shmida, A. (1990). Spatiotemporal demographic processes in plant populations: an approach and a case study. *American Naturalist*, 135, 382–97.

Kaiser, A. (1995). Estimating turnover, movements and capture parameters of resting passerines in standardized capture–recapture studies. *Journal of Applied Statistics*, 22, 1039–47.

Kaltz, O. and Shykoff, J.A. (1998). Local adaptation in host–parasite systems. *Heredity*, 81, 361–70.

Kaplan, E.L. and Meier, P. (1958). Nonparametric estimation from incomplete observations. *Journal of the American Statistical Association*, 53, 188–93.

Karanth, K.U. and Nichols, J.D. (1998). Estimation of tiger densities in India using photographic captures and recaptures. *Ecology*, 79, 2852–62.

Karban, R. and Baldwin, I.T. (1997). *Induced responses to herbivory*. University of Chicago Press, Chicago, IL.

Kareiva, P. (1990). Population dynamics in spatially complex environments: theory and data. In *Population regulation and dynamics* (ed. M.P. Hassell and R.M. May), pp. 53–68. Royal Society, London.

Karlson, R.H. and Taylor, H.M. (1992). Mixed dispersal strategies and clonal spreading of risk: predictions from a branching process model. *Theoretical Population Biology*, 42, 218–33.

Karlson, R.H. and Taylor, H.M. (1995). Alternative predictions for optimal dispersal in response to local catastrophic mortality. *Theoretical Population Biology*, 47, 321–30.

Kawecki, T.J. (1997). Sympatric speciation via habitat specialization driven by deleterious mutations. *Evolution*, 51, 1751–63.

Kawecki, T.J. and Stearns, S.C. (1993). The evolution of life histories in spatially heterogeneous environments: optimal reaction norms revisited. *Evolutionary Ecology*, 7, 155–74.

Keane, B. (1990). The effect of relatedness on reproductive success and mate choice in the white-footed mouse, *Peromyscus leucopus*. *Animal Behaviour*, 39, 264–73.

Keane, B., Creel, S.R., and Waser, P.M. (1996). No evidence of inbreeding avoidance or inbreeding depression in a social carnivore. *Behavioural Ecology*, 7, 480–9.

Keddy, P.A. (1981). Experimental demography of the sand-dune annual, *Cakile edentula*, growing along an environmental gradient in Nova Scotia. *Journal of Ecology*, 69, 615–30.

Keddy, P.A. (1982). Population ecology on an environmental gradient: *Cakile edentula* on a sand dune. *Oecologia*, 52, 348–55.

Keeley, J.E. and Fotheringham, C.J. (1998). Mechanisms for smoke-induced seed germination in a post-fire chaparral annual. *Journal of Ecology*, 86, 27–36.

Keeling, M.J. (1996). *The ecology and evolution of spatial host–parasite systems*. PhD thesis, University of Warwick, Coventry.

Kehler, D. and Bondrup-Nielsen, S. (1999). Effects of isolation on the occurrence of a fungivorous forest beetle, *Bolitotherus cornutus*, at different spatial scales in fragmented and continuous forests. *Oikos*, **84**, 35–43.

Keller, L.F. (1998). Inbreeding and its fitness effects in an insular population of song sparrows (*Melospiza melodia*). *Evolution*, **52**, 240–50.

Keller, L.F. and Arcese, P. (1998). No evidence for inbreeding avoidance in a natural population of song sparrows (*Melospiza melodia*). *American Naturalist*, **152**, 380–92.

Keller, L.F. and Ross, K.G. (1993a). Phenotypic basis of reproductive success in a social insect: genetic and social determinants. *Science*, **260**, 1107–10.

Keller, L.F. and Ross, K.G. (1993b). Phenotypic plasticity and 'cultural' transmission of alternative social organizations in the fire ant *Solenopsis invicta*. *Behavioral Ecology and Sociobiology*, **33**, 121–9.

Keller, L.F., Arcese, P., Smith, J.N.M., Hochachka, W.M., and Stearns, S.C. (1994). Selection against inbred song sparrows during a natural population bottleneck. *Nature*, **372**, 356–7.

Kempenaers, B., Adriaensen, F., and Dhondt, A.A. (1998). Inbreeding and divorce in great tits. *Animal Behaviour*, **56**, 737–40.

Kendall, B.E., Bjørnstad, O.N., Bascompte, J., Keitt, T.H., and Fagan, W.F. (2000). Dispersal, environmental correlation, and spatial synchrony in population dynamics. *American Naturalist*, **155**, 628–36.

Kendall, D.G. (1948). On the role of variable generation time in the development of a stochastic process. *Biometrika*, **35**, 316–30.

Kendall, W.L., Nichols, J.D., and Hines, J.E. (1997). Estimating temporary emigration using capture–recapture data with Pollock's robust design. *Ecology*, **78**, 563–78.

Ketterson, E.D. and King, J.R. (1977). Metabolic and behavioral responses to fasting in the white-crowned sparrow (*Zonotrichia leucophrys gambelii*). *Physiological Zoology*, **50**, 115–29.

Kiester, A.R. (1979). Conspecific as cues: a mechanism for habitat selection in the Panamanian grass anole (*Anolis auratus*). *Behavioral Ecology and Sociobiology*, **5**, 323–30.

Kimura, M. (1953). 'Stepping stone' model of population. *Annual Report of the National Institute of Genetics, Japan*, **3**, 62–3.

Kimura, M. (1957). Some problems of stochastic processes in genetics. *Annals of Mathematics and Statistics*, **28**, 882–901.

Kimura, M. and Weiss, G.H. (1964). The stepping stone model of population structure and the decrease of genetic correlation with distance. *Genetics*, **49**, 561–76.

King, K.A., Keith, J.O., and Keirans, J.E. (1977). Ticks as a factor in nest desertion of California brown pelicans. *Condor*, **79**, 507–9.

Kingman, J.F.C. (1969). Markov population processes. *Journal of Applied Probabilities*, **6**, 1–18.

Kirkpatrick, M. and Barton, N.H. (1997). Evolution of a species' range. *American Naturalist*, **150**, 1–23.

Kirkpatrick, S., Gelatt, C.D., and Vecchi, M.P. (1983). Optimization by simulated annealing. *Science*, **220**, 671–80.

Kishimoto, K. (1990). Coexistence of any number of species in the Lokta–Volterra competitive system over two-patches. *Theoretical Population Biology*, **38**, 149–58.

Kleiman, D.G. (1981). Correlations among life history characteristics of mammalian species exhibiting two extreme forms of monogamy. In *Natural selection and social behavior* (ed. R.D. Alexander and D.W. Tinkle), pp. 332–44. Chiron Press, New York.

Klopfer, P.H. and Ganzhorn, J.U. (1985). Habitat selection: behavioral aspects. In *Habitat selection in birds* (ed. M.L. Cody), pp. 435–53. Academic Press, Orlando, FL.

Knight-Jones, E.W. (1951). Gregariousness and some other aspects of the setting behaviour of *Spirorbis*. *Journal of the Marine Biological Association*, **30**, 201–22.

Kobayashi, Y. and Yamamura, N. (2000). Evolution of seed dormancy due to sib competition: effect of dispersal and inbreeding. *Journal of Theoretical Biology*, **202**, 11–24.

Koenig, W.D. (1999). Spatial autocorrelation of ecological phenomena. *Trends in Ecology and Evolution*, **14**, 22–6.

Koenig, W.D., Pitelka, F.A., Carmen, W.J., Mumme, R.L., and Stanback, M.T. (1992). The evolution of delayed dispersal in cooperative breeders. *Quarterly Review of Biology*, **67**, 111–50.

Koenig, W.D., van Vuren, D., and Hooge, P.N. (1996). Detectability, philopatry, and the distribution of dispersal distances in vertebrates. *Trends in Ecology and Evolution*, **11**, 514–17.

Kolasa, J. and Rollo, C.D. (1991). Introduction: the heterogeneity of heterogeneity: a glossary. In *Ecological heterogeneity* (ed. J. Kolasa and S.T.A. Pickett), pp. 1–23. Springer, New York.

Kollmann, J. and Pirl, M. (1995). Spatial pattern of seed rain of fleshy-fruited plants in a scrubland–grassland transition. *Acta Oecologica*, **16**, 313–29.

Komdeur, J. (1996). Influence of helping and breeding experience on reproductive performance in the Seychelles warbler: a translocation experiment. *Behavioral Ecology*, **7**, 326–33.

Komdeur, J., Daan, S., Tinbergen, J., and Mateman, C. (1997). Extreme adaptive modification in sex ratio of the Seychelles warbler's eggs. *Nature*, **385**, 522–5.

Komers, P.E. (1997). Behavioral plasticity in variable environments. *Canadian Journal of Zoology*, **75**, 161–9.

Kondrashov, A.S. and Kondrashov, F.A. (1999). Interactions among quantitative traits in the course of sympatric speciation. *Nature*, **400**, 351–4.

Koprowski, J.L. (1996). Natal philopatry, communal nesting, and kinship in fox squirrels and grey squirrels. *Journal of Mammalogy*, **77**, 1006–16.

Kortner, G. and Geiser, F. (1998). Ecology of natural hibernation in the marsupial mountain pygmy-possum (*Burramys paravus*). *Oecologia*, **113**, 170–8.

Kot, M., Lweis, M.A., and van den Driessche, P. (1996). Dispersal data and the spread of invading organisms. *Ecology*, **77**, 2027–42.

Kowalski, R. and Benson, J.F. (1978). A population dynamics approach to the wheat bulb fly *Delia coarctata* problem. *Journal of Applied Ecology*, **15**, 89–104.

Kramer, D.L., Rangeley, R.W., and Chapman, L.J. (1997). Habitat selection: patterns of spatial distribution from behavioural decisions. In *Behavioural ecology of teleost fishes* (ed. J.J. Godin), pp. 37–79. Oxford University Press, Oxford.

Krebs, C.J. (1992). The role of dispersal in cyclic rodent populations. In *Animal dispersal: small mammals as a model* (ed. N.C. Stenseth and W.Z. Lidicker), pp. 160–76, Chapman and Hall, London.

Kress, S.W. and Nettleship, D.N. (1988). Reestablishment of atlantic puffins (*Fratercula arctica*) at a former breeding site in the Gulf of Maine. *Journal of Field Ornithology*, **59**, 161–70.

Kruuk, L. (1997). *Barriers to gene flow: a Bombina (fire-bellied toad) hybrid zone and multilocus cline theory*. PhD thesis, University of Edinburgh, Edinburgh.

Kubo, T., Iwasa, Y., and Furumoto, N. (1996). Forest spatial dynamics with gap expansion: total gap area and gap size distribution. *Journal of Theoretical Biology* **180**, 229–46.

Kuhner, M.K., Yamato, J., and Felsenstein, J. (1995). Estimating effective population size and mutation rate from sequence data using Metropolis-Hastings sampling. *Genetics*, **140**, 1421–30.

Kuno, E. (1981). Dispersal and the persistence of populations in unstable habitats: a theoretical note. *Oecologia*, **49**, 123–6.

Kuusaari, M., Nieminen, M., and Hanski, I. (1996). An experimental study of migration in the Glanville fritillary butterfly *Melitaea cinxia*. *Journal of Animal Ecology*, **65**, 791–801.

Lacey, E.A. and Sherman, P.W. (1991). Social organisation of naked mole-rat colonies: evidence for divisions of labour. In *The biology of the naked mole-rat* (ed. P.W. Sherman, J.U.M. Jarvis, and R.D. Alexander), pp. 275–336. Princeton University Press, Princeton, NJ.

Lacey, E.A. and Sherman, P.W. (1997). Cooperative breeding in naked mole-rats: implications for vertebrate and invertebrate sociality. In *Cooperative breeding in mammals* (ed. N. Solomon and J.A. French), pp. 267–301. Cambridge University Press, Cambridge.

Lacey, E., Alexander, R.D., Braude, S., Sherman, P.W., and Jarvis, J.U.M. (1991). An ethogram for the naked mole-rat: non-vocal behaviors. In *The biology of the naked mole-rat* (ed. P.W. Sherman, J.U.M. Jarvis, and R.D. Alexander), pp. 209–42. Princeton University Press; Princeton, NJ.

Lack, D. (1933). Habitat selection in birds with special reference to the effect of afforestation on the breckland avifauna. *Journal of Animal Ecology*, **2**, 239–62.

Laferrère, B., Fried, S.K., Hough, K., Campbell, S.A., Thornton, J., and Pi-Sunyer, F.X. (1998). Synergistic effects of feeding and dexamethasone on serum leptin levels. *Journal of Clinical Endocrinology and Metabolism*, **83**, 3742–5.

Lamberson, R.H., McKelvey, R., Noon, B.R., and Voss, C. (1992). A dynamic analysis of northern spotted owl viability in a fragmented forest landscape. *Conservation Biology*, **6**, 505–12.

Lambin, X. (1994a). Natal philopatry, competition for resources, and inbreeding avoidance in Townsend's voles (*Microtus townsendii*). *Ecology*, **75**, 224–35.

Lambin, X. (1994b). Changes in sex allocation and female philopatry in voles. *Journal of Animal Ecology*, **64**, 945–53.

Lambin, X. (1994c). Litter sex ratio does not determine natal dispersal tendency in female Townsend's voles. *Oikos*, **69**, 353–6.

Lambin, X. (1997). Home range shifts by breeding female Townsend's voles (*Microtus townsendii*): a test of the territory bequeathal hypothesis. *Behavioural Ecology and Sociobiology*, **40**, 363–72.

Lambin, X. and Krebs, C. (1993). Influence of female relatedness on the demography of Townsend's vole populations in the spring. *Journal of Animal Ecology*, **62**, 536–50.

Lambin, X. and Yoccoz, N.G. (1998). The impact of population kin-structure on nestling survival in Townsend's voles, *Microtus townsendii*. *Journal of Animal Ecology*, **67**, 1–16.

Lance, D. and Barbosa, P. (1979). Dispersal of larval Lepidoptera with special reference to forest defoliators. *Biologist*, **61**, 90–110.

Lande, R. (1988). Demographic models of the northern spotted owl (*Strix occidentalis caurina*). *Oecologia*, **75**, 601–7.

Lande, R. (1992). Neutral theory of quantitative genetic variance in an island model with local extinction and colonization. *Evolution*, **46**, 381–9.

Lande, R. (1993). Risks of population extinction from demographic and environmental stochasticity and random catastrophes. *American Naturalist*, **142**, 911–27.

Lande, R. and Schemske, D.W. (1985). The evolution of self-fertilization and inbreeding depression in plants. I. Genetic models. *Evolution*, **39**, 24–40.

Lande, R., Engen, S., and Saether, B.-E. (1999). Spatial scale of population synchrony: environmental correlation versus dispersal and density regulation. *American Naturalist*, **154**, 271–81.

Langtimm, C.A., O'Shea, T.J., Pradel, R., and Beck, C.A. (1998). Estimates of annual survival probabilities for adult Florida manatees (*Trichechus manatus latirostris*). *Ecology*, **79**, 981–97.

Larsen, K. and Boutin, S. (1994). Movements, survival, and settlement of red squirrel (*Tamiasciurus hudsonicus*) offspring. *Ecology*, **75**, 214–23.

Lavee, D., Safriel, U.N., and Meilijson, I. (1991). For how long do trans-Sahara migrants stop over at an oasis? *Ornis Scandinavica*, **22**, 33–44.

Lavorel, S., O'Neill, R.V., and Gardner, R.H. (1994). Spatio-temporal dispêrsal strategies and annual plant species coexistence in a structured landscape. *Oikos*, **71**, 75–88.

Lavorel, S., Gardner, R.H., and O'Neill, R.V. (1995). Dispersal of annual plants in hierarchically structured landscapes. *Landscape Ecology*, **10**, 277–89.

Lawrence, W.S. (1987). Effects of sex ratio on milkweed beetle emigration from host patches. *Ecology*, **68**, 539–46.

Lawton, J.H. (1996). Population abundances, geographic ranges and conservation. 1994 Witherby lecture. *Bird Study*, **43**, 3–19.

Le Boeuf, B. and Reiter, J. (1988). Lifetime reproductive success in northern elephant seals. In *Reproductive Success: studies of individual variation in contrasting breeding systems* (ed. T. Clutton-Brock), pp. 344–62. Chicago University Press, Chicago, IL.

Le Galliard, J.F. (1999). Local competition, cooperation and mobility. A theoretical and experimental approach. Master's thesis, University of Paris VI, Paris.

Le Maho, Y., Vu Van Kha, H., Koubi, H., Dewasmes, G., Girard, J., Ferre, P., and Cagnard, M. (1981). Body composition, energy expenditure, and plasma metabolites in long-term fasting geese. *American Journal of Physiology*, **241**, E342–E354.

Lebreton, J.D., Pradel, R., and Clobert, J. (1993). The statistical analysis of survival in animal populations. *Trends in Ecology and Evolution*, **8**, 91–5.

Lee, E.T. (1980). *Statistical methods for survival data analysis*. John Wiley, New York.

Legendre, S. and Clobert, J. (1995). ULM, a software for conservation and evolutionary biologists. *Journal of Applied Statistics*, **22**, 817–34.

Lehman, C.R. and Tilman, D. (1998). Competition in spatial habitats. In *Spatial dynamics: the role of space in population dynamics and interspecific interactions*. (ed. D. Tilman and P. Kareiva), pp. 185–203. Princeton University Press, Princeton, NJ.

Lei, G. and Hanski, I. (1997). Spatial dynamics of two competing specialist parasitoids in a host metapopulation. *Oikos*, **78**, 91–100.

Leimar, O. and Norberg, U. (1997). Metapopulation extinction and genetic variation in dispersal-related traits. *Oikos*, **80**, 448–58.

Lemel, J.-Y., Bélichon, S., Clobert, J., and Hochberg, M. E. (1997). The evolution of dispersal in a two-patch system: some consequences of differences between migrants and residents. *Evolutionary Ecology*, **11**, 613–29.

Léna, J.-P. and de Fraipont, M. (1998). Kin discrimination and dispersal in the common lizard. *Behavioral Ecology and Sociobiology*, **42**, 341–7.

Léna, J. P., Clobert, J., de Fraipont, M., Lecomte, J., and Guyot, G. (1998). The relative influence of density and kinship on

dispersal in the common lizard. *Behavioral Ecology*, **9**, 500–7.

Lenormand, T., Guillemaud, T., Bourguet, D., and Raymond, M. (1998). Evaluating gene flow using selected markers: a case study. *Genetics*, **149**, 1383–92.

Lenormand, T., Bourguet, D., Guillemaud, T., and Raymond, M. (1999). Tracking the evolution of insecticide resistance in the mosquito *Culex pipiens. Nature*, **400**, 861–4.

Lens, L. and Dhondt, A.A. (1994). Effects of habitat fragmentation on the timing of crested tit *Parus cristatus* natal dispersal. *Ibis*, **136**, 147–52.

Leonard, G.H., Bertness, M.D., and Yund, P.O. (1999). Crab predation, waterborne cues and inducible defenses: higher risk mussel, *Mytilus edulis. Ecology*, **80**, 1–14.

Lesins, K., Lesins, I., and Gillies, C.B. (1970). Medicago murex. *Chromosoma*, **30**, 109–22.

Lessells, C.M., Avery, M.C., and Krebs, J.R. (1994). Nonrandom dispersal of kin: why do European bee-eater (*Merops apiaster*) brothers nest close together? *Behavioral Ecology*, **5**, 105–13.

Levene, H. (1953). Genetic equilibrium when more than one ecological niche is available. *American Naturalist*, **87**, 331–3.

Levin, S.A. (1974). Dispersion and population interactions. *American Naturalist*, **108**, 207–28.

Levin, S.A., Cohen, D., and Hastings, A. (1984). Dispersal strategies in patchy environments. *Theoretical Population Biology*, **26**, 165–91.

Levins, R. (1963). Theory of fitness in a heterogeneous environment. II. Developmental flexibility and niche selection. *American Naturalist*, **47**, 75–90.

Levins, R. (1968). *Evolution in changing environments*. Princeton University Press, Princeton, NJ.

Levins, R. (1969). Some demographic and genetic consequences of environmental heterogeneity for biological control. *Bulletin of the Entomological Society of America*, **15**, 237–40.

Levins, R. (1970). Extinction. In *Some mathematical problems in biology* (ed. M. Gustenhaver), **2**, pp. 75–107. American Mathematical Society, Providence, RI.

Levins, R. and Culver, D. (1971). Regional coexistence of species and competition between rare species. *Proceedings of the National Academy of Sciences*, **68**, 1246–8.

Levitan, D.R. and Young, C.M. (1995). Reproductive success in large populations: empirical measures and theoretical predictions of fertilization in the sea biscuit *Clypeaster roaceus*. *Journal of Experimental Marine Biology and Ecology*, **190**, 221–41.

Lewontin, R.C. (1972). The apportionment of human diversity. *Evolutionary Biology*, **6**, 381–98.

Lewontin, R.C. (1974). *The genetic basis of evolutionary change*. Columbia University Press, New York.

Lewontin, R.C., Moore, J.A., Provine, W.B., and Wallace, B. (1981). *Dobzhansky's 'Genetics of natural populations' I-XLIII*. Columbia University Press, New York.

Li, J. and Margolies, D.C. (1994). Responses to direct and indirect selection on aerial dispersal behaviour in *Tetranychus urticae*. *Heredity*, **72**,10–22.

Li, W.H. and Nei, M. (1974). Stable linkage disequilibrium without epistasis in subdivided populations. *Theoretical Population Biology*, **6**, 173–83.

Licht, P., Frank, L.G., Pavgi, S., Yalcinkaya, T.M., Siiteri, P.K., and Glickman, S.E. (1992). Hormonal correlates of 'masculinization' in female spotted hyaenas (*Crocuta crocuta*). 2. Maternal and fetal steroids. *Journal of Reproduction and Fertility*, **95**, 463–74.

Lidicker, W.Z. 1975. The role of dispersal in the demography of small mammals. In *Small mammals: their productivity and population dynamics* (ed. F.B. Golley, K. Petrusewicz, and L. Ryszkowski), pp. 103–34, Cambridge University Press, Cambridge.

Lidicker, W.Z. and Patton, J.L. (1987). Patterns of dispersal and genetic structure in populations of small rodents. In *Mammalian dispersal patterns* (ed. B.D. Chepko-Sade, T. Tang, and Z. Halpin), pp. 144–61. University of Chicago Press, Chicago, IL.

Lima, S.L. (1998). Stress and decision making under the risk of predation: recent developments from behavioral, reproductive, and ecological perspectives. *Advances in the Study of Behavior*, **77**, 215–90.

Lima, S.L. and Zollner, P.A. (1996). Towards a behavioral ecology of ecological landscapes. *Trends in Ecology and Evolution*, **11**, 131–5.

Lindberg, M.S., Sedinger, J.S., Derksen, D.V., and Rockwell, R.F. (1998). Natal and breeding philopatry in a black brant, *Branta bernicla nigricans*, metapopulation. *Ecology*, **79**, 1893–904.

Lindquist, N. and Hay, M.E. (1996). Palatability and chemical defense of marine invertebrate larvae. *Ecological Monographs*, **66**, 431–50.

Linhart, Y.B. and Grant, M.C. (1996). Evolutionary significance of local genetic differentiation in plants. *Annual Review of Ecology and Systematics*, **27**, 237–77.

Lively, C.M. and Raimondi, P.T. (1987). Desiccation, predation, and mussel–barnacle interactions in the northern Gulf of California. *Oecologia*, **74**, 304–9.

Lloyd, A.L. and May, R.M. (1996). Spatial heterogeneity in epidemic models. *Journal of Theoretical Biology*, **179**, 1–11.

Lloyd, H.G. (1980). *The red fox*. Batsford, London.

Lofdahl, K.L. (1986). A genetic analysis of habitat selection in the cactophilic species, *Drosophila mojavensis*. In *Evolutionary genetics of invertebrate behavior: progress and prospects* (ed. M.D. Huettel). Plenum Press, New York.

Lofgren, C.S. (1986). History of imported fire ants in the United States. In *Fire ants and leaf-cutting ants; biology and management* (ed. C.S. Lofgren and R.K. vander Meer), pp. 36–47. Westview Press, Boulder, CO.

Lofgren, C.S., Banks, W.A., and Glancey, B.M. (1975). Biology and control of imported fire ants. *Annual Review of Entomology*, **20**, 1–30.

Lomnicki, A. (1988). *Population ecology of individuals*. Princeton University Press, Princeton, NJ.

Long, J.C., Naidu, J.M., Mohrenweiser, H.W., Gershowitz, H., Johnson, P.L., and Wood, J.W. (1986). Genetic characterization of Gainj- and Kalam-speaking peoples of

Papua New Guinea. *American Journal of Phyiological Anthropology*, **70**, 75–96.

Loreau, M. (1998a). Biodiversity and ecosystem functioning: a mechanistic model. *Proceedings of the National Academy of Sciences*, **95**, 5632–6.

Loreau, M. (1998b). Ecosystem development explained by competition within and between material cycles. *Proceedings of the Royal Society of London B*, **265**, 33–8.

Loreau, M. and Mouquet, N. (1999). Immigration and the maintenance of local species diversity. *American Naturalist*, **154**, 427–40.

Louda, S.M. (1989). Predation in the dynamics of seed regeneration. In *Ecology of soil seed banks* (ed. M.A. Leck, V.T. Parker, and R.L. Simpson), pp. 25–51. Academic Press, New York.

Lovegrove, B.G. (1991). The evolution of eusociality in mole-rats (Bathyergidae) a question of risks, numbers, and costs. *Behavioral Ecology and Sociobiology*, **28**, 37–45.

Lovegrove, B.G. and Wissel, C. (1988). Sociality in mole-rats: metabolic scaling and the role of risk sensitivity. *Oecologia*, **74**, 600–6.

Lowry, C.A. and Moore, F.L. (1991). Corticotropin-releasing factor (CRF) antagonist suppresses stress-induced locomotor activity in an amphibian. *Hormones and Behavior*, **25**, 84–96.

Lowry, C.A., Rose, J.D., and Moore, F.L. (1996). Corticotropin-releasing factor-enhances locomotion and medullary neuronal firing in an amphibian. *Hormones and Behavior*, **30**, 50–9.

Loye, J.E. and Carroll, S.P. (1991). Nest ectoparasite abundance and cliff swallow colony site selection, nestling development, and departure time. In *Bird–parasite interactions. Ecology, behaviour and evolution* (ed. J.E. Loye and M. Zuk), pp. 222–41. Oxford University Press, Oxford.

Lubin, Y., Ellner, S., and Kotzman, M. (1993). Web relocation and habitat selection in a desert widow spider. *Ecology*, **74**, 1915–28.

Lugon-Moulin, N., Brünner, H., Wyttenbach, A., Hausser, J., and Goudet, J. (1999). Hierarchical analyses of genetic differentiation in a hybrid zone of *Sorex araneus* (Insectivora: Soricidae). *Molecular Ecology*, **8**, 419–31.

Lynch, M. (1991). Analysis of population genetic structure by DNA fingerprinting. In *DNA fingerprinting: approaches and applications* (ed. T. Burke, G. Dolf, A.J. Jeffreys, and R. Wolff), pp. 114–26. Birkhauser Basel.

Lynch, M. and Hill, W.G. (1986). Phenotypic evolution by neutral mutation. *Evolution*, **40**, 915–35.

Lynch, M. and Walsh, B. (1998). *Genetics and analysis of quantitative traits*. Sinauer Associates, Sunderland, MA.

Lynch, M., Perender, M., Spitze, K., Lehman, N., Hicks, J., Allen, D., Latta, L., Ottene, M., Bouge, F., and Colbourne, J. (1999). The quantitative and molecular genetic architecture of a subdivided species. *Evolution*, **53**, 100–10.

MacArthur, R.H. and Wilson, E.O. (1967). *The theory of island biogeography*. Princeton University Press, Princeton, NJ.

MacColl, A.D.C. (1998). *Factors affecting recruitment in red grouse*. PhD thesis, University of Aberdeen, Aberdeen.

Macdonald, D.W. (1984). *Running with the fox*. George Allen and Unwin, London.

Macdonald, D.W. (1995). Unresolved questions for the control of wildlife rabies: social perturbation and interspecific interactions. In *Rabies in a changing world*, pp. 33–48. Proceedings of the British Small Animal Veterinary Association, Cheltenham, UK.

Macdonald, D.W. and Smith, H. (1990). Dispersal, dispersion and conservation in the agricultural ecosystem. In *Species dispersal in agricultural habitats* (ed. R.G.H. Bunce and D.C. Howard), pp. 18–64. Belhaven Press, London.

Macdonald, D.W. and Smith, H. (1991). New perspectives on agro-ecology: between theory and practice in the agricultural ecosystem. In *The ecology of temperate cereal fields: 32nd symposium of the British Ecological Society* (ed. L.G. Firbank, N. Carter, J.F. Darbyshire, and G. R. Potts), pp. 413–48. Blackwell Scientific Publications, Oxford.

Macdonald, D.W. and Strachan, R. (1999). *The mink and the water vole: analyses for conservation*. Wildlife Conservation Research Unit, University of Oxford, Oxford.

Macdonald, D.W., Mace, G., and Rushton, S.P. (1998). *Proposals for the future monitoring British mammals*. Department of the Environment, Transport and the Regions, London.

Macdonald, D.W., Mace, G.M., and Barreto, G.R. (1999). The effects of predators on fragmented prey populations: a case study for the conservation of endangered prey. *Journal of Zoology of London*, **247**, 487–506.

Macdonald, D.W., Mace, G., and Rushton, S.P. (2000a). Conserving British mammals: is there a radical future? In *Priorities for conservation of mammalian biodiversity: has the panda had its day?* (ed. A. Entwhistle and N. Dunstone), 175–205. Cambridge University Press, Cambridge.

Macdonald, D.W., Tattersall, F.H., Rushton, S.P., South, A.B., Rao, S., Maitland, P., and Strachan, R. (2000b). Reintroducing the beaver (*Castor fiber*) to Scotland: a protocol for identifying and assessing suitable release sites. *Animal Conservation*, **3**, 125–33.

Macgrath, R.D. and Whittingham, L.A. (1997). Subordinate males are more likely to help if unrelated to the breeding female in cooperatively breeding white-browed scrubwrens. *Behavioral Ecology and Sociobiology*, **41**, 185–92.

MacKay, P.A. and Wellington, W.G. (1977). Maternal age as a source of variation in the ability of an aphid to produce dispersing forms. *Researches on Population Ecology*, **18**, 195–209.

Mackay, T.F.C. and Doyle, R.W. (1978). An ecological genetic analysis of the settling behaviour of a marine polychaete: 1. Probability of settlement and gregarious behaviour. *Heredity*, **40**, 1–12.

Mackenzie, A. (1996). A trade-off for host plant utilization in the black bean aphid, *Aphis fabae*. *Evolution*, **50**, 155–62.

MacNair, M.R. (1987). Heavy metal tolerance in plants: a model evolutionary system. *Trends in Ecology and Evolution*, **2**, 354–9.

Madsen, T., Stille, B., and Shine, R. (1996). Inbreeding depression in an isolated population of adders *Vipera berus*. *Biological Conservation*, **75**, 113–18.

Malécot, G. (1948). *Les Mathématiques de l'Hérédité*. Masson et Cie, Paris.

Malécot, G. (1950). Quelques schémas probabilistes sur la variabilité des populations naturelles. *Annales de l'Université de Lyon A*, **13**, 37–60.

Malécot, G. (1975). Heterozygosity and relationship in regularly subdivided populations. *Theoretical Population Biology*, **8**, 212–41.

Malendowicz, L.K., Macchi, C., Nussdorfer, G.G., and Nowak, K.W. (1998). Acute effects of recombinant murine leptin on rat pituitary–adrenocortical function. *Endocrinological Research*, **24**, 235–46.

Mallet, J. (1993). Speciation, raciation and color pattern evolution in *Heliconius* butterflies: evidence from hybrid zones. In *Hybrid zones and the evolutionary process* (ed. R.G. Harrison), pp. 226–60. Oxford University Press, Oxford.

Mallet, J., Barton, N., Lamas, L.G., Santisteban, J.C., Muedas, M.M., and Eeley, H. (1990). Estimates of selection and gene flow from measures of cline width and linkage disequilibrium in *Heliconius* hybrid zones. *Genetics*, **124**, 921–36.

Mangel, M. and Tier, C. (1993). Dynamics of metapopulations with demographic stochasticity and environmental catastrophes. *Theoretical Population Biology*, **44**, 1–31.

Mappes, T., Ylönen, H., and Viitala J. (1995). Higher reproductive success among kin groups of bank voles (*Clethrionomys glareolus*). *Ecology*, **76**, 1276–82.

Marchetti, K. (1992). Costs to host defence and the persistence of parasitic cuckoos. *Proceedings of the Royal Society of London B*, **248**, 41–6.

Markin, G.P., Dillier, J.H., Hill, S.O., Blum, M.S., and Hermann, H.R. (1971). Nuptial flight and flight ranges of the imported fire ant, *Solenopsis saevissima richteri* (Hymenoptera: Formicidae). *Journal of the Georgia Entomological Society*, **6**, 145–56.

Marrow P., Law R., and Cannings C. (1992). The coevolution of predator–prey interactions: ESSs and red queen dynamics. *Proceedings of the Royal Society of London B*, **250**, 133–41.

Martinez, J.G., Soler, J.J., Soler, M., Møller, A.P., and Burke, T. (1999). Comparative population structure and gene flow of a brood parasite, the great spotted cuckoo (*Clamator glandarius*), and its primary host, the magpie (*Pica pica*). *Evolution*, 53, 269–78.

Maruyama, T. (1970). On the fixation probability of mutant genes in a subdivided population. *Genetical Research*, 15, 221–5.

Maruyama, T. and Kimura, M. (1980). Genetic variability and effective population size when local extinction and recolonization of subpopulations are frequent. *Proceedings of the National Academy of Sciences*, 77, 6710–14.

Masaki, S. and Seno, E. (1990). Effect of selection on wing dimorphism in the ground cricket *Dianemobius fascipes* (Walker). *Boletin de Sanidad vegetal, Plagas (Fuera De Serie)*, 20, 381–93.

Massot, M. and Clobert, J. (1995). Influence of maternal food availabilty on offspring dispersal. *Behavioral Ecology and Sociobiology*, 37, 413–18.

Massot, M. and Clobert, J. (2000). Processes at the origin of similarities in dispersal behaviour among siblings. *Journal of Evolutionary Biology*, 13, 707–19.

Massot, M., Clobert, J., Pilorge, T., Lecomte, J., and Barbault, R. (1992). Density dependence in the common lizard – demographic consequences of a density manipulation. *Ecology*, 73, 1742–56.

Massot, M., Clobert, J., Lecomte, J. and Barbault, R. (1994). Incumbent advantage in common lizards and their colonizing ability. *Journal of Animal Ecology*, 63, 431–40.

Matsuda, H., Ogita, N., Sasaki, A., and Sato, K. (1992). Statistical mechanics of population: the lattice Lotka–Volterra model. *Progress in Theoretical Physics*, 88, 1035–49.

Matthiopoulos, J., Moss, R., and Lambin, X. (1998). Models of red grouse cycles. A family affair? *Oikos*, 82, 574–90.

Matthysen, E. (1987). Territory establishment of juvenile nuthatches after fledging. *Journal of Animal Ecology*, 56, 655–73.

Matthysen, E., Adriaensen, F., and Dhondt, A.A. (1995). Dispersal distances of nuthat-ches, *Sitta europaea*, in a highly fragmented forest habitat. *Oikos*, 72, 375–81.

May, R.M. (1978). Host–parasitoid systems in patchy environments: a phenomenological model. *Journal of Animal Ecology*, 47, 833–43.

May, R.M. (1979). When to be incestuous. *Nature*, 279, 192–4.

May, R.M. and Nowak, M.A. (1994). Superinfection, metapopulation dynamics, and the evolution of diversity. *Journal of Theoretical Biology*, 170, 95–114.

May, R.M. and Robinson, S.K. (1985). Population dynamics of avian brood parasitism. *American Naturalist*, 125, 475–94.

May, R.M. and Southwood, T.R.E. (1990). Introduction. In *Living in a patchy environment* (ed. B. Shorrock and I.R. Swingland), pp. 1–22. Oxford University Press, Oxford.

Mayfield, H. (1961). Nesting success calculated from exposure. *Wilson Bulletin*, 73, 255–61.

Mayfield, H. (1975). Suggestions for calculating nest success. *Wilson Bulletin*, 87, 456–66.

Mayhew, P.J. (1997). Adaptive patterns of host-plant selection by phytophagous insects. *Oikos*, 79, 417–28.

Maynard Smith, J. (1962). Disruptive selection, polymorphism, and sympatric speciation. *Nature*, 195, 60–2.

Maynard Smith, J. (1964). Group selection and kin selection. *Nature*, 201, 1145–7.

Maynard Smith, J. (1978a). *Models in ecology*. Cambridge University Press, Cambridge.

Maynard Smith, J. (1978b). Optimisation theory in evolution. *Annual Review of Ecology and Systematics*, 9, 31–56.

Maynard Smith, J. (1982). *Evolution and the theory of games*. Cambridge University Press, Cambridge.

Maynard Smith, J. (1990). The evolution of prokaryotes – does sex matter? *Annual Review of Ecology and Systematics*, 21, 1–12.

Maynard Smith, J. and Hoekstra, R. (1980). Polymorphism in a varied environment; how robust are the models? *Genetical Research*, 35, 45–57.

Mayr, E. (1961). Cause and effect in biology. *Science*, 134, 1501–6.

Mayr, E. (1982). *The growth of biological thought: diversity, evolution and inheritance*. Belknap Press, Cambridge, MA.

McCallum, H.I. (1992). Effects of immigration on chaotic population dynamics. *Journal of Theoretical Biology*, **154**, 277–84.

McCarty, M.A. (1997). Competition and dispersal from multiple nests. *Ecology*, **78**, 873–83.

McCauley, D.E. (1993). Evolution in metapopulations with frequent local extinction and recolonization. *Oxford Survey of Evolutionary Biology*, **10**, 109–34.

McCauley, D.E. and Eanes, W.F. (1987). Hierarchical population structure analysis of the milkweed beetle, *Tetraopes tetraophthalmus* (Forster). *Heredity*, **58**, 193–202.

McComb, K. and Clutton-Brock, T.H. (1994). Is mate choice copying or aggregation responsible for skewed distributions of females on leks? *Proceedings of the Royal Society of London B*, **255**, 13–19.

McCormick, M.I. (1998). Behaviorally induced maternal stress in a fish influences progeny quality by a hormonal mechanism. *Ecology*, **79**, 1873–83.

McCoy, K.D., Boulinier, T. Chardine, J.W. Danchin, E., and Michalakis, Y. (1999). Dispersal and distribution of the tick *Ixodes uriae* within and among seabird host populations: the need for a population genetic approach. *Journal of Parasitology*, **85**, 196–202.

McCullough, D.L. (1997). *Metapopulations and wildlife conservation*. Island Press, Washington, D.C.

McEwen, B.S. (1998). Protective and damaging effects of stress mediators. *New England Journal of Medicine*, **338**, 171–9.

McGee, B.L. and. Targett, N.M. (1989). Larval habitat selection in *Crepidula* (L.) and its effect on adult distribution patterns. *Journal of Experimental Marine Biology and Ecology*, **131**, 195–214.

McGregor, P.K. and Dabelsteen T. (1996). Communication networks. In *Ecology and evolution of acoustic communication in birds* (ed. D.E Kroodsma and E.H. Miller), pp. 409–25. Cornell University Press, Ithaca, NY.

McGregor, P.K. and Krebs, J.R. (1982). Mating and song types in the great tit. *Nature*, **297**, 60–1.

McGuire, B. and Getz, L.L. (1998). The nature and frequency of social interactions among free-living prairie voles (*Microtus ochrogaster*). *Behavioral Ecology and Sociobiology*, **43**, 271–9.

McLean, I.G. (1996). Teaching an endangered mammal to recognise predators. *Biological Conservation*, **75**, 51–62.

McNaughton, F.J., Dawson, A., and Goldsmith, A.R. (1992). Juvenile photorefractoriness in starlings, *Sturnus vulgaris*, is not caused by long days after hatching. *Proceedings of the Royal Society of London B*, **248**, 123–8.

McNaughton, F.J., Dawson, A., and Goldsmith, A.R. (1995). A comparison of the responses to gonadotropin-releasing hormone of adult and juvenile, and photosensitive and photorefractory European starlings, *Sturnus vulgaris*. *General and Comparative Endocrinology*, **97**, 135–44.

McNeil, J.N., Cusson, M., Delisle, J., Orchard, I., and Tobe, S.S. (1995). Physiological integration of migration in Lepidoptera. In *Insect migration: tracking resources through space and time* (ed. V.A. Drake and A.G. Gatehouse), pp. 279–302. Cambridge University Press, Cambridge.

McNutt, J.W. (1996). Sex-biased dispersal in African wild dogs (*Lycaon pictus*). *Animal Behaviour*, **52**, 1067–77.

McPeek, M.A. and Holt, R.D. (1992). The evolution of dispersal in spatially and temporally varying environments. *American Naturalist*, **140**, 1010 27.

Meadows, D. (1995). Effects of habitat geometry on territorial defence costs in a damselfish. *Animal Behaviour*, **49**, 1406–8.

Meadows, P.S. and Campbell, J.I. (1972). Habitat selection by aquatic invertebrates? In *Advances in marine Biology* (ed. F.S. Russell and M. Yonge), **10**, pp. 271–382. Academic Press, London.

Medina, E. (1996). CAM and C4 plants in the humid tropics. In *Tropical forest plant ecophysiology* (ed. S.S. Mulkey, R.L. Chazdon, and A.P. Smith). Chapman and Hall, New York.

Metz, J.A.J. and van den Bosch, F. (1995). Velocities of epidemic spread. In *Epidemic* models. *Their structure and relation to data* (ed. D. Mollison), pp. 150–84. Cambridge University Press, Cambridge.

Metz, J.A.J., de Jong, T.J., and Klinkhamer, P.G.L. (1983). What are the advantages of dispersing: a paper by Kuno explained and extended. *Oecologia*, **57**, 166–9.

Metz, J.A.J., Nisbet, R.M., and Geritz, S.A.H. (1992). How should we define 'fitness' for general ecological scenarios? *Trends in Ecology and Evolution*, **7**, 198–202.

Metz, J.A.J., Geritz, S.A.H., Meszéna, G., Jacobs, F.J.A., and van Heerwaarden, J.S. (1996). Adaptive dynamics, a geometrical study of the consequences of nearly faithful reproduction. In *Stochastic and spatial structures of dynamical systems* (ed. S.J. van Strien and S.M. Verduyn Lunel), pp. 183–231. North Holland, Amsterdam.

Meyer, W.B. and Turner, B.L. II. (ed.) (1994). *Changes in land use and land cover: a global perspective*. Cambridge University Press, Cambridge.

Meylan, S., de Fraipont, M., Clobert, J., and John-Alder, H. (1999). Offspring pjilopatry is promoted by mother stress in the common lizard (*Lacerta vivipara*). In *Proceedings of the 9th ordinary general Meeting of the Societas Europaea Herpetologica* (ed. C. Miaud and R. Guyetant), pp. 325–330. SEH, Chambéry.

Michalakis, Y. and Excoffier, L. (1996). A generic estimation of population subdivision using distances between alleles with special interest to microsatellite loci. *Genetics*, **142**, 1061–4.

Michod, R.E. (1999). *Darwinian dynamics: evolutionary transitions in fitness and individuality*. Princeton University Press, Princeton, NJ.

Michod, R.E. and Hamilton, W.D. (1980). Coefficients of relatedness in sociobiology. *Nature*, **288**, 694–7.

Midtgaard, F. (1999). Is dispersal density-dependent in carabid beetles? A field experiment with *Harpalus rufipes* (Degeer) and *Pterostichus niger* (Schaller) (Col., Carabidae). *Journal of Applied Entomology*, **123**, 9–12.

Millar, J.S. and Zwickel, F.C. (1972). Determination of age, age structure, and mortality

of the pika, *Ochotona princeps* (Richardson). *Canadian Journal of Zoology*, **50**, 229–32.

Miller, K.E. and Smallwood, J.A. (1997). Natal dispersal and philopatry of southeastern American kestrels in Florida. *Wilson Bulletin*, **109**, 226–32.

Miller, R.G. (1983). *Survival analysis*. John Wiley, New York.

Mills, M.G.L., Ellis, S., Woodroffe, R., Maddock, A., Stander, P., Rasmussen, G., Pole, A., Fletcher, P., Bruford, M., Wildt, D., Macdonald, D.W., and Seal, U. (1998). *Population and habitat viability assessment for the African wild dog (Lycaon pictus) in Southern Africa. Final Workshop Report*. IUCN/SSC Conservation Breeding Specialist Group, Apple Valley, MN.

Mills, S., Doak, D.F., and Wisdom, M.J. (1999). Reliability of conservation actions based on elasticity analysis of matrix models. *Conservation Biology*, **13**, 815–29.

Mitchell, R.J. (1993). Path analysis: pollination. In *Design and analysis of ecological experiments* (ed. S.M. Scheiner and J. Gurevitch), pp. 211–31. Chapman and Hall, London.

Miyasato, L.E. and Baker, M.C. (1999). Black-capped chickadee call dialects along a continuous habitat corridor. *Animal Behaviour*, **57**, 1311–18.

Moehlman, P.D. (1996). Ecology and cooperation in canids. In *Ecological aspects of social evolution* (ed. D.I. Rubenstein and R.W. Wrangham). Princeton University Press, Princeton, NJ.

Moilanen, A., Smith, A.T., and Hanski, I.A. (1998). Long-term dynamics in a metapopulation of the American pika. *American Naturalist*, **152**, 530–42.

Møller, A.P. (1987). Advantages and disadvantages of coloniality in the swallow, *Hirundo rustica*. *Animal Behaviour*, **35**, 819–32.

Møller, A.P. (1990). Effects of haematophagous mite on the barn swallow (*Hirundo rustica*): a test of the Hamilton and Zuk hypothesis. *Evolution*, **44**, 771–84.

Mollison, D. (1977). Spatial contact models for ecological and epidemic spread. *Journal of the Royal Statistical Society B*, **39**, 283–326.

Mollison, D. (1991). Dependence of epidemic and population velocities on basic parameters. *Mathematical Biosciences*, **107**, 255–87.

Monge, J.P. and Cortesero, A.M. (1996). Tritophic interactions among larval parasitoids, bruchids and Leguminosae seeds: influence of pre- and post-emergence learning on parasitoids' response to host and host-plant cues. *Entomologia Experimentalis et Applicata*, **80**, 293–6.

Monkkonen, M., Helle, P., and Soppela, K. (1990). Numerical and behavioural responses of migrant passerines to experimental manipulation of resident tits (*Parus* spp.): heterospecific attraction in northern breeding bird communities. *Oecologia*, **85**, 218–25.

Monkkonen, M., Forsman, J.T., and Helle, P. (1996). Mixed-species foraging aggregations and heterospecific attraction in boreal communities. *Oikos*, **77**, 127–36.

Monkkonen, M., Hardling, R., Forsman, J.T., and Tuomi, J. (1999). Evolution of heterospecific attraction: using other species as cues in habitat selection. *Evolutionary Ecology*, **13**, 91–104.

Monnat, J.-Y., Danchin, E., and Rodriguez Estrella, R. (1990). Assessment of environmental quality within the framework of prospection and recruitment: the squatterism in the kittiwake. *Comptes-Rendus de l'Académie des Sciences de Paris*, **311**, Série 3, 391–6.

Mooij, W.M. and DeAngelis, D.L. (1998). Error propagation in spatially explicit population models: a reassessment. *Conservation Biology*, **13**, 930–3.

Moore, J. and Ali, R. (1984). Are dispersal and inbreeding avoidance related? *Animal Behaviour*, **32**, 94–112.

Mori, J.G. and George, J.C. (1978). Seasonal changes in serum levels of certain metabolites, uric acid and calcium in the migratory Canada goose (*Branta canadensis interior*). *Comparative Biochemistry and Physiology*, **59B**, 362–9.

Mori, K. and Nakasuji, F. (1990). Genetic analysis of the wing-form determination of the small brown planthopper, *Laodelphax striatellus* (Hemiptera: Delphacidae). *Researches on Population Ecology*, **32**, 279–87.

Morimoto, Y., Arisue, K., and Yamamura, Y. (1977). Relationship between circadian rhythm of food intake and that of plasma corticosterone and effect of food restriction on circadian adrenocortical rhythm. *Neuroendocrinology*, **23**, 212–22.

Morita, J.G., Lee, T.W., and Mowday, R.T. (1989). Introducing survival analysis to organizational researchers: a selected application to turnover research. *Journal of Applied Psychology*, **74**, 280–92.

Morris, A.J. (1997). *Representing spatial interactions in simple ecological models*. PhD thesis, University of Warwick, Coventry.

Morris, D.W. (1987). Spatial scale and the cost of density-dependent habitat selection. *Evolutionary Ecology*, **1**, 379–88.

Morris, D.W. (1991). Fitness and patch selection by white-footed mice. *American Naturalist*, **138**, 702–16.

Morris, D.W. (1992). Scales and costs of habitat selection in heterogeneous landscapes. *Evolutionary Ecology*, **6**, 412–32.

Morris, D.W. (1994). Habitat matching: alternatives and implications to populations and communities. *Evolutionary Ecology*, **4**, 387–406.

Morse, D.H. (1985). Habitat selection in North American parulid warblers. In: *Habitat selection in birds* (ed. M.L. Cody), pp. 131–57. Academic Press, New York.

Morse, D.H. (1993). Some determinants of dispersal by crab spiderlings. *Ecology*, **74**, 427–32.

Morse, D.H. and Schmitt, J. (1985). Propagule size, dispersal ability, and seedling performance in *Asclepias syriaca*. *Oecologia*, **67**, 372–9.

Morton, E.S. (1990). Habitat segregation by sex in the hooded warbler: experiments on proximate causation and discussion of its evolution. *American Naturalist*, **135**, 319–33.

Morton, E.S., Forman, L., and Braun, M. (1990). Extrapair fertilization and the evolution of colonial breeding in purple martins. *Auk*, **107**, 275–83.

Moss, W.W. and Camin, J.H. (1970). Nest parasitism, productivity and clutch size in purple martins. *Science*, **168**, 1000–3.

Motro, U. (1982a). Optimal rates of dispersal. I. Haploid populations. *Theoretical Population Biology*, **21**, 349–411.

Motro, U. (1982b). Optimal rates of dispersal. II. Diploid populations. *Theoretical Population Biology*, **21**, 412–29.

Motro, U. (1983). Optimal rates of dispersal. III. Parent–offspring conflict. *Theoretical Population Biology*, **23**, 159–68.

Motro, U. (1991). Avoiding inbreeding and sibling competition: the evolution of sexual dimorphism for dispersal. *American Naturalist*, **137**, 108–15.

Motro, U. (1994). Evolutionary and continuous stability in asymmetric games with continuous strategy sets: the parental investment conflict as an example. *American Naturalist*, **144**, 229–41.

Mousseau, T.A. and Dingle, H. (1991). Maternal effects in insect life histories. *Annual Review of Entomology*, **36**, 511–34.

Mousseau, T.A. and Roff, D.A. (1989). Geographic variability in the incidence and heritability of wing dimorphism in the striped ground cricket, *Allonemobius fasciatus*. *Heredity*, **62**, 315–18.

Mulder, R.A. and Langmore, N.E. (1993). Dominant males punish helpers for temporary defection in superb fairy wrens. *Animal Behaviour*, **45**, 830–3.

Muller, K.L. (1998). The role of conspecifics in habitat settlement in a territorial grasshopper. *Animal Behaviour*, **56**, 479–85.

Muller, K.L., Stamps, J.A., Krishnan, V.V., and Willits, N.H. (1997). The effects of conspecific attraction and habitat quality on habitat selection in territorial birds (*Troglodytes aedon*). *American Naturalist*, **150**, 650–61.

Mulvey, M., Aho, J.M., Lydeard, C., Leberg, P.R., and Smith, M.H. (1991). Comparative population genetic structure of a parasite (*Fascioloides magna*) and its definitive host. *Evolution*, **45**, 1628–40.

Murdoch, W.W., Briggs, C.J., Nisbet, R.M., Gurney, W.S.C., and Stewart-Oaten, A. (1992). Aggregation and stability in metapopulation models. *American Naturalist*, **140**, 41–58.

Murray, D.R. (1986). *Seed dispersal*. Academic Press, Sydney.

Murray, J.D. (1980). *Mathematical Biology*. Springer, Berlin.

Myers, J.H., Boettner, G., and Elkinton, J. (1998). Maternal effects in gypsy moth: only sex ratio varies with population density. *Ecology*, **79**, 305–14.

Myles, T.G. and Nutting, W.L. (1978). Termite eusocial evolution: a re-examination of Bartz's hypothesis and assumptions. *Quarterly Review of Biology*, **63**, 1–23.

Nadler, S.A. (1995). Microevolution and the genetic structure of parasite populations. *Journal of Parasitology*, **81**, 395–403.

Nagra, B., Breitenbach, R.P., and Meyer, R.K. (1963). Influence of hormones on food intake and lipid deposition in castrated pheasants. *Poultry Science*, **42**, 770–5.

Naguib, M. and Todt, D. (1997). Effects of dyadic vocal interactions on other conspecific receivers in nightingales. *Animal Behaviour*, **54**, 1535–43.

Nagylaki, T. (1975). Conditions for the existence of clines. *Genetics*, **80**, 595–615.

Nagylaki, T. (1976). The decay of genetic variability in geographically structured populations. II. *Theoretical Population Biology*, **10**, 70–82.

Nagylaki, T. (1989). Gustave Malecot and the transition from classical to modern population genetics. *Genetics*, **122**, 253–68.

Nagylaki, T. and Moody, M. (1980). Diffusion model for genotype dependent migration. *Proceedings of the National Academy of Sciences of the USA*, **77**, 4842–6.

Nakamaru M., Matsuda H., and Iwasa Y. (1997). The evolution of cooperation in a lattice-structured population. *Journal of Theoretical Biology*, **184**, 65–81.

Narins, P.M., Lewis, E.R., Jarvis, J.U.M., and O'Riain, J. (1997). The use of seismic signals by fossorial Southern African mammals: a neuroethological gold mine. *Brain Research Bulletin*, **44**, 641–6.

Nath, H.B. and Griffiths, R.C. (1996). Estimation in an island model using simulation. *Theoretical Population Biology*, **50**, 227–53.

Nee, S. and May, R.M. (1992). Dynamics of metapopulations: habitat destruction and competitive coexistence. *Journal of Animal Ecology*, **61**, 37–40.

Nee, S., May, R.M., and Hassell, M.P. (1997). Two-species metapopulation models. In *Metapopulation biology* (ed. I. Hanski and M. Gilpin), pp. 123–48. Academic Press, San Diego, CA.

Negro, J.J., Hiraldo, F., and Donazar, J.A. (1997). Causes of natal dispersal in the lesser kestrel: inbreeding avoidance or resource competition? *Journal of Animal Ecology*, **66**, 640–8.

Neigel, J.C. and Avise, J.C. (1993). Application of a random walk model to geographic distributions of animal mtDNA variation. *Genetics*, **135**, 1209–20.

Neigel, J.E. (1997). A comparison of alternative strategies for estimating gene flow from genetic markers. *Annual Review of Ecology and Systematics*, **28**, 105–28.

Nelson, D.A. (1997). Social interaction and sensitive phases for song learning: a critical review. In *Social influences on vocal development* (ed. C.T. Snowdon and M. Hausberger), pp. 7–22. Cambridge University Press, Cambridge.

Nelson, R.J. (1994). *An introduction to behavioral endocrinology*. Sinauer, Sunderland, MA.

Nemtozov, S.C. (1997). Intraspecific variation in home range exclusivity by female green razorfish *Xyrichtys spendens* (family Labridae) in different habitats. *Environmental Biology of Fishes*, **50**, 371–81.

Nevo, E. and Reig, O.A. (1990). *Evolution of subterranean mammals at the organismal and molecular levels*. Wiley-Liss, New York.

Nichols, J.D. (1996). Sources of variation in migratory movements of animal populations: statistical inference and a selective review of empirical results for birds. In *Population dynamics in ecological space and time* (ed. O.E. Rhodes, Jr., R.K. Chesser, and M.H. Smith), pp. 147–97. The University of Chicago Press, Chicago.

Nichols, J.D. and Kaiser, A. (1999). Quantitative studies of bird movement: a methodological review. *Bird Study (Suppl.)*, **46**, S289–S298.

Nichols, J.D. and Kendall, W.L. (1995). The use of multi-state capture–recapture models to address questions in evolutionary ecology. *Journal of Applied Statistics*, **22**, 835–46.

Nichols, J.D. and Pollock, K.H. (1990). Estimation of recruitment from immigration versus *in situ* reproduction using Pollock's robust design. *Ecology*, **71**, 21–6.

Nichols, J.D., Morris, R.W., Brownie, C., and Pollock, K.H. (1986). Sources of variation in extinction rates, turnover, and diversity of marine invertebrate families during the Paleozoic. *Paleobiology*, **12**, 421–32.

Nichols, J.D., Brownie, C., Hines, J.E., Pollock, K.H., and Hestbeck, J.B. (1993). The estimation of exchanges among populations or subpopulations. In *The use of marked individuals in the study of bird population dynamics: models, methods, and software* (ed. J.-D. Lebreton and P.M. North) pp. 265–79. Birkhauser, Basel.

Nichols, J.D., Hines, J.E., Lebreton, J.-D., and Pradel, R. (2000). The relative contributions of demographic components to population growth: a direct estimation approach based on reverse-time capture–recapture. *Ecology*, **81**.

Nichols, R. and Bondrup-Nielsen, S. (1995). The effect of a single dose of testosterone propionate on activity, and natal dispersal in the meadow vole, *Microtus pennsylvanicus*. *Annales Zoologici Fennici*, **32**, 209–15.

Nicholson, A.J. (1954). An outline of the dynamics of animal populations. *Australian Journal of Zoology*, **2**, 9–65.

Nicholson, A.J. and Bailey, V.A. (1935). The balance of animal populations. Part I. *Proceedings of the Zoological Society*, **3**, 551–98.

Nilsson, J.-Å. (1989). Causes and consequences of natal dispersal in the marsh tit, *Parus palustris*. *Journal of Animal Ecology*, **58**, 619–36.

Nilsson, J.-Å. and Smith, H.G. (1985). Early fledgling mortality and the timing of juvenile dispersal in the marsh tit *Parus palustris*. *Ornis Scandinavica*, **16**, 293–8.

Nilsson, J.-Å. and Smith, H.G. (1988). Effects of dispersal date on winter flock establishment and social dominance in marsh tits *Parus palustris*. *Journal of Animal Ecology*, **57**, 917–28.

Nilsson, P., Fagerström, T., Tuomi, J., and Aström, M. (1994). Does seed dormancy benefit the mother plant by reducing sib competition? *Evolutionary Ecology*, **8**, 422–30.

Nordborg, M. and Donnelly, P. (1997). The coalescent process with selfing. *Genetics*, **146**, 1185–95.

Nordell, S.E. and Valone, T.J. (1998). Mate choice copying as public information. *Ecology Letters*, **1**, 74–6.

Norrdahl, K., Suhonen, J., Hemminki, O., and Korpimaki, E. (1995). Predator presence may benefit: kestrels protect curlew nests against nest predators. *Oecologia*, **101**, 105–9.

North, P.M. (1988). A brief review of the (lack of) statistics of bird dispersal. *Acta Ornithologica*, **24**, 63–74.

Nowak, M.A. and May, R.M. (1994). Coinfection and the evolution of parasite virulence. *Proceedings of the Royal Society of London B*, **255**, 81–9.

Nuismer, S.L., Thompson, J.N., and Gomulkiewicz, R. (1999). Gene flow and geographically structured coevolution. *Proceedings of the Royal Society of London B*, **266**, 605–9.

Nunes, M.V. and Hardie, J. (1996). Differential photoperiodic responses in genetically identical winged and wingless pea aphids, *Acyrthosiphon pisum*, and the effect of day length on wing development. *Physiological Entomology*, **21**, 339–43.

Nunes, S. and Holekamp, K.E. (1996). Mass and fat influence the timing of natal dispersal in breeding Belding's ground squirrels. *Journal of Mammalogy*, **77**, 807–17.

Nunes, S., Zugger, P.A., Engh, A.L., Reinhart, K.O., and Holekamp, K.E. (1997). Why do female Belding's ground squirrels disperse away from food resources? *Behavioral Ecology and Sociobiology*, **40**, 199–207.

Nunes, S., Co-Diem, T.H., Garrett, P.J., Mueke, E.-M., Smale, L., and Holekamp, K.E. (1998). Body fat and time of year interact to mediate dispersal behaviour in ground squirrels. *Animal Behaviour*, **55**, 605–14.

Nunney, L. (1985). Group selection, altruism and structured-deme models. *American Naturalist*, **126**, 212–30.

Nunney, L. (1999). The effective size of a hierarchically structured population. *Evolution*, **53**, 1–10.

Nürnberger, B.D., Barton, N.H., MacCallum, C., Gilchrist, J., and Appleby, M. (1995). Natural selection on quantitative traits in the *Bombina* hybrid zone. *Evolution*, **49**, 1224–38.

O'Riain, J.M., Jarvis, J.U.M., and Faulkes, C.G. (1996). A dispersive morph in the naked mole-rat. *Nature*, **380**, 619–21.

O'Riain, M.J. and Jarvis, J.U.M. (1997). Colony member recognition and xenophobia in the naked mole-rat. *Animal Behaviour*, **53**, 487–98.

O'Riain, M.J. and Jarvis, J.U.M. (1998). The dynamics of growth in naked mole-rats: the effects of litter order and changes in social structure. *Journal of Zoology*, **246**, 49–60.

O'Riain, M.J., Jarvis, J.U.M., and Faulkes, C. (1996). A dispersive morph in the naked mole-rat. *Nature*, **380**, 619–21.

Okubo, A. (1980). *Diffusion and ecological problems: mathematical models*. Springer, Berlin.

Oliveira, R.F., McGregor, P.K., and Latruffe, C. (1998). Know thine enemy: fighting fish gather information from observing conspecific interactions. *Proceedings of the Royal Society of London B*, **265**, 1045–9.

Olivieri, I. and Berger, A. (1985). Seed dimorphism and dispersal: physiological, genetic and demographical aspects. In *Genetic differentiation and dispersal in plants* (ed. P. Jacquard, G. Heim, and J. Antonovics). Springer, Berlin.

Olivieri, I. and Gouyon, P.-H. (1985). Seed dimorphism for dispersal: theory and implications. In *Structure and functioning of plant populations* (ed. J. Haeck and J.W. Woldendrop), pp. 77–89. North-Holland, Amsterdam.

Olivieri, I. and Gouyon, P.-H. (1997). Evolution of migration rate and other traits. In *Metapopulation biology, ecology, genetics, and evolution* (ed. I.A. Hanski and M.E. Gilpin), pp. 293–323. Academic Press, San Diego, CA.

Olivieri, I., Michalakis, Y., and Gouyon, P.H. (1995). Metapopulation genetics and the evolution of dispersal. *American Naturalist*, **146**, 202–28.

Olla, B.L., Davis, M.W., and Ryer, C.H. (1998). Understanding how the hatchery environment represses or promotes the development of behavioral survival skills. *Bulletin of Marine Science*, **62**, 531–50.

Olsson, M., Gullberg, A., and Tegelström, H. (1996). Malformed offspring, sibling matings, and selection against inbreeding in the sand lizard (*Lacerta agilis*). *Journal of Evolutionary Biology*, **9**, 229–42.

Oppliger, A., Vernet, R., and Baez, M. (1999). Parasite local maladaptation in the Canarian lizard *Gallotia galloti* (Reptilia: Lacertidae) parasitized by haemogregarian blood parasite. *Journal of Evolutionary Biology*, **12**, 951–5.

Orchinik, M. (1998). Glucocorticoids, stress, and behavior: shifting the timeframe. *Hormones and Behavior*, **34**, 320–7.

Orchinik, M., Murray, T.F., and Moore, F.L. (1991). A corticosteroid receptor in neuronal membranes. *Science*, **252**, 1848–51.

Ortega-Reyes, L. and Provenza, F.D. (1993). Amount of experience and age affect the development of foraging skills of goats browsing blackbush (*Coleogyne ramosissima*). *Applied Animal Behavior Science*, **36**, 169–83.

Otter, K., McGregor, P.K., Terry, A.M.R., Burford, F.R.L., Peake, T.M., and Dabelsteen, T. (1999). Do female great tits (*Parus major*) assess males by eavesdropping? A field study using interactive song playback. *Proceedings of the Royal Society of London B*, **266**, 1305–9.

Ottinger, M.A. and Abdelnabi, M.A. (1997). Neuroendocrine systems and avian sexual differentiation. *American Zoologist*, **37**, 514–23.

Ozaki, K. (1995). Intergall migration in aphids – a model and a test of ESS dispersal rate. *Evolutionary Ecology*, **9**, 542–9.

Pacala, S.W. (1997). Dynamics of plant communities. In *Plant ecology* (ed. M.E. Crawley), pp. 532–55. Blackwell, Oxford.

Pacala, S.W. and Levin, S.A. (1997). Biologically generated spatial pattern and the coexistence of competing species. In *Spatial ecology* (ed. D. Tilman and P. Kareiva), pp. 204–32. Princeton University Press, Princeton, NJ.

Pacala, S.W. and Rees, M. (1998). Models suggesting field experiments to test two hypotheses explaining successional diversity. *American Naturalist*, **152**, 729–37.

Pacala, S.W., Hassell, M.P. and May, R.M. (1990). Host–parasitoid associations in patchy environments. *Nature*, **344**, 150–3.

Packer, C. (1979). Inter-troop transfer and inbreeding avoidance in *Papio anubis*. *Animal Behaviour*, **27**, 1–36.

Packer, C. (1985). Dispersal and inbreeding avoidance. *Animal Behaviour*, **33**, 676–8.

Packer, C. and Pusey, A.E. (1993). Dispersal, kinship and inbreeding in African lions. In *The natural history of inbreeding and outbreeding* (ed. N.W. Thornhill), pp. 375–91. University of Chicago Press, Chicago, IL.

Packer, C., Herbst, L., Pusey, A.E., Bygott, J.D., Hanby, J.P., Cairns, S.J., and Borgerhoff-Mulder, M. (1988). Reproductive success of lions. In *Reproductive Success* (ed. T.H. Clutton-Brock), pp. 363–83. University of Chicago Press, Chicago, IL.

Paetkau, D., Calvert, W., Stirling, I., and Strobeck, C. (1995). Microsatellite analysis of population structure in Canadian polar bears. *Molecular Ecology*, **4**, 347–54

Pagel, M. and Payne, R.J.H. (1996). How migration affects estimation of the extinction threshold. *Oikos*, **76**, 323–9.

Palmer, J.O. and Dingle, H. (1986). Direct and correlated responses to selection among life-history traits in mikweed bugs (*Oncopeltus fasciatus*). *Evolution*, **40**, 767–77.

Palmer, J.O. and Dingle, H. (1989). Responses to selection on flight behavior in a migratory population of milkweed bug (*Oncopeltus fasciatus*). *Evolution*, **43**, 1805–8.

Pannell, J.R. and Charlesworth, B. (1999). Neutral genetic diversity in a metapopulation with recurrent local extinction and recolonization. *Evolution*, **53**, 664–76.

Paradis, E. (1995). Survival, immigration and habitat quality in the Mediterranean pine vole. *Journal of Animal Ecology*, **64**, 579–91.

Parker, G. (1979). Sexual selection and sexual conflict. In *Sexual selection and reproductive competition in insects* (ed. M.B. Blum and N.A. Blum), pp. 123–66. Academic Press, New York.

Parker, G. (1983). Mate quality and mating decisions. In *Mate choice* (ed. P.P.G. Bateson), pp. 141–64. Cambridge University Press, Cambridge.

Parker, G.A. and Stuart, R.A. (1976). Animal behaviour as a strategy optimizer: evolution of resource assessment strategies and optimal emigration thresholds. *American Naturalist*, **110**, 1055–76.

Parker, G.A. and Sutherland, W.J. (1986). Ideal free distributions when individuals differ in competitive ability: phenotype-limited ideal free models. *Animal Behaviour*, **34**, 1222–42.

Parsons, K.E. (1997). Role of dispersal ability in the phenotypic differentiation and plasticity of two marine gastropods. I. Shape. *Oecologia*, **110**, 461–71.

Parsons, K.E. (1998). The role of dispersal ability in the phenotypic differentiation and plasticity of two marine gastropods. II. Growth. *Journal of Experimental Marine Biology and Ecology*, **221**, 1–25.

Pärt, T. (1991). Philopatry pays: a comparison between collared flycatcher sisters. *American Naturalist*, **138**, 790–6.

Pärt, T. (1996). Problems with testing inbreeding avoidance: the case of the collared fly-catcher. *Evolution*, **50**, 1625–30.

Patriquin-Meldrum, K.J. and Godin, J.-G.J. (1998). Do female sticklebacks copy the mate choice of others? *American Naturalist*, **151**, 570–7.

Pawlik, J.R. (1992). Chemical ecology of the settlement of benthic marine invertebrates. *Oceanography and Marine Biology*, **30**, 273–335.

Payne, A.M. and Maun, M.A. (1981). Dispersal and floating ability of dimorphic fruit segments of *Cakile edentula* var. *lacustris*. *Canadian Journal of Botany*, **59**, 2595–602.

Payne, R.B. and Payne, L.L. (1997). Field observations, experimental design, and the time and place of learning bird songs. In *Social influences on vocal development* (ed. C.T Snowdon and M. Hausberger), pp. 57–84. Cambridge University Press, Cambridge.

Peacock, M.M. (1995). Dispersal, inbreeding and population structure of pika (*Ochotona princeps*). PhD thesis, Arizona State University.

Peacock, M.M. (1997). Determining natal dispersal patterns in a population of North American pikas (*Ochotona princeps*) using direct mark–resight and indirect genetic methods. *Behavioral Ecology*, **8**, 340–50.

Peacock, M.M. and Smith, A.T. (1997a). The effect of habitat fragmentation on dispersal patterns, mating behavior, and genetic variation in a pika (*Ochotona princeps*) metapopulation. *Oecologia*, **112**, 524–33.

Peacock, M.M. and Smith, A.T. (1997b). Nonrandom mating in pikas *Ochotona princeps*: evidence for inbreeding between individuals of intermediate relatedness. *Molecular Ecology*, **6**, 801–11.

Peck, S.L., Ellner, S.P., and Gould, F. (1998). A spatially explicit stochastic model demonstrates the feasibility of Wright's shifting balance theory. *Evolution*, **52**, 1834–9.

Pelleymounter, M.A., Cullen, M.J., Baker, M.B., Hecht, R., Winters, D., Boone, T., and Collins, F. (1995). Effects of the obese gene product on body weight reduction in *ob/ob* mice. *Science*, **269**, 540–3.

Pen, I. (2000). Reproductive effort in viscous populations. *Evolution*, **54**, 293–7.

Pepper, J.W., Braude, S., Lacey, E., and Sherman, P.W. (1991). Vocalizations of the naked mole-rats. In *The biology of the naked mole-rat* (ed. P.W. Sherman, J.U.M. Jarvis, and R.D. Alexander), pp. 243–74. Princeton University Press, Princeton, NJ.

Perrin, N. and Mazalov, V. (1999). Dispersal and inbreeding avoidance. *American Naturalist*, **154**, 282–92.

Perrin, N. and Mazalov, V. (2000). Local competition, inbreeding, and the evolution of sex-biased dispersal. *American Naturalist*, **155**, 116–27.

Philippi, T. and Seger, J. (1989). Hedging one's evolutionary bets, revisited. *Trends in Ecology and Evolution*, **4**, 41–4.

Phoenix, C.H., Goy, R.W., Gerall, A.A., and Young, W.C. (1959). Organizing action of prenatally administered testosterone propionate on the tissues mediating mating behavior in the female guinea pig. *Endocrinology*, **65**, 369–82.

Pickett, S.T.A., Kolasa, J., and Jones, C.G. (1994). *Ecological understanding*. Academic Press, San Diego, CA.

Piertney, S.B., MacColl, A.D.C., Bacon, P.J., and Dallas, J.F. (1998). Local genetic structure in red grouse (*Lagopus lagopus scoticus*): evidence from microsatellite DNA markers. *Molecular Ecology*, 7, 1645–54.

Pile, A.J., Lipcius, R.N., VanMontfrans, J., and Orth, R.J. (1996). Density dependent settler–recruit–juvenile relationships in blue crabs. *Ecological Monographs*, 66, 277–300.

Podolsky, R.H. (1990). Effectiveness of social stimuli in attracting Laysan albatross to new potential nesting sites. *Auk*, 107, 119–25.

Podolsky, R.H. and Kress, S.W. (1989). Factors affecting colony formation in leach's storm-petrel. *Auk*, 106, 332–6

Pollard, E. and Yates, T.J. (1993). *Monitoring butterflies for ecology and conservation*. Chapman and Hall, London.

Pollock, K.H. (1982). A capture–recapture design robust to unequal probability of capture. *Journal of Wildlife Management*, 46, 757–60.

Pollock, K.H., Solomon, D.L., and Robson, D.S. (1974). Tests for mortality and recruitment in a *K*-sample tag–recapture experiment. *Biometrics*, 30, 77–87.

Pollock, K.H., Winterstein, S.R., and Conroy, M.J. (1989a). Estimation and analysis of survival distributions for radio tagged animals. *Biometrics*, 45, 99–109.

Pollock, K.H., Winterstein, S.R., Bunck, C.M., and Curtis, P.D. (1989b). Survival analysis in telemetry studies: the staggered entry design. *Journal of Wildlife Management*, 53, 7–15.

Pollock, K.H., Nichols, J.D., Brownie, C., and Hines, J.E. (1990). Statistical inference for capture–recapture experiments. *Wildlife Monographs*, 107.

Pollock, K.H., Bunck, C.M., Winterstein, S.R., and Chen, C.-L. (1995). A capture–recapture survival analysis model for radio-tagged animals. *Journal of Applied Statistics*, 22, 661–72.

Porter, A.H. and Geiger, H. (1995). Limitations to the inference of gene flow at regional geographic scales – an example from the *Pieris napi* group (Lepidoptera: Pieridae) in

Europe. *Biological Journal of the Linnean Society*, 54, 329–48.

Porter, A.H., Wenger, R., Geiger, H., Scholl, A., and Shapiro, A.M. (1997). The *Pontia daplidice-edusa* hybrid zone in northwestern Italy. *Genetics*, 51, 1561–73.

Porter, S.D. (1991). Origins of new queens in polygyne red imported fire ant colonies (Hymenoptera: Formicidae). *Journal of Entomological Science*, 26, 474–8.

Potter, M.A. (1990). Movement of North Island brown kiwi (*Apteryx australis mantelli*) between forest remnants. *New Zealand Journal of Ecology*, 14, 17–24.

Potts, W.K., Manning, C., and Wakeland, E.K. (1991). Mating patterns in semi-natural populations of mice influenced by MHC genotype. *Nature*, 352, 619–21.

Poulin, R. (1994). Meta-analysis of parasite-induced behavioural changes. *Animal Behaviour*, 48, 137–46.

Poulin, R. (1995). 'Adaptive' changes in the behaviour of parasitized animals: a critical review. *International Journal of Parasitology*, 25, 1371–83.

Powell, R.A. and Fried, J.J. (1992). Helping by juvenile voles (*Microtus pinetorum*), growth and survival of younger siblings, and the evolution of pine vole sociality. *Behavioral Ecology*, 3, 325–33.

Pradel, R. (1996). Utilization of capture–mark–recapture for the study of recruitment and population growth rate. *Biometrics*, 52, 703–9.

Pradel, R., Hines, J.E., Lebreton, J.-D., and Nichols, J.D. (1997). Capture–recapture survival models taking account of transients. *Biometrics*, 53, 60–72.

Pralong, F.P., Roduit, R., Waeber, G., Castillo, E., Mosimann, F., Thorens, B., and Gaillard, R.C. (1998). Leptin inhibits directly glucocorticoid secretion by normal human and rat adrenal gland. *Endocrinology*, 139, 4264–8.

Price, K. and Boutin, S. (1993). Territorial bequeathal by red squirrel mothers. *Behavioral Ecology*, 4, 144–150.

Price, M.V. and Jenkins, S.H. (1986). Rodents as seed consumers and dispersers. In *Seed dispersal*, (ed. D.R. Murray), pp. 191–235. Academic Press, Sydney.

Price, M.V., Kelly, P.A., and Goldingay, R.L. (1994). Distances moved by the Stephens' kangaroo rat (*Dipodomys stephens*) and implications for conservation. *Journal of Mammalogy*, **75**, 929–39.

Primack, R.B. and Miao, S.L. (1992). Dispersal can limit local plant distribution. *Conservation Biology*, **6**, 513–19.

Prins, H.H.T. (1996). *Ecology and behaviour of the African buffalo: social inequality and decision making*. Chapman and Hall, New York.

Provine, W.B. (1971). *The origins of theoretical population genetics*. University of Chicago Press, Chicago, IL.

Provine, W.B. (1986). *Sewall Wright and evolutionary biology*. University of Chicago Press, Chicago, IL.

Pruett-Jones, S. (1992). Independent versus nonindependent mate choice: do females copy each others? *American Naturalist*, **140**, 1000–9.

Pulliam, H.R. (1988). Sources, sinks, and population regulation. *American Naturalist*, **132**, 652–61.

Pulliam, H.R. (1996). Sources and sinks: empirical evidence and population consequences. In: *Population dynamics in ecological space and time* (ed. O.E. Rhodes, Jr., R.K. Chester, and M. H. Smith), pp. 45–70. University of Chicago Press, Chicago, IL.

Pusey, A. (1980). Inbreeding avoidance in chimpanzees. *Animal Behaviour*, **28**, 543–82.

Pusey, A.E. (1987). Sex biased dispersal and inbreeding avoidance in birds and mammals. *Trends in Ecology and Evolution*, **2**, 295–9.

Pusey, A.E. and Packer, C. (1987). Dispersal and philopatry. In *Primate societies* (ed. B.B. Smuts, D.L. Cheney, R.M. Seyfarth, R.W. Wrangham, and T.T. Struhsaker), pp. 250–66. University of Chicago Press, Chicago, IL.

Pusey, A.E. and Wolf M. (1996). Inbreeding avoidance in animals. *Trends in Ecology and Evolution*, **11**, 201–6.

Pyler, D.B. and Proseus, T.E. (1996). A comparison of the dormancy characteristics of *Spartina patens* and *Spartina alterniflora* (Poaceae). *American Journal of Botany*, **83**, 11–14.

Raimondi, P.T. (1988). Settlement cues and determination of the vertical limit of an intertidal barnacle. *Ecology*, **69**, 400–7.

Rainey, P.B. and Travisano, M. (1998). Adaptative radiations in a heterogenous environment. *Nature*, **394**, 69–72.

Ralls, K., Harvey, P.H., and Lyles, A.M. (1986). Inbreeding in natural populations of birds and mammals. In *Conservation biology: the science of scarcity and diversity* (ed. M.E. Soulé), pp. 35–56. Sinauer Associates, Sunderland, MA.

Ramenofsky, M. (1990). Fat storage and fat metabolism in relation to migration. In *Bird migration* (ed. E. Gwinner), pp. 214–31. Springer, Berlin.

Ramsay, S.M., Otter, K., and Ratcliff, L.M. (1999). Nest-site selection by female black-capped chickadees: settlement based on conspecific attraction? *Auk*, **116**, 604–17.

Rand, D.A. (1998). Correlation equations and pair-approximations for spatial ecologies. In *Advanced ecological theory: advances in principles and applications* (ed. J.M. McClade), pp. 100–43. Blackwell, Oxford.

Rand, D.A. and Wilson, H.B. (1995). Using spatio-temporal chaos and intermediate-scale determinism in artificial ecologies to quantify spatially-extended ecosystems. *Proceedings of the Royal Society of London B*, **259**, 111–17.

Rangeley, R.W. and Kramer, D.L. (1998). Density-dependent antipredator tactics and habitat selection in juvenile pollock. *Ecology*, **7**, 943–52.

Rankin, M.A. (1989). Hormonal control of flight. In *Insect flight* (ed. G.J. Goldsworthy and C.H. Wheeler), pp. 139–63. CRC Press, Boca Raton, FL.

Rankin, M.A. (1991). Endocrine effects on migration. *American Zoologist*, **31**, 217–30.

Rankin, M.A. and Burchsted, J.C.A. (1992). The cost of migration in insects. *Annual Review of Entomology*, **37**, 533–59.

Rankin, M.A., McAnelly, M.L., and Bodenhamer, J.E. (1986). The oogenesis–flight syndrome revisited. In *Insect flight: dispersal and migration* (ed. L.W. Danthanarayama), pp. 27–48. Springer, Berlin.

Rannala, B. and Hartigan, J.A. (1995). Identity by descent in island–mainland populations *Genetics*, **139**, 429–37.

Rannala, B. and Hartigan, J.A. (1996). Estimating gene flow in island populations. *Genetical Research*, **67**, 147–58.

Rannala, B. and Mountain, J. (1997). Detecting immigration by using multilocus genotypes. *Proceedings of the National Academy of Sciences*, **94**, 9197–221.

Raver, N., Taouis, M., Dridi, S., Derouet, M., Simon, J., Robinzon, B., Djiane, J., and Gertler, A. (1998). Large-scale preparation of biologically active recombinant chicken obese protein (leptin). *Protein Expression and Purification*, **14**, 403–8.

Ray, C. (1997). *Dynamics and persistence of spatially structured populations: theoretical and empirical studies*. PhD thesis, University of California, Davis, CA.

Ray, C. (2000). Maintaining genetic diversity despite local extinctions: a spatial scaling problem. *Biological Conservation*, in press.

Ray, C. and Hastings, A. (1996). Density dependence: are we searching at the wrong spatial scale. *Journal of Animal Ecology*, **65**, 556–66.

Ray, C., Gilpin, M., and Smith, A.T. (1991). The effect of conspecific attraction on metapopulation dynamics. *Biological Journal of the Linnean Society*, **42**, 123–34.

Raymond, M. and Rousset, F. (1995a). GENEPOP version 1.2: population genetics software for exact tests and ecumenicism. *Journal of Heredity*, **86**, 248–9.

Raymond, M., and Rousset, F. (1995b). An exact test for population differentiation. *Evolution*, **49**, 1280–3.

Rayner, D.V., Dalgliesh, G.D., Duncan, J.S., Hardie, L.J., Hoggard, N., and Trayhurn, P. (1997). Postnatal development of the *ob* gene system: elevated leptin levels in suckling *fa/fa* rats. *American Journal of Physiology*, **273**, R446–R450.

Real, L. (1990). Search theory and mate choice: I. models of single sex discrimination. *American Naturalist*, **136**, 379–405.

Reboud, X. and Bell, G. (1998). Experimental evolution in *Chlamydomonas*. III. Evolution of specialist and generalist types in environments that vary in space and time. *Heredity*, **78**, 507–14.

Reed, J.M. (1999). The role of behaviour in recent avian extinctions and endangerments. *Conservation Biology*, **13**, 232–41.

Reed, J.M. and Dobson, A.P. (1993). Behavioural constraints and conservation biology: conspecific attraction and recruitment. *Trends in Ecology and Evolution*, **8**, 253–6.

Reed, J.M. and. Oring, L.W. (1992). Reconnaissance for future breeding sites by spotted sandpipers. *Behavioral Ecology*, **3**, 310–17.

Reed, J.M., Boulinier, T., Danchin, E., and Oring, L.W. (1999). Prospecting by birds for breeding site. *Current Ornithology*, **15**, 189–259.

Reeve, H.K., Westneat, D.F., Noon, W.A., Sherman, P.W., and Aquadro C.F. (1990). DNA 'fingerprinting' reveals high levels of inbreeding in colonies of the eusocial naked mole-rat. *Proceedings of the National Academy of Sciences*, **87**, 2496–500.

Regehr, H.M., Rodway, M.S., and Montevecchi, W.A. (1998). Antipredator benefits of nest-site selection in black-legged kittiwakes. *Canadian Journal of Zoology*, **76**, 910–15.

Reid, M.L. and Stamps, J.A. (1997). Female mate choice tactics in a resource-based mating system: field tests of alternative models. *American Naturalist*, **150**, 98–121.

Rendon Martos, M. and Johnson, A.R. (1996). Managment of nesting sites for Greater flamingos. *Colonial Waterbirds*, **19**, 167–83.

Renshaw, E. (1986). A survey of stepping-stone models in population dynamics. *Advances in Applied Probabilities*, **18**, 581–627.

Renshaw, E. (1991). *Modelling biological populations in space and time*. Cambridge Studies in Mathematical Biology. Cambridge University Press, Cambridge.

Reznick, D. (1981). 'Grandfather effects': the genetics of interpopulation differences in offspring size in the mosquito fish. *Evolution*, **35**, 941–53.

Reznick, D.N., Bryga, H., and Endler, J.A. (1990). Experimentally induced life-history evolution in a natural population. *Nature*, **346**, 357–9.

Ribble, D.O. (1992). Dispersal in a monogamous rodent, *Peromyscus californicus*. *Ecology*, **73**, 859–66.

Rice, W.R. and Salt, G.W. (1988). Speciation via disruptive selection on habitat preference: experimental evidence. *American Naturalist*, **131**, 911–17.

Ricklefs, R.E. and Schluter, D. (1993). Species diversity: regional and historical influences. In *Species diversity in ecological communities: historical and geographical perspectives* (ed. R.E. Ricklefs and D. Schluter), pp. 350–63. University of Chicago Press, Chicago, IL.

Riesenfeld, G., Berman, A., and Hurwitz, S. (1981). Glucose kinetics and respiratory metabolism in fed and fasted chickens. *Comparative Biochemistry and Physiology*, **70A**, 223–7.

Roberts, H.C., Hardie, L.J., Chappell, L.H., and Mercer, J.G. (1999). Parasite-induced anorexia: leptin, insulin and corticosterone responses to infection with the nematode, *Nippostrongylus brasiliensis*. *Parasitology*, **118**, 117–23.

Robertson, I.C. (1998). Flight muscle changes in male pine engraver beetles during reproduction: the effects of body size, mating status and breeding failure. *Physiological Entomology*, **23**, 75–80.

Robin, J.-P., Boucontet, L., Chillet, P. and Groscolas, R. (1998). Behavioral changes in fasting emperor penguins: evidence for a 'refeeding signal' linked to a metabolic shift. *American Journal of Physiology*, **274**, R746–R753.

Robins, G.L. and Reid, M.L. (1997). Effects of density on the reproductive success of pine engravers: is aggregation in dead trees beneficial? *Ecological Entomology*, **22**, 329–34.

Robinson, B.W. and Wilson, D.S. (1998). Optimal foraging, specialization, and a solution to Liem's paradox. *American Naturalist*, **151**, 223–35.

Rodenhouse, N.L., Sherry, T.W., and Holmes, R.T. (1997). Site-dependent regulation of population size: a new synthesis. *Ecology*, **78**, 2025–42.

Roff, D.A. (1986). The genetic basis of wing dimorphism in the sand cricket, *Gryllus firmus* and its relevance to the evolution of wing dimorphisms in insects. *Heredity*, **57**, 221–31.

Roff, D.A. (1989). Exaptation and the evolution of dealation in insects. *Journal of Evolutionary Biology*, **2**, 109–23.

Roff, D.A. (1990a). Selection for changes in the incidence of wing dimorphism in *Gryllus firmus*. *Heredity*, **65**, 163–8.

Roff, D.A. (1990b). Understanding the evolution of insect life cycles: the role of genetical analysis. In *Genetics, evolution and coordination of insect life cycles* (ed. F. Gilbert), pp. 5–27. Springer, New York.

Roff, D.A. (1990c). The evolution of flightlessness in insects. *Ecological Monographs*, **60**, 389–421.

Roff, D.A. (1992). *The evolution of life histories: theory and analysis*. Chapman and Hall, New York.

Roff, D.A. (1994a). Evidence that the magnitude of the trade-off in a dichotomous trait is frequency dependent. *Evolution*, **48**, 1650–6.

Roff, D.A. (1994b). Habitat persistence and the evolution of wing dimorphism in insects. *American Naturalist*, **144**, 772–98.

Roff, D.A. (1995). Antagonistic and reinforcing pleiotropy: a study of differences in development time in wing dimorphic insects. *Journal of Evolutionary Biology*, **8**, 405–19.

Roff, D.A. and Bradford, M.J. (1996). Quantitative genetics of the trade-off between fecundity and wing dimorphism in the cricket *Allonemobius socius*. *Heredity*, **76**, 178–85.

Roff, D.A. and Bradford, M.J. (1998). The evolution of shape in the wing dimorphic cricket, *Allonemobius socius*. *Heredity*, **80**, 446–55.

Roff, D.A. and Fairbairn, D.J. (1991). Wing dimorphisms and the evolution of migratory polymorphisms among the insecta. *American Zoology*, **31**, 243–51.

Roff, D.A. and Simons, A.M. (1997). The quantitative genetics of wing dimorphism under laboratory and 'field' conditions in the cricket *Gryllus pennsylvanicus*. *Heredity*, **78**, 235–40.

Roff, D.A., Stirling, G., and Fairbairn, D.J. (1997). The evolution of threshold traits: a quantitative genetic analysis of the physiological and life history correlates of wing dimorphism in the land cricket. *Evolution*, **51**, 1910–19.

Roff, D.A., Tucker, J., Stirling, G., and Fairbairn, D.J. (1999). The evolution of threshold traits: effects of selection on fecundity and correlated response in wing dimorphism in the sand cricket. *Journal of Evolutionary Biology*, **12**, 535–46.

Rohani, P., Lewis, T.J., Grunbaum, D., and Ruxton, G.D. (1997). Spatial self-organisation in ecology: pretty patterns or robust reality? *Trends in Ecology and Evolution*, **12**, 70–4.

Roitberg, B. and Mangel, M. (1993). Parent–offspring conflict and life-history consequences in herbivorous insects. *American Naturalist*, **142**, 443–56.

Roitt, I., Brostoff, J., and Male, D. (1996). *Immunology*, 4th edn. Mosby, London.

Ronce, O. (1999). *Histoires de vie dans un habitat fragmenté: étude théorique de la dispersion et d'autres traits*. PhD thesis, University of Montpellier, France.

Ronce, O. and Olivieri, I. (1997). Evolution of reproductive effort in a metapopulation with local extinctions and ecological succession. *American Naturalist*, **150**, 220–49.

Ronce, O., Clobert, J., and Massot, M. (1998). Natal dispersal and senescence. *Proceedings of the National Academy of Sciences of the USA*, **95**, 600–5.

Ronce, O., Perret, F., and Olivieri, I. (2000). Evolutionarily stable dispersal rates do not always increase with local extinction rates. *American Naturalist*, **155**, 485–96.

Ronfort, J. and Couvet, D. (1995). A stochastic model of selection on selfing rates in structured populations. *Genetical Research*, **65**, 209–22.

Rood, J.P. (1986). Ecology and social evolution in the mongooses. In *Ecological aspects of social evolution* (ed. D.I. Rubenstein and R.W. Wrangham). pp. 131–52 Princeton University Press, Princeton, NJ.

Rosenberg, R., Nilsson, H.C., Hollertz, K., and Hellman, B. (1997). Density-dependent migration in an *Amphiura filiformis* (Amphiuridae, Echinodermata) infaunal population. *Marine Ecology Progress Series*, **159**, 121–31.

Rosenzweig, M.L. (1981). A theory of habitat selection. *Ecology*, **62**, 327–35.

Rosenzweig, M.L. (1985). Some theoretical aspects of habitat selection. In *Habitat*

selection in birds (ed. M.L. Cody), pp. 517–40. Academic Press, New York.

Rosenzweig, M.L. (1991). Habitat selection and population interactions: the search for mechanism. *American Naturalist*, **137**, S5–S28.

Rosenzweig, M.L. (1995). *Species diversity in space and time*. Cambridge University Press, Cambridge.

Ross, K.G. (1993). The breeding system of the fire ant *Solenopsis invicta*: effects on colony genetic structure. *American Naturalist*, **141**, 554–76.

Ross, K.G. (1997). Multilocus evolution in fire ants: effects of selection, gene flow, and recombination. *Genetics*, **145**, 961–74.

Ross, K.G. and Fletcher, D.J.C. (1986). Diploid male production – a significant colony mortality factor in the fire ant *Solenopsis invicta* (Hymenoptera: Formicidae). *Behavioral Ecology and Sociobiology*, **19**, 283–91.

Ross, K.G. and Keller, L. (1995a). Ecology and evolution of social organization: insights from fire ants and other highly eusocial insects. *Annual Review of Ecology and Systematics*, **26**, 631–56.

Ross, K.G. and Keller, L. (1995b). Joint influence of gene flow and selection on a reproductively important genetic polymorphism in the fire ant *Solenopsis invicta*. *American Naturalist*, **146**, 325–48.

Ross, K.G. and Keller, L. (1998). Genetic control of social organization in an ant. *Proceedings of the National Academy of Sciences of the USA*, **95**, 14232–7.

Ross, K.G. and Shoemaker, D.D. (1993). An unusual pattern of gene flow between the two social forms of the fire ant *Solenopsis invicta*. *Evolution*, **47**, 1595–605.

Ross, K.G. and Shoemaker, D.D. (1997). Nuclear and mitochondrial genetic structure in two social forms of the fire ant *Solenopsis invicta*: insights into transitions to an alternate social organization. *Heredity*, **78**, 590–602.

Ross, K.G., Vargo, E.L., Keller, L., and Trager, J.C. (1993). Effect of a founder event on variation in the genetic sex-determining system of the fire ant *Solenopsis invicta*. *Genetics*, **135**, 843–54.

Ross, K.G., Vargo, E.L., and Keller, L. (1996). Social evolution in a new environment: the case of introduced fire ants. *Proceedings of the National Academy of Sciences of the USA*, **93**, 3021–5.

Ross, K.G., Krieger, M.J.B., Shoemaker, D.D., Vargo, E.L., and Keller, L. (1997). Hierarchical analysis of genetic structure in native fire ant populations: results from three classes of molecular markers. *Genetics*, **147**, 643–55.

Ross, K.G., Shoemaker, D.D., Krieger, M.J.B., DeHeer, C.J., and Keller, L. (1999). Assessing genetic structure with multiple classes of molecular markers: a case study involving the introduced fire ant *Solenopsis invicta*. *Molecular Biology and Evolution*, **16**, 525–43.

Ross, R.M. (1987). Sex change linked growth acceleration in coral-reef fish, *Thalassoma buperrey*. *Journal of Experimental Zoology*, **244**, 2455–61.

Rossiter, M.C. (1991). Environmentally-based maternal effects, a hidden force in insect population-dynamics. *Oecologia*, **87**, 288–94.

Rossiter, M.C. (1994). Maternal effects hypothesis of herbivore outbreak. *BioScience*, **44**, 752–63.

Rossiter, M.C. (1996). Incidence and consequences of inherited environmental effects. *Annual Review of Ecology and Systematics*, **27**, 451–76.

Rothschild, M. and Clay, T. (1952). *Fleas, flukes and cuckoos*. Collins, London.

Rothstein, S.I. (1990). A model system for the study of coevolution: avian brood parasitism. *Annual Review of Ecology and Systematics*, **21**, 481–508.

Roughgarden, J. (1979). *Theory of population genetics and evolutionary ecology: an introduction*. Macmillan, New York.

Roughgarden, J., Gaines, S., and Possingham, H. (1988). Recruitment dynamics in complex life cycles. *Science*, **241**, 1460–6.

Rousset, F. (1996). Equilibrium values of measures of population subdivision for stepwise mutation processes. *Genetics*, **142**, 1357–62.

Rousset, F. (1997). Genetic differentiation and estimation of gene flow from *F*-statistics

under isolation by distance. *Genetics*, **145**, 1219–28.

Rousset, F. (1999). Genetic differentiation within and between two habitats. *Genetics*, **151**, 397–407.

Rousset, F. (2000a). Genetic differentiation between individuals. *Journal of Evolutionary Biology*, **13**, 58–62.

Rousset, F. (2000b). Inferences from spatial population genetics. In *Handbook of statistical genetics* (ed. D. Balding, M. Bishop, and C. Cannings). John Wiley, Chichester, UK, in press.

Rousset, F. (2000c). Reproductive value in sources and sinks. *Oikos*, in press.

Rousset, F. and Billiard, S. (2000). A theoretical basis for measures of kin selection in subdivided populations: finite populations and localized dispersal. *Journal of Evolutionary Biology*, **13**, 814–25.

Ruckelshaus, M., Hartway, C., and Kareiva, P. (1997). Assessing the data requirements of spatially explicit dispersal models. *Conservation Biology*, **11**, 1298–306.

Ruckelshaus, M., Hartway, C., and Kareiva, P. (1999). Dispersal and landscape errors in spatially explicit population models: a reply. *Conservation Biology*, **13**, 1223–4.

Rushton, S.P., Barreto, G.W., Cormack, R.M., Macdonald, D.W., and Fuller, R. (2000). Modelling the effects of mink and habitat fragmentation on the water vole. *Journal of Applied Ecology*, **37**, 475–90.

Ruxton, G.D. and Rohani, P. (1998). Fitness-dependent dispersal in metapopulations and its consequences for persistence and synchrony. *Journal of Animal Ecology*, **67**, 530–9.

Ruxton, G.D., Gonzales-Andujar, J.L., and Perry, J.N. (1997). Mortality during dispersal stabilizes local population fluctuations. *Oikos*, **66**, 289–92.

Sabelis, M.W., van Baalen, M., Bakker, F.M., Bruin, J., Drukker, B., Egas, M., Janssen, A., Lesna, I., Pels, B., van Rijn, P., and Scutereanu, P. (1999). The evolution of direct and indirect defence against herbivorous arthropods. In *Herbivores: between plants and predators* (ed. H. Olff, V. Brown, and R.H. Drenth), pp. 109–66. Blackwell Science, London.

Saccheri, I., Kuussaari, M., Kankare, M., Vikman, P., Fortelius, W., and Hanski, I. (1998). Inbreeding and extinction in a butterfly metapopulation. *Nature*, **392**, 491–4.

Sale, P.F. (1982). Stock recruit relationship and regional coexistence in a lottery competitive system. *American Naturalist*, **120**, 139–59.

Sanderson, N. (1989). Can gene flow prevent reinforcement? *Evolution*, **43**, 1223–35.

Santiago, E. and Caballero, A. (1998). Effective size and polymorphism of linked neutral loci in populations under directional selection. *Genetics*, **149**, 2105–17.

Sarrazin, F. and Barbault, R. (1996). Reintroduction: challenges and lessons for basic ecology. *Trends in Ecology and Evolution*, **11**, 474–8.

Sarrazin, F., Bagnolini, C., Pinna, J.L., and Danchin, E. (1996). Breeding biology during establishment of a reintroduced griffon vulture *Gyps fulvus* population. *Ibis*, **138**, 315–25.

Satô, K. and Konno, N. (1995). Successional dynamic models on the 2-dimensional lattice space. *Journal of the Physical Society of Japan*, **64**, 1866–9.

Satô, K., Matsuda H., and Sasaki, A. (1994). Pathogen invasion and host extinction in lattice structured population. *Evolutionary Ecology*, **6**, 352–6.

Savage, A.A. (1996). Density dependent and density independent relationships during a twenty-seven year study of the population dynamics of the benthic macroinvertebrate community of a chemically unstable lake. *Hydrobiologia*, **335**, 115–31.

Savage, I.R. (1956). Contributions to the theory of order statistics – the two sample case. *Annals of Mathematics and Statistics*, **27**, 590–615.

Sawyer, S. (1977). Asymptotic properties of the equilibrium probability of identity in a geographically structured population. *Advances in Applied Problems*, **9**, 268–82.

Schippers, P., Verboom, J., Knaapen, J.P., and van Apeldoorn, R.C. (1996). Dispersal and habitat connectivity in complex heterogeneous landscapes: an analysis with a GIS-based random walk model. *Ecography*, **19**, 97–106.

Schjørring, S., Gregersen, J., and Bregnballe, T. (1999). Prospecting enhances breeding success of first-time breeders in the great cormorant *Phalacrocorax carbosinensis*. *Animal Behaviour*, **57**, 647–54.

Schjørring, S., Gregersen, J., and Bregnballe, T. (2000). Sex difference in criteria determining fidelity towards breeding sites in the great cormorant. *Journal of Animal Ecology*, **69**, 214–23.

Schlichting, C.D. and Pigliucci, M. (1998). *Phenotypic evolution: a reaction norm perspective*. Sinauer Associates, Sunderland, MA.

Schlupp, I. and Ryan, M.J. (1997). Male sailfin mollies (*Poecilia latipinna*) copy the mate choice of other males. *Behavioral Ecology*, **8**, 104–7.

Schlupp, I., Marler, C., and Ryan, M.J. (1994). Benefit to male sailfin mollies of mating with heterospecific females. *Science*, **263**, 373–4.

Schmitt, R.J. and Holbrook, S.J. (1996). Local-scale patterns of larval settlement in a planktivorous damselfish: do they predict recruitment? *Marine and Freshwater Research*, **47**, 449–63.

Schmitt, R.J. and Holbrook, S.J. (1999). Mortality of juvenile damselfish: implications for assessing processes that determine abundance. *Ecology*, **80**, 35–50.

Schneider, J.E., Goldman, M.D., Tang, S., Bean, B., Ji, H., and Friedman, M.I. (1998). Leptin indirectly affects estrous cycles by increasing metabolic fuel oxidation. *Hormones and Behavior*, **33**, 217–28.

Scholz, D., Borgemeister, C., Markham, R.H., and Poehling, H.M. (1997). Flight initiation in *Prostephanus truncatus*: influence of population density and aggregation pheromone. *Entomologia Experimentalis et Applicata*, **85**, 237–45.

Schumacher, P., Weber, D.C., Hagger, C., and Dorn, S. (1997a). Heritability of flight distance for *Cydia pomonella*. *Entomologia Experimentalis et Applicata*, **85**, 169–75.

Schumacher, P., Weyeneth, A., Weber, D.C., and Dorn, S. (1997b). Long flights in *Cydia pomonella* L. (Lepidoptera: Totricidae) measured by a flight mill: influence of sex, mated status and age. *Physiological Entomology*, **22**, 149–60.

Schwabl, H. (1993). Yolk is a source of maternal testosterone for developing birds. *Proceedings of the NationalAcademy of Sciences*, **90**, 11466–70.

Schwabl, H. (1996a). Environment modifies the testosterone levels of a female bird and its eggs. *Journal of Experimental Zoology*, **276**, 157–63.

Schwabl, H. (1996b). Maternal testosterone in the avian egg enhances postnatal growth. *Comparative Biochemistry and Physiology*, **114A**, 271–6.

Schwabl, H. (1997a). Maternal steroid hormones in the egg. In *Perspectives in avian endocrinology* (ed. S. Harvey and R.J. Etches), pp. 3–13. Society of Endocrinology, Bristol.

Schwabl, H. (1997b). The contents of maternal testosterone in house sparrow *Passer domesticus* eggs vary with breeding conditions. *Naturwissenschaften*, **84**, 406–8.

Schwartz, O.A. and Armitage, K.B. (1980). Genetic variation in social mammals: the marmot model. *Science*, **207**, 665–7.

Schwarz, C.J. (1993). Estimating migration rates from tag-recovery data. In *Marked individuals in the study of bird population* (ed. J.-D. Lebreton and P.M. North), pp. 255–64. Birkhauser, Basel.

Schwarz, C.J. and Arnason, A.N. (1990). Use of tag–recovery information in migration and movement studies. *American Fisheries Society Symposium*, **7**, 588–603.

Schwarz, C.J. and. Stobo, W.T. (1997). Estimating temporary migration using the robust design. *Biometrics*, **53**, 178–94.

Schwarz, C.J., Schweigert, J.F., and Arnason, A.N. (1993). Estimating migration rates using tag recovery data. *Biometrics*, **49**, 177–93.

Scribner, K.T., Arntzen, J.W., and Burke, T. (1994). Comparative analysis of intra- and interpopulation genetic diversity in *Bufo bufo*, using allozyme, single-locus microsatellite, minisatellite, and multilocus minisatellite data. *Molecular Biology and Evolution*, **11**, 737–48.

Seabloom, R.W., Iverson, S.L., and Turner, B.N. (1978). Adrenal responses in a wild *Microtus* population: seasonal aspects. *Canadian Journal of Zoology*, **56**, 1433–40.

Seeley, T.D. and Buhrman, S.C. (1999). Group decision making in swarms of honey bees. *Behavioral Ecology and Sociobiology*, **45**, 19–31.

Seger, J. and Brockmann, H.J. (1987). What is bet-hedging? *Oxford Surveys in Evolutionary Biology*, **4**, 182–211.

Selander, R.K. (1970). Behaviour and genetic variation in natural populations. *American Zoologist*, **10**, 53–66.

Semlitsch, R.D., Harris, R.N., and Wilbur, H.M. (1990). Paedomorphosis in *Ambystoma talpoideum*: maintenance of population variation and alternative life-history pathways. *Evolution*, **44**, 1604–13.

Severaid, J.H. (1955). *The natural history of the pika (mammalian genus Ochotona)*. PhD, thesis, University of California, Berkeley, CA.

Sharp, P.M. (1995). Cross-species transmission and recombination of 'AIDS' viruses. *Philosophical Transactions of the Royal Society B*, **349**, 41–47.

Shaw, M.P.J. (1970a). Effects of population density on alienicolae of *Aphis fabae* Scop. II. The effects of crowding on the expression of the migratory urge among alatae in the laboratory. *Annals of Applied Biology*, **65**, 197–203.

Shaw, M.J.P. (1970b). Effects of population density on alienicolae of *Aphis fabae* Scop. III. The effect of isolation on the development of form and behaviour of alatae in a laboratory clone. *Annals of Applied Biology*, **65**, 205–12.

Shaw, M.W. (1995). Simulation of population expansion and spatial pattern when individual dispersal distributions do not decline exponentially with distance. *Proceedings of the Royal Society of London B*, **259**, 243–8.

Shepherdson, D. (1994). The role of environmental enrichment in the captive breeding and reintroduction of endangered species. In *Creative conservation: interactive management of wild and captive animals* (ed. P.J.Olney, G.M. Mace, and A.T. Feistner), pp. 167–77. Chapman and Hall, London.

Shettleworth, S.J. (1998). *Cognition, evolution and behavior*. Oxford University Press, Oxford.

Shields, W.M. (1982). *Philopatry, inbreeding and the evolution of sex.* State University of New York Press, Albany, NY.

Shields, W.M. (1983). Optimal inbreeding and the evolution of philopatry. In *The ecology of animal movement* (ed. I.R. Swingland and P.J. Greenwood), pp. 132–59. Clarendon, Oxford.

Shields, W.M. (1987). Dispersal and mating systems: investigating their causal connections. In *Mammalian dispersal patterns: the effects of social structure on population genetics* (ed. B.D. Chepko-Sade and Z.T. Halpin), pp. 3–24. University of Chicago Press, Chicago, IL.

Shields, W.M., Crook, J.R., Hebblethwaite, M.L., and Wiles-Ehmann, S.S. (1988). Ideal free coloniality in swallows. In *The ecology of social behavior.* (ed. C.N. Slobodchikoff), pp. 189–228. Academic Press, San Diego, CA.

Shirai, Y. (1995). Longevity, flight ability and reproductive performance of the diamond-back moth, *Plutella xylostella* (L.) (Lepidoptera, Yponomautidae), related to adult body size. *Researches on Population Ecology*, **37**, 269–77.

Shmida, A. and Ellner, S. (1984). Coexistence of plant species with similar niches. *Vegetatio*, **58**, 29–55.

Shmida, A. and Wilson, M.V. (1985). Biological determinants of species diversity. *Journal of Biogeography*, **12**, 1–20.

Shoemaker, D.D. and Ross, K.G. (1996). Effects of social organization on gene flow in the fire ant *Solenopsis invicta. Nature*, **383**, 613–16.

Siegel, H.S. (1980). Physiological stress in birds. *Bioscience*, **30**, 529–33.

Sillero-Zubiri, C., Gottelli, D., and McDonald, D.W. (1996). Male philopatry, extra-pack copulations and inbreeding avoidance in Ethiopian wolves (*Canis simensis*). *Behavioral Ecology and Sociobiology*, **38**, 331–40.

Silverin, B. (1997). The stress response and autumn dispersal behaviour in willow tits. *Animal Behaviour*, **53**, 451–9.

Silverin, B., Viebke, P.A. and Westin, J. (1989). Hormonal correlates of migration and territorial behavior in juvenile willow tits during autumn. *General and Comparative Endocrinology*, **75**, 148–56.

Simberloff, D. (1988). The contribution of population and community biology to conservation science. *Annual Review of Ecology and Systematics*, **19**, 473–511.

Simberloff, D., Farr, J.A., Cox, J., and Moehlman, D.W. (1992). Movement corridors: conservation bargains or poor investments. *Conservation Biology*, **6**, 493–504.

Sites, J.W., Barton, N.H., and Reed, K.M. (1995). The genetic structure of a hybrid zone between two chromosome races of the *Sceloporus grammicus* complex (Sauria, Phrynosomatidae) in central Mexico. *Evolution*, **49**, 9–36.

Slagsvold, T. (1980). Habitat selection in birds: on the presence of other bird species with special regard to *Turdus pilaris. Journal of Animal Ecology*, **49**, 523–36.

Slagsvold, T. (1986). Nest site settlement by the pied flycatcher: does the female chose her mate for the quality of his house or himself? *Ornis Scandinavica*, **17**, 210–20.

Slatkin, M. (1973). Gene flow and selection in a cline. *Genetics*, **75**, 733–56.

Slatkin, M. (1974). Competition and regional coexistence. *Ecology*, **55**, 128–34.

Slatkin, M. (1977). Gene flow and genetic drift in a species subject to frequent local extinctions. *Theoretical Population Biology*, **12**, 253–62.

Slatkin, M. (1978a). On the equilibration of fitnesses by natural selection. *American Naturalist*, **112**, 845–59.

Slatkin, M. (1978b). Spatial patterns in the distribution of polygenic characters. *Journal of Theoretical Biology*, **70**, 213–28.

Slatkin, M. (1981). Estimating levels of gene flow in natural populations. *Genetics*, **99**, 323–35.

Slatkin, M. (1985a). Gene flow in natural populations. *Annual Review of Ecology and Systematics*, **16**, 393–430.

Slatkin, M. (1985b). Rare alleles as indicators of gene flow. *Evolution*, **39**, 53–65.

Slatkin, M. (1991). Inbreeding coefficients and coalescence times. *Genetical Research*, **58**, 167–75.

Slatkin, M. (1993). Isolation by distance in equilibrium and non-equilibrium populations. *Evolution*, **47**, 264–79.

Slatkin, M. (1994). Gene flow and population structure. In *Ecological genetics* (ed. L.A. Real), pp. 3–17. Princeton University Press, Princeton, NJ.

Slatkin, M. (1995). A measure of population subdivision based on microsatellite allele frequencies. *Genetics*, **139**, 457–62.

Slatkin, M. and Maddison, W.P. (1989). A cladistic measure of gene flow inferred from the phylogenies of alleles. *Genetics*, **123**, 603–13.

Slooten, E., Dawson, S.M., and Lad, F. (1992). Survival rates of photographically identified Hector's dolphins from 1984 to 1988. *Marine Mammal Science*, **8**, 327–43.

Smith, A.T. (1974a). The distribution and dispersal of pikas: consequences of insular population structure. *Ecology*, **55**, 1112–19.

Smith, A.T. (1974b). The distribution and dispersal of pikas: influences of behavior and climate. *Ecology*, **55**, 1368–76.

Smith, A.T. (1978). Comparative demography of pikas (*Ochotona*): effect of spatial and temporal age-specific mortality. *Ecology*, **59**, 133–9.

Smith, A.T. (1980). Temporal changes in insular populations of the pika (*Ochotona princeps*). *Ecology*, **6**, 8–13.

Smith, A.T. (1987). Population structure of pikas: dispersal versus philopatry. In *Mammalian dispersal patterns: the effects of social structure on population genetics*. (ed. B.D. Chepko-Sade and Z.T. Halpin), pp. 128–43. University of Chicago Press, Chicago, IL.

Smith, A.T. (1993). The natural history of inbreeding and outbreeding in small mammals. In *The natural history of inbreeding and outbreeding* (ed. N.W. Thornhill), pp. 329–51. University of Chicago Press, Chicago, IL.

Smith, A.T. and Gilpin, M. (1997). Spatially correlated dynamics in a pika metapopulation. In *Metapopulation biology: ecology, genetics and evolution* (ed. I.A. Hanski and M.E. Gilpin), pp. 407–28. Academic Press, San Diego, CA.

Smith, A.T. and Ivins, B.L. (1983). Colonization in a pika population: dispersal vs philopatry. *Behavioral Ecology and Sociobiology*, **13**, 37–47.

Smith, A.T. and Ivins, B.L. (1984). Spatial relationships and social organization in adult pikas: a facultatively monogamous mammal. *Zeitschrift für Tierpsychologie*, **66**, 289–308.

Smith, A.T. and Peacock, M.M. (1990). Conspecific attraction and the determination of metapopulation colonization rates. *Conservation Biology*, **4**, 320–3.

Smith, A.T. and Weston, M.L. (1990). *Ochotona princeps*. *Mammal Species*, **352**, 1–8.

Smith, C.A.B. (1969). Local fluctuations in gene frequencies. *Annals of Human Genetics*, **32**, 251.

Smith, J.N.M. and Arcese, P. (1989). How fit are floaters? Consequences of alternative territorial behaviours in a nonmigratory sparrow. *American Naturalist*, **133**, 830–45.

Smith, R.J.F. (1986). Evolution of alarm signals: role of benefits of retaining group members or territorial neighbors. *American Naturalist*, **128**, 604–10.

Smith, S.M. (1984). Flock switching in chickadees. Why be a winter floater? *American Naturalist*, **123**, 81–98.

Smouse, P.E. and Peakall, R. (1999). Spatial autocorrelation analysis of individual multiallele and multilocus genetic structure. *Heredity*, **82**, 561–73.

Smouse, P.E., Vitzthum, V.J., and Neel, J.V. (1981). The impact of random and lineal fission on the genetic divergence of small human groups: a case study among the Yanomama. *Genetics*, **98**, 179–97.

Sokal, R.R. and Wartenberg, D.E. (1983). A test of spatial autocorrelation analysis using an isolation by distance model. *Genetics*, **105**, 219–37.

Sokal, R.R., Oden, N.L., Legendre, P., Fortin, M.J., Kim, J., Thomson, B.A., Vaudor, A., Harding, R.M., and Barbujani, G. (1990). Genetics and language in European populations. *American Naturalist*, **135**, 157–75.

Solbreck, C. 1978. Migration, diapause and direct development as alternative life histories in the seed bug, *Neacoryphus bicrucis*. In *Evolution of insect migration and diapause*. (ed. H. Dingle), pp. 195–217. Springer, New York.

Soler, M., Soler, J.J., Martinez, J.G., Perez-Contreras, T., and Møller, A.P. (1998).

Micro-evolutionary change and population dynamic of a brood parasite and its primary host: the intermittent arms race hypothesis. *Oecologia*, **117**, 381–90.

Solomon, N.G., Vandenberg, J.G., and Sullivan, W.T.J. (1998). Social influences on intergroup transfer by pine voles (*Microtus pinetorum*). *Canadian Journal of Zoology*, **76**, 2131–6.

Sonerud, G.A., Solheim, R., and Prestrud, K. (1988). Dispersal of Tengmalm's owl *Aegolius funereus* in relation to prey availability and nesting success. *Ornis Scandinavica*, **19**, 175–81.

Sorci, G. and Clobert, J. (1995). Effects of maternal parasite load on offspring life-history traits in the common lizard (*Lacerta vivipara*). *Journal of Evolutionary Biology*, **8**, 711–23.

Sorci, G., Massot, M., and Clobert, J. (1994). Maternal parasite load increases sprint speed and philopatry in female offspring of the common lizard. *American Naturalist*, **144**, 153–64.

Sorci, G., Møller, A.P., and Boulinier, T. (1997). Genetics of host–parasite interactions. *Trends in Ecology and Evolution*, **12**, 196–9.

Sorensen, A.E. (1986). Seed dispersal by adhesion. *Annual Review of Ecology and Systematics*, **17**, 443–63.

Soulsby, E.J.L. (1986). *Helminths, arthropods and protozoa of domesticated animals*, 7th edn. Baillière Tindall, London.

South, A. (1999). Dispersal in spatially explicit population models. *Conservation Biology*, **13**, 1039–46.

South, A.B., Rushton, S.P., and Macdonald, D.W. (2000). Simulating the proposed re-introduction of the European beaver (*Castor fiber*) to Scotland. *Biological Conservation*, **93**, 103–16.

Southwick, C.H., Golian, S.C., Whitworth, M.R., Halfpenny, J.C., and Brown, R. (1986). Population density and fluctuations of pikas (*Ochotona princeps*) in Colorado (USA). *Journal of Mammalogy*, **67**, 149–53.

Southwood, T.R.E. (1962). Migration of terrestrial arthropods in relation to habitat. *Biological Review*, **37**, 171–214.

Southwood, T.R.E. (1977). Habitat, the templet for ecological strategies? *Journal of Animal Ecology*, **46**, 337–65.

Spendelow, J.A., Nichols, J.D., Nisbet, I.C.T., Hays, H., Cormons, G.D., Burger, J., Safina, C., Hines, J.E., and Gochfeld, M. (1995). Estimating annual survival and movement rates of adults within a metapopulation of roseate terns. *Ecology*, **76**, 2415–28.

Stamps, J.A. (1987). Conspecifics as cues to territory quality: a preference of juvenile lizard (*Anolis aeneus*) for previously used territories. *American Naturalist*, **129**, 629–42.

Stamps, J.A. (1988). Conspecific attraction and aggregation in territorial species. *American Naturalist*, **131**, 329–47.

Stamps, J.A. (1991). The effect of conspecifics on habitat selection in territorial species. *Behavioral Ecology and Sociobiology*, **28**, 29–36.

Stamps, J.A. (1994). Territorial behavior: testing the assumptions. *Advances in the Study of Behavior*, **23**, 173–232.

Stamps, J.A. (1995). Motor learning and the value of familiar space. *American Naturalist*, **146**, 41–58.

Stamps, J.A. and Krishnan, V.V. (1999). A learning-based model of territory establishment. *Quarterly Review of Biology*, **74**, 1–28.

Stattersfield, A.J., Crosby, M.J., Long, A.J., and Wege, D.C. (1998). *Endemic bird areas of the world: priorities for biodiversity conservation*. Birdlife International, Cambridge.

Stearns, S.C. (1992). *The evolution of life-histories*. Oxford University Press, Oxford.

Stenseth, N.C. (1983). Causes and consequences of dispersal in small mammals. In *The ecology of animal movement* (ed. I.R. Swingland and P.J. Greenwood), pp. 63–101, Clarendon Press, Oxford.

Stenseth, N.C. and Lidicker, W.Z., Jr. (ed.) (1992). *Animal dispersal: small mammals as a model*. Chapman and Hall, London.

Stephens, D.W. (1987). On economically tracking a variable environment. *Theoretical Population Biology*, **32**, 15–25.

Stephens, D.W. (1991). Change, regularity and value in the evolution of animal learning. *Behavioral Ecology*, **2**, 77–89.

Stephens, D.W. and Krebs, J.R. (1986). *Foraging theory*. Princeton University Press, Princeton, NJ.

Stephens, J.C., Gilbert, D.A., Yuhki, N., and O'Brien S.J. (1992). Estimation of heterozygosity for single-probe multilocus DNA fingerprints. *Molecular Biology and Evolution*, **9**, 729–43.

Stephens, P.A. and Sutherland, W.J. (1999). Consequences of the Allee effect for behaviour, ecology and conservation. *Trends in Ecology and Evolution*, **14**: 401–5.

Stiles, E.W. (1992). Animals as seed dispersers. In *Animals as seed dispersers* (ed. M. Fenner), pp. 105–56. CAB International, Wallingford, UK.

Stockley, P., Searle, J.B., Macdonald, D.W., and Jones, C.S. (1993). Female multiple mating behaviour in the common shrew as a strategy to reduce inbreeding. *Proceedings of the Royal Society of London B*, **254**, 173–9.

Stoner, D.S. (1990). Recruitment of a tropical colonial ascidian: relative importance of pre-settlement vs. post-settlement processes. *Ecology*, **71**, 1682–90.

Strickland, D. (1991). Juvenile dispersal in gray jays: dominant brood member expels siblings from natal territory. *Canadian Journal Zoology*, **69**, 2935–45.

Strong, D.R., Jr., Simberloff, D. Abele, L.G., and Thistle, A.B. (ed.). (1984). *Ecological communities: conceptual issues and the evidence*. Princeton University Press, Princeton, NJ.

Sugg, D.W., Chesser, R.K., Dobson, F.S., and Hoogland, J.L. (1996). Population genetics meets behavioural ecology. *Trends in Ecology and Evolution*, **11**, 338–42.

Suhonen, J., Norrdahl, K., and Korpimaki, E. (1994). Avian predation risk modifies breeding bird community on a farmland area. *Ecology*, **75**, 1626–34.

Sukhdeo, M.V.K. and Sukhdeo, S.C. (1994). Optimal habitat selection by helminths within the host environment. *Parasitology*, **106**, S41–S55.

Sultan, S.E. (1995). Phenotypic plasticity and plant adaptation. *Acta Botanica Neerlandica*, **44**, 363–83.

Sultan, S.E., Wilczek, A.M., Hann, S.D., and Brosi, B.J. (1998a). Contrasting ecological breadth of co-occurring annual *Polygonum* species. *Journal of Ecology*, **86**, 363–83.

Sultan, S.E., Wilczek, A.M., Bell, D.L., and Hand, G. (1998b). Physiological response to complex environments in annual *Polygonum* species of contrasting ecological breadth. *Oecologia*, **115**, 564–78.

Sundström, L. (1995). Dispersal polymorphism and physiological condition of males and females in the ant, *Formica truncorum*. *Behavioural Ecology*, **6**, 132–9.

Sutcliffe, O. L., Thomas, C. D., and Moss, D. (1996). Spatial synchrony and asynchrony in butterfly population dynamics. *Journal of Animal Ecology*, **65**, 85–95.

Sutherland, O.W.R. (1969a). The role of crowding in the production of winged forms by two strains of the pea aphid *Acyrthosiphon pisum*. *Journal of Insect Physiology*, **15**, 1385–410.

Sutherland, O.W.R. (1969b). The role of the host plant in the production of winged forms by two strains of the pea aphid, *Acyrthosiphon pisum*. *Journal of Insect Physiology*, **15**, 2179–201.

Sutherland, W.J. (1996). *From individual behaviour to population ecology*. Oxford University Press, Oxford.

Swingland, I.R. (1983). Intraspecific differences in movement. In *The ecology of animal movement* (ed. I.R. Swingland and P.J. Greenwood), pp. 102–15. Clarendon Press, Oxford.

Swinton, J., Tuyttens, F.A.M., Macdonald, D.W., and Cheeseman, C.L. (1997). Social perturbation and bovine tuberculosis in badgers: fertility control and lethal control compared. *Philosophical Transactions of the Royal Society of London*, **352**, 619–31.

Switzer, P.V. (1993). Site fidelity in predictable and unpredictable habitats. *Evolutionary Ecology*, **7**, 533–55.

Switzer, P.V. (1997). Past reproductive success affects future habitat selection. *Behavioral Ecology and Sociobiology*, **40**, 307–12.

Szymczak, M.R. and Rexstad, E.A. (1991). Harvest distribution and survival of a gadwall population. *Journal of Wildlife Management*, **55**, 592–600.

Szymura, J.M. and Barton, N.H. (1991). The genetic structure of the hybrid zone between the fire-bellied toads *Bombina bombina* and *B. variegata*: comparisons between transects and between loci. *Evolution*, **45**, 237–61.

Tallamy, D.W. and Schaefer, C. (1997). Maternal care in the Hemiptera: ancestry, alternatives, and current adaptive value. In *The evolution of social behavior in insects and arachnids* (ed. J.C. Choe and B.J. Crespi), pp. 94–115. Cambridge University Press, Cambridge.

Taouis, M., Chen, J.-W., Daviaud, C., Dupont, J., Derouet, M., and Simon, J. (1997). Cloning the chicken leptin gene. *Gene*, **208**, 239–42.

Tapper, S.C. (1973). *The spatial organization of pikas (Ochotona), and its effect on population recruitment*. PhD thesis, University of Alberta, Edmonton.

Taylor, A.D. (1988). Large-scale spatial structure and population dynamics in arthropod predator–prey systems. *Annales Zoologici Fennici*, **25**, 63–74.

Taylor, A.D. (1990). Metapopulations, dispersal and predator–prey dynamics: an overview. *Ecology*, **71**, 429–36.

Taylor, A.D. (1991). Studying metapopulation effects in predator–prey systems. *Biological Journal of the Linnean Society*, **42**, 305–23.

Taylor, P.D. (1981). Intra-sex and inter-sex sibling interactions as sex-ratios determinants. *Nature*, **291**, 64–6.

Taylor, P.D. (1988). An inclusive fitness model for dispersal of offspring. *Journal of Theoretical Biology*, **130**, 363–78.

Taylor, P.D. (1989). Evolutionary stability in one-parameter models under weak selection. *Theoretical Population Biology*, **36**, 125–43.

Taylor, P.D. and Bulmer, M.G. (1980). Local mate competition and the sex ratio. *Journal of Theoretical Biology*, **86**, 409–19.

Taylor, P.D. and Crespi, B.J. (1994). Evolutionarily stable strategy sex ratios when correlates of relatedness can be assessed. *American Naturalist*, **143**, 297–316.

Taylor, P.D. and Frank, S.A. (1996). How to make a kin selection model. *Journal of Theoretical Biology*, **180**, 27–37.

Taylor, P.D. and Merriam, G. (1995). Wing morphology of a forest damselfly is related to landscape structure. *Oikos*, **73**, 43–8.

Telenius, A. and Torstensson, P. (1989). The seed dimorphism of *Spergularia marina* in relation to dispersal by wind and water. *Oecologia*, **80**, 206–10.

Temple, S.A. (1978). Manipulating behavioral patterns of endangered birds: a potential management technique. In *Endangered birds: management techniques for presenting threatened species* (ed. S.A. Temple), pp. 435–43. University of Winsconsin Press, Madison, WI.

Templeton, A.R. (1986). Coadaptation and outbreeding depression. In *Conservation biology* (ed. M. Soulé), pp. 105–16. Sinauer Associates, Sunderland, MA.

Templeton, A.R. (1992). Human origins and analysis of mitochondrial DNA sequences. *Science*, **255**, 737.

Templeton, A.R. and Read, B. (1994). Inbreeding: one word, several meanings, much confusion. In *Conservation genetics* (ed. J. Loeschke, J. Tomiuk, and S.K. Jain), pp. 91–105. Birkhäuser, Basel.

Templeton, J.J. and Giraldeau, L.A. (1995). Patch assessment in foraging flocks of European starlings: evidence for the use of public information. *Behavioral Ecology*, **6**, 65–72.

Templeton, J.J. and Giraldeau, L.A. (1996). Vicarious sampling: the use of personal and public information by starlings foraging in a simple patchy environment. *Behavioral Ecology and Sociobiology*, **38**, 105–14.

Teuschl, Y., Taborsky, B., and Taborsky, M. (1998). How do cuckoos find their hosts? The role of habitat imprinting. *Animal Behaviour*, **56**, 1425–33.

Thomas, C.D. (1994). Local extinctions, colonizations and distributions: habitat tracking by British butterflies. In *Individuals, populations and patterns in ecology* (ed. S.R. Leather, A.D. Watt, N.J. Mills, and K.F.A. Walters), pp. 319–36. Intercept, Andover.

Thomas, C.D. and Hanski, I. (1997). Butterfly metapopulations. In *Metapopulation biology: ecology, genetics and evolution* (ed. I.A. Hanski and M.E. Gilpin), pp. 359–86. Academic Press, San Diego, CA.

Thomas, C.D. and Jones, T.M. (1993). Partial recovery of a skipper butterfly (*Herspera comma*) from population refuges: lessons for conservation in a fragmented landscape. *Journal of Animal Ecology*, **62**, 472–81.

Thomas, C.D., Hill, J.K., and Lewis, O.T. (1998). Evolutionary consequences of habitat fragmentation in a localized butterfly. *Journal of Animal Ecology*, **67**, 485–97.

Thornhill, R. and Alcock, J. (1983). *The evolution of insect mating systems*. Harvard University Press, Boston, MA.

Thrall, P.H. and Burdon, J.J. (1997). Host–pathogen dynamics in a metapopulation context: the ecological and evolutionary consequences of being spatial. *Journal of Ecology*, **85**, 743–53.

Tilman, D. (1990). Constraint and tradeoffs: toward a predictive theory of competition and succession. *Oikos*, **58**, 3–15.

Tilman, D. (1994). Competition and biodiversity in spatially structured habitats. *Ecology*, **75**, 2–16.

Tilman, D. (1997). Community invasibility, recruitment limitation, and grassland biodiversity. *Ecology*, **78**, 81–92.

Tilman, D. and Kareiva, P. (1997). *Spatial ecology: the role of space in population dynamics and interspecific interactions. Monographs in Population Biology*. Princeton University Press, Princeton, NJ.

Tilman, D. and Pacala, S. (1993). The maintenance of species richness in plant communities. In *Species diversity in ecological communities: historical and geographical perspectives* (ed. R.E. Ricklefs and D. Schluter), pp. 13–25. University of Chicago press, Chicago, IL.

Tilman, D., May, R.M., Lehman, C.L., and Nowak, M.A. (1994). Habitat destruction and the extinction debt. *Nature*, **371**, 65–6.

Tinbergen, N. (1963). On aims and methods of ethology. *Zeitschrift für Tierpsychologie*, **20**, 410–33.

Tinkle, D.W., Dunham, A.E., and Congdon, J.D. (1993). Life history and demographic variation in the lizard *Sceloporus graciosus*: a long term study. *Ecology*, **74**, 2413–29.

Tobias, J. (1997). Asymmetric territorial contests in the European robin: the role of settlement costs. *Animal Behaviour*, **54**, 9–21.

Toft, C.A., Aeschlimann, A., and Bolis, L. (ed.) (1991). Parasite–host association. *Coexistence or conflict?* Oxford Scientific Publications, Oxford.

Tordoff, H.B., Martell, M.S., and Redig, P.T. (1998). Effect of fledge site on choice of nest site by midwestern peregrine falcons. *The Loon*, **70**, 127–9.

Travis, J.M.J. and Dytham, C. (1998). The evolution of dipsersal in a metapopulation: a spatially explicit, individual-based model. *Proceedings of the Royal Society of London B*, **265**, 17–23.

Tregenza, T. (1995). Building on the ideal free distribution. *Advances in Ecological Research*, **26**, 253–307.

Trent, T.T. and Rongstad, O.J. (1974). Home range and survival of cottontail rabbits in southwestern Wisconsin. *Journal of Wildlife Management*, **38**, 459–72.

Trottier, G., Koski, K.G., Brun, T., Toufexis, D.J., Richard, D., and Walker, C.D. (1998). Increased fat intake during lactation modifies hypothalamic–pituitary–adrenal responsiveness in developing rat pups: a possible role for leptin. *Endocrinology*, **139**, 3704–11.

Trzcinski, M.K., Fahrig, L., and Merriam, G. (1999). Independent effects of forest cover and fragmentation on the distribution of forest breeding birds. *Ecological Applications*, **9**, 586–93.

Tsai, K. (1996). *Survival analysis for telemetry data in animal ecology*. PhD thesis, North Carolina State University, Raleigh, NC.

Tschinkel, W.R. (1998). The reproductive biology of fire ant societies. *BioScience* **48**, 593–605.

Tufto, J., Engen, S. and Hindar, K. (1996). Inferring patterns of migration from gene frequencies under equilibrium conditions. *Genetics*, **144**, 1911–21.

Tuljapurkar, S. (1990). *Population dynamics in variable environments*. Springer, Berlin.

Tuljapurkar, S. (1997). Stochastic matrix models. In *Structured-population models in marine, terrestrial, and freshwater systems* (ed. S. Tuljapurkar and H. Caswell), pp. 59–88. Chapman and Hall, London.

Tuljapurkar, S. and Caswell, H. (ed.) (1997). *Structured-population models in marine, terrestrial, and freshwater systems*. Chapman and Hall, London.

Tuljapurkar, S. and Istock, C. (1993). Environmental uncertainty and variable diapause. *Theoretical Population Biology*, **43**, 251–80.

Turchin, P. (1998). *Quantitative analysis of movement: measuring and modeling population redistribution in animals and plants*. Sinaeur Associates, Sunderland, MA.

Turchin, P., Reeve, J.D., Cronin, J.T., and Wilkens, R.T. (1997). Spatial pattern formation in ecological systems: bridging theoretical and empirical approaches. In *Modeling spatiotemporal dynamics in ecology* (ed. J. Bascompte and R.V. Solé), pp. 199–214. Springer, New York.

Turner, B.N., Iverson, S.L., and Severson, K.L. (1980). Effects of castration on open-field behaviour and aggression in male meadow voles (*Microtus pennsylvanicus*). *Canadian Journal of Zoology*, **58**, 1927–32.

Turner, B.N., Iverson, S.L., and Severson, K.L. (1983). Seasonal changes in open-field behavior in wild male meadow voles (*Microtus pennsylvanicus*). *Behavioral and Neural Biology*, **39**, 60–77.

Tuyttens, F.A.M. and Macdonald, D.W. (1998). Sterilization as an alternative strategy to control wildlife diseases: bovine tuberculosis in European badgers as a case study. *Biodiversity and Conservation*, **7**, 705–23.

Tuyttens, F.A.M. and Macdonald, D.W. (2000). Consequences of social perturbation for wildlife management and conservation. In *Behaviour and Conservation* (ed. L.M. Gosling and W.J. Sutherland), pp. 315–329. Cambridge University Press, Cambridge.

Tyson, J.J. (1984). Evolution of eusociality in diploid species. *Theoretical Population Biology*, **26**, 283–95.

Urbanek, M., Goldman, D., and Long, J.C. (1996). The apportionment of dinucleotide repeat diversity in native Americans and Europeans: a new approach to measuring gene identity reveals asymmetric patterns of divergence. *Molecular Biology and Evolution*, **13**, 943–53.

Uyenoyama, M.K., Holsinger, K.E., and Waller, D.M. (1993). Ecological and genetic factors directing the evolution of self-fertilization. *Oxford Surveys in Evolutionary Biology*, **9**, 327–81.

Valone, T.J. (1989). Group foraging, public information, and patch estimation. *Oikos*, **56**, 357–63.

Valone, T.J. (1991). Bayesian and prescient assessment: foraging with pre-harvest information. *Animal Behaviour*, **41**, 569–77.

Valone, T.J. (1993). Patch information and estimation: a cost of group foraging. *Oikos*, **68**, 258–66.

Valone, T.J. (1996). Food-associated calls as public information about patch quality. *Oikos*, **77**, 153–7.

Valone, T.J. and Giraldeau, L.A. (1993). Patch estimation by group foragers: what information is used? *Animal Behaviour*, **45**, 721–8.

van Baalen, M. (2000). Pair approximations for different spatial geometries. In *The geometry of ecological interactions: simplifying spatial complexity* (ed. U. Dieckmann, R. Law, and J.A.J. Metz), pp. 359–87. Cambridge University Press, Cambridge.

van Baalen, M. and Rand, D.A. (1998). The unit of selection in viscous populations and the evolution of altruism. *Journal of Theoretical Biology*, **193**, 631–48.

van Baalen, M. and Sabelis, M.W. (1993). Coevolution of patch selection strategies of predators and prey and the consequences for ecological stability. *American Naturalist*, **142**, 646–70.

van Baalen, M. and Sabelis, M.W. (1995). The milker–killer dilemma in spatially structured predator–prey interactions. *Oikos*, **74**, 391–400.

van Baalen, M. and Sabelis, M.W. (1999). Non-equilibrium dynamics of 'ideal and free' prey and predators. *American Naturalist*, **154**, 69–88.

van den Bosch, F., Hengeveld, R., and Metz, J.A.J. (1992). Analyzing the velocity of animal range expansion. *Journal of Biogeography*, **19**, 135–50.

van der Pijl, L. (1982). *Principles of dispersal in higher plants*. Springer, Berlin.

van der Schaar, W., Alonso-Blanco, C., Leon-Kloosterziel, K.M., Jansen, R.C., van Ooijen, J.W., and Koornneef, M. (1997). QTL analysis of seed dormancy in *Arabidopsis* using recombinant inbred lines and MQM mapping. *Heredity*, **79**, 190–200.

van Dijken, F.R. and Scharloo, W. (1979). Divergent selection on locomotor activity in *Drosophila melanogaster*. I. Selection response. *Behaviour Genetics*, **9**, 543–53.

van Hinsberg, A. (1998). Maternal and ambient environmental effects of light on germination in *Plantago lanceolata*: correlated responses to selection on leaf length. *Functional Ecology*, **12**, 825–33.

van Horne, B., Olson, G.S., Schooley, R.L., Corn, J.G., and Burnham, K.P. (1997). Effects of drought and prolonged winter on Townsend's ground squirrel demography in shrubsteppe habitats. *Ecological Monographs*, **67**, 295–315.

van Noordwijk, A.J. (1995). On bias due to observer distribution in the analysis of data on natal dispersal in birds. *Journal of Applied Statistics*, **22**, 683–94.

van Tienderen, P.H. (1997). Generalists, specialists, and the evolution of phenotypic plasticity in sympatric populations of distinct species. *Evolution*, **51**, 1372–80.

van Valen, L. (1971). Group selection and the evolution of dispersal. *Evolution*, **25**, 591–8.

van Vuren, D. (1996). Ectoparasites, fitness and social behaviour of yellow-bellied marmots. *Ethology*, **102**, 686–94.

van Vuren, D. and Armitage K.B. (1994). Survival of dispersing and philopatric yellow-bellied marmots: what is the cost of dispersal? *Oikos*, **69**, 179–81.

Vander Wall, S.B. (1990). *Food hoarding in animals*. University of Chicago Press, Chicago, IL.

Vargo, E.L. (1996). Sex investment ratios in monogyne and polygyne populations of the fire ant, *Solenopsis invicta*. *Journal of Evolutionary Biology*, **9**, 783–802.

Vargo, E.L. and Fletcher, D.J.C. (1986). Evidence of pheromonal queen control over the production of male and female sexuals in the fire ant, *Solenopsis invicta*. *Journal of Comparative Physiology A*, **159**, 741–9.

Vargo, E.L. and Laurel, M. (1994). Studies on the mode of action of a queen primer pheromone of the fire ant *Solenopsis invicta*. *Journal of Insect Physiology* **40**, 601–10.

Vargo, E.L. and Porter, S.D. (1989). Colony reproduction by budding in the polygyne form of the fire ant, *Solenopsis invicta* (Hymenoptera: Formicidae). *Annals of the Entomological Society of America*, **82**, 307–13.

Veen, J. (1977). Functional and causal aspects of nest distribution in colonies of the sandwich tern (*Sterna sandvicensis* Lath.). *Behaviour (Suppl.)*, **20**, 1–193.

Veiga, J.A.S., Roselino, E.S., and Migliorini, R.H. (1978). Fasting, adrenalectomy, and gluconeogenesis in the chicken and a carnivorous bird. *American Journal of Physiology*, **234**, R115–R121.

Veldhuis, H.D., De Kloet, E.R., Van Zoest, I., and Bohus, B. (1982). Adrenalectomy reduces exploratory activity in the rat: a specific role of corticosterone. *Hormones and Behavior*, **16**, 191–8.

Venable, D.L. and Brown, J.S. (1988). The selective interactions of dispersal, dormancy, and seed size as adaptations for reducing risk in variable environments. *American Naturalist*, **131**, 360–84.

Venable, D.L. and Burquez, A. (1989). Quantitative genetics of size, shape, life-history, and fruit characteristics of the seed-heteromorphic composite *Heterosperma pinnatum*. I. Variation within and among populations. *Evolution*, **43**, 113–24.

Venable, D.L., Dyreson, E., Pinero, D., and Becerra, J.X. (1998). Seed morphometrics and adaptive geographic differentiation. *Evolution*, **52**, 344–54.

Venier, L.A. and Fahrig, L. (1996). Habitat availability causes the species abundance–distribution relationship. *Oikos*, **76**, 564–70.

Verboom, J., Metz, J.A.J., and Meelis, E. (1993). Metapopulation models for impact assessment of fragmentation. *IALE Studies in Landscape Ecology*, **1**, 172–91.

Verboom, J., Schotman, A., Opdam, P., and Metz, J.A.J. (1991). European nuthatch metapopulations in a fragmented agricultural landscape. *Oikos*, **61**, 149–56.

Verhulst, S., Perrins, C.M., and Riddington, R. (1997). Natal dispersal of great tits in a patchy environment. *Ecology*, **78**, 864–72.

Vet, L.E.M. and Dicke. M. (1999). Plant–carnivore interactions: evolutionary and ecological consequences for plant, herbivore and carnivore. In *Herbivores: between plants and predators* (ed. H. Olff, V.K. Brown, and R.H. Drenth) pp. 483–520. Blackwell Science, London.

Vet, L.E.M., Wäckers, F.L., and Dicke, M. (1991). How to hunt for hiding hosts: the reliability–detectability problem in foraging parasitoids. *Netherlands Journal of Zoology*, **41**, 202–13.

Vinson, S.B. and Greenberg, L. (1986). The biology, physiology, and ecology of imported fire ants. In *Economic impact and control of social insects* (ed. S.B. Vinson), pp. 193–226. Praeger, New York.

Visser, M.E. (1991). Prey selection by predators depleting a patch: an ESS model. *Netherlands Journal of Zoology*, **41**, 63–80.

Vitousek, P.M., Mooney, H.A., Lubchenko, J., and Melillo, J.M. (1997). Human domination of Earth's ecosystems. *Science*, **277**, 494–9.

Vleeschouwers, L.M., Bouwmeester, H.J., and Karssen, C.M. (1995). Redefining seed dormancy: an attempt to integrate physiology and ecology. *Journal of Ecology*, **83**, 1031–7.

vom Saal, F.S. (1984). The intrauterine position phenomenon: effects on physiology, aggressive behaviour and population dynamics in house mice. In *Biological perspectives on aggression* (ed. K. Flannelly, R. Blanchard, and D. Blanchard), pp. 135–79. Alan R. Liss, New York.

Vyas, N.B., Kuenzel, W.J., and Hill, E.F. (1995). Acephate affects migratory orientation of the white-throated sparrow (*Zonotrichia albicollis*). *Environmental Toxicology and Chemistry*, **14**, 1961–5.

Wade, G.N., Lempicki, R.L., Panicker, A.K., Frisbee, R.M., and Blaustein, J.D. (1997). Leptin facilitates and inhibits sexual behavior in female hamsters. *American Journal of Physiology*, **272**, R1354–R1358.

Wade, M.J. and Breden, F.J. (1987). Kin selection in complex groups: mating structure, migration structure, and the evolution of social behaviors. In *Mammalian dispersal patterns: the effects of social structure on population genetics* (ed. B.D. Chepko-Sade and Z.T. Halpin), pp. 273–83. University of Chicago Press, Chicago, IL.

Wade, M.J. and McCauley, D.E. (1988). Extinction and recolonization: their effects on the genetic differentiation of local populations. *Evolution*, **42**, 995–1005.

Wagner, R.H. (1991). The use of extrapair copulations for mate appraisal by razorbills, *Alca torda*. *Behavioral Ecology*, **2**, 198–203.

Wagner, R.H. (1993). The pursuit of extrapair copulations by female birds: a new hypothesis of colony formation. *Journal of Theoretical Biology*, **163**, 333–46.

Wagner, R.H. (1997). Hidden leks: sexual selection and the clumping of avian territories. In *Extra-pair mating tactics in birds. Ornithological monographs* (ed. P.G. Parker and N. Burley), pp. 123–45. American Ornithologists' Union, Washington, DC.

Wagner, R.H. (1999). Sexual selection and colony formation. In *Proceedings of the 22nd International Ornithological Congress* (ed. N. Adams and R. Slotow), pp. 1304–13. University of Natal Press, Durban.

Wagner, R.H., Schug, M.D., and Morton, E.S. (1996). Condition-dependent control of paternity by females purple martins: implications for coloniality. *Behavioral Ecology and Sociobiology*, **38**, 379–89.

Wahlberg, N. (1997). The life history and ecology of *Melitaea diamina* (Nymphalidae) in Finland. *Nota Lepidopterologica*, **20**, 70–81.

Wahlberg, N., Moilanen, A., and Hanski, I. (1996). Predicting the occurrence of endangered species in fragmented landscapes. *Science*, **273**, 1536–8.

Wakeley, J. (1998). Segregating sites in Wright's island model. *Theoretical Population Biology*, **53**, 166–74.

Wallace, A.R. (1889). *Darwinism*. Macmillan, London.

Wang, J. and Caballero, A. (1999). Developments in predicting the effective size of subdivided populations. *Heredity*, **82**, 212–26.

Wang, R.L., Wakeley, J., and Hey, J. (1997). Gene flow and natural selection in the origin of *Drosophila pseudoobscura* and close relatives. *Genetics*, **147**, 1091–106.

Ward, I.L. (1972). Prenatal stress feminizes and demasculinizes the behavior of males. *Science*, **175**, 82–4.

Ward, I.L. and Stehm, K.E. (1991). Prenatal stress feminizes juvenile play patterns in male rats. *Physiological Behaviour*, **50**, 601–5.

Ward, R.D., Woodward, M., and Skibinski, D.O.F. (1994). A comparison of genetic diversity levels in marine, freshwater, and anadromous fishes. *Journal of Fish Biology*, **44**, 213–32.

Ward, S.A. (1987). Optimal habitat selection in time-limited dispersers. *American Naturalist*, **129**, 568–79.

Warner, R.R. (1985). Mating behavior and hermaphroditism in coral reef fishes. *American Scientist*, **72**, 128–36.

Waser, N.M. (1994). Crossing-distance effects in *Delphinium nelsonii*: outbreeding and inbreeding depression in progeny fitness. *Evolution*, **48**, 842–52.

Waser, N.M. and Price, M.V. (1994). Crossing-distance effects in *Delphinium nelsonii*: outbreeding and inbreeding depression in progeny fitness. *Evolution*, **48**, 842–52.

Waser, P.M. (1985). Does competition drive dispersal? *Ecology*, **66**, 1170–5.

Waser, P.M. (1987). A model predicting dispersal distance distribution. In *Mammalian dispersal patterns: the effects of social structure on population genetics* (ed. B.D. Chepko-Sade and Z. Tang Halpin), pp. 251–56, University of Chicago Press, Chicago, IL.

Waser, P.M. and Elliott L.F. (1991). Dispersal and genetic structure in kangaroo rats. *Evolution*, **45**, 935–43.

Waser, P.M. and Jones, W. (1983). Natal philopatry among solitary mammals. *Quarterly Review of Biology*, **58**, 355–90.

Waser, P.M. and Strobeck, C. (1998). Genetic signatures of interpopulation dispersal. *Trends in Ecology and Evolution*, **13**, 43–4.

Waser, P.M. Austad, S.N., and Keane, B. (1986). When should animals tolerate inbreeding? *American Naturalist*, **128**, 529–37.

Watkinson, A.R. and Sutherland, W.J. (1995). Sources, sinks and pseudo-sinks. *Journal of Animal Ecology*, **64**, 126–30.

Watson, A. (1985). Socia-class, socially-induced loss, recruitment and breeding of red grouse. *Oecologia*, **67**, 493–8.

Watson, A., Moss, R., Parr, R. Mountford, M.D., and Rothery, P. (1994). Kin landownership, differential aggression between kin and non-kin, and population fluctuations in red grouse. *Journal of Animal Ecology*, **63**, 39–50.

Watt, C., Dobson, A.P., and B.T. Grenfell. (1995). Glossary. In *Ecology of infectious diseases in natural populations* (ed. B.T. Grenfell and A.P. Dobson), pp. 510–21. Cambridge University Press, Cambridge.

Wauters, L., Matthysen, E., and Dhondt, A. (1994). Survival and lifetime reproductive success in dispersing and resident red squirrels. *Behavioral Ecology and Sociobiology*, **34**, 197–201.

Webb, N.J., Ibrahim, K.M., Bell, D.J., and Hewitt G.M. (1994). Natal dispersal and genetic structure in a population of the European rabbit (*Oryctolagus cuniculus*). *Molecular Ecology*, **4**, 239–47.

Weber, J.C. and Epifanio, C.E. (1996). Response of mud crab (*Panopeus herbstii*) megalopae to cues from adult habitat. *Marine Biology*, **126**, 655–61.

Wecker, S.C. (1963). The role of early experience in habitat selection by the prairie deer mouse, *Peromyscus maniculatus bairdi*. *Ecological Monographs*, **33**, 307–25.

Weddell, B.J. (1986). *The effect of patch area, isolation, and habitat quality on the distribution and dispersion of Columbian ground squirrels (Spermophilus columbianus) in meadow steppe*. DPhil dissertation, Washington State University, Pullman, WA.

Wehausen, J.D. (1999). Rapid extinction of mountain sheep populations revisited. *Conservation Biology*, **13**, 378–84.

Weir, B.S. and Cockerham, C.C. (1984). Estimating *F*-statistics for the analysis of population structure. *Evolution*, **38**, 1358–70.

Weiss, G.H. and Kimura, M. (1965). A mathematical analysis of the stepping stone model of population genetics. *Journal of Applied Probabilities*, **2**, 129–49.

Weisser, W.W. and Hassell, M.P. (1996). Animals 'on the move' stabilise host–parasitoid systems. *Proceedings of the Royal Society of London B*, **263**, 749–54.

Weisser, W.W., Braendle, C., and Minoretti, N. (1999). Predator-induced morphological shift in the pea aphid. *Proceedings of the Royal Society London B*, **266**, 1175–81.

Weisz, J. and Ward, I.L. (1980). Plasma testosterone and progesterone titers of pregnant rats, their male and female fetuses, and neonatal offspring. *Endocrinology*, **106**, 306–16.

Wennergren, U., Ruckelshaus, M., and Kareiva, P. (1995). The promise and limitations of spatial models in conservation biology. *Oikos*, **74**, 349–56.

Wenny, D.G. and Levey, D.J. (1998). Directed seed dispersal by bellbirds in a tropical cloud forest. *Proceedings of the National Academy of Sciences of the USA*, **95**, 6204–7.

Werren, J.H. (1993). The evolution of inbreeding haplodiploid organsisms. In *The natural history of inbreeding and outbreeding* (ed. N.W. Thornhill), pp. 42–59. University of Chicago Press, Chicago, IL.

West, M.J., King, A.P., and Freeberg, T.M. (1997). Building a social agenda for the study of bird song. In *Social influences on vocal development* (ed. C.T. Snowdon and M. Hausberger), pp. 41–56. Cambridge University Press, Cambridge.

Westemeier, R.L., Brawn, J.D., Simpson, S.A., Esker, T.L., Jansen, R.W., Walk, J.W., Kerschner, E.L., Bouzat, J.L., and Paige, K.N. (1998). Tracking the long-term decline and recovery of an isolated population. *Science*, **282**, 1695–8.

Westneat, D.F. (1990). Genetic parentage in the indigo bunting: a study using DNA fingerprinting. *Behavioral Ecology and Sociobiology*, **27**, 67–76.

White, G.C. and Garrott, R.A. (1990). *Analysis of wildlife radio-tracking data*. Academic Press, New York.

Whitlock, M.C. (1997). Founder effects and peak shifts without genetic drift: adaptive peak shifts occur easily when environments fluctuate slightly. *Evolution*, **51**, 1044–8.

Whitlock, M.C. (2000). Neutral additive genetic variance in a metapopulation. *Genetical Research*, in press.

Whitlock, M.C. and Barton, N.H. (1997). The effective size of a subdivided population. *Genetics*, **146**, 427–41.

Whitlock, M.C. and McCauley, D.E. (1990). Some population genetic consequences of colony formation and extinction: genetic correlations within founding groups. *Evolution*, **44**, 1717–24.

Whitlock, M.C. and McCauley, D.E. (1998). Indirect measures of gene flow and migration: F_{st} not equal to $1/(4Nm + 1)$. *Heredity*, **82**, 117–25.

Whitmore, T.C. (1997). Tropical rainforest disturbance, disappearance and species loss. In *Tropical forest remnants: ecology, management and conservation of fragmented communities* (ed. W.F. Laurence and R.O. Bierregaard), pp. 3–12. Chicago University Press, Chicago, IL.

Wicklund, C.G. (1996). Determinants of dispersal in breeding merlins (*Falco columbarius*). *Ecology*, **77**, 1920–7.

Wiener, P. and Feldman, M.W. (1993). The effects of the mating system on the evolution of migration in a spatially heterogeneous population. *Evolutionary Ecology*, **7**, 251–69.

Wiens, J.A. (1991). The ecology of desert birds. In *The ecology of desert communities* (ed. G.A. Polis), pp. 278–310. University of Arizona Press, Tucson, AZ.

Wiens, J.A. (1992). Ecological flows across landscape boundaries: a conceptual overview. In *Landscape boundaries: consequences for biotic diversity and ecological flows* (ed. A.J. Hansen and F. diCastri), pp. 217–35. Springer, New York.

Wiens, J.A. (1994). Habitat fragmentation: island v landscape perspectives on bird conservation. *Ibis*, **137 (suppl.)**, 97–104.

Wiens, J.A. (1995a). Landscape mosaics and ecological theory. In *Mosaic landscapes and ecological processes* (ed. L. Hansson, L. Fahrig, and G. Merriam), pp. 1–26. Chapman and Hall, London.

Wiens, J.A. (1995b). Habitat fragmentation: island v landscape perspectives on bird conservation. *Ibis*, **137**, S97–S104.

Wiens, J.A. (1997a). The emerging role of patchiness in conservation biology. In *Enhancing the ecological basis of conservation: heterogeneity, ecosystem function, and biodiversity* (ed. S.T.A. Pickett, R.S. Ostfeld, M. Shachak, and G.E. Likens), pp. 93–107. Chapman and Hall, New York.

Wiens, J.A. (1997b). Metapopulation dynamics and landscape ecology. In *Metapopulation biology: ecology genetics and evolution* (ed. I. Hanski and M.E. Gilpin), pp. 43–62. Academic Press, San Diego, CA.

Wiens, J.A. (1999). The science and practice of landscape ecology. In *Landscape ecological analysis: issues and applications* (ed. J.M. Klopatek and R.H. Gardner), pp. 371–83. Springer, New York.

Wiens, J.A. (2000). Ecological heterogeneity: an ontogeny of concepts and approaches. In *Ecological consequences of habitat heterogeneity*. (ed. M.J. Hutchings, E.A. John, and A.J.A. Stewart), pp. 9–31. Blackwell Science, Oxford, in press.

Wiens, J.A. and Johnston, R.F. (1977). Adaptive correlates of granivory in birds. In *Granivorous birds in ecosystems* (ed. J. Pinowski and S.C. Kendeigh), pp. 301–40. Cambridge University Press, Cambridge.

Wiens, J.A. and Milne, B.T. (1989). Scaling of 'landscapes' in landscape ecology, or, landscape ecology from a beetle's perspective. *Landscape Ecology*, **3**, 87–96.

Wiens, J.A., Stenseth, N.C., Ban Horne, B., and Ims, R.A. (1993). Ecological mechanisms and landscape ecology. *Oikos*, **66**, 369–80.

Wiens, J.A., Crist, T.O., With, K.A., and Milne, B.T. (1995). Fractal patterns of insect movement in microlandscape mosaics. *Ecology*, **76**, 663–6.

Wiggett, D.R. and Boag, D.A. (1993). Annual reproductive success in three cohorts of Columbian ground squirrels: founding immigrants, subsequent immigrants, and natal residents. *Canadian Journal of Zoology*, **71**, 1577–84.

Williams, T.D., Dawson, A., Nicholls, T.J., and Goldsmith, A.R. (1987). Reproductive endocrinology of free-living nestling and juvenile starlings, *Sturnus vulgaris*: an altricial species. *Journal of Zoology*, **212**, 619–28.

Willson, M.F. (1992). The ecology of seed dispersal. In *The ecology of regeneration in plant communities* (ed. M. Fenner), pp. 61–86. CAB International, Wallingford, UK.

Wilson, E.O. (1992). *The diversity of life*. Penguin, London.

Wilson, H.B. and Hassell, M.P. (1997). Host–parasitoid spatial models: the interplay of demographic stochasticity and dynamics. *Proceedings of the Royal Society of London B*, **264**, 1189–95.

Wilson, H.B. and Keeling, M.J. (1999). Spatial scales and low-dimensional deterministic dynamics. In *The geometry of spatial interactions: simplifying spatial complexity* (ed. U. Dieckmann, R. Law, and J.A.J Metz), pp. 209–26. Cambridge University Press, Cambridge.

Wingfield, J.C. and Farner, D.S. (1978). The endocrinology of a natural breeding population of the white-crowned sparrow (*Zonotrichia leucophrys pugetensis*). Physiological Zoology, **51**, 188–205.

Wingfield, J.C. and Ramenofsky, M. (1997). Corticosterone and facultative dispersal in response to unpredictable events. *Ardea*, **85**, 155–66.

Wingfield, J.C. and Silverin, B. (1986). Effects of corticosterone on territorial behavior of free-living male song sparrows *Melospiza melodia*. *Hormones and Behavior*, **20**, 405–17.

Wingfield, J.C., Schwabl, H., and Mattocks, P.W., Jr. (1990). Endocrine mechanisms of migration. In *Bird migration* (ed. E. Gwinner), pp. 232–56. Springer, Berlin.

Wingfield, J.C., Breuner, C.W., and Jacobs, J.D. (1997). Corticosterone and behavioral responses to unpredictable events. In *Perspectives in avian endocrinology* (ed. S. Harvey and R.J. Etches), pp. 267–78. Society of Endocrinology, Bristol.

Wingfield, J.C., Maney, D.L., Breuner, C.W., Jacobs, J.D., Lynn, S., Ramenofsky, M., and Richardson, R. D. (1998). Ecological bases of hormone–behavior interactions: the 'emergency life history stage'. *American Zoologist*, **38**, 191–206.

Wisenden, B.D. and Sargent, C. (1997). Antipredator behaviour and suppressed aggression by convict cichlids in response to injury–released chemical cues of conspecifics but not to those of an allopatric heterospecific. *Ethology* **103**, 283–91.

With, K.A. (1994). Using fractal analysis to assess how species perceive landscape structure. *Landscape Ecology*, **9**, 25–36.

With, K.A. (1999). Is landscape connectivity necessary and sufficient for wildlife management? In *Forest fragmentation: wildlife and management implications* (ed. J.A. Rochelle, L.A. Lehmann, and J. Wisniewski). Brill Academic Publishers, Leiden, The Netherlands.

With, K.A. and Crist, T.O. (1995). Critical thresholds in species' responses to landscape structure. *Ecology*, **76**, 2446–59.

With, K.A. and King, A.W. (1997). The use and misuse of neutral landscape models in ecology. *Oikos*, **79**, 219–29.

With, K.A. and King, A.W. (1999). Dispersal success on fractal landscapes: a consequence of lacunarity thresholds. *Landscape Ecology*, **14**, 73–82.

With, K.A., Cadaret, S.J., and Davis, C. (1999). Movement responses to patch structure in experimental fractal landscapes. *Ecology*, **80**, 1340–53.

Witte, K. and Ryan, M.J. (1998). Male body length influences mate-choice copying in the sailfin molly *Poecilia latipinna*. *Behavioral Ecology*, **9**, 534–9.

Witter, M.S. and Swaddle, J.P. (1997). Mass regulation in juvenile starlings: response to change in food availability depends on initial body mass. *Functional Ecology* **11**, 11–15.

Wolff, J.O. (1992). Parents suppress reproduction and stimulate dispersal in opposite-sex juvenile white-footed mice. *Nature*, **359**, 409–10.

Wolff, J.O. (1994). More on juvenile dispersal in mammals. *Oikos*, **71**, 349–52.

Wolff, J.O. (1997). Population regulation in mammals: an evolutionary perspective. *Journal of Animal Ecology*, **66**, 1–13.

Wolff, J.O. and Plissner, J.H. (1998). Sex biases in avian natal dispersal: an extension of the mammalian model. *Oikos*, **83**, 327–30.

Wolff, J.O., Lundy, K., and Baccus, R. (1988). Dispersal, inbreeding avoidance and reproductive success in white-footed mice. *Animal Behaviour*, **36**, 456–65.

Wood, D.L., Akers, R.P., Owen, D.R., and Parmeter, J.R.J. (1986). The behaviour of bark beetles colonizing ponderosa pine. In *Insects and the plant surface*, (ed. B. Juniper and R. Southwood), pp. 91–103. Edward Arnold, London.

Wood, J.W. (1987). The genetic demography of the Gainj of Papua New Guinea. 2. Determinants of effective population size. *American Naturalist*, **129**, 165–87.

Wood, J.W., Smouse, P.E., and Long, J.C. (1985). Sex-specific dispersal patterns in two human populations of highland New Guinea. *American Naturalist*, **125**, 747–68.

Woodroffe, R.B. (2000). Strategies for carnivore conservation: lessons from contemporary extinctions. In *Carnivore conservation* (ed. J.L. Gittleman, R.K. Wayne, D.W. Macdonald, and S. Funk). Cambridge University Press, Cambridge, in press.

Woodroffe, R.B. and Ginsberg, J.R. (1998). Edge effects and the extinction of populations inside protected areas. *Science*, **280**, 2126–8.

Woodroffe, R.B. and Ginsberg, J.R. (2000). Ranging behaviour and vulnerability to extinction in carnivores. In *Behaviour and conservation* (ed. L.M. Gosling and W.J. Sutherland), pp. 125–40. Cambridge University Press, Cambridge.

Woodroffe, R.B. and MacDonald, D. (1995). Female/female competition in European badgers *Meles meles*: effects on breeding success. *Journal of Animal Ecology*, **64**, 12–20.

Woodroffe, R.B., MacDonald, D.W., and da Silva, J. (1993). Dispersal and philopatry in the European badger, *Meles meles*. *Journal of Zoology*, **237**, 227–39.

Woodroffe, R.B., Ginsberg, J.R. and Macdonald, D.W. (1997). *The African wild dog: status, survey and conservation action plan*. IUCN/SSC Canid Specialist Group, Gland, Switzerland.

Woolfenden, G.E. and Fitzpatrick, J.W. (1990). Florida scrub jay after 19 years of study. In *Cooperative breeding in birds* (ed. P.B. Stacey and W.D. Koenig), pp. 241–66. Cambridge University Press, New York.

Wright, S. (1921). Systems of mating. *Genetics*, **6**, 111–78.

Wright, S. (1931). Evolution in Mendelian populations. *Genetics*, **16**, 97–159.

Wright, S. (1932). The roles of mutation, inbreeding, crossbreeding and selection in evolution. *Proceedings of the Sixth International Congress of Genetics*, **1**, 356–66.

Wright, S. (1939). *Statistical genetics in relation to evolution. Actualités scientifiques et industrielles 802. Exposés de biométrie et de la statistique biologique XIII*. Hermann et Cie, Paris.

Wright, S. (1940). Breeding structure of populations in relation to speciation. *American Naturalist*, **74**, 232–48.

Wright, S. (1943). Isolation by distance. *Genetics*, **28**, 114–38.

Wright, S. (1946). Isolation by distance under diverse systems of mating. *Genetics*, **31**, 39–59.

Wright, S., Dobzhansky, T., and Hovanitz, W. (1942). Genetics of natural populations VII. The allelism of lethals in the third chromosome of *Drosophila pseudoobscura*. *Genetics*, **27**, 363–94.

Yahner, R.H. (1993). Old nests as cues for nest-site selection by birds: an experimental test in small even-aged forest plots. *Condor*, **95**, 239–41.

Yamada, S.B., Navarrete, S.A., and Needham, C. (1998). Predation induced changes in behavior and growth rate in three populations of intertidal snail, *Littorina sitkana* (Philippi). *Journal of Experimental Marine Biology and Ecology*, **20**, 213–26.

Yoccoz, N.G., Engen, S., and. Stenseth, N.C. (1993). Optimal foraging: the importance of environmental stochasticity and accuracy in parameter estimation. *American Naturalist*, **141**, 139–57.

Yoerg, S.I. and Shier, D.M. (1997). Maternal presence and rearing condition affect responses to a live predator in kangaroo rats (*Dipodomys hermanni arenae*). *Journal of Comparative Psychology*, **111**, 362–9.

Yoshimura, J. and Jansen, V.A.A. (1996). Evolution and population dynamics in stochastic environments. *Researches on Population Ecology*, **38**, 165–82.

Young, C.M. (1989). Selection of predator-free settlement sites by larval ascidians. *Ophelia*, **30**, 131–40.

Zakharov, A.A. and Tompson, L.C. (1998). Tunnels and territorial structure in polygyne fire ants, *Solenopsis wagneri* (Hymenoptera, Formicidae). *Zoologiceskij Zhurnal*, **77**, 911–22.

Zangerl, A.R. and Bazzaz, F.A. (1983). Plasticity and genotypic variation in photosynthetic behavior of an early and a late successional species of *Polygonum*. *Oecologia*, **57**, 270–3.

Zera, A.J. and Denno, R.F. (1997). Physiology and ecology of dispersal polymorphism in insects. *Annual Review of Entomology*, **42**, 207–31.

Zera, A.J. and Tanaka, S. (1996). The role of juvenile hormone and juvenile hormone esterase in wing morph determination in *Modicogryllus confirmatus*. *Journal of Insect Physiology*, **42**, 909–15.

Zera, A.J. and Tiebel, K.C. (1989). Differences in juvenile hormone esterase activity between presumptive macropterous and brachypterous *Gryllus rubens*: implications for the hormonal control of wing polymorphism. *Journal of Insect Physiology*, **35**, 7–17.

Zera, A.J., Strambi, C., Tiebel, K.C., Strambi, A., and Rankin, M.A. (1989). Juvenile hormone and ecdysteroid titers during critical periods of wing morph determination in *Gryllus rubens*. *Journal of Insect Physiology*, **35**, 501–11.

Zera, A.J., Sall, J., and Grudzinski, K. (1997). Flight-muscle polymorphism in the cricket *Gryllus firmus*: muscle characteristics and their influence on the evolution of flightlessness. *Physiological Zoology*, **70**, 519–29.

Zhang, J. (1995). Differences in phenotypic plasticity between plants from dimorphic seeds of *Cakile edentula*. *Oecologia*, **102**, 353–60.

Zhang, Y., Proenca, R., Maffei, M., Barone, M., Leopold, L., and Friedman, J.M. (1994). Positional cloning of the mouse obese gene and its human homologue. *Nature*, **372**, 425–32.

Zicus, M.C. and Hennes, S. (1991). Nest prospecting by common goldeneyes. *Condor*, **91**, 807–12.

Zielinski, W.J., vom Saal, F.S., and Vandenberh, J.G. (1992). The effect of intrauterine position on the survival, reproduction and home range size of female house mice (*Mus musculus*). *Behavioural Ecology and Sociobiology*, **30**, 185–91.

Zimmerman, J.L. (1971). The territory and its density dependent effect in *Spiza americana*. *Auk*, **88**, 591–612.

Zobel, M. (1992). Plant species coexistence – the role of historical, evolutionary and ecological factors. *Oikos*, **65**, 314–20.

Zobel, M. (1997). The relative role of species polls in determining plant species richness: an alternative explanation of species coexistence. *Trends in Ecology and Evolution*, **12**, 266–9.

Zollner, P.A. and Lima, S.L. (1997). Landscape-level perceptual abilities in white-footed mice: perceptual range and the detection of forested habitat. *Oikos*, **80**, 51–60.

Zollner, P.A. and Lima, S.L. (1999). Search strategies for landscape-level interpatch movements. *Ecology*, **80**, 1019–30.

Index

active dispersal   265–6
adaptation, evolution of dispersal   356
adaptive dynamics   57–79
   canonical equation   59–60
   dispersal and altruism   73–6
   models   xviii
adaptive landscapes   278–9
African hunting dogs   369
age-dependent dispersal   346
aggregative response   303
aggression   116, 235
Agrostis tenuis   332
Allee effects   233, 360
allele frequency   334, 335
alleles   332
   fixation of beneficial   277–8
allelic identity   19
altruism   58, 71–8, 137, 325, 342
American mink   367
among-patch movement   48–9
anchored pairs   64, 65–7
androgens   221, 222, 228
animal movement   16
antagonistic interactions   xx, 299–309
ants   231
aphids   118, 207, 210, 351, 352
Arctic foxes   369
assessment, habitats   237–8
assignment   22
asymmetrical migration   281
auto-correlation   22

backward dispersal   18, 24
balanced dispersal   86, 88
band-sharing   48, 49, 51
bark beetles   210, 240
basic suitability   *see* habitat quality
beetles   102–3, 104
behavioural development   234–5
behavioural flexibility   300, 302–3
behavioural mechanisms   360
behavioural models   303
behavioural modification   264–5
behavioural phenotype   205
behavioural profile, naked mole-rats   149–50
behaviour patterns   237–9
   dispersal   350–54, 360–61
   information gathering   250–54
Belding ground squirrels   115, 205, 206, 207, 220
bellbirds   185
bet-hedging strategy   342, 343
between-form differentiation   38–9
between-form gene flow   39–40
between-population variation   165

birds   112, 183, 218, 219, 222–4, 225–6, 234
body condition   218–20, 228
body mass   220
body size   205
breeding dispersal   xvii, 248, 351–2
breeding habitat selection   243–58
   ESS (evolutionarily stable strategy)   246, 254–6
   ideal free distribution (IFD)   254
   models   254
   philopatry   251
   strategies   255–6
brood parasitic birds   177–8
brown kiwis   106

Cakile edentula   271–2
canonical equation, adaptive dynamics   59–60
carp   265
choosiness   137, 141
cliff swallows   171
climate change   366
clines   22–3, 332, 333, 339–40
Clusia species   264
coadaptation, dispersal and altruism   71–8
co-ancestry   124
co-evolution   177–8, 187
coexistence of species   xx, 292–3, 312, 319, 324, 325, 349
collared flycatchers   253
colonial ascidians   184
colonial birds   171–2
colonization   xx, 281
colony multiplication   31–2
Columbian ground squirrels   186
combined data   5, 15–16
common lizards   115, 118, 173, 346, 351, 353–4
communities
   competitive   *see* competitive communities
   dynamics   xix
   ecology   349
   organization   xx
   properties   317–19
comparisons, genetic and demographic
   estimates   23–4, 28
competition   xviii, 50, 55, 110–11
competition–colonization trade-off   313
competitive ability   205
competitive communities   311–26
   ESS (evolutionarily stable strategy)   320, 322, 323, 324
   evolution   319–24
   fitness   322
   metapopulations   312
competitive inhibition   212
competitors   293, 319–20, 322

complex spatial dynamics 293–6
computer modelling 109
conditional dispersal xviii, xix, 141, 182, 203–16, 254, 342, 346–7, 355–6
conservation 240–41, 258
  applications 368–9
  dispersal 358–9
  management 361, 362, 364, 365, 368, 371–72
  models 359–60, 362, 364–5, 372
  planning 364
conservation biology xx, 358–72
conspecific attraction 231–2, 239–40
conspecific cueing 238
conspecifics 232, 233, 257
constraints, evolution of dispersal 356
convergence stability 129
cooperation 137, 148
correlated response 197–201
correlation equations 58, 64–70
corridors 368
corticosterone 223, 224, 227
cost-benefit perspective 100–101
costs
  dispersal 155–6, 207–8, 262, 305
  inbreeding 136–7, 158–9, 161–4, 165
  information 247
  search 236–7
  settlement 235–6
covariates, methods to include 13–15
Cox regression *see* proportional hazards model
Crepis sancta 351
crested tits 185, 206
crickets 195, 218
crisis management 371–2
cues xix, 246, 247, 353–4
  external 248–9
  indirect 238, 241
curlews 185

Daphnia 265
definitions, dispersal 226
demes 332
demographic analyses 55
demographic consequences 356–7
demographic equilibrium 84
demographic instability 24
demographic stochasticity 95, 309
demography xviii, 166, 347, 348
density 233
density-dependent dispersal 113, 121, 204, 208–9
density-dependent emigration xx, 287–8
determinism, dispersal behaviour 350–54
deterministic models 343–4
differentiation 35, 36, 41
diffusion-limited dynamics 301
diffusive instability 309
dimorphic seeds 193
dimorphism 193
direct fitness 71, 157
direct life cycles 174
direct pathways 205–7, 209–210

dispersal 3, 40–42, 97
  active 265–6
  age-dependent 346
  altruism 71–8
  backward 18, 24
  balanced 86, 88
  behaviour 350–54, 360–61
  breeding xvii, 248, 351–2
  causes 342–7
  coalitions 121, 152
  conditional 141, 182, 203–216, 254, 342, 346–7
  conservation 358–9
  continuous habitat 47–50
  costs 155–6, 207–8, 262, 305
  decisions xix, 301
  definitions 226
  density-dependent 113, 121, 204, 208–9
  development of ability 182–3
  environment 99
  environmental 205
  event 4
  evolution *see* evolution of dispersal
  facultative 225
  fire ants 33–9
  fragmented habitat 50–51
  fruit 185
  genetic approach 40–42, 191–202
  hosts 170–74
  inferred from demographic methods 45–6
  innate 205
  limited 105, 302
  long-distance 26–7, 105
  mode of 266
  mortality 370
  naked mole-rats 149
  natal xvii, 217–28, 248, 346, 351
  parasites 173–6
  passive 266, 307
  pathways 98, 100, 101
  phenotype xix, 350–53, 355–6
  plants 183
  population dynamics 333–4
  population genetics 29
  predation 180–81
  pre-saturation 115
  propensity 57
  random 332
  seeds 184, 185
  sex biased 111–13, 147–9
  stage-dependent 211–13, 352
  strategies 301, 305
  unconditional 182
  variable 276–80
dispersal distance 345
dispersal-distance function 97, 99, 106, 107, 108
dispersal rates
  co-evolution xx, 349–50
  ESS (evolutionarily stable strategy) 156, 164, 267–72, 343
  estimation 18–28
  evolution 297, 332, 344

measurement 285–7
  naked mole-rats 145–6
  predator and prey 187
divergence 330–31
DNA fingerprints 47
DNA sequencing 337
dominant Lyapunov exponent 87, 92
dormancy 267
dormice 360
Drosophila pseudoobscura 339

ecological succession 105
ectoparasites 171
effective migration rate 274–5
effective population size 52, 276–7, 280
egg discrimination 177
eggs 222–3
emigration 181–4, 284, 287, 288
EMS (experimental model systems) 102–4
endocrine system 226
energy reserves 149, 205, 212
environmental autocorrelation 250
environmental change 365–8
environmental conditions 204–5
environmental dispersal 205
environmental effects 351–2
environmental factors 182
environmental gradients 306–8
environmental variation 256
equilibrium, propagule rain model 314, 316
equitability 314, 317
ESS (evolutionarily stable strategy)
  breeding habitat selection 246, 254–6
  competitive communities 320, 322, 323, 324
  dispersal rates 156, 164, 267–72, 343
  genetic drift 344
  invasion fitness 61
  kin competition 111, 133
  models 354
  predator-prey systems 307
establishment, immigrants 186
estimation
  dispersal probabilities 3–17
  dispersal rates 18–28
  methods 11–13
  selfing rate 25
European badgers 116–17, 121, 221
European kestrels 114
European nuthatches 108
evolution, competitive communities 319–24
evolutionarily stable strategy *see* ESS
evolutionary biology xx
evolutionary branching 61, 349
evolutionary dynamics, spatial invasion fitness 77
evolutionary feedback 347–50
evolutionary modelling approach 123
evolutionary singularities 61
evolutionary stable dispersal patterns 129–39
evolution of dispersal xix, xx, 56, 155–67, 341–57
  adaptation 356

altered environments 370–71
antagonistic interactions 308
constrains 356
extinction 158, 159–61
ideal free interpretation 85
maternal perspective 267
models 254
predation 187
spatial heterogeneity xviii
temporally constant environments 86–8
experimental model systems (EMS) 102–4
exploration 213, 235–6
external cues 248–9
extinction
  conservation biology 358, 359, 361, 363, 367, 369
  evolution of dispersal 158, 159–61
  local population 281, 297, 343, 358, 367
  metapopulations 368
  rates 166
  risk 288–9

factorial experiments 215
facultative dispersal 225
fairy wrens 264
fat 149–50, 205, 212, 213, 219, 220
fecundity 197–201
feedback 105, 308, 363
fidelity function 11
fire ants 29–42
  dispersal 33–9
  gene flow 33–40
  genetic structure 33–8
  social biology 30–33
fitness 3, 84, 85
  competitive communities 322
  direct 71, 157
  dominant Lyapunov exponent 87, 92
  gene flow 339–40
  habitat selection 232–3, 237
  immigrant xix–xx
  individual 262
  invasion 57–79
  local *see* local fitness
  maximal 301
  neighbour-modulated fitness 71
  offspring 267–72
  parental 267
  partial mixing 88
  population dynamics 333, 334
  spatial invasion 58, 62–71, 73, 77, 78
fitness equilibration 85, 92, 95
fixation of beneficial alleles 277–8
food chains 180
food-hoarding 183, 184, 185
food limitation 204
food resources 206, 219
fragmentation 50–51, 106, 107, 108, 297, 366, 369, 370–71
frequency dependence 269–70
fruit, dispersal 185

F_{ST} 331, 334, 335
F_{ST} analogues   20, 41
F statistics   19, 35–6, 41
fungus beetles   108

game-theoretical approach   127
gap-crossing   99
genealogies   337–9
gene flow   xix, 29, 40–42
  between-form   39–40
  fire ants   33–40
  fitness   339–40
  local adaption   329–40
  social behaviour   30
gene frequency data   16
generalists   266
genetically-based dispersal   191–202
genetic analyses   55
genetic correlation   195–7
genetic data   16
genetic drift   276–7, 280, 344
genetic effects   350–51
genetic identities   348
genetic inference   44, 46, 51–4, 55, 56
genetic load   126, 139, 144, 165
genetic methods   285
genetics   xviii
genetic similarity   55
genetic structure   29, 347–8, 362–3
  fire ants   33–8
  pikas   51
  statistical approaches   41
genetic variance   279–80
genetic variation   330, 350
  migratory traits   192–3
  naked mole-rats   152–4
  pikas   51
genotype   xix, 32, 346, 350
geographical isolation   332
geometric mean local fitness   89, 90–92, 94
GIS technologies   365
Glanville fritillary butterflies   289–90, 296
global competition   110
global temporal variance   156
glucocorticoids   223, 224–5
goldenrod beetles   113–14
gonads   225
grain-extent window   104
grandmaternal effect   351
grasshoppers   103, 104
gray jays   117
great spotted cuckoos   178
great tits   115, 173
grey-sided voles   205, 221
Gryllus firmus   *see* sand crickets
gypsy moths   209

habitat cueing   238–9
habitat imprinting   231–2, 240, 241, 242
habitats
  altered   371

  assessment   237–8
  destruction   297–8, 366
  fragmentation   *see* fragmentation
  heterogeneity   xviii
  indirect assessment   238
  loss   107
  quality   234–5, 353–4
  saturation   116
  sex-specific requirements   xviii
  sink   88–94
  stability   204
  see also patches
habitat selection   230–42
  antagonistic interactions   305
  behavioural flexibility   300, 302–3
  breeding   *see* breeding habitat selection
  definition   342
  dispersal behaviour   345
  evolutionary consequences   353
  fitness   232–3, 237
  ideal free theory   84–5
  mate choice   248–9
  models   236
  optimal   xix
  predation   185–6
habitat training   234–5
Hamilton's rule   76–7
hamsters   228
haplotype   39
haystack model   320–21
hazard function   12
hazel dormice   367
Heisey–Fuller estimator   12
Heliconius butterflies   332
helping behaviour   264
herbivores   184
heritability   192, 193
  estimates   194
heterosis   275
heterospecifics   233–4
Heterosperma pinnatum   271
hibernation   265
homegeneity, population   165
honeybees   239
hormones   220–23
horseshoe crabs   265
host-parasite associations   169–70, 303
hosts   349
  activity behaviour   176
  dispersal   170–74
human activity   105, 240–41, 297, 365, 366, 369
hybridizations   336
hydrodynamics limit models   62

ideal despotic habitat selection model   235
ideal free distribution (IFD)   xviii, 83–95, 302
  breeding habitat selection   254
  simultaneous   303–4
  temporally constant environments   86–8
ideal free theory   84–5
IFD   *see* ideal free distribution

immigrants
  establishment 186
  fitness xix–xx
immigration xix, 185–6, 284, 289, 336
immune defences 177
inbreeding xviii, 123–42, 143–54, 370
  costs 136–7, 158–9, 161–4, 165
    pedigree 153
    pikas 46
    system-of-mating 153
    without competition 131–2
inbreeding avoidance xviii, 55, 111, 123, 125–7,
    133, 139, 140
inbreeding coefficient 124
inbreeding depression 125–6, 132, 139, 140, 141
  avoidance 156
  Glanville fritillary butterflies 289
  kin competition 132–3
  naked mole-rats 144
indirect assessment, habitats 238
indirect cues 238, 241
indirect parasitic life cycles 175
indirect pathways 205–7, 209–210
individual fitness 262
individuals 274
individual variation 274–5
induced defences 265
inferences 4, 16, 334–9
  genetic 44, 46, 51–4, 55, 56
information 352–3
  costs 247
  gathering 247, 250–54
  pooling 239
  private 246, 248
  public 243–58
  value of 246–7
innate dispersal 205
insects 209–210, 218
interactive effects 215
internal conditions 204, 205
interspecific interactions 168
intrasexual competition 111–13
intraspecific competition 113–17
intraspecific interactions 168
introgression 336
invasion fitness 57–79
island biogeography 99
island model 21, 25, 273, 275–6, 330, 331, 332
isolation
  geographical 332
  reproductive 333
isolation by distance 21–2, 25
  models 337
iteropar species 250, 251

joint likelihoods 16
juvenile hormone 218

kangaroo rats 116, 120
Kaplan–Meier estimator 12–13

killer whales 30
kin competition xviii, 57, 95, 140, 156
  avoidance 111, 117–19
  ESS (evolutionarily stable strategy) 111, 133
  evolution of dispersal 157–8, 159–64
  inbreeding depression 132–3
  naked mole-rats 148
kin cooperation 119, 137–9, 140
  philopatry 119, 120, 121, 138
kin interaction xviii, 269, 270, 271
kin recognition 139
kin selection 57, 128, 342, 344
  models 77
kin structures 138
kittiwakes 171–2
known-status data 5, 10–15

labile modifying factors 225
landscape 96–109, 369
  change 104–8
  structure 104
land-use changes 105–6
larvae, marine organisms 114
larval ascidians 185
larval barnacles 237, 238
lattice models 62, 63
leptin 227–8
liability 197, 198
life cycles 157, 165
  direct 174
  indirect parasitic 175
life history theory 186–7
life history traits *see* traits
light 263–4
likelihood 20–21, 25, 28, 106, 337
  joint 16
  maximum 20, 21, 23
limited dispersal 105, 302
limpets 114
linkage disequilibria 336
lions 121
living museum 370
local adaptation 176–7
local competition 111
  without inbreeding 128–31
local demographic structure 204–5
local dynamics 311–26
local fitness 84, 85
  geometric mean 89, 90–92, 94
  non-equilibration 88–94
local mate competition xviii, 135–6
local populations 283–98
  dynamics 287–91, 293–4, 305
  exploitation 304–6
  extinction 281, 297, 343, 358, 367
  size 287–91, 343
local resource competition (LRC) 120,
    135–6
local resource enhancement (LRE) 120
local species richness 325

local survival probabilities 9
local temporal variance 156
locomotor behaviour 226–7
logistic regression 14–15
long-distance dispersal 26–7, 105
long-term selective forces 344
Lotka–Volterra model 299
LRC *see* local resource competition
LRE *see* local resource enhancement

magpies 178, 247
mainland–island model 273
mammals 220, 221–2
management, conservation 361, 362, 364, 365,
    368, 371–2
manna ash trees 354
marginal value theorem (MVT) 305
marginal value threshold 305
marine ascidian 114
marine bivalves 184
marine gastropods 266–7
marked animals 4
mark-recapture studies 43–4
mark-release-recapture (MRR) 285–7
marsh tits 115, 219
mate choice 137
    habitat selection 248–9
    naked mole-rats 150–52
maternal effects 207, 210–11, 223, 352
maternal genes 351
maternal perspective 267–72
maternal phenotypic plasticity 262
mate search 237
mating
    behaviours 265
    patterns 45–6
    systems 133–7
maximal fitness 301
maximum likelihood 20, 21, 23
Medicago species 351
Membranipora 265
metacommunity xx, 314–17
metapopulations xviii, 54
    competitive communities 312
    conservation biology 359, 363
    contour xx
    evolutionary feedback 347–8
    extinction 368
    genetic properties 273–82
    models 62, 106, 319
    persistence xx, 291–3
    pikas 50, 51–2
    population dynamics 283–98, 334
    predator–prey 305–6
    survival 285
methodological tools, estimating parameters 4
mice 228
microsatellite markers 336
migrants 201, 274–5, 344
    number of 334, 335
migration xix, 191–202, 225–6, 274

asymmetrical 281
    matrix 336
    model 335
migratory traits, genetic variation 192–3
milker strategy 306
milking 306
milkweed 193
milkweed bugs 207
mobile animals 264–5
models 109
    adaptive dynamics xviii
    assumptions 164–6
    behavioural 303
    breeding habitat selection 254
    computer modelling 109
    conservation 359–60, 362, 364–5, 372
    deterministic 343–4
    EMS (experimental model systems) 102–4
    ESS dispersal 267–72
    evolutionary modelling approach 123
    evolution of dispersal 254
    habitat selection 236
    haystack 320–21
    hydrodynamics limit 62
    ideal despotic habitat selection 235
    island 21, 25, 273, 275–6, 330, 331, 332
    isolation by distance 337
    kin selection 77
    lattice 62, 63
    Lotka–Volterra 299
    mainland-island 273
    metacommunity 314–17
    metapopulation 62, 106, 319
    migration 335
    multisite 7, 8
    mutation 28
    Nicholson–Bailey 299
    open-population modelling approach 7
    pair-dynamics 64–70
    population structure 273–4
    propagule rain 313–14, 316
    proportional hazards 13–14
    quantitative genetic 199, 202, 354
    reaction-diffusion 62
    Ricker 293–4
    scalar 86
    social attraction 258
    source–sink 325
    spatially explicit dispersal 109
    spatial population 57–79
    stage-structured matrix 87
    stepping-stone 62, 274
    stochastic 343–4
    threshold 192, 193
moles 185
moment equations 62
monogamous species 134
monogyne social form 31, 32
morphology 182
    naked mole-rats 149–50
mortality risk 186

mother–daughter competition 118
movement xvii, 99, 100, 101–4
  among-patch 48–9
  animal 16
  probabilities 8
MRR *see* mark-release-recapture
mtDNA 34–5, 36
  variance 35
multi-causal approach xviii
multiple condition dependence 214–15
multiple equilibria 295–6
multisite studies 5
  models 7, 8
mussels 265
mutant invasion rate 60–61
mutants 58, 59, 76, 77, 309, 319, 321–3
mutation 280
  model 28
  rates 24–5
mutations 330
mutual exclusion 59
MVT *see* marginal value theorem

naked mole-rats 143–54, 221, 271
  behavioural profile 149–50
  dispersal 149
  dispersal coalitions 152
  dispersal rates 145–6
  genetic variation 152–4
  inbreeding depression 144
  kin competition 148
  mate choice 150–52
  morphology 149–50
  nascent colonies 146–7, 151, 152, 153
  philopatry 148
  sex-biased dispersal 147–9
nascent colonies, naked mole-rats 146–7, 151, 152, 153
natal dispersal xvii, 217–28, 248, 346, 351
natural populations 276
neighbourhood size 233, 241
neighbour-modulated fitness 71
neighbours 233
neutral alleles 330
neutrality, genetic markers 24
Nicholson–Bailey model 299
non-equilibrium persistence 303
non-equilibrium population structure 276–80
non-migrants 201
nuclear markers 37, 39
nut-hatches 241

observations, animals 4
offspring 262–7
  fitness 267–72
  parent–offspring competition 118, 351
Oncopeltus fasciatus 195
ontogeny, dispersal behaviour 350–54
open-population modelling approach 7
optimization 100, 101
outbreeding 143–54

oyster-catchers 247

pair approximation 69
pair-dynamics models 64–70
panmictic populations 330
parameters 365
parasites 170–74, 349
  dispersal 173–6
  population 174
parasitic hymenopterans 231
parasitic life styles 174
parasitism xviii, 168–79
parasitoids 302, 303
parental fitness 267
parent–offspring competition 118, 351
partial mixing 87–8
partitioning, genetic variance 35–6
passerine birds 185
passive dispersal 266, 307
patch dynamics 105
patches 100, 304
  movement among 48–9
  prey 306
  refuge 304
  see also habitats
patchiness 308–9
patch sample information *see* private information
patch suitability 248
paternal investment 134
path-analytical approaches 215
pea aphids 182, 351
pedigree inbreeding 153
pelagic dipterans 185
peregrine falcons 231
persistence, metapopulations xx, 291–3
phenotypes xix, 59, 253–4
  behavioural 205
  dispersal xix, 350–53, 355–6
  philopatry 262–3
phenotypic plasticity 261–72
phenotypic variation 346
philopatry xviii
  breeding habitat selection 251
  kin cooperation 119, 120, 121, 138
  mating systems 134, 135
  naked mole-rats 148
  phenotypic responses 262–3
  settlement pattern 47, 51
  sex-biased dispersal 112, 113
  social species 369
  strategy 255, 256
photoperiod 204
physiology 217, 228
pikas 45–54, 241
  genetic structure 51
  genetic variation 51
  inbreeding 46
  metapopulations 50, 51–2
  natural history 45
pine martens 221, 367
pine voles 115

Piper arieianum 264
Piper sanctifelicis 263
planning, conservation 364
plant-herbivore systems 183
planthoppers 113
plants 309
  dispersal 183
  mortality 186
  productivity 318
plasticity 332
  dispersal-distance function 108
Polygonum species 266
polygyne social form 31, 32
polygynous species 134, 148
population density 204, 233
population dynamics xix, xx, 283–98
  antagonistic interactions 300, 301, 303, 304, 305
  dispersal 333–4
  fitness 333, 334
  metapopulations 283–98, 334
  spatial scale 312
population ecology 3
population lattices, spatial dynamics 63–4
population persistence xix, 364
population regulation 287
population size 283
population structure 124
  evolutionary consequences 330, 331
  metapopulations 274–5, 278, 279
  models 273–4
  non-equilibrium 276–80
population synchrony 294–5
population viability analysis (PVA) 360, 364–5
post-dispersal seed predation 186
predation xviii, 168, 180–87, 367
predator–prey metapopulation 305–6
predator–prey systems 182–3, 300, 305, 306, 307, 349
predator-related mortality 184
predators 301, 305, 306
pre-emigration seed predation 183
pre-saturation dispersal 115
presence strategy 255
prey 306
prey patches 306
primary productivity 318
private information 246, 248
probability density function 12
probability of fixation 277–8
propagule rain model 313–14, 316
proportional hazards model 13–14
prospecting 247, 250–54
proximate mechanisms 217–28, 231–2
pseudo-sinks 290
public information 243–58
PVA *see* population viability analysis

quality strategy 255, 256
quantitative genetic models 199, 202, 354

radio-telemetry 10–11

random dispersal 332
random drift 330–31, 334–6
random strategy 255
ranging 226
razorfish 265
reaction-diffusion models 62
recapture/resighting data 5, 6–10
recessive deleterious alleles 126
recovery data 5–6
red foxes 117, 361
red grouse 119
red squirrels 107–8, 118
refuge patches 304
refuge use 304
regional dynamics 311–26
regional similarity rule 316
reintroduction 368
relatedness 76, 77, 124
reobservation data 5
reproductive isolation 333
rescue effect 289, 295
residency 232–5
resistance 177, 178
resource-competition hypothesis 134–5
Ricker model 293–4
risk-spreading 291–2, 344
robustness 25, 28
root voles 120

salamanders 271
sampling 28, 41
sand crickets 195, 197–201, 352
sandwich terns 257
scalar models 86
scale 300–302
  landscape 104
  spatial *see* spatial scale
  timescales 301, 342, 344, 356
seabird ticks 171–2, 174–5
search costs 236–7
searching 232, 236–7
seeds
  dimorphic 193
  dispersal 184, 185
  dormancy 263
  germination 263
  size 193
seismic vibrations 151
selecting new site 4
selection 197–201, 331
  experiments 354–5
  sexual 137
  spatial variation 331–3
selection derivative 61
selfing rates 166
semelpar species 250, 251
sequential prospecting 251, 253
sessile species 251, 263
settlement xix, 232
  costs 235–6
sex asymmetries 140

sex-biased dispersal 111–13
  naked mole-rats 147–9
sex hormones 221, 222
sex ratio 33, 34, 120
sexual selection 137
Seychelles warblers 120
shifting balance theory 278–9, 331
Shorea species 264
short-term selective forces 344
sibling competition 117
simultaneous prospecting 251
sink habitats 88–94
sink populations 290
sinks 281, 369
snail kites 13
sneaking 265
snow crabs 114
social attraction, models 258
social behaviour 30, 41, 233
social biology, fire ant 30–33
sociality 369
socially breeding species 249
social structures 137–9
Solenopsis invicta *see* fire ants
solitary species 249
song sparrows 114–15
source populations 290
sources 281
source-sink dynamics 290–91, 313–19
source-sink models 325
space xx, 300
space occupation 319
Spanish imperial eagles 219
spatial auto-correlation 213
spatial dynamics, population lattices 63–4
spatial heterogeneity xviii, 99, 268, 303,
  308–9
spatial invasion fitness 58, 62–71, 73, 77, 78
spatially explicit dispersal models 109
spatial patterns 295, 334
spatial populations
  distribution 300
  models 57–79
  structure 273
spatial problems xviii
spatial reproductive value 89, 93–4
spatial scale 300, 342, 357, 361
  dependence 213–14
  dispersal distance 345
  local environment 312
  population dynamics 312
spatial variation 165, 331–3
specialists 266
specialized dispersal morphs 183–4
speciation 332, 333
species
  coexistence xx, 292–3, 312, 319, 324,
    325, 349
  interaction between 349–50
  iteropar 250, 251
  monogamous 134

polygynous 134, 148
richness *see* species richness
semelpar 250, 251
sessile 251, 263
socially breeding 249
solitary 249
threatened 359
species richness 312–13, 317
  local 325
spider monkeys 120
spotted hyenas 120–21, 221
spruce budworms 205
stage-dependent dispersal 211–13, 352
stage-structured matrix models 87
standard pair approximation 69
statistical approaches, genetic structure 41
stay-or-leave 4, 300, 305
stepping-stone models 62, 274
stochastic models 343–4
strategies, breeding habitat selection 255–6
stress 224–5
study sites 5
subpopulations 269
success strategy 255
suitability 84, 185
swallow bugs 171
system-of-mating inbreeding 153

temporal heterogeneity xviii, 268, 308–9
temporal scales *see* timescales
temporal variation 88–94, 156
territory bequeathal 118
territory ownership 118
testosterone 221, 222
theoretical unification xviii, xx, 354
threatened species 359
threshold models 192, 193
threshold population sizes 369
threshold traits 192
time lags 181, 208, 209, 211
timescales 301, 342, 344, 356
time-scale separation 59
timing, natal dispersal 219
Townsend voles 114, 116, 120
tracking 362
trade-offs 186–7, 202
  competition–colonization 313
  habitat selection 237, 246, 250–54
  offspring fitness 271
traits xix, 202, 348–9
  migratory 192–3
  threshold 192
transit 184–5
transition probabilities 9–10
transition rates, anchored pairs 65–7
transmission 174, 179
Tribolium 347
Tundra voles 116
turkeys 237
turnover 48, 50
tussock moths 295

unconditional dispersal   182
unifying principles   *see* theoretical unification

variable dispersal   276–80
virulence   176

water voles   361, 367
waves of advance   27
western screech-owls   219, 223

where to settle   300, 301
white-footed mice   117, 242
widow spiders   184
willow tits   226
wing-dimorphic insects   182, 183, 186, 193
within-population patterns   47–8

yellow-bellied marmots   173

Breinigsville, PA USA
28 October 2009
226635BV00004B/9/A

9 780198 506591